4th Global Conference on Materials Science and Engineering (CMSE 2015)

IOP Conference Series: Materials Science and Engineering
Volume 103

Macau, China
3 – 6 August 2015

Editors:

Harry E. Ruda
Alexander Khotsianovsky

ISBN: 978-1-5108-1812-5
ISSN: 1757-8981

Printed from e-media with permission by:

Curran Associates, Inc.
57 Morehouse Lane
Red Hook, NY 12571

CURRAN ASSOCIATES INC. proceedings.com

Some format issues inherent in the e-media version may also appear in this print version.

Printed by Curran Associates, Inc. (2016)

Institute of Physics
Dirac House, Temple Back
Bristol BS1 6BE UK

Phone: 44 1 17 929 7481
Fax: 44 1 17 920 0979

techtracking@iop.org

Additional copies of this publication are available from:

Curran Associates, Inc.
57 Morehouse Lane
Red Hook, NY 12571 USA
Phone: 845-758-0400
Fax: 845-758-2634
Email: curran@proceedings.com
Web: www.proceedings.com

Table of contents

Volume 103

4th Global Conference on Materials Science and Engineering (CMSE 2015)
3–6 August 2015, Macau, China

Accepted papers received: 2 November 2015
Published online: 9 December 2015

IOP Conf. Series: Materials Science and Engineering **103** (2015) 011001 doi:10.1088/1757-899X/103/1/011001

Preface

IOP Conference Series: Materials Science and Engineering is publishing a volume of conference proceedings that contains a selection of papers presented at the 4th Global Conference on Materials Science and Engineering (CMSE 2015), which is an annual event started from 2012. CMSE 2015, technically supported by Institute of Applied Physics and Materials Engineering of University of Macau, organized by Wuhan Advance Materials Society, was successfully held at the University of Macau-new campus located on Hengqin Island during August 3rd-6th, 2015. It aims to bring together leading academic scientists, researchers and scholars to exchange and share their experience and research results on all aspects of Materials Science and Engineering, and to discuss the practical challenges encountered and the solutions adopted.

Macau, one of the two special administrative regions of the People's Republic of China, where East meets West, turned out to be an ideal meeting place for domestic and overseas participants of this annual international conference. The conference program included keynote presentations, special sessions, oral and poster contributions. From several hundred submissions, 52 most promising and IOP mainstream-relevant contributions were included in this volume, which present original ideas or results of general significance supported by clear reasoning and compelling evidence, and methods, theories and practices relevant to the research. The authors state clearly the problems and the significance of their research to theory and practice. Being a successful conference, this event gathered more than 200 qualified and high-level researchers and experts from over 40 countries, including 10 keynote speakers from 6 countries, which created a good platform for worldwide researchers and engineers to enjoy the academic communication.

Taking advantage of this opportunity, we would like to thank all participants of this conference, and particularly the authors of all accepted papers for their high quality and fruitful contributions. Special thanks are due to all reviewers for their carefully critical reading of the manuscripts and useful comments and suggestions. We do hope that this volume will be beneficial for readers to their future research endeavours and careers. We also gratefully acknowledge tremendous efforts and dedication of many individuals, especially CMSE Conference Secretary Ms. Liu Qin, Editor Anete Ashton and all the Editorial Board members in IOP Publishing for their support in producing the proceedings of this event.

Guest Editors:

Prof. Harry E. Ruda
University of Toronto, Canada
Dr. Alexander Khotsianovsky
Pisarenko Institute of Problems of Strength of the National Academy of Sciences of Ukraine, Ukraine

Peer review statement

All papers published in this volume of *IOP Conference Series: Materials Science and Engineering* have been peer reviewed through processes administered by the proceedings Editors. Reviews were conducted by expert referees to the professional and scientific standards expected of a proceedings journal published by IOP Publishing.

IOP Conf. Series: Materials Science and Engineering **103** (2015) 012001 doi:10.1088/1757-899X/103/1/012001

Influence of specimen size on autogenous volume deformation of concrete with magnesium oxide

S L Yan[1,2] and C L Chen[1,3]

[1]School of Materials and Architectural Engineering, Guizhou Normal University, Guiyang, 550025, China
[2]School of Civil Engineering, Guizhou University, Guiyang, 550025, China

E-mail: ccl@gznu.edu.com

Abstract. In order to investigate the influence of specimen size on autogenous volume deformation of concrete with magnesium oxide (MgO), the role of autogenous volume deformation of MgO concrete was tested. The results show that the specimen size has an obvious influence on autogenous volume deformation of MgO concrete.

1. Introduction

Cracks appear in mass concrete to different extents during their life time, decreasing the crack resistance of the concrete. Some penetrating cracks may endanger the safety of a project. Mehta and Pirtz mixed MgO into concrete as an expanding agent, and they studied the expansion character of MgO , the results showed that expansion deformation generated by the hydration of MgO could compensate for temperature-induced shrinkage [1, 2]. This had also been proved by engineering practice in China [3]. There are many factors that influence autogenous volume deformation of MgO concrete. Fang studied the influence of factors such as the curing temperature of test-piece and amount of admixture of MgO on deformation characteristics of concrete [4]. Chen pointed out that fly ash leading to the autogenous volume deformation of additive MgO concrete increased in the early stages and decreased later [5]. Li pointed out that with increased water-cement ratio, the autogenous volume deformation of MgO concrete also increased [6].

However, there are no relevant studies on the influence of the size of test specimen on autogenous volume deformation of MgO concrete, let alone taking the size effect of autogenous volume deformation of MgO concrete into industrial deformation. In accordance with "Test Code for Hydraulic Concrete (SL352-2006)" [7], the standard size of test specimen for testing autogenous volume deformation of the concrete is 200mm with its diameter, and 500 mm-600 mm with its height; the aggregate whose particle size is greater than 40 mm will be filtered out when the test specimen formed (in short, standard test specimen). Obviously, there exist differences in construction. So, if it is assumed that aggregate whose particle size is greater than 40 mm is not filtered out, how does the autogenous deformation volume of MgO concrete proceed? This research tests MgO concrete samples with different sizes to measure the autogenous volume deformation and MIP of drilling core sample, aged of 1 year, for revealing the effects of specimen sizes on the autogenous volume deformation of MgO concrete.

[3] Address for correspondence: C L Chen, School of Materials and Architectural Engineering, Guizhou Normal University, Guiyang, 550025, China. E-mail: ccl@gznu.edu.com.

2. Experiment

2.1. Raw materials

- Cement: Ordinary Portland cement of Lafarge produced in Shuicheng County, Guizhou province. The density is 3.05 $g \cdot cm^{-3}$, specific surface area is 271 $m^2 \cdot kg^{-1}$, water requirement of normal consistency is 26.3% and the stability is qualified, the main chemical component of the cement is shown in table 1.

Table 1. The chemical composition of Cement (mass fraction) /%.

CaO	S_iO_2	Al_2O_3	Fe_2O_3	MgO	SO_3	Na_2O	K_2O	Loss
58.15	21.23	6.34	3.81	3.01	2.70	0.19	0.30	3.78

- Fly ash: Class F and II fly ash processed in Guizhou Yemazhai Power Plant. The water content ratio is 91.2%, the loss on ignition is 6.20%, fineness (residue on sieve filtered out by 0.045mm) is 16.55% and the density is 2.46 $g \cdot cm^{-3}$, which meets the requirement of "Technical specification of fly ash for use in hydraulic concrete (DL/T5050-2007)", the testing result of its main chemical component is shown in table 2.

Table 2. The main chemical composition of fly ash (mass fraction) /%.

SiO_2	Al_2O_3	Fe_2O_3	CaO	MgO	SO_3	Na_2O	K_2O	TiO_2	MnO	Loss
45.77	24.72	12.80	3.86	2.48	0.60	0.68	1.31	2.90	0.08	6.20

- Magnesium oxide: The magnesium oxide is produced by Dongfang Huamei Co. Ltd, in Haicheng city, Liaoning province. The density is 3.27 $g \cdot cm^{-3}$, its quality is in accordance with "Quality technical requirements of light magnesia materials for water conservancy and hydropower projects (for Trial Implementation)". The main chemical component is shown in table 3.

Table 3. The chemical composition of MgO (mass fraction) /%.

MgO	SiO_2	CaO	Fe_2O_3	Al_2O_3	SO_3	Loss
93.36	2.05	0.77	0.62	0.36	0.12	1.74

- Aggregate: The aggregate is manufactured sand and macadam in a hydropower station in Guizhou province. The fineness modulus of manufactured sand is 2.98, which belongs to class II (medium sand),and the grain composition is qualified.
- Superplasticizer: The superplasticizer is produced by Chongqing Sansheng Special Building Materials Co. Ltd; the quality conforms to the current standard.

Table 4. The mixture proportion of concrete/ $kg \cdot m^{-3}$.

Items	Water	Cement	Fly ash	Sand	Aggregate(mm)			MgO	Superplasticizer
					5-20	20-40	40-80		
55-30-0	110	140	60	706.2	430.14	430.14	573.52	0.00	1.6
55-30-6	110	140	60	706.2	430.14	430.14	573.52	12.00	1.8
55-30-8	110	140	60	706.2	430.14	430.14	573.52	16.00	2.0
55-30-10	110	140	60	706.2	430.14	430.14	573.52	20.00	2.4

[a] 55-30-6 shows that water-binder ratio is 0.55, fly ash is 30%, the content of MgO is 6% and so on.

2.2. Mixture proportion of concrete

The concrete compositions are shown in Table 4. This mixture proportion is the same raw materials, water-binder ratio is 0.55, fly ash replacement is 30%, the mixing amount of MgO is 0%, 0.6%, 8%, 10% and adjusted the mixing amount of super-plasticizer to make sure the mixture reaches the same

IOP Conf. Series: Materials Science and Engineering **103** (2015) 012001 doi:10.1088/1757-899X/103/1/012001

slump from 20 mm to 50 mm.

2.3. Conditions and methods

Test specimens of the concrete are as follows: the standard specimen with 200 mm diameter and 500 mm height; medium specimen with 250 mm diameter and 600 mm height; and large specimen with 250 mm diameter and 600 mm height.

The test conforms with "Test Code for Hydraulic Concrete (SL352-2006)". When the volume of molding concrete turns into the standard test specimen, the coarse aggregate whose grain diameter is greater than 40 mm has been filtered out; when it comes to mold medium test specimen and large test specimen, the coarse aggregate whose grain diameter is greater than 40 mm has not been filtered out. Before sealing the test specimen, embedding vertically differential strain gauge into the central position of specimen barrel to read data. After sealing the test specimen, it is placed in the environment of constant temperature (20±2℃) and water-proofed.

Core and samples from drilling in the test specimen of concrete after 1 year of aging, are used to test the pore structure of the concrete.

3. Results and analysis

3.1. Results

Figs 1-4 show the volume deformation of MgO concrete from the test results for the specimens of different sizes. Also, the testing results of MIP specimens are summarized in Table 5.

3.2. Analysis

- As it can be seen from figure 1, no matter what sizes of the test specimens are, the volume deformation of concrete without MgO appears contraction state. In addition, before the age of 60 days, all the curves decreased sharply; after the age of 60 days, the contraction rate of volume deformation of the concrete dropped obviously, it is namely that it is appeared stable shrinkage trend. As a whole, the shrinkage value of the standard specimen is smaller than that of the medium and large specimen, but the value is smaller than that of the large specimen, only 5×10^{-6}.

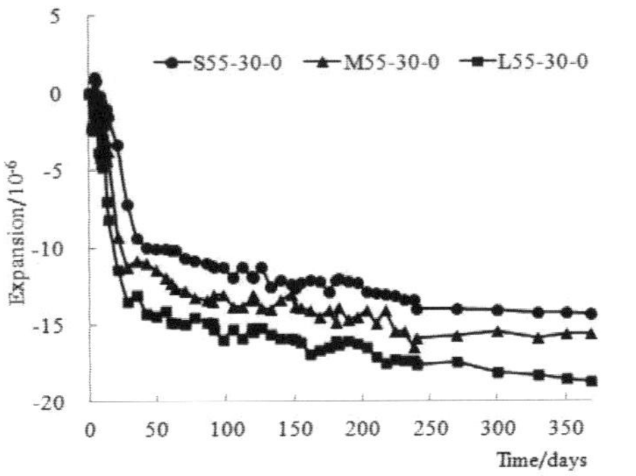

Figure 1. Effect of different specimen sizes on autogenous volume deformation of concrete.
[a] S, M and L show that the standard specimen, medium specimen, large specimen and so on.

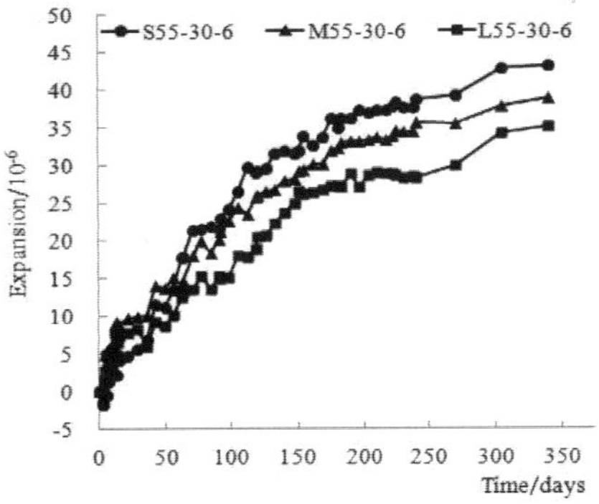

Figure 2. Effect of different specimen sizes on autogenous volume deformation of concrete with 6% of MgO.

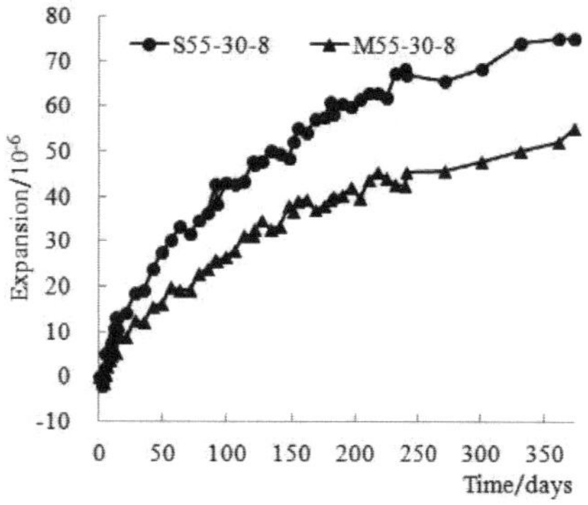

Figure 3. Effect of different specimen sizes on autogenous volume deformation of concrete with 8% of MgO.

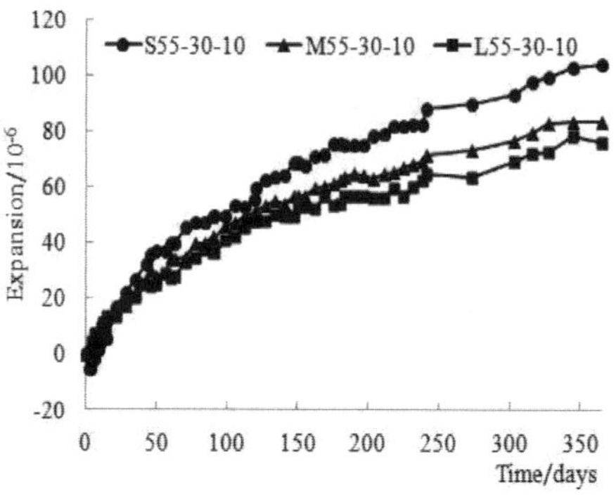

Figure 4. Effect of different specimen sizes on autogenous volume deformation of concrete with 10% of MgO.

Table 5. The MIP results of different specimens.

Items	Porosity /%	Average pore size/nm	Pore size (nm) distribution /%						
			<20	20～50	50～100	10^2 ～ 10^3	10^3 ～ 10^4	10^4 ～ 10^5	>10^5
S55-30-0	1.46	50.9	17.92	16.91	7.33	20.72	7.58	8.19	21.35
M55-30-0	9.65	19.10	36.14	37.33	9.71	7.30	2.81	3.35	3.36
L55-30-0	11.47	23.5	21.65	28.76	27.35	8.42	3.77	3.41	6.64
S55-30-6	13.04	22.2	23.56	29.13	25.7	10.06	3.96	3.67	3.92
M55-30-6	14.51	14.5	43.07	30.43	7.25	8.02	1.93	3.12	6.18
L55-30-6	13.56	24.00	22.02	32.54	28.16	9.77	2.63	2.95	1.93
S55-30-8	13.52	19.80	37.29	34.54	8.35	10.2	3.63	2.29	3.70
M55-30-8	15.14	19.40	34.98	32.13	10.95	11.56	4.15	2.32	3.91
S55-30-10	9.83	25.80	21.07	26.68	22.19	10.47	5.70	4.47	9.42
M55-30-10	17.30	15.2	46.97	30.30	5.97	7.69	3.45	2.89	2.73
L55-30-10	17.22	21.30	24.08	32.35	22.93	10.51	3.06	3.32	3.75

- As it can be seen from the results of Figs 2-4, the sizes of test specimens have obvious effects on the value of volume deformation of mixture MgO concrete. The general trend is that the measurement value of volume deformation of mixture MgO concrete decreases with the increase of sizes of the test specimen, standard specimen>medium specimen>large specimen. For instance, when it is with 6% of MgO, the measurement values of volume deformation of standard specimen, medium specimen, and large specimen at the age of 180 days are 36×10^{-6}, 32×10^{-6}, 27×10^{-6}, comparing large specimen with standard specimen and medium specimen, the values decreased by 15.6% and 25.0% respectively; similarly, when it is with 10% of MgO, the measurement values are 76×10^{-6}, 62×10^{-6}, 54×10^{-6} and the values decrease by 12.9% and 28.9% respectively.

Moreover, for 8% of MgO, as compared with the standard specimen, the medium test specimen volume deformation of the concrete increases by $(10\sim15)\text{x}10^{-6}$ before 200 days, and increases by $15\text{x}10^{-6}$ after 200 days. When it is with 10% of MgO, the measurement value of volume deformation of the concrete for the standard specimen increases by $(5\sim12)\text{x}10^{-6}$ than that of medium specimen before 200 days, and increases by $20\text{x}10^{-6}$ after 200 days; Similarly, the measurement value of volume deformation of the concrete of medium specimen increases by $(3\sim8)\text{x}10^{-6}$ than that of large test specimen before 200 days and increases by $10\text{x}10^{-6}$ after 200 days, which means that the growth between standard specimen and medium specimen is significantly higher than that between the medium specimen and large specimen. The differences between standard specimen and medium specimen are that the length of radical direction is increased by 50mm and the height is the same and the differences between medium specimen and large specimen are that the length of vertical direction is increased 100mm, the length of radical direction is the same, which prove that changes in the length of radical direction have a greater effect than that of the length of vertical direction on the measurement value of volume deformation of the concrete; on one hand, the changes of the length of radical direction may have larger constraining force on the volume deformation of the concrete than that of the length of vertical direction; on the other hand, the aggregate whose grain diameter is greater than 40mm has not been filtered out when the standard specimen has been molded, whereas, when the medium specimen and large specimen are molded, the aggregate whose grain diameter is greater than 40mm has been filtered out, then the MgO content in unit volume of standard specimen is more than that of the medium specimen and large specimen, resulting in more swelling deformations of standard specimen than that of the medium specimen and large specimen [8]. Of course, there also exists the other reasons that the grain diameter of aggregate in medium specimen and large test specimen is larger than that of in standard specimen, the compatibility between aggregate and cement paste interface decreased, consequently, swelling capacity caused by the hydration of a part of MgO absorbed, leading to the effective swelling capacity decreased [9].

- As it can also be seen from the results of Figs 2-4, the sizes of test specimen have no obvious effect on the deformation rate of the MgO concrete. The deformation rates of the three sizes are the same on the whole, but the deformation rate is a little larger before 180 days, after that, the deformation rate begins to decrease with the age. For example, the standard specimen mixed 8% of MgO, when the concretes are at 60 days, 180 days, 300 days, the corresponding measurement values of volume deformation are 33×10^{-6}, 60×10^{-6}, 68×10^{-6}, the deformation rate has been decreased from 0.23×10^{-6}/day before 180 days to 0.07×10^{-6}/day after 180 days.
- The results of Figs 2-4 prove again that no matter the size is large or small, the aggregate whose grain diameter is larger than 40mm or not, the volume deformation of MgO concrete increases with the increase of MgO mixture, and increases slowly with the age. Lu pointed out that the growth of $Mg(OH)_2$ crystal had been filled the holes of the concrete, which lead to the dense structure of concrete [10]. Before 28 days, the concrete is dramatically swelled. Between the age of 60 days and 200 days, the slope of swelling process of MgO concrete in three sizes reaches a maximum; the rates of deformation reduced after the age of 200 days, the deformation of concrete does not occur reduced with ages [11]. For instance, at the age of 360 days, when 6% of MgO is mixed, the measurement values of volume deformation of standard specimen, medium specimen and large specimen are 50×10^{-6}, 44×10^{-6} and 38×10^{-6}; when 10% of MgO is mixed, the measurement values of volume deformation of stand specimen, medium specimen and large specimen are 104×10^{-6}, 84×10^{-6} and 76×10^{-6}, which respectively increases by 108%, 91%, 100% than that of 6% of MgO.

This reason is that the increase of MgO mixture increases the amount of $Mg(OH)_2$ crystal created by hydration reaction, and the corresponding pressure and swelling force of crystal growth lead to the deformation of concrete. As for the expansive action, Chartterji believed that the expansion power of cement paste originates from the pressure of crystal growth created by MgO hydration reaction [12], the hydration reaction of the early age of MgO was rather intensive, the reaction speed was fast, elasticity modulus of the concrete at this time is rather low, which all contributed to the increase of deformation rate of MgO concrete; at the late age of MgO, once the hydration reaction came to an end, the dilatational strain was over, so the deformation of MgO concrete increased slowly with the extend of the age, but it would gradually become stable eventually.

- The influence of sizes of the test specimen on the interior pore structure of MgO concrete. The volume deformation of the test specimen of MgO concrete is not only related to the MgO mixture, but also relevant with porosity, including the average pore size and pore size distribution. It is always the outcome of multifactor. The larger proportion that the harmful pores and multiple damage pores inside concrete take, the bigger of the interior porosity and average pore size, then the degradation of interior structure of the concrete [13]. Al-Tabbaa pointed out that the influence of MgO concrete deformation mainly depended on the microstructure inside the concrete [14]. According to the pore sizes, the Wu indicated that the interior pore of the concrete was as follows: damage pores (>200 nm), harmful pores (50 nm～200 nm), less harmful pores (20 nm～50 nm) and harmless pores (<20 nm) [15].

As can be seen in Table 5, when the MgO has not been mixed, the size order of the interior structure of the concrete is as follow: large specimen (11.47%) > medium specimen (9.65%) > standard specimen (1.46%). Accordingly, the measurement values of shrinkage of the large specimen, medium specimen and standard specimen at the age of 360 days are 18×10^{-6}, 15×10^{-6}, 14×10^{-6}, which has proved that the shrinkage of the volume deformation become bigger with the increasing of interior porosity of concrete ,When mixed with 6% of MgO, the interior porosity of the standard specimen of the concrete (13.04%) is smaller than that of medium specimen (14.51%), the measurement value of dilatational strain of the volume of the test specimen is smaller than that of standard specimen, which is shown in Fig2. This is because the larger interior porosity of the concrete can absorb more MgO than that of the smaller interior porosity of the concrete, which results in the reducing of the ultimately valid swell capacity of MgO concretes. Meanwhile, when mixed with 6% of MgO, the proportion of the harmful pore and multiple damage pore inside the large specimen and medium specimen is 45.44% and 26.5%, both the harmful pore and multiple damage pore absorb the hydration products of MgO, leading to the measurement value of the volume deformation decreases with the increase of the size, which is shown in Fig2. When mixed with 8% of MgO, the interior porosity of the concrete of the standard specimen and medium specimen is 13.52% and 15.14%, the proportion of harmful pore and multiple damage pore is 28.17% and 32.89%, leading to the measurement value of the volume deformation decreases with the increase of the specimen size; which is shown in Fig3. When mixed with 10% of MgO, the interior porosity of the concrete of the standard specimen (9.83%) is the smallest, the medium specimen (17.30%) and large specimen (17.22%) are much larger than that of the standard specimen, and the porosity of medium specimen and large test specimen are very close. Therefore, Fig4 also shows that the measurement values of the volume deformation decrease with the increase of the specimen sizes.

4. Conclusions

The specimen size has an obvious influence on the autogenous volume deformation of MgO concrete. It is largest for the standard specimen, then the medium specimen and the third the large specimen. Besides, changes in radical dimension of the test specimen have a more obvious influence on the measurement value of autogenous volume deformation of MgO concrete than that of the vertical dimension.

The specimen size also has an influence on the interior the pore structure of MgO concrete. The larger size of the specimen, the worse parameter values of the pore structure of the concrete; otherwise, the smaller size of the specimen, the better parameter values of the pore structure of the concrete. Changes of the parameter values of the porosity character inside the test specimen are the internal causes for autogenous volume deformation of MgO concrete with different sizes.

Once again, the autogenous volume deformation of MgO concrete increases with the increase of MgO mixture is verified; the deformation rate of MgO concrete is larger at the early age, then it becomes slower at the later age and stable eventually.

Acknowledgments

This research is supported by the National Natural Science Foundation of China (Grant no. 51269003) and Guizhou Province "125" major scientific and technological project (Grant no. [2012] 004).

References

[1] Mehta P K and Pirtz D 1980 Magnesium oxide additive for producing self stress in mass

concrete *J. Proceedings of the 7th International Congress on the Chemistry of Cement* **3** 6-9
[2] Mehta P K 1973 *J. Cement and Concrete Research* **3** 1-6
[3] Li C M and Yuan M D 2003 *J. Advances in Science and Technology of Water Resources* **23** 57-63
[4] Fang K H 2004 *J. Hydroelectric Engineering* **24** 46-9
[5] Chen C L 2013 *J. Hydroelectric Engineering* **32** 253-6
[6] Li W W, Chen C L, Fang K H and Du Y 2011 *J. Water Resources and Power* **29** 58-9
[7] China Institute of Water Resources and Hydropower, Nanjing Hydraulic Research Institute, Chang Jiang and Water Resources Commission Academy of Sciences 2006 Hydraulic Concrete Testing Procedures (SL352-2006) (Beijing: China Water Power Press) **4** 124-6
[8] Zhu B F 2003 *J. Hydroelectric Engineering* **2** 2-12
[9] Gan C C, Li J T, Nv W Q, Guan Z W and Mai R 2007 *J. Concrete* **9** 16-7
[10] Lu X L 2013 *J. Construction and Building Materials* **40** 998-1001
[11] Liu S H and Fang K H 2007 *J. Hydroelectric Engineering* **2** 23-4
[12] Chartterji S 1995 *J. Cement and Concrete Research* **1** 51-6
[13] Chen H X, Ma X W and Chen B L 2003 *J. Materials Science and Engineering* **21** 361-3
[14] Al-Tabbaa A 2014 *J. Cement and Concrete Research* **1** 1-12
[15] Wu Z W and Lian H Z 1999 *M . High performance concrete* (Beijing: China Railway Press) **2** 24-5

IOP Conf. Series: Materials Science and Engineering **103** (2015) 012002 doi:10.1088/1757-899X/103/1/012002

Experimental study on the microstructure evolution of 55SiMnMo

Y Q Zhao[1,4], H Y Qi[2], F F Liu[1], X G Wang[1] and Z H Ma[3]

[1] Northeastern University at Qinhuangdao, Qinhuangdao, China
[2] Yanching Institute of Technology, Beijing, China
[3] College of Information Science and Engineering, Northeastern University, Shenyang, China

E-mail: zhaoyanyuqian@163.com

Abstract. Isothermal compressive experiments on 55SiMnMo steel were carried out with the Gleeble 3500 hot-simulation machine. High temperature flow stress-strain curves were measured over the deformation temperature range of 950 to 1050℃ and a strain rate from 0.01 to 10 s^{-1}. Experimental results revealed that the peak stress decreases with increasing deformation temperature and decreasing strain rate. In addition, when the deformation temperature T ≥ 1000℃, and the strain rate $\dot{\varepsilon} \leq 0.1$ s^{-1}, the dynamic recrystallization of 55SiMnMo steel occurs. The stress and strain constitutive models and austenite recrystallization model were constructed to form the foundation for studying the forming process of drill rods.

1. Introduction

Drill steel is a kind of tool steel that is often used in the mining industry to produce drill rods, which are connected to rock drills and drill bits. Usually, drill rods have a hexagonal outer cross section that transfers torque from the rock drill in addition to a hollow center that allows high-pressure water to pass through and wash away rock powder. During operation, the rods are subjected to impact loads that cause considerable alternating stresses in the metal. In addition, the rods are subject to corrosion due to the water flowing through the hollow center. Therefore, it is necessary to construct rods with high-quality surface metals and accurate geometrical dimensions.

The drill rod is a performance component that transfers impact energy. In modern impact drilling, a drill rod ordinarily functions under fatigue load; therefore, the performance requirements for drilling steel should take into consideration the stress state of steel under fatigue loading, and in addition, a wet drill should also have good resistance to water corrosion. Furthermore, drill rods are most likely to fail due to fatigue fractures. If the requirements are not correctly implemented, then the drill experiences a fatigue-induced crack; then, it undergoes the fatigue arc zone development process, and finally, the final fracture occurs. The maximum stress that is borne by drill rods is the superposition of axial tensile stress and bending stress. Therefore, the requirements for drill steel include high purity, lower notch sensitivity, and fatigue resistance. In addition, the effectiveness of the drill steel production process and brazing process also should be considered. In a heat treatment, the drill rod is a

[4] Address for correspondence: Y Q Zhao, Northeastern University at Qinhuangdao, Qinhuangdao, China. E-mail: zhaoyanyuqian@163.com.

IOP Conf. Series: Materials Science and Engineering **103** (2015) 012002 doi:10.1088/1757-899X/103/1/012002

long rod with a great slenderness ratio. Moreover, if the whole drill rod needs to undergo a quenching heat treatment, the equipment is complex, and deflection is inevitable. Furthermore, there are more than fifty types of drilling steel grades. Most carbon tool steel has 0.702% to 1% carbon content, such as Z708 in Sweden and LWS2 in Germany, and bearing steel and alloy structural steel with Ni, Cr, and Mo—such as 95CrMo and 30CrNi3Mo—are also used in drilling steel.

Drilling steel has been extensively researched. For example, Belusov reported the manufacturing experience with hollow drilling steel at various metallurgical plants. The U7 carbon steel billet was employed with a square cross section, with a hole through the center with an EI94 austenite steel core [1]. Furthermore, Larkin developed a technique to manufacture drilling rods by using high-frequency current in the final heat treatment [2]. Postonogov designed a pass system to roll the hexagon drilling steel, which is an oval -- edging oval -- special leader pass -- open finishing process [3]. Moreover, Petrov analyzed force and torque and determined that drilling rod breakdown is not caused by stability loss but is instead caused by the bending deformation of its free end that occurs as a result of deviation of the drill bit from the set direction due to unevenness in the rock [4]. Song studied the stress wave theory and strength design principle as well as discussed the dynamic responses of the fatigue strength of different rod steels under a typical technological process [5]. Finally, Xing acquired the fatigue test parameters of the tapering connection drilling rod through analyzing the combination of stress and strain wave [6].

Understanding material flow behavior in hot deformation by modeling is important for the analysis and optimization of metal-forming processes [7, 8]. In this paper, the hot compression test and the plastic finite element method were used to investigate the microstructure evolution of 55SiMnMo steel in the hot deformation process.

2. Experimental procedure

2.1. Preparation of test specimen

55SiMnMo was designed based on an expected microstructure and based upon certain limited resources in China [9]. (There is a lack of Ni and Cr resources in China.) The chemical composition and the material properties of the 55SiMnMo steel are shown in tables 1 and 2, respectively. The drilling steel production process is a major factor in determining the overall quality and performance. Therefore, a study of the microstructure evolution process of 55SiMnMo steel has both important theoretical and practical significance [10].

Table 1. Chemical composition of 55SiMnMo.

Grade	C	Si	Mn	Mo	S	P	Cu
ZK55SiMnMo	0.50-0.60	1.10-1.40	0.60-0.90	0.40-0.55	≤0.030	≤0.030	≤0.025

Table 2. Material properties of 55SiMnMo.

Modulus of elasticity E/Pa	Poisson's ratio	Density kg/m^3	Yield limits /MPa	Coefficient of thermal expansion /K	Specific heat capacity J/(kg×K)	Heat conduction coefficient $W/(m^2\times K)$
2.0×10^{11}	0.3	7.85×10^3	634	1.1×10^{-5}	452	48

The 55SiMnMo steel was studied through hot compression tests that were conducted by the Gleeble 3500 thermal simulation-testing machine in order to simulate the austenite evolution process of specimen compression. The results were measured at different temperatures so as to determine deformation rates under various conditions and to obtain true stress data.

A wire-cutting machine obtained Φ6×9mm pieces of H22 drilling steel, and the specimen surfaces were polished with alcohol and a wipe.

In the isothermal compression experiment, the sample's friction surface is the main factor that affects the test precision. Theoretically speaking, the flow stress could only accurately reflect the real situation of plastic deformation when the following conditions were met: (1) when the compressed samples did not have a central bulge and (2) when their axial strain and transverse strain were equal. Therefore, the sample surface friction had to be reduced in order to guarantee the accuracy of the compression test. For the test, a tantalum chip of a 0.1 thickness was placed between the plug and the specimen, and the tantalum chip was isolated and lubricated so as to reduce friction. In addition, deformation prevented oxidation.

2.2. *Hot compression test*

Figure 1 shows the simulation process. First, the specimens were heated to 1150°C at a rate of 10°C/s. After the maximum temperature was achieved, the temperature was maintained for 3 min in order to obtain a uniform austenite. Next, the specimens were cooled to 1050°C, 1000°C, and 950°C at a rate of 5°C/s so as to eliminate the temperature gradient inside the specimens. Thirdly, they were respectively compressed to 50% with strain rates of $0.1s^{-1}$, $1s^{-1}$, and $10s^{-1}$, with online measuring and recording of the true stress data. Real-time acquisition of the specimen temperature, strain, and stress data was obtained according to the deformation time setting and the appropriate sampling period. Then, the quench heat treatment was performed on 35 specimens, and the sample photos are shown in figure 2. Finally, metallographic specimens were made so as to obtain an average grain size using quantitative metallography.

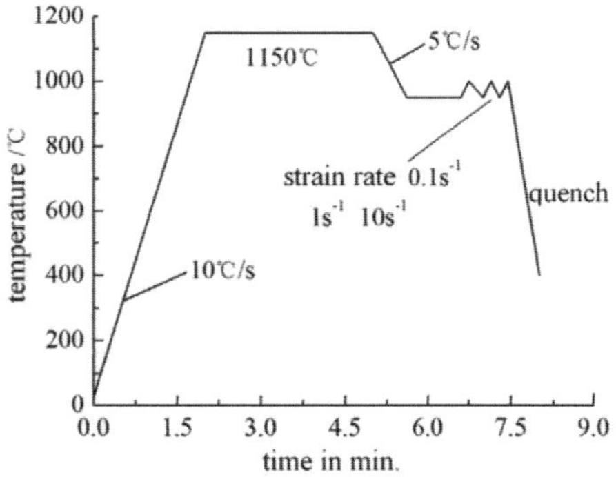

Figure 1. Schematic diagram of simulation process.

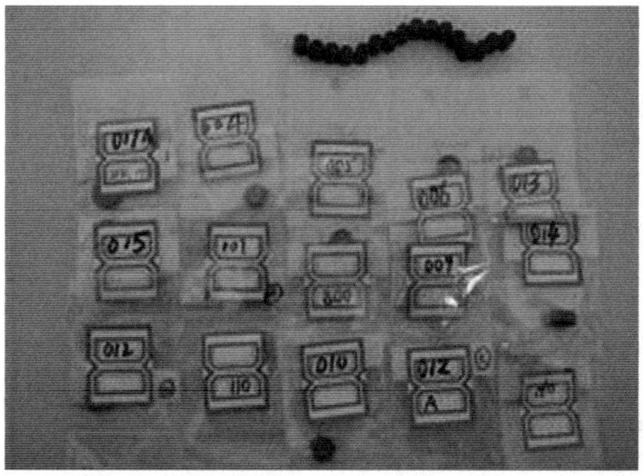

Figure 2. Deformation specimens.

3. Results and discussion

3.1. *Stress-strain curve of hot deformation*

High temperature flow stress is characteristic of a basic data of metal plastic processing performance. The research of materials in pressure processing on rheological stress has important academic significance and engineering value. Figure 3 details 55SiMnMo compression tests at different strain rates and deformation temperatures of the true stress-strain curve.

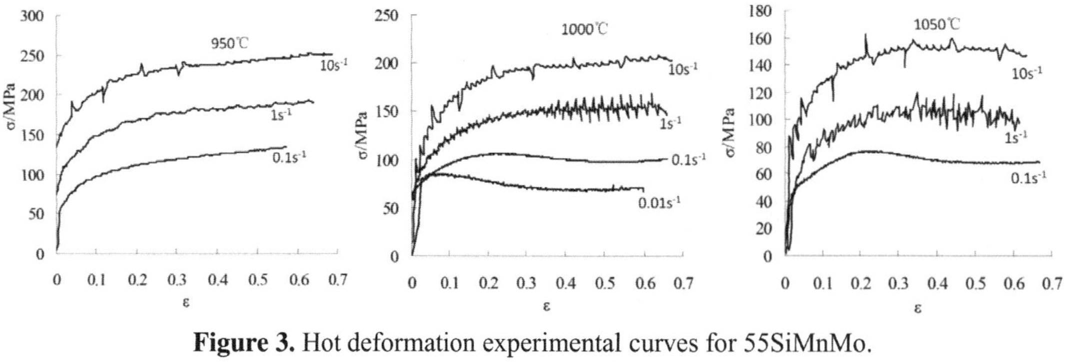

Figure 3. Hot deformation experimental curves for 55SiMnMo.

Analysis of the stress-strain curves shows that:

- Peak stress decreases with increasing deformation temperatures and decreasing strain rate.
- The curves can be classified as one of two types: dynamic recrystallization and dynamic recovery. In figure 3, those classified as dynamic recrystallization are stress-strain curves $0.1s^{-1}$, $0.01s^{-1}$ at 1000℃ and $1s^{-1}$, $0.1s^{-1}$ at 1050℃. The dynamic recovery stress curve is shown as a strain curve. Initially, the flow stress increased rapidly; however, when hardening and dynamic recovery reached the equilibrium state, the increase began to level off, and the stress remained close to a constant value.
- When the deformation temperature was greater than or equal to 1000℃ and the strain rate was less than or equal to 1/s, 55SiMnMo experienced dynamic recrystallization.
- The thermal simulation machine collected the experimental data. Each curve in the deformation area has a sampling point between 280 and 350. When the strain rate was larger than 0.1/s, the flow stress curves exhibited white noise. Matlab software was used to eliminate noise and obtain peak stress and peak strain in different deformations.

3.2. The flow stress constitutive equation

In addition to the metallurgical factors of the material itself, flow stress also depends on the temperature, deformation, and strain rate. The constitutive equation was built to describe the relationship between the flow stress and other influences by physical simulation and numerical modeling. In the flow stress constitutive equations, equation (1) is suitable for high temperatures and low strain rate conditions, and the process is controlled by diffusion. On the other hand, equation (2) is suitable for low temperatures and high strain rate conditions, and the process is controlled by slip.

$$\dot{\varepsilon} = A\sigma^{n} \exp\left(\frac{-Q}{RT}\right) \tag{1}$$

$$\dot{\varepsilon} \exp\left(\frac{Q}{RT}\right) = B\exp(\beta\sigma) \tag{2}$$

where T is the thermodynamic temperature; Q is the activation energy; R is the gas constant, 8.314 J/(K·mol); and A, n, B, β, C, α, and m are all material constants.

$$\beta = \alpha n \tag{3}$$

Peak strain (ε_p) constitutive equation:

$$\varepsilon_p = 4.362\times10^{-4} d_0^{0.3} \dot{\varepsilon}^{0.1131} \exp\left(\frac{54283}{RT}\right) \tag{4}$$

We got the parameters using above medals by experimental data and show as table 3.

Table 3. Coefficient of flow stress constitutive equation.

A	n	Q	R	B	β	α	m	C
314.0	6.43	314219.8	8.314	9.417e12	0.0396	0.00616	5.22	2.55e15

3.3. Recrystallization model

In dcformation conditions, the deformation temperature was 1000℃, and the strain rate was $0.1s^{-1}$. As can be seen in figures 4(a)-4(f), six different strains were tested, ranging from 0 to 0.693. Figure 4 shows the core photomicrographs. The true strain of every specimen was calculated by using formula (5):

$$\varepsilon = \ln \frac{h}{h_0} \tag{5}$$

where h is the height (mm) of the specimen after deformation, and h_0 is the height (mm) of the specimen before deformation.

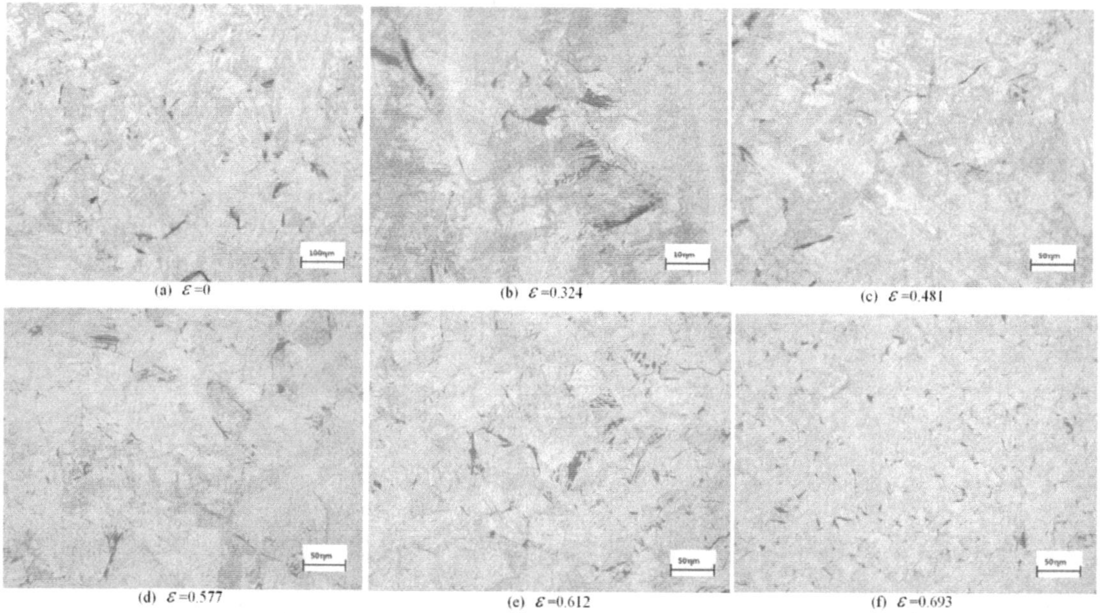

Figure 4. Grain size at different strains (1000℃, 0.1 s^{-1}).

To determine the austenite grain size, saturated picric acid and a small amount of Seagull brand shampoo were used as the etchant. The specimens were placed into an HHS-112 electric heated water bath whose temperature ranged from 50 to 60 ℃, and then, they were immersed in the etchant for 5 to 7 minutes. Afterwards, the surfaces were wiped and dried. Finally, the Austenite grain boundaries were observed using an optical microscope, as are shown in figure 4.

In figure 4(a), the specimen strain is 0. This specimen was heated to 1150ºC at a rate of 10ºC/s and then kept at this temperature for 3min. It was subsequently cooled to 1000ºC at 5ºC/s, maintained at this temperature for 1 min, and then quenched. This specimen's initial grains were in a non-deformation condition. The subsequent specimens underwent the same heating process as this specimen.

In figure 4(b), the specimen strain is 0.324. Its final height was 6.51 mm. The figure shows that there are some small dynamic recrystallization grains in large crystal grain boundaries. Recrystallization primarily occurred in the original location of the austenite grain boundaries. The dynamic recrystallization grain size was small (<10 μm), whereas the grain size of those with no recrystallization was still large.

In figure 4(c), the strain of the specimen is 0.481. This specimen was also deformed by a strain rate of 0.1s^{-1}, and its final height was 5.54 mm. The figure shows a fine recrystallized grain size and large grains with incomplete grain boundaries. The increase in deformation caused the initial large grains to rupture.

From figures 4(c)-4(f), the specimen strains increased, and dynamic recrystallization continued to occur and accumulate with increasing strain. At the largest tested strain (0.693), the grain size was significantly uniform and refined.

We measured the dynamic recrystallization percentage (Xd) and the average austenite grain size (d) of the specimens, which were deformed by the same temperature but at a different strain rate and strain.

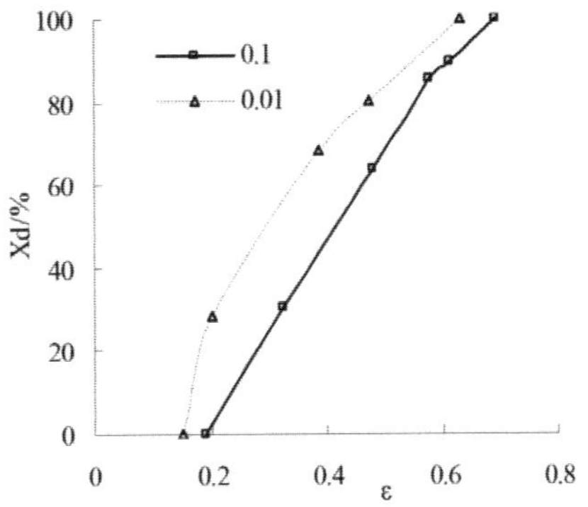

Figure 5. Curve of Xd.

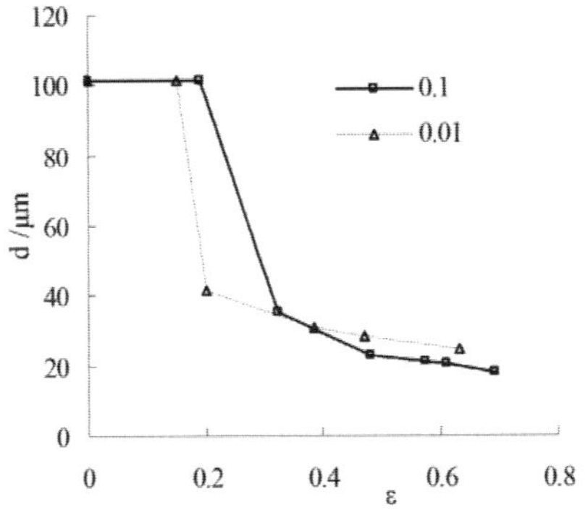

Figure 6. Crain size complied with strain.

Figure 5 graphs the change in the dynamic recrystallization percentage (Xd) complied with the strain; the faint line represents the specimens that were deformed by the 0.01 s^{-1} strain rate, and the dark, solid line represents the specimens that were deformed by the 0.1 s^{-1} strain rate. This figure shows that: 1) after reaching critical, the dynamic recrystallization percentage accumulated as the strain increased; 2) when the strain reached 0.7, the dynamic recrystallization percentage increased to 100%; and 3) in the same strain condition, when the strain rate was small, the recrystallized percentage was large.

Figure 6 shows the change in grain size complied with strain. Again, the faint line represents specimens deformed by a 0.01 s^{-1} strain rate, and the dark, solid line represents specimens deformed by a 0.1 s^{-1} strain rate. The figure shows that the dynamic recrystallization size (d) is smaller when the strain rate is larger. Therefore, within the scope of dynamic recrystallization, improving the strain rate increases grain refinement.

Using linear regression calculation, both the dynamic recrystallization mathematical model and the dynamic recrystallization grain size model can be obtained:

$$X_d = 1 - \exp\left[-0.693\left(\frac{\varepsilon - 0.83\varepsilon_p}{\varepsilon_{0.5}}\right)^{3.15}\right] \tag{6}$$

$$\varepsilon_{0.5} = 0.037 d_0^{0.2} \dot{\varepsilon}^{0.058} \exp\left(\frac{1.2\times10^5}{RT}\right) \tag{7}$$

$$d = 1.301\varepsilon^{0.61}\dot{\varepsilon}^{-0.14}\exp\left(1.2\times10^5 / RT\right) \tag{8}$$

where Xd is the dynamic recrystallization percentage, $\varepsilon_{0.5}$ is the strain when Xd is at 50%, and d is dynamic recrystallization size.

4. Conclusions

Studies on hot deformation behavior and critical strains for dynamic recrystallization are important in controlling the metal-forming processes and for improving production quality. This paper studied the hot deformation behavior and microstructure evolution of 55SiMnMo steel, which is a bainitic carbon alloy steel used for drilling rods.

The stress-strain curves of the 55SiMnMo steel in a wide range of temperatures and strain rates were measured by hot isothermal compression tests. The experimental results reveal that the peak stress decreases with increasing deformation temperature and decreasing strain rate. Moreover, the dynamic recrystallization of 55SiMnMo steel occurs when the strain $\dot{\varepsilon}$=0.01s^{-1}, or when $\dot{\varepsilon}$=0.1s^{-1} and the deformation temperature T ≥ 1000℃.

The constitutive equation was built to describe the relationship between the flow stress and temperature (T), activation energy (Q), gas constant (R), and material constants by physical simulation and numerical modeling.

Under the deformation condition of T=1000℃, $\dot{\varepsilon}$=0.1s^{-1}, and strain from 0 to 0.7, the dynamic recrystallization percentage (Xd) and the average austenite grain size (d) of the specimens were measured by metallurgical tests. Figure 6 shows that some small dynamic recrystallization grains first appear in large crystal grain boundaries. When the strain of the specimen was 0.693, the grain size was significantly uniform and refined. Using linear regression calculation, the dynamic recrystallization mathematical model and the dynamic recrystallization grain size model were obtained.

The model's correctness was tested and validated by combining the microstructure model with the finite element method.

Acknowledgments

This work was supported in part by the National Natural Science Foundation of China (31170956), the Fundamental Research Funds for the Central Universities (N120223001), the Natural Science Foundation of the Hebei Province (H2015501133), and the Science and Technology Research & Development Project Funds of Shenzhen (JCYJ20120618142137681).

References

[1] Belousov A S and Konshin P P 1957 Improving the quality of hollow drill steel *Metallurgist* **1** 103-11
[2] Larkin F R, Kudrya N A, Chubilin A M and Artemenko N S 1974 Technique of manufacturing drill rods *Metal Science and Heat Treatment* **16** 134-5
[3] Postonogov Yu A and Nikiforov V V 1980 New shape of leader pass for rolling hexagons *Metallurgist* **24** 285-7
[4] Petrov N G and Pasynkov R E 1993 Deformation and stability of drill rod *Journal of Mining Science* **29** 132-5
[5] Song S Z and Rong L 1996 Dynamic response of fatigue strength of drill rod to theoretical load spectrum of rock drill *Journal of Northeastern University* **17** 579-83
[6] Xing J Z 2011 The study of the parameter on the hollow steel fatigue life test *Modern Machinery* **1** 89-91
[7] Pater Z, Kazanecki J and Barmicki J 2006 Three dimensional thermo-mechanical simulation of the tube forming process in Diescher's mill *Journal of Materials Processing Technology* **177** 210-5
[8] Tieu A K, Jiang Z Y and Lu C A 2002 3D finite element analysis of the hot rolling of strip with lubrication *Journal of Materials Processing Technology* **125** 638-44
[9] Tao Y, Yu E L and Zhao Y Q 2013 Modeling of hot deformation behavior of 55SiMnMo medium-carbon steel *Journal of Iron and Steel Research, International* **20** 125-30
[10] Iankov R 2003 Finite element simulation of profile rolling of wire *Journal of Materials Processing Technology* **142** 355-61

IOP Conf. Series: Materials Science and Engineering **103** (2015) 012003 doi:10.1088/1757-899X/103/1/012003

Research on geometrical model and mechanism for metal deformation based on plastic flow

H P An[1,2,3], Z Y Rui[2] and X Li[1]

[1] School of Bailie Engineering Technology, Lanzhou City University, Lanzhou 730070, China
[2] College of Mechano-Electronic Engineering, Lanzhou Univ. of Tech., Lanzhou 730050, China

E-mail: ahp2004@126.com

Abstract. Starting with general conditions of metal plastic deformation, it analyses the relation between the percentage spread and geometric parameters of a forming body with typical machining process are studied. A geometrical model of deforming metal is set up according to the characteristic of a flowing metal particle. Starting from experimental results, the effect of technological parameters and friction between workpiece and dies on plastic deformation of a material were studied and a slippage deformation model of mass points within the material was proposed. Finally, the computing methods for strain and deformation energy and temperature rise are derived from homogeneous deformation. The results can be used to select technical parameters and compute physical quantities such as strain, deformation energy, and temperature rise.

1. Introduction

Metal deformation or machining is the process of shaping materials by either large plastic deformation or material removal through the relative motion of the tool and workpiece. There is a common nature for various craft processes that is the slipping of metallic material until cracks initiate, grow and fracture under extrusion. Past research on machining characteristics were used to set up the empirical formulas of the cutting force, the cutting temperature, and the cutter life by means of testing [1, 2]. To obtain the reasonable cutting parameters, a lot of experiments were needed to establish the cutting database, because of the variety of different of workpiece and cutter materials, as well as technical conditions. In fact, when using these cutting parameters they must still be modified to specific processing requirement: Besides, many empirical equations of cutting theory are the results of mathematical treatment of test data for a certain conditions. This information can guide understanding the law of metal deformation, but it is insufficient to predict the deformation mechanism. So this paper sets up the models of metal plastic deformation via studying its physical essential so as to probe into the mechanism of plastic deformation and influence of the geometrical factors. A reference to obtain some parameters at high speed can be offered for high efficiency.

2. Nature of metal plastic deformation

[3] Address for correspondence: H P An, School of Bailie Engineering Technology, Lanzhou City University, Lanzhou 730070, China. E-mail: ahp2004@126.com

Machining of metal depends on plastic deformation to get the required shape and dimension. Flow regularity of a metal particle is the most basic law. According to the relation equation between strain and displacement or geometric equation, the displacement field, and velocity field in the deforming body are set up to determine the strain value. It is helpful to see the deformation of an object and choose reasonable technical parameter for good processing quality. Meanwhile, it can provide a reference for the design and manufacture of tool and mold.

2.1. Metal flow law in plastic deformation

Keeping constant volume in plastic machining, the mass point has to move in the direction of the lowest resistance that is called the law of minimum resistance. In other words, the mass point will go the shortest path requiring the least energy. For this reason, we can control deformation by altering process conditions to adjust the magnitude of resistance in the certain direction [3]. The rectangle area of heavy line shown in Fig 1 is the geometrical model of moving masses point when a prism with rectangular cross-section is being compressed. Because of the effect of friction between the product and the forging dies, the resistance exerted on the mass point is proportional to the distance from side. Therefore, the mass points will certainly flow along the nearest normal direction of the boundary as material deforms. The rectangle shown in Fig 1 is divided into four areas of two triangles and two trapezoids marked with dot-dash line, in which mass points flow to the nearest border. It turns out that the rectangular cross-section becomes a polygon marked in double dot-dash line. The cross-section will finally become an oval that tends to minimize the perimeter of it. In the next change, all mass points flow to its radius direction. Circular cross section has the smallest perimeter in the same area of all shape section. So the law of the smallest resistance is equal to that of the smallest rim. It can be demonstrated in the following three models.

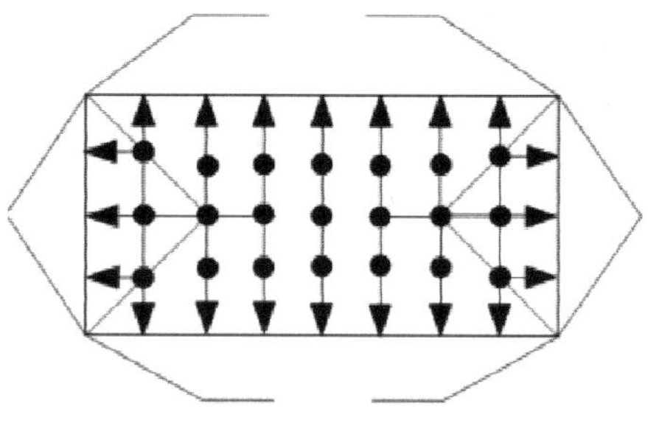

Figure 1. Diagram of minimum rim law.

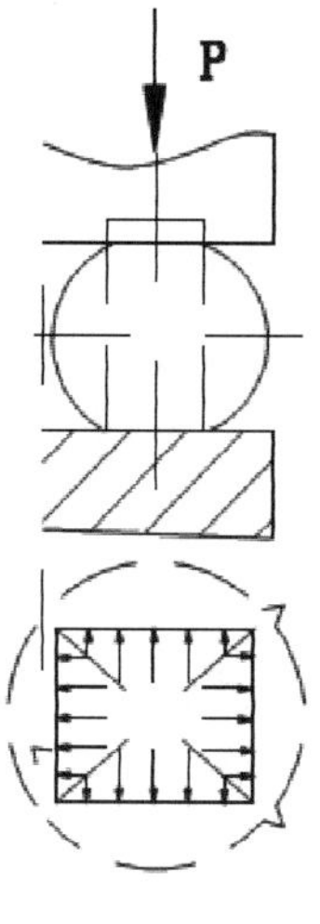

Figure 2. Deformation of square section.

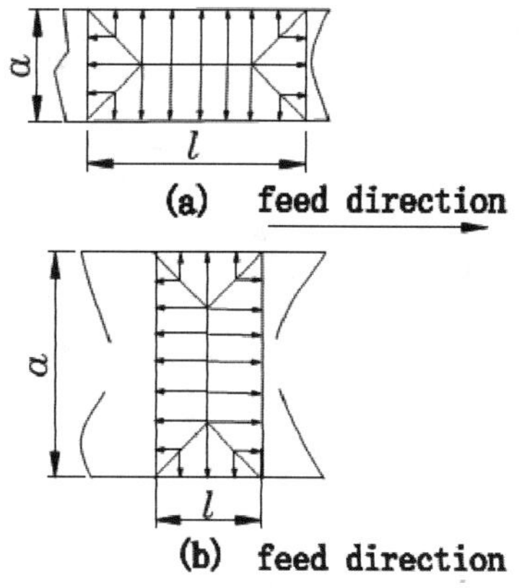

Figure 3. Deformation of swaging rectangle section.

- First is about the plastic deformation of square section material. According to the law of minimum rim, the deformation of square section material between up and down flat hammer tends to the imaginary line circle showed in figure 2.
- Next we consider the plastic deformation of a rectangular section of material. As the rectangular section of material is being elongated between anvils without regard to external impact, the transformation model relates to the relative ratio of feed (l) and width (a) of the material. When l>a most mass points transverse flow, or the wide increases more than the long showed in figure 3(a). As l<a most mass points longitudinal flow, the long (l) elongates more the wide showed in figure 3(b). The area of two triangles marked in dot-dash line represents

number of mass points flowing along the length in figure 3. From above we can know that the percentage spread of a narrow board is greater than that of a wide board when rolling plate. Considering the impact of unshaped portion, displacement direction of mass points will change comparing with those mentioned above.

- The third example considered concerning the deformation of a rolling plate. Deformation of a rolling plate is being connected with roller diameter. If keeping the same rolling reduction and width, a large diameter roller has a bigger percentage spread than a small one. In Figs 4(a), 4(b), H and h are the thicknesses of the plate before and after rolling, respectively, the width before rolling is B_1. Because of the difference of diameters of the rollers, that is $R_2>R_1$, the rolled width of the big diameter roller is larger than that of the smaller, i.e. $B'_2>B_2$. Therefore, we can control percentage spread to use cluster mill with small diameter when finish rolling is performed.

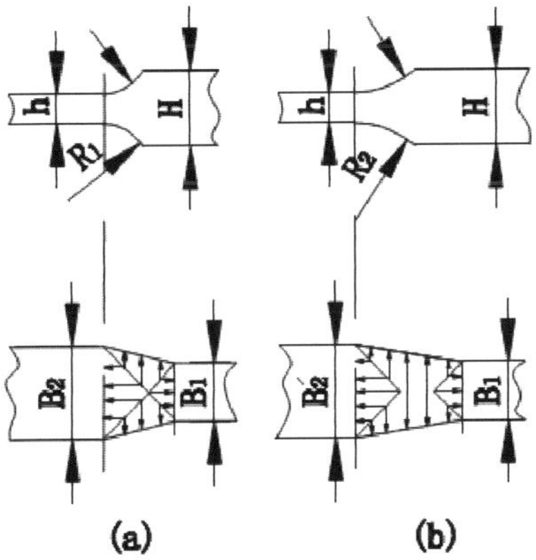

Figure 4. Effect of roller's diameter on vertical and horizontal resistance.

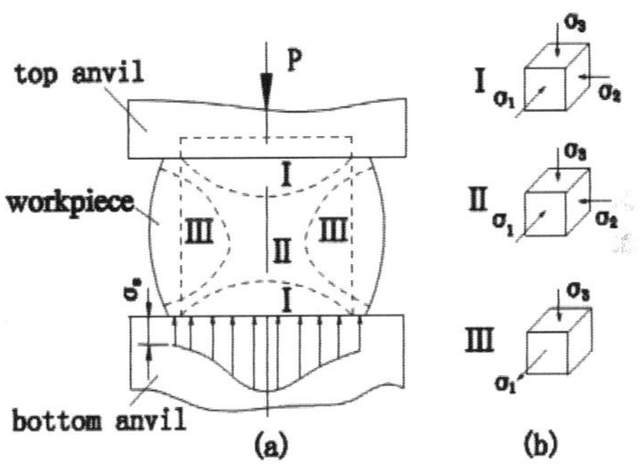

Figure 5. Effect of friction on deforming and stress distribution of pressing cylinder.

2.2. Effect of friction on plastic deformation

All factors acting on mass point flow will also incur plastic deformation [4].

- This concerns the impact of contact friction on the deformation of a solid. Under tool compression, the friction on the interface will have a certain effect on deformable body. The reason why cylindrical material changes into drum-type under pressure between up and down anvils is because of the friction effect. Friction forces make it more difficult to flow for the material nearby the contact interface than that near center section. The geometrical model showed in figure 5(a) has been divided into three different areas. It is the most difficult to deform in zone I due to the external frictional influence on the material while it is the easiest in zone II for the force acting at an angle of 45° with the material zone. The zone III is a free deformation circular area which is outside of the zone II. The stress state corresponding three areas are being showed in Fig 5(b). The zone I and II are three-way pressure stress state while the zone III is two-dimensional stress state of pulling stress and pressure stress. It shows that the friction affects not only uneven stress on the contact surface but also the deformation of the material. Such as the stress on an edge of the sample is equal to the yield limit (σ_s) of the material but it becomes bigger and bigger nearby the central area. Besides, the stress distributed from the interface to the middle of the body reduces [5-8].
- This concerns the impact of contact friction on deformation of ring unit. Friction will locally change the direction of flowing of a mass point when a ring body is pressed down. A ring body before pressing is being showed in figure 6(a), R_o is an external radius and R_i is an inner

radius, and H is a height of the annulus. It shows the geometrical model of the ring body being pressed in figure 6(b) when the friction coefficient (μ) is tiny. According to the condition of constant volume of plastic deformation, each material point flows outward along a radial direction. Then both inner diameter (r_i) and outer diameter (r_o) enlarge, that is, $r_o > R_o$, $r_i > R_i$. When the friction coefficient on interface increases (or $\mu \neq 0$), the prevention of its force of friction to the flow of metal also increases. Furthermore, as the friction coefficient increases to a critical value, the resistance of the material points received near the inner diameter of ring body is so big that they change moving direction from outward to inward. Therefore, a shunt surface with radius (R_n) appears in an annulus so that the inner material flows inward, and the outer material flows outward. The inner diameter reduces and external diameter increases after deformation, thus, $r_o > R_o$, $r_i > R_i$. The radius (R_n) of the shunt surface increases with the friction coefficient on interface and it is less than the mid value of the wall thickness of the annulus. The reason of it is an inward flow of metal bringing about decrease of the inner diameter and increase of pressure stress, respectively. The direction of a radial component of the pressure stress fits that of the frictional resistance of shunt surface of the annulus.

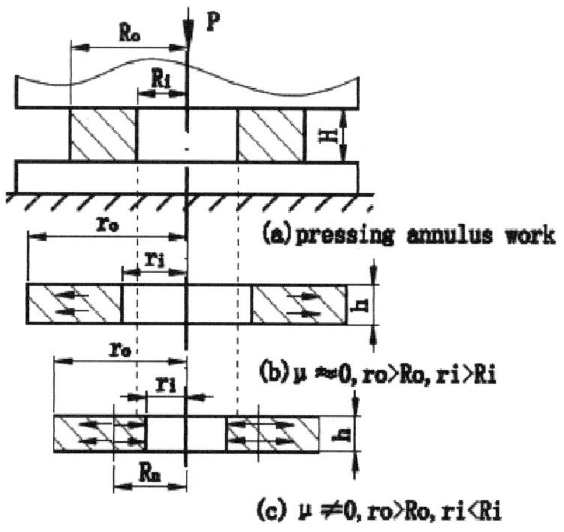

Figure 6. Model of material flow for pressing annulus.

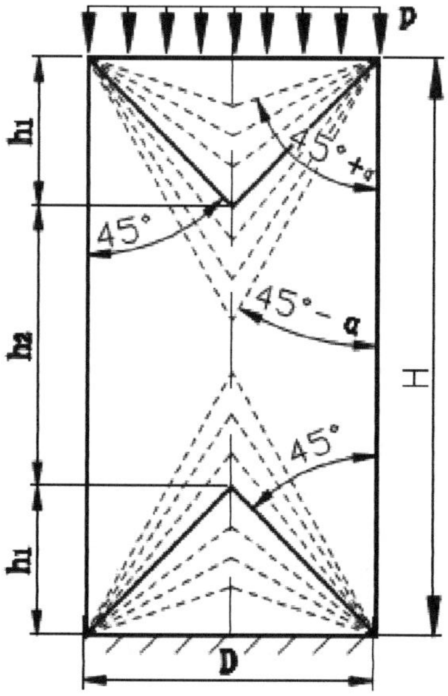

Figure 7. Model of slipping mass point for pressing material.

2.3. Geometrical factors influencing plastic deformation

The interior material slips and extend when a pressure is being exerted on the metal. According to plastic yielding criteria, slip band is some orthogonal mesh line. Initially, a slip line is at an angle of 45° to the acting force. Slipping enlarges gradually as the rolling reduction increases. At this point, when a cylinder with height (H) and diameter (D) between hammer flatteners is forged or pressed, a model of the slip line is being showed in Fig 7. We make an isosceles right triangle with a bottom of diameter (D), and a height of the half bottom as a heavy line showed in Fig 7. The triangle is called a basic cone or a main cone, in which two sides of the triangle are at angle of 45° to the acting force. A plastic deformation firstly occurs near the main cone when the cylinder is forged because its shear stress at angle of 45° to force is the largest and easy to slip. With deformation going on, the slip appears inside and outside the main cone. Slip lines inside the main cone near the friction interface is hard to move and it consumes more energy to move because there is high hydrostatic pressure here. Although lying in the zone of low hydrostatic pressure, the slip line outside the main cone needs enough energy to extend toward deep and outward because the distance is big. It depends on the distance (h_2) of the two-cone tops for which slip line of inside and outside cone is dominant [9-12].

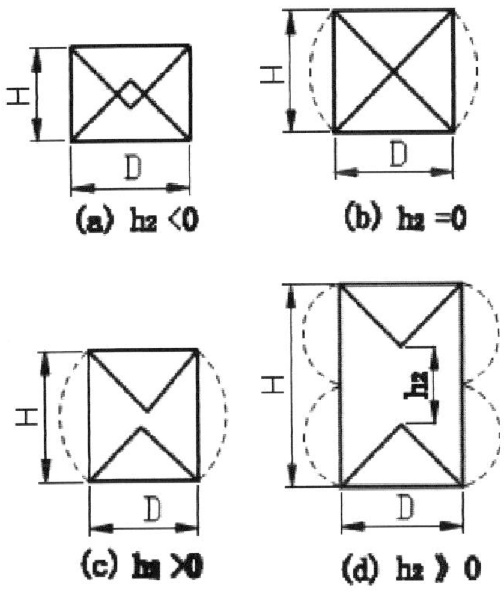

Figure 8. Effect of h_2 on slipping deformation.

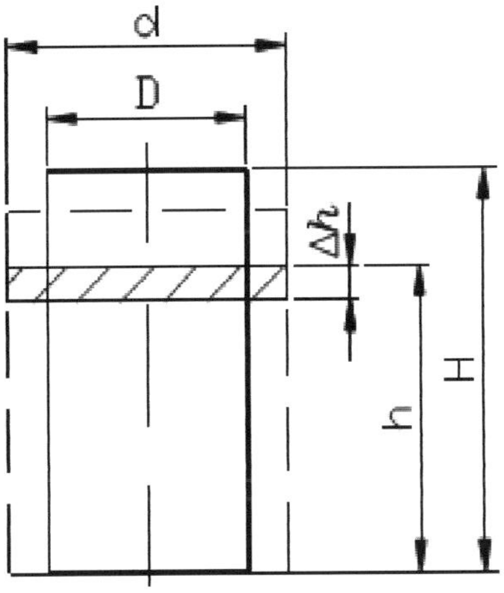

Figure 9. Homogeneous deforming of cylinder.

- When h_2<0, $H/D<1$, and its model is shown in Fig 8(a). Two cones insert each other, outside slip lines disturb severely and inside slip lines increase. Now slipping spreads the entire and deformation tends uniformly. Besides, the adhering zone of the contact surface will reduce greatly. It is possible to slide overall for small cylinder, but great external force is needed.
- When h_2=0, $H/D=1$, its model is shown in Fig 8(b). Slip lines of both inside and outside cone will appear and focus nearby angle of 45° line. Slip lines outside the two cones disturb mutual and forms a single-drum shape obviously. Currently, adhering zone of the contract surface decreases and slip begins. So total deformation is relatively uniform, but more pressure is needed. It is reasonable to select these parameters for high cylindricity.
- When h_2>0, $H/D>1$, its model is shown in Fig 8(c). The two-cone tops are out of contact but not far from. Now that there is better slip condition outside the main cone, more slip lines outside the main cone appear and also form obvious single-drum shape, which has been proved in test of forging short cylinder.
- When h_2≫0, H/D≫1, its model is shown in Fig 8(d). The two main-cone tops are far from each other; the slip lines outside the main cone are liable to appear but not going deeply. Because the deformation occurs near the two main cones and contact surface, double-drum shape appears. Besides, the contact surface is severely adhered. This phenomenon often happens in forging long cylinder.

Analysis above models shows that the condition of the slip of material throughout the whole volume is H/D<1. We can qualitatively explain the reason that deformation force increases with decreasing H/D by using these models.

3. Analysis of the mechanism of a plastic deformation on metal

Many factors act on metal plastic deformation, in which deformation is the result of coupling action of various factors. In fact, it is always expected that metal deformation is uniform and stable to keep high product quality and efficiency. To obtain the relations of some parameters, such as strain and energy consumption and temperature variation, we assume the deformation process is homogeneous, and the simplified model is showed in Fig 9.

3.1. Strain of uniform forging cylinder

Let's assume that the cylinder material is isotropic and without initial stress, which is being forged between up and down rigid anvils. The anvils are smooth and parallel. The process of deforming is homogeneous. Besides reducing height of the material and increasing its diameter, we can keep round cross section and increase its diameter at the same speed in the course of forging. Suppose the ratio of initial height (H_0) to the diameter (D_0) is less than two without axial bend.

Natural strain increment ($d\varepsilon_z$) in the axial direction is given by

$$d\varepsilon_z = -dh/h \tag{1}$$

Where h is a height of any cross section of the cylinder and *-dh* is a tiny decrement of h in compression process. The total strain of the material from H_0 to H_1 is

$$\varepsilon_z = \int_{H_0}^{H_1}(-dh/h) = \ln(H_0/H_1) \tag{2}$$

The axial strain can be calculated by equation (2) so long as giving an initial and final height of the cylinder. For isotropy and incompressible material, we have $\varepsilon_r + \varepsilon_\theta + \varepsilon_z = 0$. Additionally, the radial strain equals the circumferential strain, and they are tension strains, then we get $\varepsilon_r = \varepsilon_\theta = -\varepsilon_z/2$.

3.2. *Energy consumed in forging cylinder material*

Suppose that the axial height of metal material reduces a micro amount (-dh) when the plastic deformation occurs under acting force (P). Then the tiny consumed energy (dE) is

$$dE = P(-dh) = \sigma_z A(-dh) \tag{3}$$

Where σ_z is an axial stress, and A is any instant cross-sectional area.

When the material is compressed from an initial height (H_0) to a final height (H_1), the total energy consumed is

$$E = \int_{H_2}^{H_1} P(-dh) = \int_{H_2}^{H_1} \sigma_z A(-dh) \tag{4}$$

For incompressible material, the volume keeps invariable, then $A = V/h$. The energy consumed in compressing unit volume is

$$E/V = \int_{H_0}^{H_1} \sigma_z(-dh/h) = \int_0^{\varepsilon_z} \sigma_z d\varepsilon_z \tag{5}$$

Where the integral ($\int_0^{\varepsilon} \sigma_z d\varepsilon_z$) is obtained from an area surrounded by true stress-strain curve, the curve is decided by a test. The range of strain in practice is 0~ε.

- For ideal strong-plastic material, the one-way yield stress is σ_s. Then energy of per unit volume is

$$E/V = \int_0^{\varepsilon} \sigma_z d\varepsilon_z = \sigma_s \int_{H_0}^{H_1}(-dh/h) = \sigma_s \ln(H_0/H_1) = \sigma_s \varepsilon_z \tag{6}$$

Equation (6) shows that the energy consumed in unit volume for axial yield stress is proportional to strain when we deform a rigid-plastic material.

- For strain-hardening material, relation of true stress to natural strain is approximate obey $\sigma = A\varepsilon^n$, and the energy consumed in unit volume deformation is

$$E/V = \int_0^{\varepsilon_z} \sigma_z d\varepsilon_z = \int_0^{\varepsilon_z} A\varepsilon_z^n d\varepsilon_z = A\varepsilon_z^{n+1}/(n+1) \tag{7}$$

Where both A and n in equation (7) are material constants.

For big plastic strain, an expression corresponding with test data of engineering material is $\sigma = C(C+\varepsilon)^q$, the energy consumed per unit volume deformation is

$$E/V = \int_0^{\varepsilon_z} C(D+\varepsilon)^q d\varepsilon = \left[(D+\varepsilon_z)^{q+1} - D^{q+1}\right]C/(q+1) \tag{8}$$

Where q, C, D, and q in equation (8) are the given material constants, in which we have $0 \leq q \leq 1$.

Above statement are about strain and consuming energy based on homogeneous deformation. A homogeneous deformation is one of the most effective means for forming metal, which can be used to estimate the value in other forming process.

3.3. *Peak temperature rise of pressing cylinder material*

In a compression process of material, most energy consumed in plastic deformation converts to quantity of heat, and causing temperature rise. Assume that the compression process is adiabatic. In this case, all plastic work turns into thermal energy without spreading away from the material. In line with the energy conservation law, an energy consumed per unit volume deformation can be expressed as $E/V = J\rho c\Delta\theta$, and then we obtain

$$\Delta\vartheta = (E/V)/(J\rho c) \tag{9}$$

Where $\Delta\theta$ is average temperature rise, ρ is a density of the material, c is a specific heat of the material, and J is the heat equivalent of the work [13, 14].

Taking advantage of equation (9), we can obtain the temperature rise of plastic deformation. But the actual temperature rise is less than the calculated value.

4. Conclusion

Plastic forming and machining of metal belongs to a large deformation. The process of deformation is extremely complex, and influence factors are numerous. In this paper, several geometrical models of plastic deformation have been sets up by studying the flow characteristic of a mass point. The effect of primary geometrical parameters and function of friction have been discussed, the main conclusion is as follow:

- Besides invariant volume in plastic deformation, the movement of mass points follows the law of smallest resistance force, which means mass points move in the direction of least resistance.
- In the course of plastic deformation, each mass point moves in the direction of its nearest border so as to consume the least energy.
- Force of friction and geometric parameters of deformation have a great influence on the flow of the mass point. The force of friction has a significant influence on the deformation near the contact surface. The shorter is a deforming piece, the more homogeneous is the deformation.

Acknowledgments

This work is supported by the Natural Science Foundation of Gansu Province (145RJZA134) and the Scientific Research Innovation Funds of Lanzhou City University President (LZCU-XZ2014-01).

References

[1] An H P, Rui Z Y, Wang R F and Zhang Z M 2012 Model of shear slipping deformation and it's cutting energy consumed for high-speed machining *Advanced Materials Research* **602** 1967-70
[2] An H P, Rui Z Y, Wang R F and Zhang Z M 2014 Research on cutting-temperature field and

distribution of heat rates among a workpiece, cutter, and chip for high-speed cutting based on analytical and numerical methods *Strength of Materials* **46** 289-95

[3] Peng D S 2004 *Metal Plastic Machining Theory* (Chang Sha:Press of Central South University)

[4] Li M J 2003 *Plastic Process Technology* (Beijing: China Machine Press)

[5] An H P 2014 Summary of equations used to solve primary stress by means of measuring strain under special plane stress state *Automation and Instrument* **7** 181-4

[6] An H P 2009 The slip-line theory and its applications in the process of metal cutting *Chinese Journal of Construction Machinery* **7** 308-11

[7] Sun J L 2007 *Friction and Lubrication of Material Formation* (Bei Jing: National Defense Industry Press)

[8] Shaw M C and Vyas A 1993 Chip formation in machining hardened steels *Annals of the CIRP* **42** 29-32

[9] Shaw M C and Vyas A 1998 The mechanism of chip formation with hard turning steel *Annals of the CIRP* **47** 72-82

[10] Komanduri R and Brown R 1981 On the mechanics of chip segmentation in machining *J. of Eng. for Ind. Trans. ASME* **103** 33-51

[11] Barry J and Byme G 2002 The mechanisms of chip formation in machining hardened steels *Journal of Engineering* **124** 528-35

[12] Filice L, Micari F, Rizzuti S and Umbrello D 2007 A critical analysis on the friction modeling in orthogonal machining *Journal of Machine Tools and Manufacture* **47** 709-14

[13] Zhang X P, Gao E and WandLiu C R 2009 Optimization of process parameter residual stresses for hard turned surfaces *Journal of Materials Processing Technology* **209** 4286-91

[14] Benardos P G and Vasniakos G C 2003 Predicting surface roughness in machining: A review *International Journal of Machine Tool & Manufacture* **43** 833-44

Analysis of the influence of advanced materials for aerospace products R&D and manufacturing cost

A W Shen[1], J L Guo and Z J Wang

Aeronautics and Astronautics Engineering College, Air Force Engineering University, Xi'an, 710038, China

E-mail: zjsaw@foxmail.com

Abstract. In this paper, we pointed out the deficiency of traditional cost estimation model about aerospace products Research & Development (R&D) and manufacturing based on analyzing the widely use of advanced materials in aviation products. Then we put up with the estimating formulas of cost factor, which representing the influences of advanced materials on the labor cost rate and manufacturing materials cost rate. The values ranges of the common advanced materials such as composite materials, titanium alloy are present in the labor and materials two aspects. Finally, we estimate the R&D and manufacturing cost of F/A-18, F/A-22, B-1B and B-2 aircraft based on the common DAPCA IV model and the modified model proposed by this paper. The calculation results show that the calculation precision improved greatly by the proposed method which considering advanced materials. So we can know the proposed method is scientific and reasonable.

1. Introduction

As science and technology develop, advanced materials are being widely applied in the fields of aerospace, power generation equipment, numerical control machining, weapon systems, and especially aviation, a field that struggles to reduce every gram of weight [1]. Composite materials, titanium alloy, and other advanced materials are playing an increasingly important role in both civil and military aircraft [2]. For example [3], the American fifth-generation fighter F/A-22, the world's most advanced aircraft, uses composite materials in the most important parts of the fuselage, wings, and tail. In fact, titanium alloy makes up 40% of the total weight of this military aircraft, and composite materials make up 34%. Among civilian aircraft, composite materials make up more than 40% of the total weight of the European Airbus Company's A380, the world's largest airliner [4-7].

In the early days, metals were the main materials in aircraft manufacturing; later, advanced materials such as titanium alloy and composite materials were incorporated into nonbearing structures. With the continuous development of material properties, modern aircraft use titanium alloy and composite materials more and more widely. Manufacturers have gradually broken the original material-use restrictions and transferred the load bearing structure, moving closer to an all-composite aircraft. Table 1 shows the contrast in material application proportion between the F/A-22 fighter and the early F-15 fighter. Composite material use has great increased in the F/A-22, and the amount of titanium alloy used has also risen [8, 9].

[1] Address for correspondence: A W Shen, Aeronautics and Astronautics Engineering College, Air Force Engineering University, Xi'an, 710038, China. E-mail: zjsaw@foxmail.com.

Table 1. Contrast in material application proportions of F-15, F/A-22 fighter and B-2 bomber.

Type	Aluminum alloy	Titanium alloy	Alloy steel	Composite materials	Other materials
F-15	0.50	0.34	0.02	0.02	0.12
F/A-22	0.15	0.40	0.06	0.34	0.05
B-2	0.24	0.18	0.1	0.38	0.01

Material applications in the airframes of modern aircraft represent a significant change from those in early aircraft. This may create a sizeable effect for research and development (R&D) and manufacturing costs of airframes.

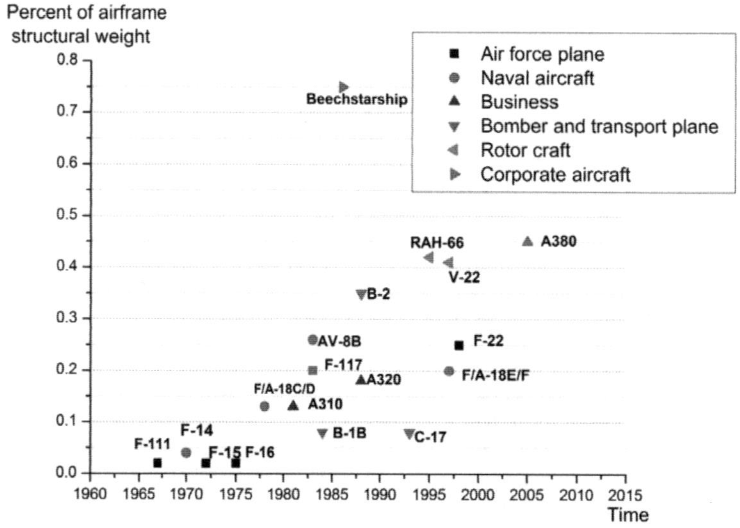

Figure 1. Development of composite materials in aircraft.

When estimating R&D and manufacturing costs of new aircraft, we must consider the different amounts of composite materials used in order to modify traditional aircraft cost models appropriately. Figure 1 reflects the evolution of applying composite materials in aircraft. The engineering development prototype of the F/A-22 fighter included 40% titanium alloy, broken down into 37% titanium 6-4 and 3% titanium 6-22-22; 34% composite materials, broken down into 23% thermosetting composite materials, 10% toughened epoxy composite materials, and 1% thermoplastic materials; 15% aluminum alloy; 6% alloy steel; and 5% other materials. With a large number of new technologies, new materials, and new manufacturing processes, aircraft performance has improved rapidly. At the same time, new technologies increase aircraft structures' complexity and need for precision. Further advances will lead to continued growth in development and production costs. Therefore, when estimating the development and production costs of a model airplane body, we must consider these changes and make corresponding amendments to cost models [10-12].

At present, the best-known cost estimation model is the Development and Procurement Costs of Aircraft (DAPCA) established by America's Rand Corporation [13], the latest version of which is known as the DAPCA IV model. In this model, the general expression for cost estimation is $y = a_0 x_1^{a_1} x_2^{a_2}$, where y is the cost, x_1 and x_2 are the weight and speed of the aircraft, respectively, and a_0, a_1, and a_2 are the correlation coefficients. The parameter estimation can be achieved by regression using the same aircraft data. This method is simple and practical, but it produces large deviations when used to estimate the cost of an advanced aircraft. The advanced materials' contribution to the cost represents a major reason behind this error [14, 15].

The German Airbus Company once analyzed the limitations of the DAPCA II model and introduced a series of cost factors such as materials, stealth, difficulty, and computer-aided design to aid in estimation. Their attempts have strong practical value; however, the correction factor is relatively rough, and the definitions are not sufficiently clear. Also, the use of factors derived from subjective judgment presents obvious defects.

After studying the rules for incorporating advanced materials into planes, this paper puts forward a series of material cost factors using statistical analysis. These factors represent the influence of advanced materials on aircraft development and production costs. We finish by analyzing the validity of material cost factors proposed in this paper through estimating the cost of U.S. military aircraft in service.

2. Calculation of material cost factor

Labor hours and materials costs are the two main elements making up the advanced material cost of an aircraft. The influence of labor hours can be calculated using a working hour's ratio of advanced materials to whole aluminum alloys. The influence of material costs acts mainly to increase the price of the material itself and decrease the utilization rate.

When analyzing the influence of labor hours, we consider the main materials to be aluminum alloy, aluminum lithium alloy, titanium alloy, alloy steel, carbon-epoxy, carbon-bismaleimide, and carbon thermoplastics, the latter three of which are composite materials. Similarly, we consider the main working hours to be categorized into Nonrecurring Engineering, Nonrecurring Tooling, Recurring Engineering, Recurring Tooling, Recurring Manufacturing, and Recurring Quality Assurance.

The concept of material cost factors plays a role in evaluating the influence of advanced materials for each type of labor hour. These factors are mainly divided into the labor hour cost coefficient and the material cost coefficient. Material cost is the ratio of aircraft cost involving many advanced materials to the cost of a whole aluminum aircraft of the same weight.

2.1. Labor hour cost coefficient

The labor hour cost coefficient can be described by two components: manufacturing process and geometric complexity.

2.1.1. The manufacturing process. Below are the formulas for estimating labor hour costs based on use of advanced materials, derived by statistical analysis.

$$WMCF_l = \sum_{l=1}^{L}\sum_{m=1}^{M} S_m \gamma_l^m \tag{1}$$

where $WMCF_l$ is the cost factor of l labor hour and S_m is the structural weight percentage of material m in the airframe. Detailed values for S_m are shown in table 2.

Table 2. Structure weight percentages of airframe materials.

Type	F-15	F-16	F/A-18	F/A-22	B-1B	B-2	C-17
Aluminum alloy	50%	75%	51%	15%	42.5%	34%	69.3%
Titanium alloy	34%	1%	14%	40%	17.6%	18%	10.3%
Alloy steel	2%	11%	12%	6%	7%	10%	12.3%
Composite material	2%	2%	11%	34%	32.9%	38%	8.1%
Other materials	12%	11%	12%	5%	—	—	—

In turn, γ_l^m represents the cost ratio of m material to l labor hours. It can be calculated using the following equation:

$$\gamma_l^m = C_l * \sigma_{l,m} + (1 - C_l) \tag{2}$$

where C_l is the proportion of airframe cost within total cost. The total cost includes not only the airframe cost but also aircraft subsystem costs such as the flight control system and final assembly costs. Values of C_l differ for different types of aircraft and different labor hours. Table 3 shows various values for C_l.

Table 3. Values of C_l.

Materials	Nonrecurring Engineering	Nonrecurring Tooling	Recurring Engineering	Recurring Tooling	Recurring Manufact-uring	Material Cost	Recurring Quality Assurance
A-6	47	80	47	80	33	—	—
A-7	20	94	16	92	81	83	83
A-10	45	88	20	24	53	—	—
B-52	24	88	23	92	88	—	—
C-5	63	93	63	93	75	57	57
C-130	61	80	61	80	83	83	83
C-141	53	80	53	80	73	63	63
F-4	51	94	51	94	50	69	69
F-14	58	94	50	94	66	35	35
F-15	43	87	43	87	74	—	—
F-16	12	68	19	69	45	16	16
KC-135	58	100	57	100	88	—	—
Average	45	87	42	82	67	58	69

[a] "—" in the table means the data are not available.

[b] Average values present the average values for U.S. active-duty aircraft rather than the average values of the data listed in the table.

Table 4. Values of $\sigma_{l,m}$ for different materials in late 1980's.

Materials	Nonrecurring Engineering	Nonrecurring Tooling	Recurring Engineering	Recurring Tooling	Recurring Manufacturing	Recurring Quality Assurance
Aluminum alloy	1.0	1.0	1.0	1.0	1.0	1.0
Aluminum lithium alloy	1.1	1.2	1.1	1.1	1.1	1.1
Titanium alloy	1.1	1.4	1.4	1.9	1.6	1.6
Alloy steel	1.1	1.1	1.1	1.4	1.2	1.4
Carbon-epoxy	1.4	1.6	1.9	2.2	1.8	2.4
Carbon-bismaleimide	1.5	1.7	2.1	2.3	2.1	2.5
Carbon thermoplastics	1.7	2.0	2.9	2.4	1.8	2.6

In equation (2), $\sigma_{l,m}$ represents the cost factor of l labor hour and m material. These factors are the rates of different materials' features, price, and utilization to the costs of ordinary aluminum alloy. For example, Table 4 shows values of $\sigma_{l,m}$ as they stood in the late 1980's.

The category "Other materials" from Table 1 mostly represents nonmetal composite materials. In the following calculations, we consider these to be carbon-epoxy.

When estimating the cost of aircraft, we consider composite materials as a whole because of their great variety. We can derive $\sigma_{l,m}$ for composite materials using weighted summations from table 4. Table 5 shows values of $\sigma_{l,m}$ for composite materials.

Table 5. Values of $\sigma_{l,m}$ for composite materials.

Time	Nonrecurring Engineering	Nonrecurring Tooling	Recurring Engineering	Recurring Tooling	Recurring Manufacturing	Recurring Quality Assurance
Late 1980's	1.53	1.77	2.30	2.30	1.90	2.50
Mid 1990's	1.30	1.50	1.50	2.17	1.63	1.97
Late 1990's	1.10	1.30	1.16	1.43	1.26	1.44
Mid 2000's	1.15	1.37	1.18	1.47	1.26	1.55

The above expressions combine to yield the labor hour materials cost coefficients for different decades of the last century.

2.1.2. Geometric complexity of component. The geometric complexity of components has a marked impact on costs beyond the influence of materials and manufacturing processes. In fact, for some types of labor hour, geometric complexity is the most important cost-driven factor. However, statistical data on components' geometric complexity by aviation product manufacturer remain relatively sparse, especially given the contribution of different materials and different accounts to the geometric complexity of the components. Additional statistical data concerning manufacturing labor hours and other labor hours is incomplete as well, preventing comprehensive and effective analysis. Therefore, we propose a more detailed formula regarding manufacturing labor hours.

Table 6. Aircraft body parts by complexity level.

Complexity	Name of main components in aircraft
Simple	skin, doors, dashboard, accessories, surface smooth skin
Medium	beam, lateral wall of oil tank, equipment tray, ceiling block, rigid skin, tank floor
Complex	horizontal rails, separator, air-intake duct, cabin keel, variable camber skin, pylon, stringers, ribs
Extremely complex	bearing, air-intake ports, guide edge

First, we must quantify the effects of geometric complexity on manufacturing labor hours. Components can be divided into four types: simple components, medium components, complex components, and extremely complex components. Simple components are single blocks or flat surface parts. Medium components incorporate a medium radius of curvature, reinforcing ribs, or a moderate number of unit parts. Complex components include pieces with complex curvature or the main internal structures; they may also involve a large number of subcomponents. Finally, extremely complex components are characterized by complex geometry, with widespread use of multi-dimensional control and high tolerance parts. Table 6 shows examples of this classification system.

Cost ratios of manufacturing labor hours are not identical across different materials, different processes, and different complexities. Table 7 shows the cost ratios of manufacturing labor hours based on medium-complexity aluminum alloy parts. The data in table 7 suggest the following rules:

- Cost ratios will increase with geometric complexity because the processing time of more complex components is longer than that of simpler components.

- Cost ratios will decrease with automation, especially for composite materials, in which the processing time of automatic fiber placement is shorter than that of manual molding.

Table 7. Cost ratios of material, manufacturing process, and complexity.

Material / Manufacturing Process	Simple	Medium	Complex	Extremely complex
Aluminum alloy / Conventional processing	0.7	1.0	1.5	2.3
Aluminum lithium alloy / Conventional processing	0.7	1.0	1.6	2.5
Titanium alloy / Conventional processing	0.7	1.2	1.7	2.9
Alloy steel / Conventional processing	0.7	1.1	1.8	2.9
Aluminum alloy / High speed machining	0.6	0.8	1.0	1.5
Aluminum lithium alloy / High speed machining	0.6	0.8	1.1	1.6
High performance titanium alloy / Processing	0.7	1.1	1.5	2.5
Composite / Hand laying	1.3	1.8	2.2	3.0
Composite / Automatic fiber placement	0.7	1.3	1.7	—
Composite / Resin transfer molding	—	—	1.4	2.3

Therefore, in the manufacturing process of airframes, the labor hour cost factor of the manufacturing process is

$$WMCF_{LAB} = \sum_{l=1}^{L}\sum_{m=1}^{M} S_l^m \gamma_l^m \tag{3}$$

where S_l^m is the percentage of airframe structural weight with l class manufacturing process and m complexity, and γ_l^m is the percentage of labor hours with l class manufacturing process and m complexity. At the same time, we can derive other labor hour cost factors when sufficient statistical data is provided.

2.2. Effect of material costs

Advanced materials costs proceed mainly from two aspects, labor hours and materials costs. The preceding discussion studied labor hours in detail; we will now focus on materials costs, which result largely from increasing prices of materials and declines in utilization rates.

Raw materials costs are related to different types of fiber and resin and their uses in manufacturing and processing. At the same time, material costs are affected by market factors like commercial dosage, purchase quantity, and competition within the industry.

Scrap rate is another important factor that influences the cost of materials. The scrap rate is the weight ratio between the raw materials and the actual manufactured aircraft components. Sources of material loss during manufacturing include material handling loss, processing and cutting loss, scrapping loss, and other operating losses. The scrap rate for automatic fiber placement is lower than for hand operation. The scrap rate of composite materials is related to processing, while the scrap rate of metal materials is closely related to the complexity of the components. The more complex the geometry of the parts, the higher the scrap rate.

Therefore, for K type of material and F method of manufacturing, the formula for estimating airframe material cost is

$$C_m = W_s * \sum_{f=1}^{F}\sum_{k=1}^{K} M_f^k S_f^k * P_k , \tag{4}$$

where W_s is the weight of the airframe structure; M_f^k is the weight percentage of *k* type material and *f* type manufacturing method; S_f^k is the scrap rate of *k* type material and *f* type manufacturing method; and P_k is the unit price of *k* type material.

When analyzing the material factor, we find that material costs grow based on the factor $\sum_{f=1}^{F}\sum_{k=1}^{K} M_f^k S_f^k$. Therefore, we define $\sum_{f=1}^{F}\sum_{k=1}^{K} M_f^k S_f^k$ as the material cost factor within manufacturing costs of materials. In order to facilitate statistical analysis of data, we simplify the formula as follow:

$$WMCF = \sum S_m \gamma_m \tag{5}$$

where γ_m is the average scrap rate of *m* material. Table 8 shows specific values:

Table 8. Scrap ratios of different materials.

Materials	Aluminum alloy	Aluminum lithium alloy	Titanium alloy	Alloy steel	Composite materials
Late 1980's	2.5	4.2	3.0	2.1	1.9
Mid 1990's	2.2	2.7	3.0	2.1	1.8

Table 8 shows that the scrap rate remained about the same for the common aircraft structure materials over the period covered. Therefore, we can adopt the value from the mid 1990's to estimate the scrap rate after the 1990's.

3. Case analysis

In the previous section, we analyzed the impact of advanced materials on labor hour and manufacturing material costs. We then defined and analyzed the material cost factors. In this section, we will consider the use of advanced materials to estimate aircraft R&D and manufacturing costs.

In the case of the B-2 Bomber, the empty weight is 166,579 lb and the maximum speed at the optimum altitude is 489.2 kn. In this aircraft, 34% of the total weight is aluminum alloy and other materials, 10% is alloy steel, 18% is titanium alloy, and 38% is composite materials. The material cost factor can be calculated as follows:

$$\begin{aligned}
WMCF_{NRENGR} &= \sum_{l=1}^{L}\sum_{m=1}^{M} S_m \gamma_l^m \\
&= 0.45\times(1\times0.34+1.1\times0+1.1\times0.18+1.1\times0.1+1.7\times0.38)+(1-0.45)=1.132 \\
WMCF_{NRTOOL} &= 0.87\times(1\times0.34+1.2\times0+1.4\times0.18+1.1\times0.1+2\times0.38)+(1-0.87)=1.402 \\
WMCF_{ENGR} &= 0.42\times(1\times0.34+1.1\times0+1.4\times0.18+1.1\times0.1+2.9\times0.38)+(1-0.42)=1.337 \\
WMCF_{TOOL} &= 0.82\times(1\times0.34+1.1\times0+1.9\times0.18+1.4\times0.1+2.4\times0.38)+(1-0.82)=1.602 \\
WMCF_{LAB} &= 0.67\times(1\times0.34+1.1\times0+1.6\times0.18+1.2\times0.1+1.8\times0.38)+(1-0.67)=1.289 \\
WMCF_{MAT} &= 0.58\times(1\times0.34+2.7\times0+2.8\times0.18+0.7\times0.1+6.5\times0.38)+(1-0.58)=2.383 \\
WMCF_{QUAL} &= 0.69\times(1\times0.34+1.1\times0+1.6\times0.18+1.4\times0.1+2.6\times0.38)+(1-0.69)=1.521
\end{aligned}$$

Using the DAPCA IV model, the nonrecurring engineering labor hour can be estimated using the following formula:

$$\begin{aligned} WHR_{NRENGR} &= 0.0168 \times W^{0.747} \times S^{0.8} \times \sec rt \times labor \times 10^{-3} \\ &= 0.0168 \times 166579^{0.747} \times 489.2^{0.8} \times 1.25 \times 161.66 \times 10^{-3} \\ &= 3827.92 \end{aligned} \tag{6}$$

However, considering the use of advanced materials, the nonrecurring engineering labor hour can instead be estimated by

$$\begin{aligned} WHR_{NRENGR} = WHR_{NRENGR} &= 0.0168 \times W^{0.747} \times S^{0.8} \times \sec rt \times labor \times WMCF_{NRENGR} \times 10^{-3} \\ &= 3827.92 \times 1.132 = 4334.36 \end{aligned} \tag{7}$$

We can take the same approach for other types of labor hours, although we will not detail these in the present article. Table 9 displays the results.

Table 9. Labor hour costs from different calculation models.

Cost	DAPCA IV model (millions)	Material cost factor	New model (millions)
Nonrecurring Engineering	3827.92	1.132	4334.36
Nonrecurring Tooling	2009.73	1.402	2817.52
Recurring Engineering	734.54	1.337	982.58
Recurring Tooling	264.38	1.602	423.51
Recurring Manufacturing	1767.05	1.289	2278.50
Material Cost	433.28	2.383	1032.38
Recurring Quality Assurance	229.72	1.521	349.55
Airframe total	9266.62	—	12218.40

Table 10. Aircraft R&D and manufacturing costs considering advanced materials.

Type	R&D cost (billons, only airframe)					
		Actual value	DAPCA IV model		New model	
			value	deviation	value	deviation
	F/A-18	1.56	1.226	21.41%	1.336	14.36%
	F/A-22	7.47	4.92	34.14%	5.978	19.97%
	B-1B	7.04	5.172	26.53%	6.372	9.49%
	B-2	14.23	9.26	34.93%	12.21	14.20%
	Unit manufacturing cost (millions)					
		Actual value	DAPCA IV model		New model	
			value	deviation	value	deviation
	F/A-18	24.2	23.52	2.81%	24.73	2.19%
	F/A-22	130	116.472	10.41%	139.284	7.14%
	B-1B	283.1	239.568	15.38%	283.74	0.23%
	B-2	840	637.764	24.08%	717.576	14.57%

Table 10 shows the results of estimating R&D and manufacturing costs of the F/A-18, F/A-22, B-1B, and B-2 using the new formula. Figures 2 and 3 provide the estimation error of the DAPCA IV model and the new model.

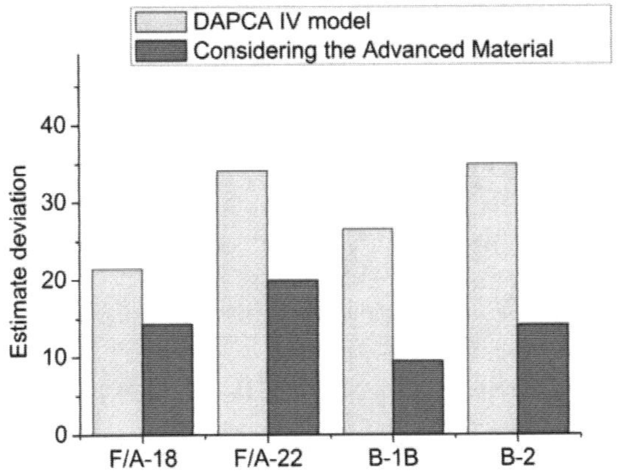

Figure 2. R&D cost estimate deviations of common aircraft.

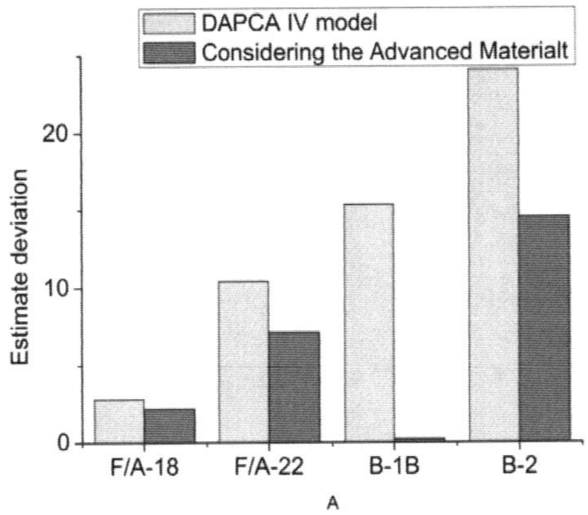

Figure 3. Manufacturing cost estimation errors of common aircraft.

Based on the analysis above, the new model appears to offer more accurate estimates than the common DAPCA IV model. We can conclude that is necessary to consider advanced materials when estimating the cost of a new aircraft.

4. Conclusion

In this article, we studied the relationship between advanced materials and aircraft R&D and manufacturing costs. Ultimately, we proposed a modified formula based on the commonly-used DAPCA IV model to estimate costs while considering advanced materials. Examples performed using US Air Force aircraft data suggest that the estimation accuracy of the new formula exceeds that of the DAPCA IV. For example, the estimation error for the B-2 bomber's R&D cost improved from 34.93% to 14.2%, and the estimated error in manufacturing cost improved from 24.08% to 14.57%. At the same time, we note that the new estimation model contains some estimation errors, which require further research.

References

[1] Rappold P M, Kline D E, Bond B H and Wiedenbeck J K 2009 Reciprocal estimation of the raw material cost of producing hardwood lumber busing the principles of activity-based costing *Forest Product Journal* **59** 84-90

[2] Younossi O, Mark V A, Moore R M, Lorell M, Mason J, John C G 2001 *Military Airframe Cost: The Effects of Advanced Materials and Manufacturing Processes* (Santa Monica, Calif.: The Rand Corporation, MR1370-AF)

[3] [2] Younossi O, Kennedy M and John C G 2005 *Lessons Learned from the F/A–22 and F/A–18E/F Development Programs* (Santa Monica, Calif.: The Rand Corporation, MG-276)

[4] Mark V A, Robert S L and Sheila M E 2006 *Historical Cost Growth of Completed Weapon System Programs* (Santa Monica, Calif.: The Rand Corporation, TR-343)

[5] Raman R, John Graser C and Younossi O 2003 *The Effects of Advanced Materials on Airframe Operating and Support Costs* (Santa Monica, Calif.: The Rand Corporation)

[6] Cook R, Cynthia R and John C G 2001 *Military Airframe Acquisition Costs: The Effects of Lean Manufacturing* (Santa Monica, Calif.: The Rand Corporation, MR-1325-AF)

[7] Jamie J K 2009 Predicting fatigue failures *Science* **325** 156-8

[8] Adrian C O, Iligo Z A, Rodney S T and Javid B 2008 Compression and post-buckling damage growth and collapse analysis of flat composite stiffened panels *Composites Science and Technology* **68** 3150-60

[9] Mittelstedt C 2008 Closed-form buckling analysis of stiffened composite plates and identification of minimum stiffener requirements *International Journal of Engineering*

Science **46** 1011-34

[10] Jedidi J, Jacquemin F and Vautrin A 2005 Design of accelerated hygrothermal cycles on polymer matrix composites in the case of a supersonic aircraft *Composite Structures* **68** 429-37

[11] Cagno E, Micheli G J L and Trucco P 2012 Eco-efficiency for sustainable manufacturing: An extended environmental costing method *Production Planning & Control* **23** 134-44

[12] Navarro H J, Doran C M and Shakeshaft A P 2011 Measuring costs of alcohol harm to others: A review of the literature *Drug and Alcohol Dependence* **114** 87-99

[13] Jia L, Guo J and Shen A 2013 The effects of advanced materials on aircraft production cost *Advanced Materials Research* **710** 798-803

[14] Woerden H J M, Sinke J and Hooijmeijer P A 2003 Maintenance of glare structures and glare as riveted or bonded repair material *Applied Composite Materials* **10** 307-29

[15] Jiang D M, Cao Z Y, Sun X, Guo L and Liu J G 2013 Effect of Yttrium addition on microstructure and mechanical properties of Mg-Zn-Ca alloy *Materials Research Innovations* **17** 33-8

IOP Conf. Series: Materials Science and Engineering **103** (2015) 012005 doi:10.1088/1757-899X/103/1/012005

A study on the influence of ultrasonic processing on microstructure during laser welding phases

H D Liu[1,3], F Y Hu[1], A Y Cui[1] and H B Li[2]

[1] Naval Aeronautical and Astronautical University, Yantai, 264001, China
[2] China University of petroleum, Qingdao, 266580, China

E-mail: mdlhd@sina.com

Abstract. This paper proposes new welding technology that combines ultrasonic processing across different phases based on laser welding (UPPLW) and laser processing technology. The welding experiment used a 1.5 mm thick titanium alloy. The specimen was made metallographically prepared and the microstructural grain size of the welded joint was rated by metallurgical processing software, verifying that this new process can refine grains and improve joint properties.

1. Introduction

Aircrafts on aircraft carriers are susceptible to collision damage as wells as corrosion fatigue cracks while occur readily under marine condition. Consequently, it is often necessary to perform aircraft maintenance on-ship where conditions are less than ideal. This can lead to residual stress from welding, leading to deformities in the aircraft and reduced stability. In fact, repeated repair can create new cracks at the weld toe, thereby making the repair process more harmful than productive. As a result, it is important to find alternative repair technology for aircraft [1].

Titanium alloys continue to play an increasingly important strategic role in the field of modern aerospace and defence. It has high specific strength and good corrosion resistance[2-5]. The alloy has excellent comprehensive performance, so it can be welded, machined, and worked at temperatures of 400°C for long periods of time. The alloy is often used as important bearing components of aircrafts, including the strengthening frame, beam, joint bolts, and wall board[6-10].

Laser welding has been widely used in manufacturing aircrafts, automobiles, and ships because it has a small heat affected zone, it is stable, it produces small welding deformations, it can be easily automated, and it can be used to weld difficult welding material [11]. However, high power heating causes high local temperature gradients in the material. This leads to rapid stratified solidification of the molten pool; restriction of the base material will lead to welding deformations and residual stress. In addition, a very short laser molten pool solidification time does not promote the molten pool flowing during solidification, which leads to uneven weld solidification microstructure and to micro cracks. Due to these factors, we studied welding with power ultrasonic trailing vibration, which proved to be suitable for shipboard maintenance and airfield emergency maintenance.

2. Experimental equipment

[3] Address for correspondence: H D Liu, Naval Aeronautical and Astronautical University, Yantai, 264001, China. E-mail: mdlhd@sina.com.

Fig. 1 shows a system schematic. The process was mainly composed of an ultrasonic vibration system, laser welding system, and specimen constraint system; the main parts were the ultrasonic vibration system, pressure measuring modules, a digital ultrasonic generator that automatically tracks frequency, an ultrasonic transducer, and an ultrasonic amplitude transformer. The specifications of the main working mechanism are as follows: a 20 kHz ultrasonic vibration frequency was subsidiarily exerted in the nearby scan location on the same side of the area when laser welding was used to repair damaged metal. The ultrasonic vibration was imported by the tool head and lead to the specimen forming a resonance effect with the ultrasonic generator frequency tracing, thus affecting several key transition stages including the formation, flow, and solidification of the welding pool and the high temperature superplastic solid-state phase transformation of the welding area. Ultimately, the technology can improve weld solidification microstructures, control welding deformation, slow-release residual joint stress, improve joint mechanical properties, and reduce defect occurrences.

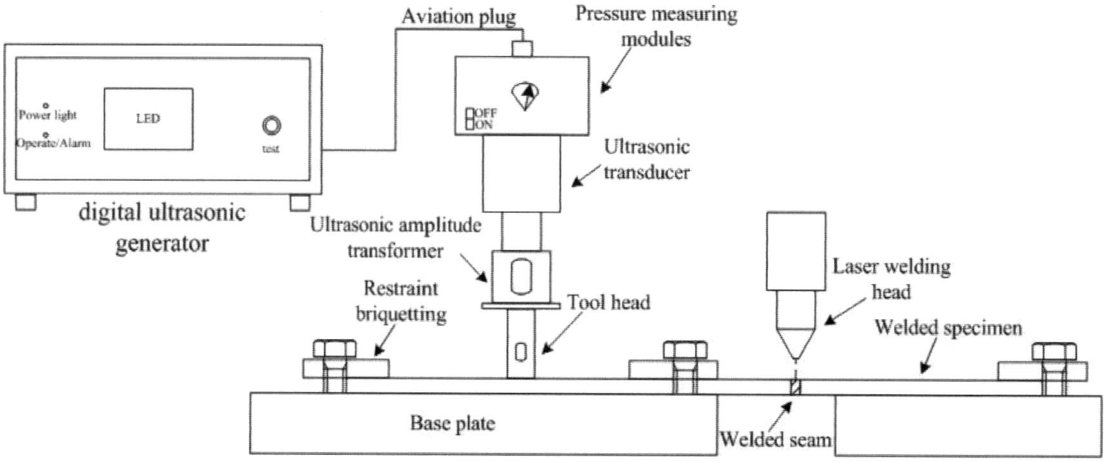

Figure1. Welding process schematic with the ultrasonic system.

3. Experimental material and methods

TC4 titanium alloy is close-packed hexagonal structure martensite alloy, which has good comprehensive properties. It has characteristics that promote high limit and ultimate strength, exceptional deformation resistance, a large ratio of yield limit and ultimate strength, and low plastic deformation at room temperature. Thus which belongs to the difficult deformation material, so it was difficult for welded titanium alloy thin-walled structure to be achieved plastic extension.

Slip plane of close-packed hexagonal structure will substantially increase when TC4 alloy was being at high temperature, so plasticity significantly improved and deformation resistance decreased sharply. On the other hand, we can analyze from the aspect of the high temperature dynamic behavior of titanium alloy: the main influence factors of the thermal deformation including deformation temperature, strain rate and deformation amount, the influence of the deformation temperature and strain rate for the thermal deformation process were more significant than deformation amount, the influence relation among deformation temperature, strain rate and the deformation resistance was shown in equation (1) [12],

$$\sigma = \sigma_0 e^{(-bT/T_0+b_2)} \left(\frac{\varepsilon}{\varepsilon_0}\right)^{b_3} \left(\frac{\dot{\varepsilon}}{\dot{\varepsilon}_0}\right)^{(b_4+b_5T/T_0)} \tag{1}$$

In the equation, σ denotes the deformation resistance, σ_0 denotes the deformation resistance reference value, T denotes the deformation temperature, $\dot{\varepsilon}$ denotes the deformation rate, ε denotes the deformation degree, and b_1 to b_5 denote the regression coefficient. Through the previous equation,

we know that when the component deformation reached a certain amount, deformation resistance dropped along as the temperature increased under the same strain rate. TD4 titanium alloy thermal deformation tests showed that the peak stress fell sharply and temperature increased when the strain rate was maintained at 1 s^{-1}; the peak stress of 890°C was 200 MPa, and the peak stress of 950°C was only about 90 MPa. These peak stresses differed more than 2 times at 60°C temperature intervals.
Table 1 shows the material composition, which was detected by third-party professional testing institutions. The specimen size was processed for 50 mm×30 mm×1.5 mm by wire-cutting.

Table 1. TC4 chemical compositions (mass fraction %).

Al	V	Fe	O	H
6.15	4.13	0.08	0.009	0.002

Table 2. TC4 laser welding parameters.

Current (A)	Impulse width (ms)	Frequency (Hz)	Welding speed ($mm \cdot s^{-1}$)	Spot diameter (mm)
175	9	6	2	0.3

The welding equipment was a pulsed laser repair machine with independent intellectual property rights. A Japan ERCR HP3 - AA00 robot was used as an actuator. Table 2 presents the welding parameters. Before welding, the sample was smoothed with metallographic sand paper and then cleaned with acetone; specimens were welded within 24 hours of this preparation. Metallographic specimens were prepared after welding was completed; a corrosion liquid with a composition of 3 ml HF + 30 ml HNO_3 + 67 ml H_2O was used to etch metallographic specimens about 8~10 min at room temperature. Then, the specimens were cleaned with alcohol. The metallographic microscope (OM) was used to observe the specimens' microstructures, and the three circles cutoff point method was used to measure the microstructural grain sizes of conventional laser welding versus this process [13].

As shown in Fig 2, the stepped tool head was used to oscillate the welded specimen, the vibration position was 100 mm from the molten pool, the downward pressure was the vibrator weight, the vibration frequency was 20 kHz, and the ultrasonic generator power was 1200 W.

4. Results and discussion

Table 3 presents the grain size measurement data, which was observed by a metallographic microscope with 100 magnification. Team A and B respectively present samples of our proces and conventional welding joints. SRMAS metallographic analysis software was used to evaluate the level of microstructural grain sizes via the three circles cutoff point method. The experiment shows that reliable accuracy can be obtained when the cut point count of each sample was 500.

The measure grid of 500 mm was used to cutoff points statistics in 8 field, the measuring data was tested by X^2 (Kai). The results showed that the cutoff point count was a normal distribution; thus, the measure value can be processed according to the statistical methods of the normal distribution. According to the relevant calculation methods of the grain size, the grain sizes of A and B contrast test group and can be calculated.

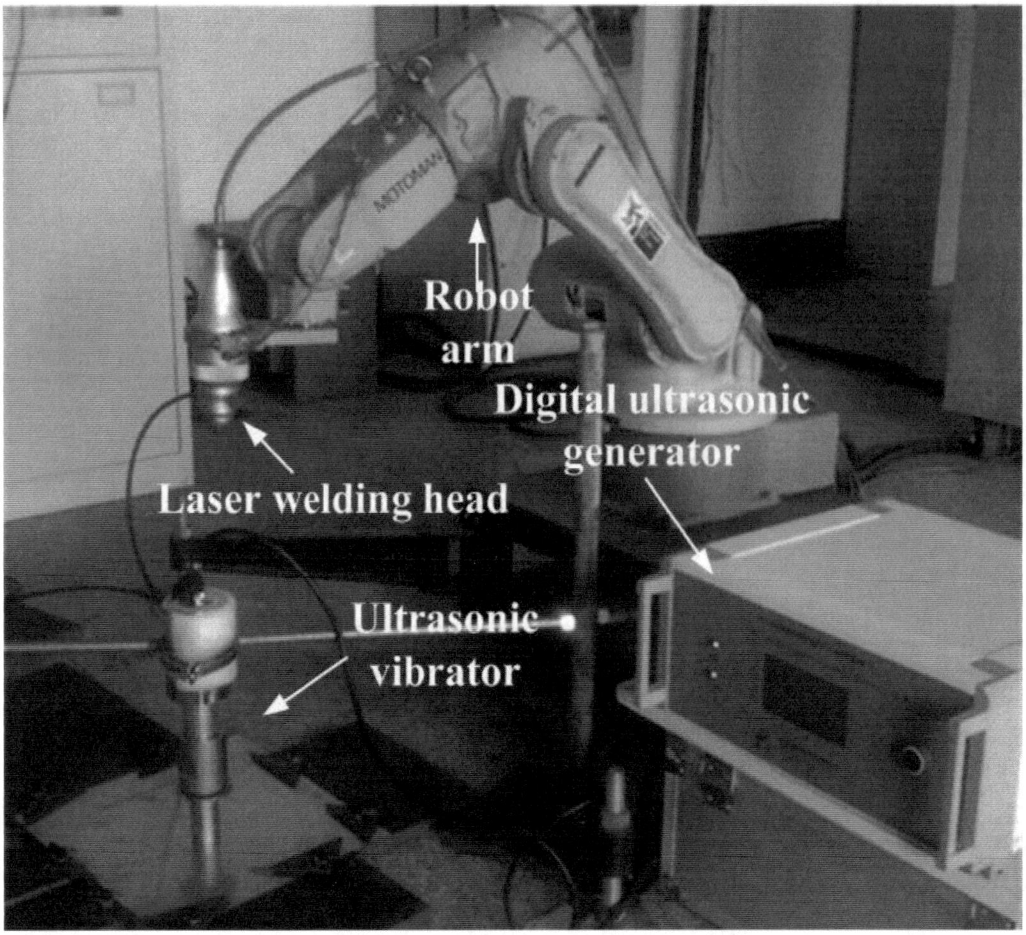

Figure 2. The ultrasonic laser welding system.

Table 3. Specimen statistics.

Team A	Sample number	1	2	3	4	5
	Level of grain size G	8.21	8.19	8.15	8.23	8.20
Team B	Sample number	1	2	3	4	5
	Level of grain size G	7.30	7.22	7.36	7.39	7.28

Table4. Specimen logarithm statistic.

Team A $x_{ai} = \lg G_{ai}$	Team B $x_{bi} = \lg G_{bi}$	$x_{ai} - x_{bi}$	$(x_{ai} - x_{bi})^2$
0.9143	0.8633	0.0510	0.0026
0.9133	0.8585	0.0548	0.0030
0.9112	0.8669	0.0443	0.0020
0.9154	0.8686	0.0468	0.0022
0.9138	0.8621	0.0517	0.0027
$\sum$		0.2486	0.0125

The grain size value was taken the logarithm, and the two experimental datasets were contrastively analyzed in accordance with the group comparison test of the fatigue reliability theory, as shown in table 4.

The average value $\overline{x}_{a-b}$ and standard deviation s_{a-b} can be obtained from table 4; they were respectively 0.0497 and 0.0059. Then, t can be calculated as 18.8360. The significant level was identified as $\alpha = 1\%$, degrees of freedom was known as $\nu = 4$, $t_a = 4.604$ can be obtained with the consulting related schedule [14] because $t > t_a$. Consequently, there were obvious differences between the two, namely, because the two experimental samples came from different normal population samples, and ultrasonic laser processing significantly influences grain size refinement. Therefore our process proved to be an effective and feasible technology in improving welding joint properties.

5. Conclusions

The study provides support the utility of combined ultrasonic and laser processing combined with conventional laser welding, and we reach the following conclusions:

- The combined process was efficient in refining welding joint microstructures and in increasing
- welding joint grain size by about 1 grade.

Acknowledgments

This material is based upon work supported by the Natural Science Foundation of China (grant no. 50804053) and the Innovation Program of Naval Aeronautical Engineering Academy (grant no. 20121009).

References

[1] Hu F Y and Dai J T 2010 *Aircraft Structural Damage Repairing Technology* (Beijing: Weapon Industry Press)
[2] Oh J, Kim N J and Lee S 2003 Correlation of fatigue properties and microstructure in investment cast Ti-6Al-4V welds *Materials Science and Engineering A* **340** 232-42
[3] Yang W L, Yang X H, Li X Y, Dui W Z, Lin J H, Li J Wand Gong S L2013 Analysis on fatigue property and fracture mechanism of TC4 titanium alloy joint during laser welding *Transaction of the China Welding Institution* **34** 92-6
[4] Ning X R 1998 Application of welding titanium in Russian aviation industry *Titanium Industry Progress* **1** 22-6
[5] Xu F 2010 On microstructure and mechanical properties of titanium alloy TC4 joints by capacitor discharge welding *Journal Shaanxi University of Technology (Natural Science Edition)* **26** 1-4
[6] ChenM, ChenW, GuanY P and ZhaoZ H2013Investigation on dynamic constitutive model of titanium TC4 *Journal of Mechanical Strength* **35** 406-501
[7] LinebergerL1998Titanium aerospace alloy*Advanced Materials and Processes* **153** 45-6
[8] Qian J H 2000 Application and development of new titanium alloys for aerospace *Chinese Journal of Rare Metals* **24** 218-21
[9] Ruan X Q, Lin X, Huang C P, YiN, MengY L and Huang W D 2015 Effect of microstructure of laser solid forming TC4 titanium alloy on ultrasonic parameters *Chinese Journal of Lasers* **42** 0103005
[10] Yang X H, Yang W L, Dui W Z and Li X Y 2014 Microstructure and microhardness of laser welding joints in TC4 alloy *Transcations of Materials and Heat Treatment* **35** 70-4
[11] Cai H and Xiao R S 2013 Statistic analysis on spatter characteristics in high power CO_2 laser and fiber laser welding of thin sheet aluminum alloy *Transaction of the China Welding Institution* **34** 28-31
[12] Cui G Z, Zhu F X, Gao D F, Liu C and Gao X L1999 Study on characteristics of hot deformation of steel T91 *Journal of Plasticity Engineering* **6** 13-6
[13] Walker C A 2010 Relief for various Tin grain structures *Journal of Electronic Materials* 39

1315-8
[14] Gao Z T and Xiong J J 2000 *Fatigue Reliability* (Beijing: Beihang University Press, Beijing)

IOP Conf. Series: Materials Science and Engineering **103** (2015) 012006 doi:10.1088/1757-899X/103/1/012006

Structural and magnetic properties of $Pr_{0.3}Tb_{0.7}Fe_{1.9-x}Mn_x$ Ribbons

Z Ren[1], L Chen[1], S Li[1,3], H Liu[2] and Z Lu[2]

[1] School of Mathematics and Physics, North China Electric Power University, Baoding 071003, China

[2] School of Material Science and Engineering, Hebei University of Technology, Tianjin 300130, China

E-mail: songtaoli2001@126.com

Abstract. The structural and magnetic properties of $Pr_{0.3}Tb_{0.7}Fe_{1.9-x}Mn_x$ ribbons have been investigated using X-ray diffraction, vibrating sample magnetometry and the standard strain gauge technique. According to the XRD spectra, all ribbons of $Pr_{0.3}Tb_{0.7}Fe_{1.9-x}Mn_x$demonstratea pure Laves phase, which benefits from the melt-spun technology. The Curie temperature of the $Pr_{0.3}Tb_{0.7}Fe_{1.9-x}Mn_x$ alloys decreases gradually from 366°C for x=0.1 to 328°C for x=0.3. Meanwhile, magnetostriction at room temperature demonstrates a maximum at x=0.1.

1. Introduction

Magnetostrictive materials have been the topic of much recent research for their potential applications on sensors and micro-drivers. According to the single-ion model [1], $CeFe_2$ and $PrFe_2$ compounds demonstrate greater magnetostriction than $TbFe_2$ and $DyFe_2$ at 0K; therefore, Ce-based and Pr-based RFe_2 ("R" refers to rare earth elements) compounds may have great potential as giant magnetostrictive materials. Moreover, the light rare earth elements Pr and Ce are much more cost effective than the heavy rare earth elements Tb and Dy. As a result, considerable studies have focused on Pr-based or Ce-basedcompounds [2-6]. However, the large radius of the Pr^{3+} ion presents difficulties for the synthesis of $PrFe_2$ with a pure Laves phase at ambient pressure [7, 8]. Only a small portion of Pr content is possible in stable RFe_2 compounds. When Pr content exceeds 20-25%, a single Laves phase material cannot be obtained in $(R,Pr)Fe_2$ compounds. Recently, great effort has been exerted to enhance Pr content in $(R,Pr)Fe_2$ compounds [9]. For instance, researchers have used a melt-spinning technique, or introduced substitute/interstitial atoms such as B, Co, etc., into the $(R,Pr)Fe_2$ system [10-12].

In this paper, Mn atoms were introduced into $(Tb,Pr)Fe_2$ compounds to substitute for Fe, and melt-spun technology was applied in the synthesis of $Pr_{0.3}Tb_{0.7}Fe_{1.9-x}Mn_x$ ribbons to obtain potential magnetostrictive materials. The structural and magnetic properties of the obtained $Pr_{0.3}Tb_{0.7}Fe_{1.9-x}Mn_x$ ribbons were then investigated.

[3] Address for correspondence: S Li, School of Mathematics and Physics, North China Electric Power University, Baoding 071003, China. E-mail: songtaoli2001@126.com.

2. Experimental procedure

The $Pr_{0.3}Tb_{0.7}Fe_{1.9-x}Mn_x$ (x=0.1, 0.2, 0.3) alloys were prepared by arcmelting constituent elements in a highly purified argon atmosphere using nonconsumable tungsten electrodes and a water-cooled copper bottom.The purities of rare earth elements are 99.9% and the metals of Fe and Mn have a the purities of 99.99%. The alloy buttons were melted four times to ensure homogenization. Next, ingots of approximately 3 g were annealed at 850°C for 72 hours; ingots of approximately 5 g portions were conducted in an atmosphere of pure argon by ejection of the molten metal onto a rotating copper roll moving at a speed of 50 m/s, which was then cooled by water. Thus, ribbons of $Pr_{0.3}Tb_{0.7}Fe_{1.9-x}Mn_x$ were obtained, and the ribbons were then annealed in an argon atmosphere at 500°C for 30 minutes. Phase identification was conducted by X-ray powder diffraction methods (Philips X'Pert MPD). The magnetization at room temperature and the Curie temperature of annealed ribbons were determined by a vibrating sample magnetometer (Lake Shore 7407 model). The magnetostriction was measured at room temperature using a standard strain gauge in directions parallel ($\lambda_{//}$) and perpendicular ($\lambda_{\perp}$) to the magnetic field.

3. Results and discussion

The X-ray diffraction spectra for the annealed $Pr_{0.3}Tb_{0.7}Fe_{1.9-x}Mn_x$ alloys are shown in figure 1. All arc-melt alloys were composed of a matrix of cubic-structured RFe_2 compound, with a small amount of the R-rich phase and RFe_3 phase. In the formula of $Pr_{0.3}Tb_{0.7}Fe_{1.9-x}Mn_x$, the rare earth rich stoichiometry was often chosen to offset the volatilization of rare earth metals in the melting process and to inhibit the formation of the RFe_3 phase. Generally, the Pr content in an $(R,Pr)Fe_2$ system cannot exceed 20-25% for rare earth materials, without the appearance of the second phases (R-rich phase and RFe_3 phase) [8]. In order to fabricate a single Laves phase with Pr content greater than 20%, we employed melt-spun technology and investigated the XRD spectrato confirm the Laves phase and the second phases.

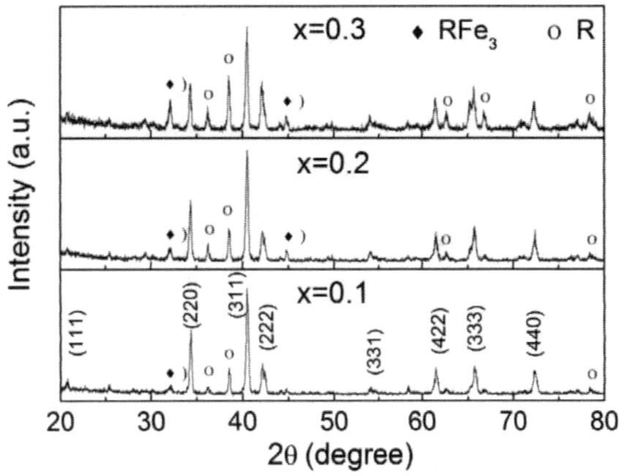

Figure 1. XRD patterns of homogenized $Pr_{0.3}Tb_{0.7}Fe_{1.9-x}Mn_x$alloys at 850°C for 72 hs.

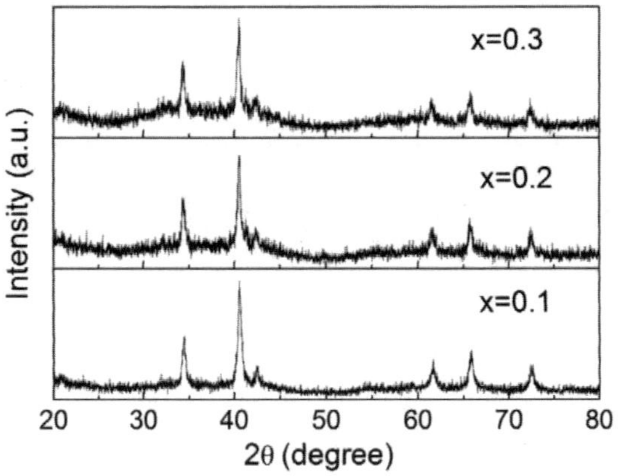

Figure 2. XRD patterns of as-spun $Pr_{0.3}Tb_{0.7}Fe_{1.9-x}Mn_x$ ribbons.

XRD patterns of $Pr_{0.3}Tb_{0.7}Fe_{1.9-x}Mn_x$ alloys homogenized at 850°C for 72 hours are shown in figure 1. The X-ray diffraction spectra for the $Pr_{0.3}Tb_{0.7}Fe_{1.9-x}Mn_x$ ribbons (as-spun and as-annealed) are shown in figures 2 and 3, respectively. In figure 1, the orientation index shows the presence of second phases in the alloys. The peaks of the R-rich phase and the RFe_3 phase are clear in the XRD spectra for the $Pr_{0.3}Tb_{0.7}Fe_{1.9-x}Mn_x$ alloys, which were annealed at 850°C for 72 hours. The high temperature and the long annealing time cannot eliminate the second phases in the $Pr_{0.3}Tb_{0.7}Fe_{1.9-x}Mn_x$ alloys. However, as shown in figure 2, which represents the as-spun $Pr_{0.3}Tb_{0.7}Fe_{1.9-x}Mn_x$ ribbons (which were not annealed), the peaks of the R-rich phase and RFe_3 phase disappear. All peaks in the diffraction patterns are indexed to the characteristics of the $MgCu_2$ crystal structure in figure 2. In figure 2, the broad crystalline peak and the broad hump indicate that a mixture of ultra-fine grains and the amorphous phase exist in the $Pr_{0.3}Tb_{0.7}Fe_{1.9-x}Mn_x$ ribbons. In order to investigate the effect of the annealing process on the as-spun $Pr_{0.3}Tb_{0.7}Fe_{1.9-x}Mn_x$ ribbons, the ribbons were annealed at 500°C for 30 minutes. The XRD of the as-annealed $Pr_{0.3}Tb_{0.7}Fe_{1.9-x}Mn_x$ ribbons are shown in figure 3. Results indicate that all diffraction peaks become sharper in figure 3 than those in figure 2, indicating the formation of the Laves phase in the as-annealed $Pr_{0.3}Tb_{0.7}Fe_{1.9-x}Mn_x$ ribbons. Thus, the melt-spin technique is determined to be an effective method for the synthesis of Laves phase alloys with high Pr content.

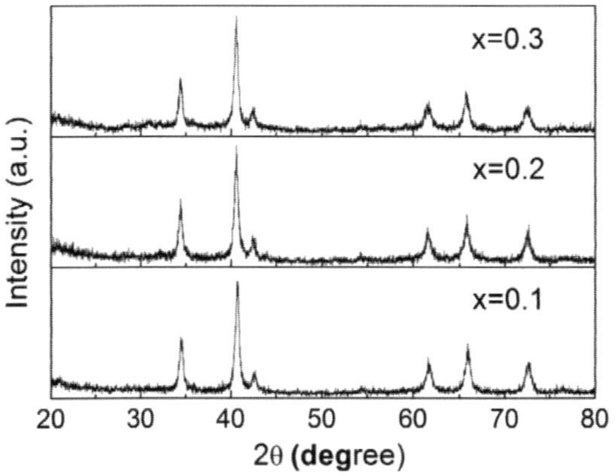

Figure 3. XRD patterns of as-annealed $Pr_{0.3}Tb_{0.7}Fe_{1.9-x}Mn_x$ribbons.

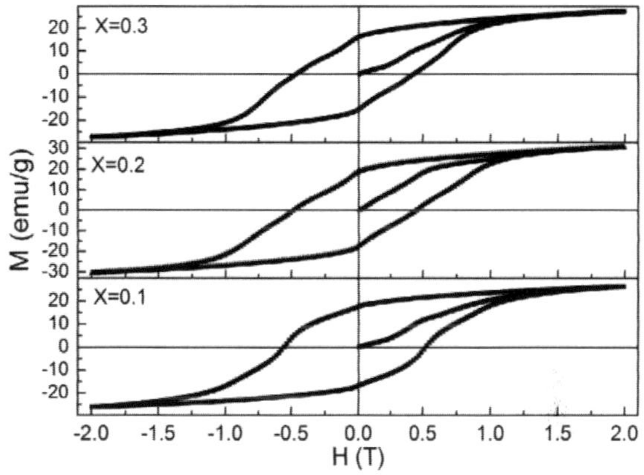

Figure 4. Magnetization of $Pr_{0.3}Tb_{0.7}Fe_{1.9-x}Mn_x$ ribbons (annealed) at room temperature.

The as-annealed $Pr_{0.3}Tb_{0.7}Fe_{1.9-x}Mn_x$ ribbons demonstrate a pure Laves phase. The magnetic properties of the as-annealed $Pr_{0.3}Tb_{0.7}Fe_{1.9-x}Mn_x$ ribbon powder were then investigated. The magnetization dependence on the external applied magnetic field is shown in figure 4. All samples are soft magnetic materials which easily reach saturation under the external applied magnetic field. The saturated magnetization and the remnant magnetization show a maximum at x=0.2, but the intrinsic coercivity decreases with increasing Mn content. Table 1 lists the values of the saturated magnetization M_s, the remnant magnetization M_r, the intrinsic coercivity H_{ci} and the Curie temperature for the $Pr_{0.3}Tb_{0.7}Fe_{1.9-x}Mn_x$ ribbons. As summarized in table 1, The Curie temperature of the as-annealed $Pr_{0.3}Tb_{0.7}Fe_{1.9-x}Mn_x$ ribbonpowder decreases from 366°C (x=0.1) to 328°C (x=0.3) as the Mn content increases. It is well-known that for R-T compounds (T representing a transition metal element), the Curie temperature is primarily determined by T-T exchange interaction. Thus, the decline of the Curie temperature may be due to the substitution of Mn for Fe in $Pr_{0.3}Tb_{0.7}Fe_{1.9-x}Mn_x$ ribbons, which disturbs the 3d-3d exchange interaction [13].

Table 1. Saturated magnetization M_s, remnant magnetization M_r, intrinsic coercivity H_{ci} and Curie temperature T_c for $Pr_{0.3}Tb_{0.7}Fe_{1.9-x}Mn_x$ ribbons (annealed).

x	M_s (Am^2/kg)	M_r (Am^2/kg)	H_{ci} (A/m)	T_c (°C)
0.1	26.099	17.219	5.363	366
0.2	30.553	18.141	4.664	357
0.3	27.316	15.658	4.458	328

The dependence of magnetostriction $\Delta\lambda = \lambda_{//} - \lambda_{\perp}$ on the external applied magnetic field is shown in figure 5. The as-annealed $Pr_{0.3}Tb_{0.7}Fe_{1.9}Mn_x$ ribbons were milled into powder and mixed with a phenol binder, with a weight ratio of powders to binder of 100:6. The composites were compacted at a pressure of 80MPa, and then solidified at room temperature in a free state. The magnetostriction of all samples was far from saturation at the external applied magnetic field of 11 kOe. With the Mn content increasing from 0.1 to 0.3, the magnetostriction at the largest magnetic field decrease from 400 ppm (x=0.1) to 300 ppm (x=0.3). This interesting magnetostrictive behavior may be ascribed to the complex coupling between R (a rare earth element) and Mn. Because the Mn-Mn interatomic spacing in the R-Fe-Mn alloy is just below the critical value d_c (0.266 nm), the itinerant 3d moment of the Mn atom becomes unstable [14]. Therefore, the substitution of Fe by Mn demonstrates a complex effect on the magnetostrictive behavior of $Pr_{0.3}Tb_{0.7}Fe_{1.9-x}Mn_x$ ribbons.

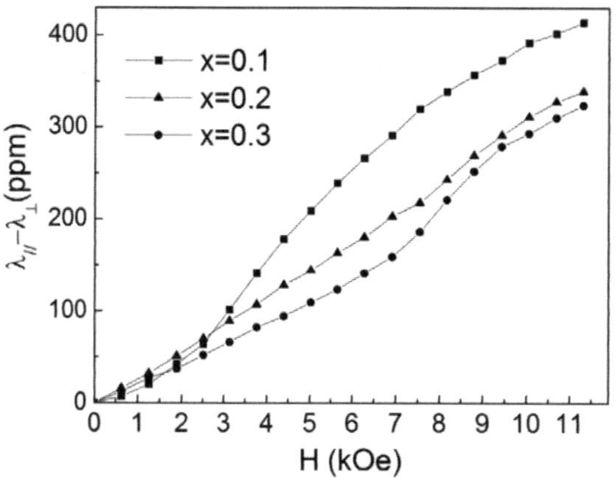

Figure 5. Magnetostriction of $Pr_{0.3}Tb_{0.7}Fe_{1.9-x}Mn_x$ ribbons (annealed) at room temperature.

4. Conclusions

In this work, $Pr_{0.3}Tb_{0.7}Fe_{1.9-x}Mn_x$ ribbons with a Laves phase structure were fabricated. The melt-spin technique proved to be a good method for the synthesis of alloys with high Pr content in the Laves phase. The substitution of Mn for Fe has a major effect on the magnetic properties and magnetostriction of $Pr_{0.3}Tb_{0.7}Fe_{1.9-x}Mn_x$ ribbons. The saturated magnetization and the remnant magnetization demonstrate maximum values at x=0.2, but the intrinsic coercivity and Curie temperature decrease with increasing Mn content. The magnetostrictionof $Pr_{0.3}Tb_{0.7}Fe_{1.9-x}Mn_x$ ribbons at a magnetic field of 11 kOe decrease with increasing Mn content, due to the complex coupling between R (rare earth elements) and Mn.

Acknowledgments

This work has been supported by the National Science Foundation of China (grant nos. 11004055 and 50901028), the Fundamental Research Funds for the Central Universities (grant nos. 12MS142, 13ZD23, 2014MS162 and 2014MS168), the Beijing-funded Joint Development Program of the Central Universities in Beijing and the Program for New Century Excellent Talents in University (grant no. NCET-12-0844).

References

[1] Clark A E 1980 *Ferromagnetic Materials* (Amsterdam: North Holland Publications) p 531

[2] Li Y, Tang C, Du J, Wu G, Zhan G, Yan W, Xu G and Yang Q 1998 Magnetostrictive and magnetic properties of the pseudobinary compounds $Pr_xTb_{1-x}Fe_2$ and $Pr_{0.15}Tb_xDy_{0.85-x}Fe_2$ *J. Appl. Phys.* **83** 7753-6

[3] Ma T, Yan M, Chen X, Jiang C and Xu H 2008 Structure and magnetic properties of magnetostrictive compounds $Tb_{0.36}Dy_{0.64}(Fe_{0.85}Co_{0.15})_{2-x}B_x$ *J. Magn. Magn. Mater.* **320** 2368-72

[4] Liu H Y, Meng F B, Li S T, Li Y X, Chen J L and Wu G H 2004 Structure, magnetic properties and magnetostriction of $Pr_{0.15}Tb_{0.3}Dy_{0.55}Fe_{1.85-x}Si_x$ alloys *J. Physica B* **351** 102-5

[5] Tang C, Du J, Li Y, Wang F, Wu G and Zhan W 1998 Magnetostriction in $(Ce_xTb_{1-x})_{0.5}Pr_{0.5}Fe_2$ compounds *Appl. Phys. Lett.* **73** 692-4

[6] Zhang S, Zhang M, Qiao Y, Liu M, Wang H and Xu Y 2013 Magnetostriction, composition and microstructure of $Tb_{0.29}(Dy_{1-x}Pr_x)_{0.71}Fe_{1.97}$ alloys with heat treatment *J. Rare Earth* **31** 577-81

[7] Zhao X, Li J, Liu S, Ji S and Jia K 1997 Magnetic properties and thermal stability of $PrFe_2$ compound *J. Alloys Comp.* **258** 39-41

[8] Lv X K, Liu W, Ren W J, Li B, Zhao X G, Li D and Zhang Z D 2008 Structure, magnetic properties and magnetostriction of laves compounds $Nd_{1-x}Pr_x(Fe_{0.35}Co_{0.55}B_{0.1})_2$ *J. Magn. Magn. Mater.* **320** 2373-5

[9] Shi Y G, Tang S L, Huang Y J, Nie B, Qian B, Lv L Y and Du Y W 2007 Structure, thermal stability and magnetostrictive properties of $PrFe_2$ alloys *J. Alloys Comp.* **443** 11-4

[10] Liu J J, Xiong D K, Ren W J, Zhang Z D and Or S W 2007 High magnetostriction at low fields of epoxy/$Tb_{1-x}Pr_x(Fe_{0.4}Co_{0.6})_{1.9}$ composites *J. Alloys Comp.* **427** 271-4

[11] Ren W J, Zhang Z D, Zhao X G, Song X P, Liu W and Geng D Y 2004 Structure and magnetostriction of $Tb_{1-x}Pr_xFe_{1.93}B_{0.15}$ alloys *J. Magn. Magn. Mater.* **269** 281-5

[12] Liu H Y, Li Y X, Meng F B, Lu Z M, Xu X W, Zhang M and Qu J 2008 Structure and magnetostriction of melt-spun $Pr_{0.3}Tb_{0.7}Fe_2B_x$ alloys *J. Alloys Comp.* **469** 310-3

[13] Guo Z J, Zhang Z D, Zhao X G, Jin X M and Geng D Y 2000 Effect of Mn substitution on structure, magnetic properties and magnetostriction of $Sm_{0.9}Pr_{0.1}Fe_2$ *J. Alloys Comp.* **308** 319-22

[14] Wu C H, Yang C P, Chuang Y C, Jin X M and Li J Y 1997 Structure and magnetostriction of $R(Fe_{1-x}Mn_x)_{1.85}$ alloys ($R=Dy_{0.6}Tb_{0.3}Pr_{0.1}$) *J. Magn. Magn. Mater.* **166** 249-52

Effect of rapid set binder on early strength and permeability of HES latex modified road repair pre-packed concrete

J W Han[1], S K Lee[1], C Yu[2] and C G Park[3,4]

[1] Department of Bio-Industry Mechanical Engineering, Kongju National University, Yesan, Republic of Korea
[2] Major of Agricultural Engineering (Inst. of Agri. & Life Sci.), Gyeongsang National University, Jinju, 660-701, Republic of Korea
[3] Department of Rural Construction Engineering, Kongju National University, Yesan, Republic of Korea

E-mail: cgpark@kongju.ac.kr

Abstract. The early strength development characteristics and permeability resistance of high early strength (HES) pre-packed road repair concrete incorporating a rapid-set binder material were evaluated for emergency repairs to road pavement. The rapid-set binder is a mixture of rapid-set cement and silica sands whose fluidity improves with the addition of styrene butadiene latex (latex). The resulting mixture has a compressive strength of 21 MPa or higher and a flexural strength of greater than 3.5 MPa after 4 hours, the maximum curing age allowed for emergency repair materials. This study examines the strength development properties and permeability resistance of HES latex-modified pre-packed road repair concrete using a rapid-set binder as a function of the latex-to-binder mixing ratio at values of 0.40, 0.33, 0.29 and 0.25. Both early strength development properties and permeability resistance increased as the ratio of latex to rapid-set binder decreased. The mixture showed a compressive strength of 21 MPa or higher after 4 hours, which is the design standard of emergency repair concrete, only when this ratio was 0.29 or lower. A flexural strength of 3.5 MPa or greater was observed after 4 hours only when this ratio was 0.33 or lower. The standard for permeability resistance, less than 2,000 C of chloride after 7 days of curing, was satisfied by all ratios. The ratio of latex to rapid-set binder satisfying all of the conditions for an emergency road repair material was 0.29 or less.

1. Introduction

Reparations to old paved roads are in continuous demand. As traffic considerations require that roads be reopened as soon as possible following a repair, many diverse studies have led to design standards for materials used in emergency road repairs [1-7]. Various methods have been explored to ensure minimal closure; for example, a precast concrete pavement as a replacement for the damaged concrete pavement can be used [8-12]. With precast concrete pavement structures, immediate re-opening to traffic is possible once the precast sections are in place [8-12]. However, with precast concrete pavement structures, rough handling during transportation and installation can lead to cracking, which can increase the water permeability [11, 12]. Another method is to use ASTM Type I cement with a

[4] Address for correspondence: C G Park, Department of Rural Construction Engineering, Kongju National University, Yesan, Republic of Korea. E-mail: cgpark@kongju.ac.kr.

hardening accelerator [13, 14]. Pavement repair materials specify a minimum curing time of 72 hours (3 days) as the standard curing period for ASTM Type I cement before re-opening to traffic and requires the compressive strength to exceed 22 MPa and the flexural strength to exceed 4.5 MPa [8, 11, 12]. However, ASTM Type I cement alone does not meet these requirements; a hardening accelerator is needed [11, 12]. The third approach is to use rapid-set cement concrete [11, 12, 15, 16]. Rapid-set cement concrete is the primary material used in such repairs, but rapid-set cement generates a high heat hydration to attain its early strength, which heats the surrounding road structure and increases the migration of vapours [5, 6]. Under actual field conditions, this migration is often restricted to the repair structure itself, resulting in tensile stress and the formation of micro cracks [5, 6]. These cracks increase the water permeability of the repair structure, which leads to a diverse range of destruction mechanisms [5, 6]. Thus, the repair loses function and raises the necessity of another repair within a short time. To solve this problem, this study evaluates the performance of a HES road repair material consisting of pre-packed concrete incorporating styrene butadiene latex (latex) and rapid-set binder materials. In a typical road repair using this material, the old road pavement to be repaired is removed and the void is filled with coarse aggregates. The rapid-set binder material is then mixed with latex, the mixture is poured into the aggregate-filled void and the repair is infused via vibration compaction. At this point, the mixture of latex and rapid-set binder should have sufficient fluidity to be self-levelling. After infusion, the surface is finished to end the road repair. Mixing latex into the rapid-set binder secures the early strength and sufficient workability while maintaining or improving durability [17-20]. The addition of latex increases the workability of unhardened concrete and reduces the ratio of water to cement [17-20]. It can also prevent the separation of materials due to the viscosity of latex itself and can increase the flexural strength, tensile strength and deformation capacity after the concrete hardens, thereby improving adhesion, water-tightness, resistance to freezing/melting, anti-abrasion properties and chemical resistance [17-20]. In this study, the early strength development properties and permeability resistance of HES latex-modified pre-packed concrete for emergency road repair were evaluated.

2. Materials and methods

2.1. Materials

The rapid-set binder material used in this study was a mixture of rapid-set cement and silica sands manufactured by Jung-Ang Polytec in Korea. The mixing ratio of rapid-set cement to silica sands was 1:1. The chemical compositions of the rapid-set cement are shown in table 1. Coarse aggregates were crushed stone with a maximum dimension of 25 mm and a density of 2.62 g/mm^3. The physical properties of the latex, made by Korean Jung-Ang Polytec, are listed in table 2.

Table 1. Chemical compositions of rapid set cement.

SiO_2 (%)	Al_2O_3 (%)	Fe_2O_3 (%)	CaO (%)	MgO (%)	K_2O (%)	SO_3 (%)
13±3	17.5±3	3>	50±3	2.5>	0.21	14±3

Table 2. Properties of styrene butadiene latex.

Solids (%)	Content Styrene Content (%)	Butadiene Content (%)	Surface Tension (dyne/cm)	Particle Size (Å)	Viscosity (cps)
8	34±1.5	66±1.5	30.57	1700	42

2.2. Mix proportions and manufacturing of test specimens

For road pavement repairs made with rapid-set cement, AASHTO and the road traffic offices of each American state, as well as the Korean Highway Corporation, have defined 4 hours as the minimum curing period prior to reopening traffic over the repair [8, 21]. The standards for traffic reopening are a compressive strength of 21 MPa or higher and a flexural strength of 3.5 MPa or more [8, 21]. This study focuses on road repairs using rapid-set cement, the goals were to obtain a material that meets these standards. Permeability has the greatest effect on the life span and durability of road pavement. The Korean Highway Corporation has suggested a limit of 2000 C or lower after 7 days of curing, based on tests of chloride ion permeation in accordance with ASTM C1202 [21]. The mixing ratios of latex to binder used in this study are presented in table 3.

Table 3. Mix Proportions of HES latex modified road repair pre-packed concrete.

Type of mix	Rapid set binder materials (kg/m^3)	Latex (kg/m^3)	Corse aggregate (kg/m^3)
0.40	575	218	1450
0.33	627	210	1450
0.29	690	198	1450
0.25	732	184	1450

Specimens for compressive strength tests was made by packing coarse aggregates into a cylindrical frame with a diameter of 100 mm and a height of 200 mm. Rapid-set cement and silica sands were mixed together. The latex was then poured into this rapid-set binder material and mixed for 1 minute and 30 seconds. After mixing, the cylindrical form packed with coarse aggregates was put on the vibration table, and the rapid-set binder materials were poured into it while the table was vibrating. Flexural strength specimens were produced in the same way except in a 100 × 100 × 400-mm rectangular frame.

2.3. Test methods

2.3.1. Strength tests. Compressive strength tests were performed in accordance with ASTM C 39, flexural tests were conducted in accordance with ASTM C 78/C78M were conducted in accordance with ASTM C 496/C496M-11. Tests were performed after 4 hours of curing.

2.3.2. Chloride ion penetration tests. Chloride ion penetration tests were conducted in accordance with ASTM C 1202-94. Specimens measuring 150 × 50 mm were tested after 4 hours and 7 days of curing. The test apparatus for chloride ion penetration tests is shown in figure 1.

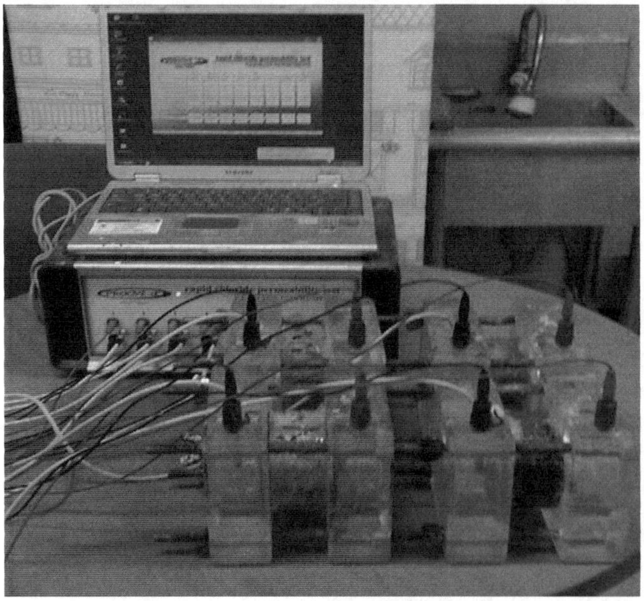

Figure 1. Chloride ion penetration test set-up.

3. Test results and discussion

3.1. Compressive strength

The compressive strength characteristics of HES latex-modified pre-packed road repair concrete for emergency road repair are shown in figure 2. As the mixing ratio of latex to rapid set binder material got smaller, the compressive strength tended to increase. When this ratio was below 0.29, the resulting material satisfied the compressive strength requirement for an emergency road repair material (21 MPa) after 4 hours, which suggests that the addition of latex retarded the manifestation of early strength. Furthermore, the lower proportion of rapid-set binder material likely did not allow sufficient early strength to develop. To satisfy both of these goals, the proportion of rapid-set binder must be higher. Thus, a latex-to-binder ratio of 0.29 or lower was deemed sufficient to satisfy the standards of an emergency road repair material.

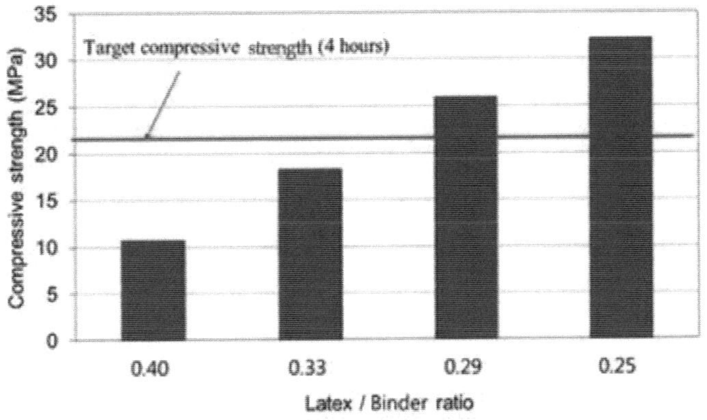

Figure 2. Compressive strength of HES latex modified road repair pre-packed concrete.

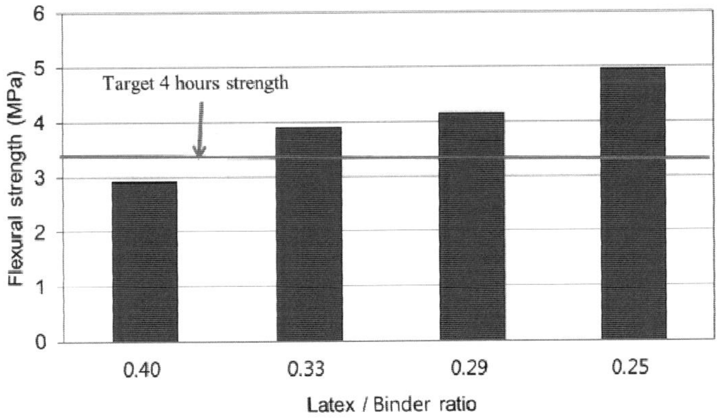

Figure 3. Flexural strength of HES latex modified road repair pre-packed concrete.

3.2. Flexural strength

The flexural strength test results are given in figure 3. The standard for flexural strength is 3.5 MPa after 4 hours. These standards were met at a latex-to-binder ratio of 0.33 or lower. These data show that the addition of latex to concrete has a greater influence on tensile and flexural strength than on compressive strength. As latex increases the binding forces between materials when the concrete receives flexural or tensile stresses, unlike compressive strength, the flexural strength requirements were satisfied even at a latex-to-binder ratio of 0.33. Flexural strength tended to increase as this binder ratio decreased.

3.3. Chloride ion penetration

The chloride ion permeability resistance of the evaluated road repair concrete is shown in figure 4, and the permeability standard according to ASTM C 1202 is presented in table 4. Chloride ion permeability tended to decrease as the latex-to-binder ratio decreased. Relative to the standard set in ASTM C 1202, the chloride ion permeability after 4 hours for the 0.40 and 0.33 mixtures was 'high', showing up to 4000 C. Conversely, the chloride permeability for the 0.29 and 0.25 mixtures was deemed 'moderate', measuring between 2000 and 4000 C. After 7 days of curing, the permeability of the 0.40 mixture was 'low', measuring 1000–2000 C, while the permeability of the 0.33, 0.29 and 0.25 mixtures was 'very low', measuring 1000 C or less. Furthermore, after 7 days, the standard of 2000 C was met by all of the evaluated mixtures. Thus, in general, the addition of latex reduced the permeability of the concrete blend by filling gaps and forming a thick latex film within the concrete. Therefore, all of the specimens exhibited excellent permeability resistance.

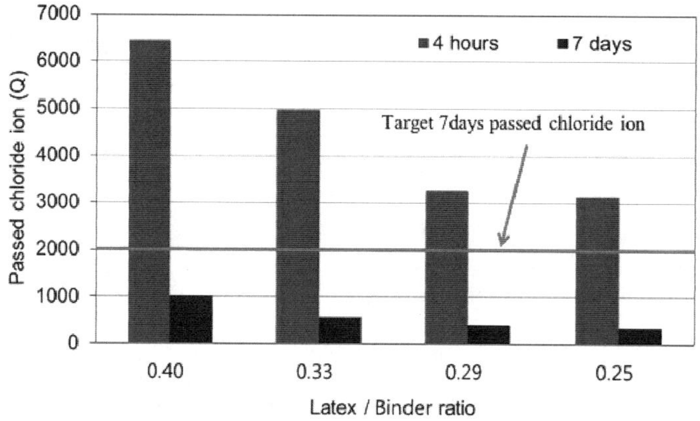

Figure 4. Chloride ion penetration of HES latex modified road repair pre-packed concrete.

Table 4. Chloride permeability based on charge passed.

Charge passed (Coulombs)	Chloride permeability
4000>	High
2000-4000	Moderate
1000-2000	Low
100-1000	Very low
<100	Negligible

4. Conclusions

This study evaluated the early strength development and permeability resistance characteristics of HES latex-modified pre-packed road repair concrete as a function of the ratio of latex to rapid-set binder. The test results can be summarized as follows:

- The HES latex-modified pre-packed road repair concrete formulations evaluated herein met the design standards for an emergency repair concrete only when the ratio of latex to rapid-set binder was 0.29 or lower.
- In the flexural strength tests, the mixing ratio of latex to rapid-set binder needs to be 0.33 or less to satisfy the conditions of 3.5 MPa after 4 hours.
- All of the HES latex-modified road repair concrete formulations met the chloride ion permeability standard of 2000 C set by ASTM C 1202. Chloride permeability decreased as the ratio of latex to rapid-set binder decreased.

Acknowledgment

This research was supported by a grant (code-14TBIP-C073993-01-000000) from Technology Business Innovation Program (TBIP) funded by Ministry of Land, Infrastructure and Transport of Korean government.

References

[1] Sprinkel M 2005 *Latex-Modified Concrete Overlay Containing Type K Cement* (Virginia Transportation Research Council, VA, FHWA/VTRC 05-R26)

[2] Sprinkel M 1998 *High Early Strength Latex Modified Concrete Overlay* (Virginia Transportation Research Council: VA, Report VTRC 88-R12)

[3] Sprinkel M 1998 *Very-Early-Strength Latex-Modified Concrete Over-Lays* (Virginia Transportation Research Council, VA, TAT99-TAR3)

[4] N Buch, T J V Dam, K Peterson and L Sutter 2008 Evaluation of high-early strength PCC mixtures used in full depth repairs *Constr. Build. Mater* **22** 162-74

[5] Won J P, Kim J M, Lee S J, Lee S W and Park S K 2011 Mix proportion of high-strength, roller-compacted, latex-modified rapid-set concrete for rapid road repair *Constr. Build. Mater.* **25** 1796-800

[6] Won J P, Kim J H, Park C G and Kang J W 2009 Shrinkage cracking of styrene butadiene polymeric emulsion-modified concrete using rapid-hardening cement *J. Applied Polymer Sci.* **112** 2229-34

[7] Ricardo Z J 2013 Development of rapid, cement-based repair materials for transportation Structures (The University of Texas at Austin May: MS thesis)

[8] AASHTO 1998 *Standard Specification of Rigid Pavement* (Washington, DC: American Association of State Highway and Transportation Officials)

[9] Bull J W and Woodford C H 1997 Design of precast concrete pavement units for rapid maintenance of runways *Computer. Structu.* **64** 857-64

[10] Ackroyd R F and Bull J W 1985 The design of precast concrete pavements on low bearing capacity subgrades *Computer. Geotech.* **1** 279-91

[11] Kim D H and Park C G 2013 Permeability, abrasion, and impact resistance of latex-modified fibre reinforced concrete for precast concrete pavement application *Progress. Rubb. Plast. Recycl. Techno.* **29** 239-54

[12] Oh R O, Kim D H, Park C G, Chan Gi 2014 Durability performance of latex modified nylon fiber reinforced concrete for precast concrete pavement applications *Indian J. Eng. Mater. Sci.* **21** 49-56

[13] Buch N, Dam T J V, Peterson K and Sutter L 2008 Evaluation of high-early strength PCC mixtures used in full depth repairs *Constru. Build. Mater.* **22** 162-74

[14] Ghafoori N, Diawara H and Beasley S 2008 Resistance to external sodium sulfate attack for early-opening-to-traffic Portland cement concrete *Cem. Concre. Compos.* **30** 444-54

[15] Won J P, Kim J H, Lee S W and Park C G 2009 Durability of low-heat, ultra rapid-hardening, latex modified polymer concrete *Progress. Rubb. Plast. Recycl. Techno.* **25** 91-102

[16] Yun K K and Choi P 2014 Causes and controls of cracking at bridge deck overlay with very-early strength latex –modified concrete *Constr. Build. Mater.* **56** 53-62

[17] Kim D H and Park C G 2013 Strength, permeability and durability of hybrid fiber-reinforced concrete containing styrene butadiene latex *J. Applied Polymer Sci.* **129** 1499-505

[18] Park C G and Lee J H 2013 Effect of styrene butadiene latex polymer contents on the bond properties of macro polypropylene fiber in polymer-modified cement-based composites *J. Applied Polymer Sci.* **126** E330-7

[19] Oh R O and Park C G 2014 Interface pullout properties of macrosynthetic fibres in latex modified cement based nanosilica and silica fume composites *Mater. Res. Innova.* **15** 220-4

[20] Kim D H, Lee S K and Park C G 2014 Bonding characteristics of macro polypropylene (PP) fibre-reinforced styrene butadiene latex-modified cement-based composites *Progress in Rubb. Plast. Recycli Techno.* **30** 1-17

[21] Korea Expressway Corporation 2005 *Construction Material Quality and Standard for Highway Construction* (Seongnam, Republic of Korea: Korea Expressway Corporation)

IOP Conf. Series: Materials Science and Engineering **103** (2015) 012008 doi:10.1088/1757-899X/103/1/012008

Facile synthesis of silver-tin oxide electrical contact materials by in-situ decomposition of a uniform mixture precursor

Z J Lin[1], X D Sun[1,4], S HLiu[1,2], M Xie[2], J L Chen[2] and J G Li[3]

[1] Key Laboratory for Anisotropy and Texture of Materials (Ministry of Education), School of Materials and Metallurgy, Northeastern University, Shenyang, Liaoning 110819, China

[2] State Key Laboratory of Advanced Technologies for Comprehensive Utilization of Platinum Metals, Kunming Institute of Precious Metals, Kunming 650106, China

[3] Advanced Materials Processing Unit, National Institute for Materials Science, 1-2-1 Sengen, Tsukuba, Ibaraki 305-0044, Japan

E-mail: xdsun@mail.neu.edu.cn

Abstract. A uniform mixture precursor containing Ag_2O and Ag_2SnO_3 was synthesized via a facile ion exchange reaction method. The influence of pH value on composition and precipitation rate of the precursor was investigated systematically. The Ag-SnO_2 composite powder was obtained by decomposition of the precursor at 600°C, and uniform mixing of elements was inherited from the precursor and fresh and intimate Ag-SnO_2 interfaces were created. The Ag-SnO_2 electrical contact material was prepared using this composite powder by hot-pressing method. It was found that nano-sized SnO_2 particles (about 20.6 nm) are homogeneously dispersed in the Ag-matrix. The as-prepared Ag-SnO_2 electrical contact material shows excellent performances, such as high hardness (154 MPa), high density (9.90 g/cm^3) and good electrical conductivity (72.24 %IACS).

1. Introduction

As eco-friendly electrical contact materials, Ag-SnO_2 composite materials have been widely concerned [1-3]. It is believed that the Ag-SnO_2 samples reinforced by nano-sized SnO_2 would have great anti-arc performances due to the fast moving of arc-foot and easy decomposition of SnO_2 [4, 5]. However, traditional PM method reaches its limit for homogeneously mixing nano-sized SnO_2 and Ag powders [6]. It is urgently need to obtain an effective and convenient method for preparation of Ag-SnO_2 electrical contact materials reinforced by nano-sized SnO_2.

In 1997, Linke [7] reported the formation of Ag_2SnO_3 crystals by the solid state reaction method. The modulated structure of Ag_2SnO_3 was then investigated by Takeo [8]. Feng et al. [9, 10] calculated the thermodynamic, mechanical and electronic properties of Ag_2SnO_3, revealing its positive effect on Ag-SnO_2 electrical contact materials. An ion exchange method was used to synthesize nano-sized Ag_2SnO_3 by Yin et al. [11]. The as-prepared Ag_2SnO_3 shows excellent photocatalytic performance. The Ag-SnO_2 heterogeneous nanomaterial [12] was then in situ produced by thermal decomposition of

[4] Address for correspondence: X D Sun, Key Laboratory for Anisotropy and Texture of Materials (Ministry of Education), School of Materials and Metallurgy, Northeastern University, Shenyang, Liaoning 110819, China. E-mail: xdsun@mail.neu.edu.cn.

Ag_2SnO_3. The heterogeneous nanomaterial has a unique advantage of good photocatalytic properties, because Ag and SnO_2 nanoparticles contact tightly with each other.

In this work, the ion exchange method was used to synthesize a mixture precursor containing Ag_2O and Ag_2SnO_3. We investigated systematically the effect of pH value on phase composition and precipitation rate. The thermal behaviors of the as-prepared precursor were then studied. The Ag-SnO_2 composite powder was then in-situ produced from the precursor by direct thermal decomposition. Finally, the Ag-SnO_2 electrical contact material with good mechanical and electrical properties was made by hot pressing.

2. Experiment procedure

The starting materials including silver nitrate ($AgNO_3$), sodium stannate (Na_2SnO_3), sodium hydroxide (NaOH, analytical reagent), and nitric acid (HNO_3) were all analytical reagent and purchased from Sinopharm Chemical Reagent Co., and used as received without further purification.

0.816 mol silver nitrate and 0.080 mol sodium stannate were added into de-ionized (DI) water to make 1000 mL solutions, respectively. The two solutions were then mixed and magnetically stirred at room temperature. Sodium hydroxide and nitrate acid solution was used to adjust pH values. After 60 min aging, the precursor was collected via centrifugation, washed several times with DI water, and dried at 100°C for 24 h in air. The obtained precursors were calcined in a muffle furnace at 600°C. The Ag-SnO_2 composite powders were thus obtained and then used to prepared the electrical contact materials (Φ12×10 mm) by hot-pressed at 700°C with a pressure of 60 MPa.

3. Characterization techniques

Density of the Ag-SnO_2 contact materials was measured by the Archimedes principle. A Vickers hardness tester was used to measure the Hardness. The electric conductivity of samples was measured by an FQR 7051 vortex conductivity apparatus. DSC-TG analysis of the precursor was taken under flowing air with a heating rate of 10°C/min (QMS 403D; NETZSCH Sleb, Germany). Phase identification was performed by XRD (Model PW 3040/60; PANALYTICAL B.V, Netherlands) operated at 40 KV per 40 mA, using nickel-filtered CuKα radiation. The contents of Ag and Sn in supernatant liquid were determined using ICP/AES (ProdigyXP; LEEMAN LABS, USA). FE-SEM was used to observed microstructures of the Ag-SnO_2 composite powders and electrical contact materials with an acceleration voltage of 15 kV (Ultra Plus, Carl Zeiss AG, German).

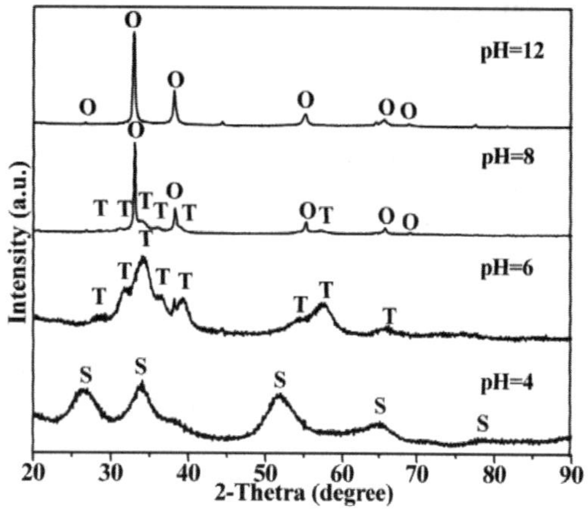

Figure 1. XRD patterns of the precursors prepared at various pH values. Letters S, T, and O represent $Sn(OH)_4$, Ag_2SnO_3, and Ag_2O, respectively.

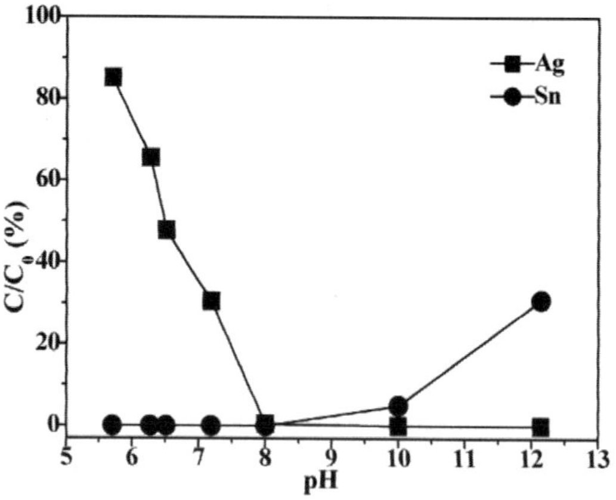

Figure 2. The residual ratios of Ag and Sn elements in the supernatant liquid as a function of pH value.

4. Results and discussion

4.1. Preparation and characterization of the precursors

Figure 1 shows XRD patterns of the precursors synthesized at different pH values. It was found that the phase components of the precursors are highly dependent on pH value of the reaction solution. The Ag_2SnO_3 compound can only be obtained in a certain range of pH value (6~8). When the pH value is lower than 4, the Sn element exists mainly as $Sn(OH)_4$. While the pH value is increased to 12, Ag^+ is captured by the extra OH^- to form Ag(OH), which will decompose into Ag_2O in a short time, as shown by equation (1).

$$Ag_2SnO_3 + 2OH^- \rightarrow Ag_2O + SnO_3^{2-} + H_2O \quad (1)$$

Figure 2 shows the residual ratios of Ag and Sn elements in the supernatant liquid at different pH values. The results show that both Ag and Sn rarely exist (over range of measurement) in the supernatant liquid when pH=8, indicating complete precipitation of the elements.

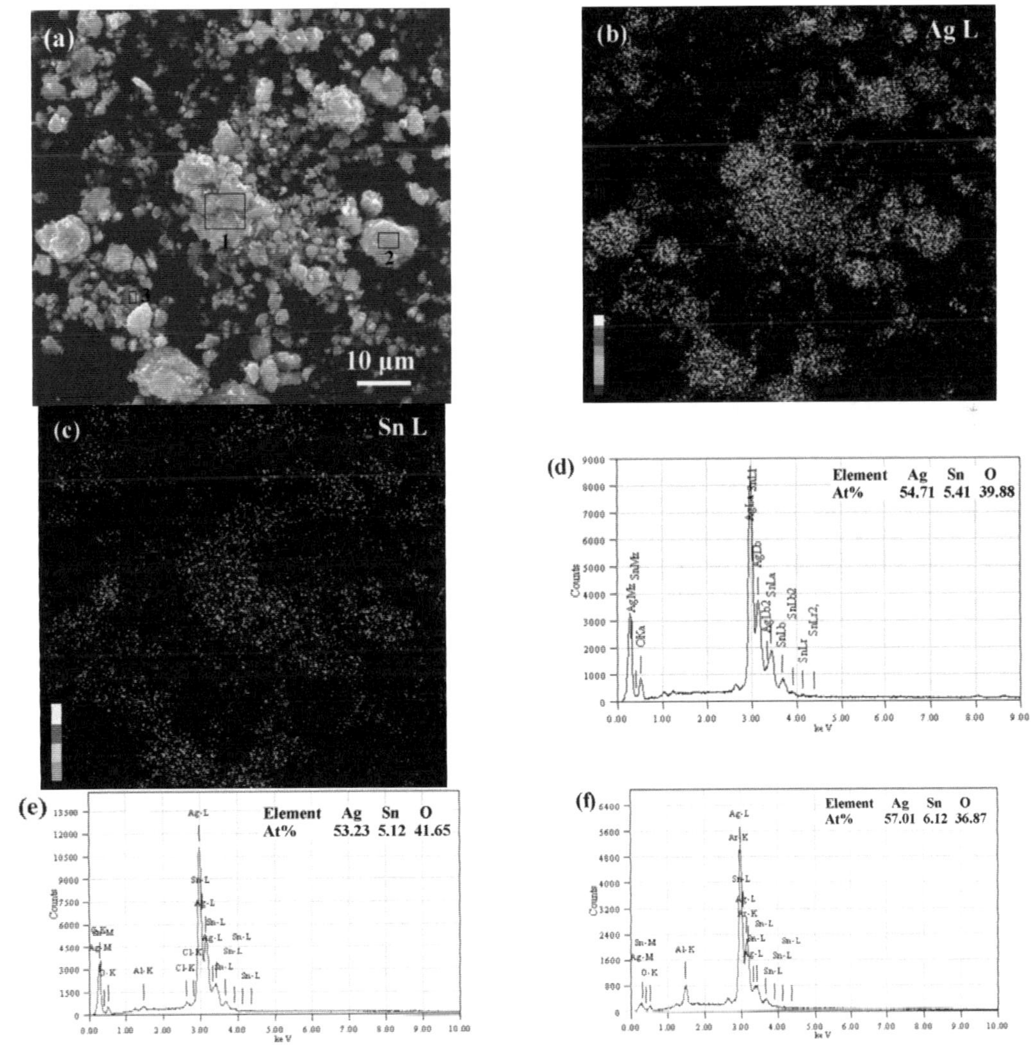

Figure 3. Morphology of the precursor synthesized by the in-situ reaction method (SEM) (a); the EDS maps for Ag (b) and Sn (c); EDS analysis for the three selected zones in (a): (d) zone 1, (e) zone 2 and (f) zone 3.

Figure 3 shows morphology and element distribution of the precursor synthesized at a pH value of 8. The concentration of Ag and Sn elements varies synchronously in the EDS maps. The EDS quantitative analyses were taken for three selected areas in figure 3(a). The molar ratio of Sn/Ag varies from 0.096 to 0.107, indicating that Ag and Sn elements are basically homogenously mixed in the precursor.

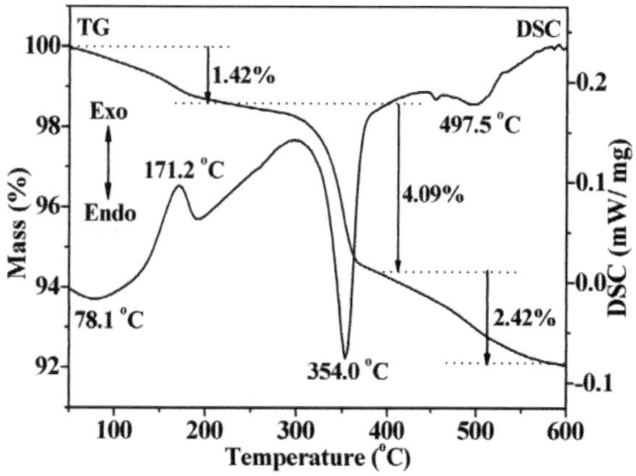

Figure 4. DTATG test of the Ag-Sn-O precursors at a heating rate of 10°C/ min in flowing air.

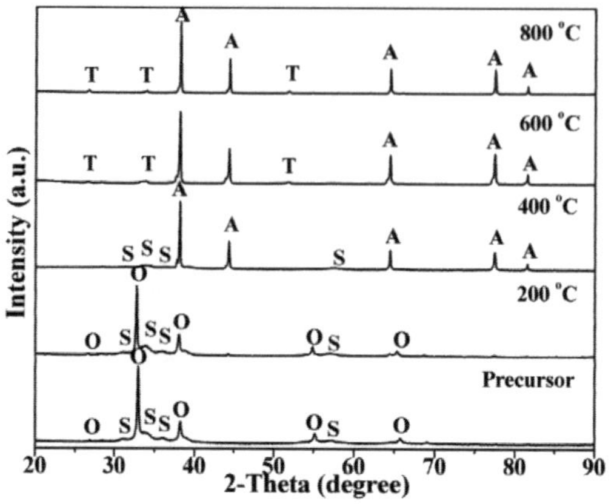

Figure 5. XRD patterns of the precursors calcined at various temperatures. Letters O, S, A and T represent Ag_2O, Ag_2SnO_3, Ag and SnO_2, respectively.

4.2. Preparation and characterization of the Ag-SnO_2 composites

The thermal decomposition behaviors were shown in figure 4. The TG curve can be described as a four-step weight loss, claimed by a series of endothermic processes, respectively. The first endothermic process below 200°C, with a weight loss of about 1.42%, is largely due to the evaporation of surface adsorbed water. The exothermic peak located at 171.2°C on the DSC curve is due to the crystallization of compounds. A sharp endothermic peak located at 354°C, accompanied by a main mass loss about 4.09%, can be attributed to the decomposition of Ag_2O.The last step weight loss starts at 400°C, with a series of fluctuate endothermic peaks in the DSC curves, which may be due to the decomposition of Ag_2SnO_3, as shown in equation (2).

$$2Ag_2SnO_3 \rightarrow 4Ag + 2SnO_2 + O_2 \uparrow \quad (2)$$

XRD analysis also demonstrates the above results, as shown in figure 5. The existence of silver metal phase for the products calcined at 200°C may be attributed to the photodecomposition of silver compounds. The precursor was transformed into Ag-SnO_2 composite at 600°C. Crystallinity of the SnO_2 was further improved at 800°C.

The morphology and element distribution of the obtained Ag-SnO_2 composite powders calcined at 600°C are shown in figure 6. The concentration of silver and tin elements varies synchronously in the EDX maps. During the thermal decomposition, both silver and tin oxides are in-situ created from the precursor. Therefore, the Ag-SnO_2 composites inherit the uniform mixing from the precursor.

4.3. Preparation and characterization of the Ag-SnO_2 electrical contact materials

XRD pattern of the Ag-SnO_2 electrical contact material reveals that the sample is composed of metal Ag and SnO_2. No other phases exist, as shown in figure 7. The size of the SnO_2 particles is about 20.6 nm calculated by the Scherrer equation.

Microstructure of the Ag-SnO_2 electrical contact materials are shown in figure 8. Higher magnification observation indicates that the SnO_2 particles (black color) are distributed uniformly in the Ag-matrix. Lower magnification observation reveals that the uniform distribution exists in a large

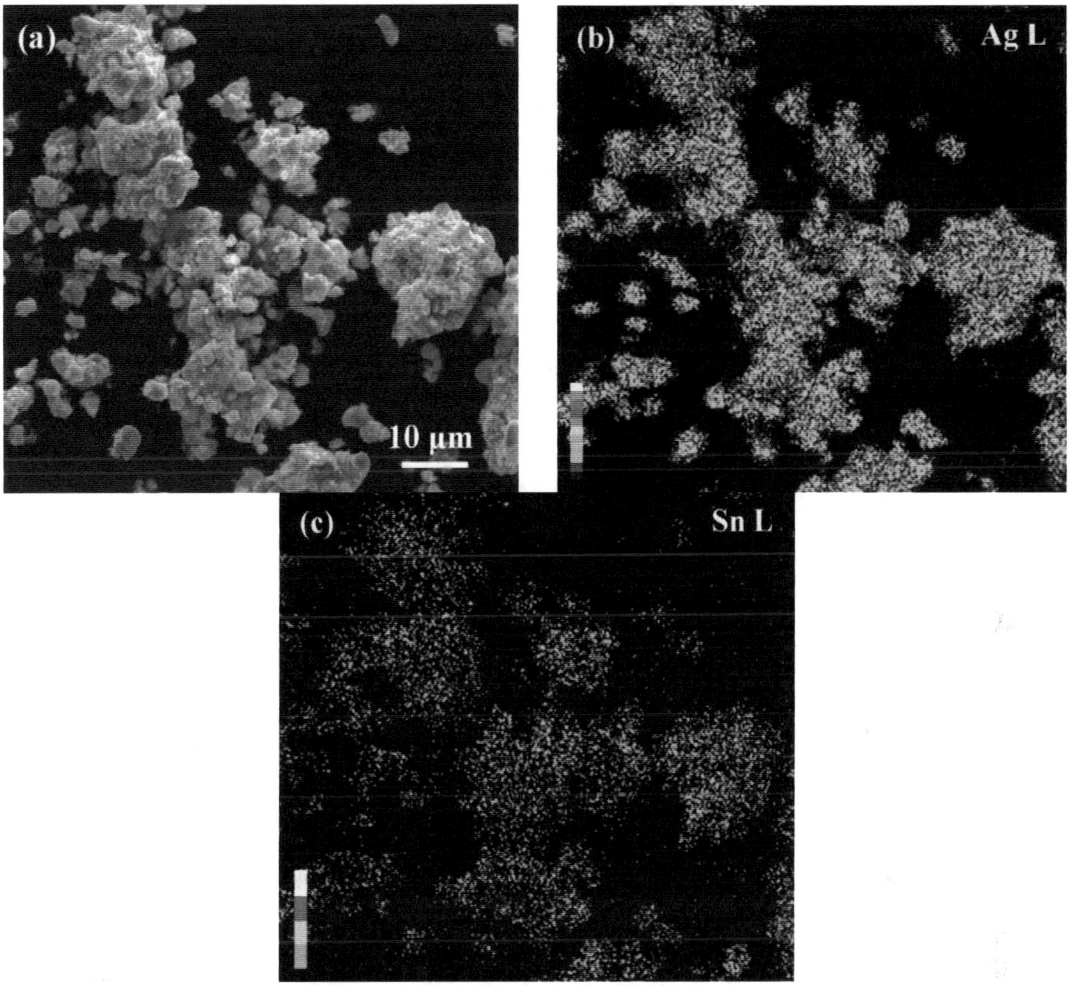

Figure 6. SEM image of the Ag-SnO_2 composite powders calcined at 600°C (a), and the EDS maps for Ag (b) and Sn (c).

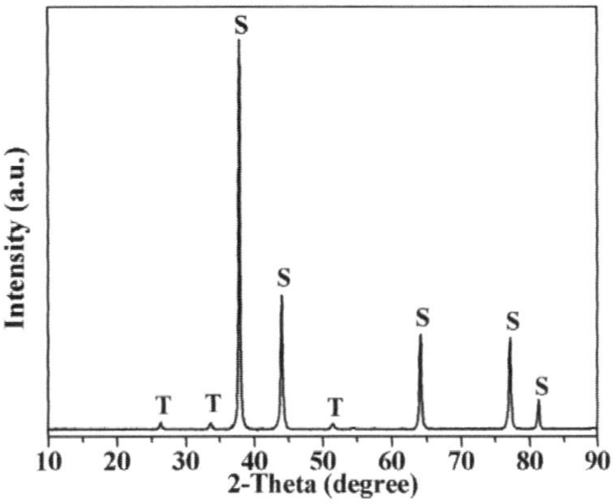

Figure 7. XRD pattern of the Ag-SnO_2 electrical contact materials prepared by hot-pressing. Letters S and T represent Ag and SnO_2, respectively

area. In our previous work [13], nano-sized SnO_2 reinforced Ag-SnO_2 electrical contact materials were prepared by the sol-gel method. The physical and electrical properties of the Ag-SnO_2 electrical

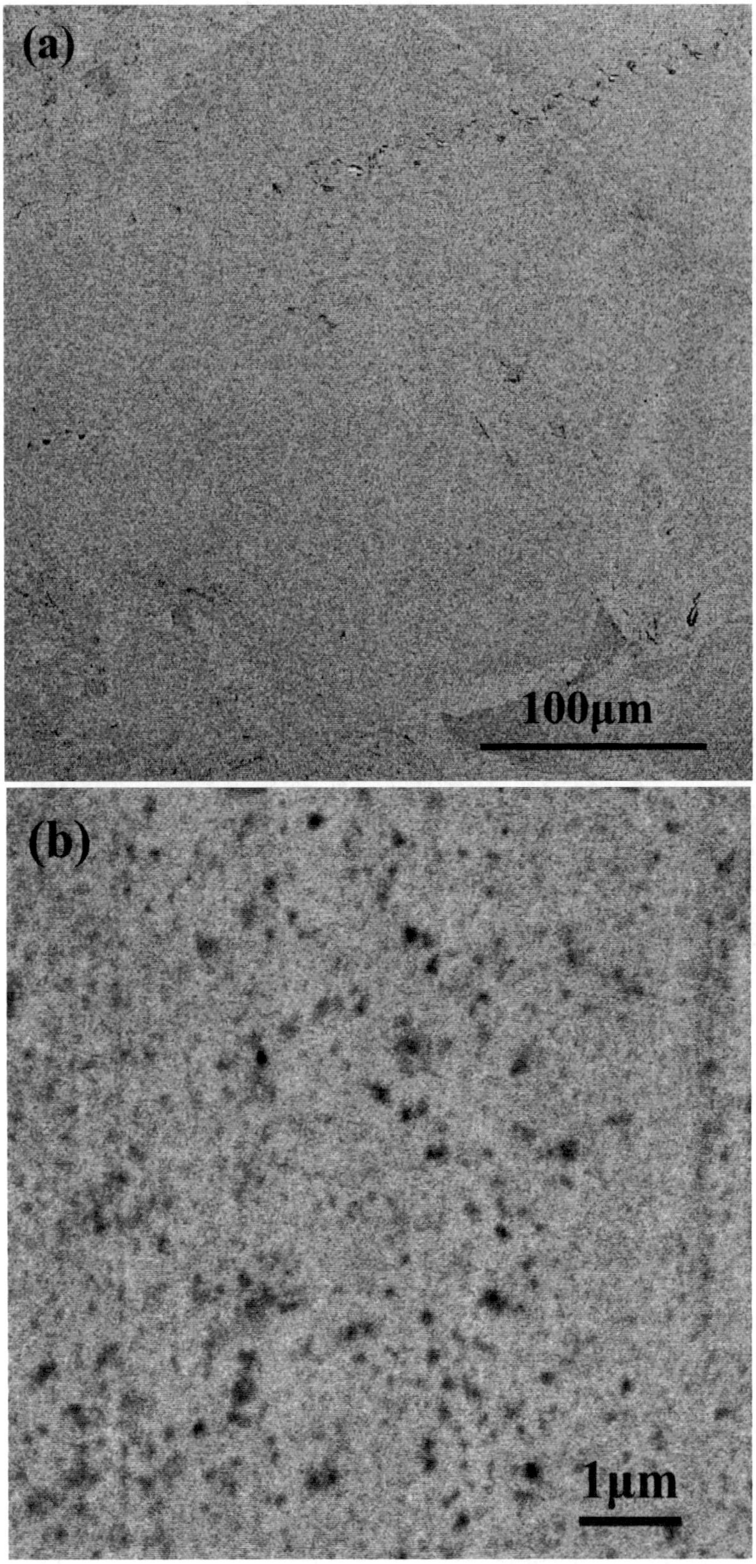

Figure 8. Microstructure of the Ag-SnO_2 electrical contact materials prepared by hot-pressing (SEM), with (a) lower magnification and (b) higher magnification.

contact materials are compared in table 1. The relative density of both samples is higher than 99.5% (the theoretical density is about 9.92 g/cm^3), indicating the effective densification by the hot-pressing method. In contrast to the sample prepared by the sol-gel method, the sample prepared by the present ion exchange method shows a higher values of hardness, density and electrical conductively, which may be due to the uniform mixture and strong bonding between tin oxide and silver matrix. The contact resistance between the SnO_2 particles and Ag matrix is therefore significantly reduced in the sample prepared by the ion exchange method.

Table 1. Physical and electrical properties of the Ag-SnO_2 electrical contact materials.

Sample	Hardness (MPa)	Density (g/cm^3)	Electrical conductively (%IACS)
In-situ method	154	9.90	72.24
Sol-gel method	123	9.87	68.97

5. Conclusion

In this work, a mixture precursor containing Ag_2O and Ag_2SnO_3 was synthesized via an ion exchange method using sodium stannous and silver nitrate as the starting materials. A homogeneous precursor mixture of Ag_2O and Ag_2SnO_3 was obtained. The precursor was decomposed into a mixture of Ag and SnO_2 after being calcinated at 600°C. During the calcination process, the Ag and SnO_2 particles were in-situ created from the precursor and inherited the uniform mixing of different phases. The Ag-SnO_2 electrical contact material was fabricated by hot-pressing. The nano-sized (about 20.6 nm) SnO_2 particles are homogeneously dispersed in the Ag-matrix. Fresh and stable interfaces between the SnO_2 particles and the Ag-matrix may be formed by the in-situ reaction. The contact resistance between the SnO_2 particles and the Ag-matrix is therefore significantly reduced. As a consequence, the as-prepared Ag-SnO_2 electrical contact material shows good performance in hardness, density and electrical conductively.

Acknowledgments

This work was supported by the Joint Funds of the National Natural Science Foundation of China (U1302272), the Fundamental Research Funds for the Central Universities (grants N141006002), and the fund of the State Key Laboratory of Advanced Technologies for Comprehensive Utilization of Platinum Metals (SKL-SPM-201506, SKL-SPM-201505, SKM-SPM-201214 and SKL-SPM-201206).

References

[1] Lawson C, Shen Q H, Zhu S M, Li H Y, Mu C F, Wu X H, Chen X T, Qi G X, He Q, Qiao X Q, Fan X P and Yang H 2012 Ag/(SnO_2) 12 electrical contact material with fibre-like arrangement of reinforcing nanoparticles: preparation, formation mechanism and properties *Proc. 26th Int. Con. on Electrical Contacts* (China: Beijing) pp 346-52

[2] Hossein-Babaei F, Moghadam S and Masoumi S 2015 Forming ohmic Ag/SnO_2 contacts *Mater. Lett.* **141** 141-4

[3] Braumann P, Koffler A 2004 The influence of manufacturing process, metal oxide content, and additives on switching behavior of Ag/SnO_2 in DC and AC relays *Proc. 23rd Int.Con. on Electrical Contacts* (Japan: Sendai) pp 90-97

[4] Bourda C, Jeannot D, Pinard J and Ramoni P 1992 Properties and effects of doping agents used in Ag-SnO_2 contact materials *Proc. 16th Int. Con. on Electrical Contacts* (England: Loughborough) pp 377-82

[5] Jeannot D, Pinard J, Ramoni P and Jost E M 1993 The effects of metals oxide additions or dopants on the electrical performance of Ag-SnO_2 contact materials *Proc. 39th Holm Conference on Electrical Contacts* (USA: Pittsburgh) pp 51-9

[6] Heringhaus F, Braumann P, Rühlicke D, Susnik E and Wolmer R 2000 On the improvement of dispersion in Ag-SnO_2-based contact materials *Pro. 20th Int. con. on Electrical contact* vol 199 (Stockhom: IEEE) pp 204.

[7] Linke C and Jansen M 1997 On Ag_2SnO_3, the first silver stannate *Z. Anorg. Allg. Chem.* **623** 1441-6

[8] Takeo O, Anna C, Jan-Olov B, Christer S, Reine W L, Christoph L and Martin J 2000 Modulated structure of Ag_2SnO_3 studied by high-resolution electron microscopy *Acta Crystallogr. B* **56** 363-8

[9] Feng J, Chen J C, Xiao B, Zhou C T, Hong Z J and Zhou R 2009 Stability, thermodynamic and mechanical properties of the compounds in the Ag-Sn-O system *Physica B* **404** 2461-7

[10] Feng J, Xiao B, Chen J C and Zhou C T 2009 Theoretical study on the stability and electronic property of Ag_2SnO_3 *Solid State Sci.* **11** 259-64

[11] Yin K, Wang T, Zhang J, Gao A M and Shao M W 2013 The effects of gamma ray irradiation on the photocatalytic efficiency: A way to enhance the Lewis acid-base interaction between Ag_2SnO_3 and Congo Red *Mater. Lett.* **113** 17-9

[12] Yin K, Shao M W, Zhang Z S and Lin Z Q 2012 A single-source precursor route to Ag/SnO_2 heterogeneous nanomaterials and its phot-catalysis in degradation of Conco Red *Mater. Res. Bull.* **47** 3704-8

[13] Lin Z J, Liu S H, Sun X D, Xie M, Li J G, Li X D, Chen Y T, Chen J L, Huo D, Zhang M, Zhu Q and Liu M M 2014 The effects of citric acid on the synthesis and performance of silver-tin oxide electrical contact materials *J. Alloy. Compd.* **588** 30-5

IOP Conf. Series: Materials Science and Engineering **103** (2015) 012009 doi:10.1088/1757-899X/103/1/012009

Tensile properties of glass/natural jute fibre-reinforced polymer bars for concrete reinforcement

J W Han[1], S K Lee[1], K W Kim[2] and C G Park[2,3]

[1] Department of Bio-Industry Mechanical Engineering, Kongju National University, Yesan, Republic of Korea

[2] Department of Rural Construction Engineering, Kongju National University, Yesan, Republic of Korea

E-mail: cgpark@kongju.ac.kr

Abstract. The tensile performance of glass/natural jute fibre-reinforced polymer (FRP) bar, intended for concrete reinforcement was evaluated as a function of volume fraction of natural jute fibre. Natural jute fibre, mixed at a ratio of 7:3 with vinyl ester, was surface-treated with a silane coupling agent and used to replaced glass fibre in the composite in volume fractions of 0%, 30%, 50%, 70%, and 100%. The tensile load–displacement curve showed nearly linear elastic behaviour up to 50% natural jute fibre, but was partially nonlinear at a proportion of 70%. However, the glass/natural jute FRP bars prepared using 100% natural jute fibre showed linear elastic behaviour. Tensile strength decreased as the natural jute fibre volume fraction increased because the tensile strength of natural jute fibre is much lower than that of glass fibre (about 1:8.65). The degree of reduction was not proportional to the natural jute fibre volume fraction due to the low density of natural jute fibre (1/2 that of glass fibre). Thus, as the mix proportion of natural jute fibre increased, the amount (wt%) and number of fibres used also increased.

1. Introduction

Environmental pollution and global warming concerns have prompted research and development of new biocomposite materials for various applications [1-4]. An example of this is the replacement of synthetic fibres made from petroleum products (e.g. polymers) with inexpensive, lightweight, naturally occurring fibres from plant and animal resources [5-10]. Natural fibres used in biocomposite materials, such as jute, flax, kenaf, and hemp, offers many advantages [4, 11]. Natural fibres are biodegradable, easier to recycle than synthetic fibre, and may be effective in reducing overall unit production cost by increasing the efficiency of production, thus reducing carbon emissions.4 Reinforcement bars, widely used in concrete structures to increase tensile strength are prone to corrosion [12-16]. To address this issue, research has been conducted on the tensile reinforcing properties of fibre-reinforced polymer (FRP) composites based on glass, carbon, and aramid fibres [17-21] FRP materials, applied to concrete in the form of bars, plates, and sheets, are resistant to corrosion and highly durable in the long term. Additionally, FRP reinforcing bars have a similar form

[3] Address for correspondence: C. G. Park, Department of Rural Construction Engineering, Kongju National University, Yesan, Republic of Korea. E-mail: cgpark@kongju.ac.kr.

to steel bars. The use of FRP as a reinforcement material has steadily increased, though it is more expensive than steel [12-21]. This study developed a tensile reinforcing bar made from biocomposite materials as a replacement for tensile reinforcing bars that use a glass fibre (GFRP) composite. To date, most of the research on concrete reinforcement has focused on the sheet/plate form of biofibre-based composite materials; few studies have investigated the bar form [3, 4, 22, 23]. In general, the price of biofibres is about 1/3 to 1/5 that of glass fibre [3, 4]. Despite the low cost, biofibre has a low density (1.2–1.5 g cm^{-3}) compared with that of glass fibre (2.56 g cm^{-3}), as well as lower strength and elasticity; however, the specific strength and specific elasticity modulus of natural fibre and glass fibre are comparable [3, 4, 22, 23]. Previous research indicated that 100% replacement of glass fibre with natural fibres would degrade the reinforcement strength of FRP bars, but partial replacement with natural biofibres may offer some benefit. In this study, glass/natural jute FRP bars were manufactured for reinforcing concrete structures. The tensile performance of FRP containing 0%, 30%, 50%, 70%, and 100% natural jute within the fibre portion, where the remaining volume was glass, was evaluated.

2. Experimental methods

2.1. Materials

Glass fibre was obtained from Hankuk Fibre Company (Milyang, Republic of Korea). The natural jute fibre used in this research was obtained from Nycontech Company (Seoul, Republic of Korea). Vinyl ester resin was purchased from Ashland Chemical (Covington, KY, USA). The properties of the material components, as specified by the manufacturers, are listed in table 1. Natural jute fibre, having a lower tensile strength than E-glass fibre, is 52% as dense and similar elastic modulus in comparison with glass fibre (table 1). Because natural jute fibre is less expensive than glass fibre and has little effect on the human body, natural fibre can replace synthetic fibre in certain applications. A major weakness of natural fibres is their hydrophilicity and resulting water sensitivity and weak binding/adhesion to most hydrophobic polymeric resins [3, 4, 6]. Thus, the surface of natural fibres must be conditioned for use in FRP bars. Accordingly, in this study, a silane coupling agent was used to treat the natural jute fibre surface [4, 6]. The polymer composite matrix is commonly composed of epoxy, vinyl ester, or polyester resins. Though vinyl ester resin, like polyesters, is degraded by OH^- ions and is thus not an ideal tensile reinforcement material in concrete, it was nonetheless used here, as it is inexpensive and has been shown to be durable in previous studies of FRP composite materials [12].

Table 1. Mechanical properties of glass fiber and matrix resin.

Mechanical properties	Vinyl ester resin	E-glass	Natural jute fiber
Tensile stress (MPa)	90	3,400	393
Elastic modulus (GPa)	3.4	71	55
Specific modulus	-	28	38
Strain (%)	-	4.79	0.71
Fiber density (g/cm^3)	-	2.62	1.3

2.2. Pre-treatment of natural fibre

A silane coupling agent, γ-methacryloxypropyltrimethoxysilane, was used to strengthen the interfacial binding between the natural jute fibres and vinyl ester resin. γ-methacryloxypropyltrimethoxysilane was hydrolysed for 1 h after dilution with a cosolvent of methanol and distilled water (weight ratio 95:5). The pH of the solution was adjusted to 4.0 with acetic acid and the concentration of the silane coupling agent was fixed at 0.8 wt%. Natural jute fibre was then soaked in the hydrolysed solution for 30 min, removed, dried in an oven at 110°C, and allowed to dry completely at room temperature.

2.3. Glass/natural jute FRP bars bar preparation

Five kinds of glass/natural jute FRP bars were prepared at various volume ratios of natural jute fibre, glass fibre, and vinyl ester resin (table 2). A combination of pultrusion and braiding process was used to manufacture the tests specimens. PVA fibre was used in the braiding process to improve the bonding with concrete. Figure 1 shows the shape of the glass/natural jute FRP bars.

Table 2. Mixture ratio of glass/natural jute fiber reinforced polymer bar.

No. of mix	Natural jute fiber (Vol. %)	Glass fiber (Vol. %)
No. 1	0	100
No. 2	30	70
No. 3	50	50
No. 4	70	30
No. 5	100	0

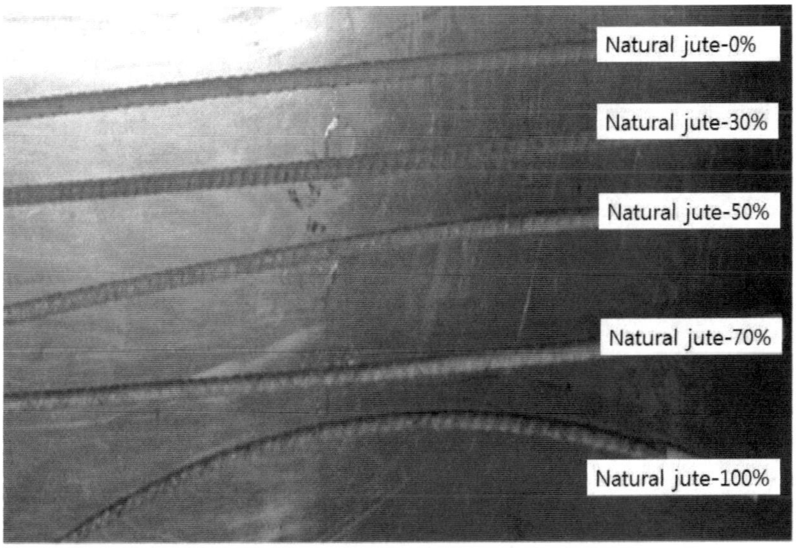

Figure 1. Photograph of the glass/natural jute fiber reinforced polymer bar.

2.4. Test method

The tensile strength of the FRP bars was tested in accordance with ACI Subcommittee 440 K guidelines [17]. As FRP bars were difficult to secure in the test apparatus and they were easily broken by slippage between the specimen grip and FRP bar, the centre of the grip point was reinforced with a silica–epoxy mix to increase the surface friction between the grip and the composite bar. The grip measured 300 mm by 30 mm, and the glass/natural jute FRP bars were 800 mm by 5 mm. Tensile tests were performed using a displacement-adjustable universal tensile machine (UTM; Han Sin Gum Pung Company, Seoul, Republic of Korea), with a capacity of 1,000 kN at a loading speed of 5 mm min^{-1}. The strain on the concrete FRP reinforcement bar was measured using a linear variable differential transformer (LVDT). The test was performed on five specimens. Figure 2 shows the tensile test apparatus.

3. Results and discussion

3.1. Tensile load–displacement behaviour

Generally, FRP reinforcing materials are brittle and exhibit linear elastic behaviour. Figure 3 shows the tensile load–displacement curves for the composite bars; the tensile–load behaviour exhibited nearly linear elasticity, up to a fibre volume fraction of 50% natural jute fibre. FRPs in which 70% of

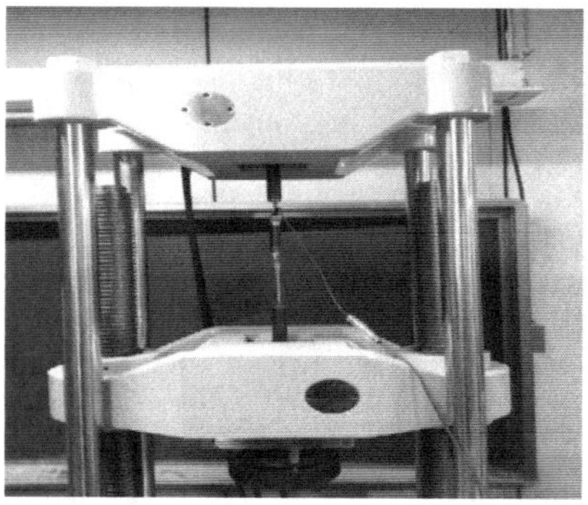

Figure 2. The tensile test set-up.

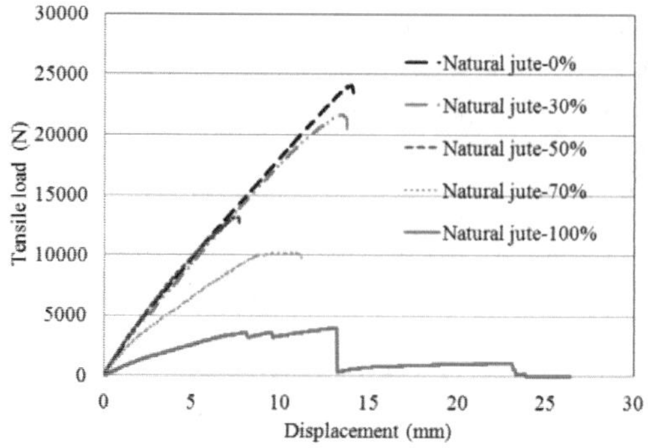

Figure 3. Load-displacement behavior of the glass/natural jute fiber reinforced polymer bars.

the fibre was natural jute initially showed linear elastic behaviour, followed by plastic deformation. Glass/natural jute FRP bars prepared using 100% natural jute fibre displayed linear elastic behaviour. Numerous studies have examined the use of hybrid FRP bars for concrete reinforcement applications to promote ductile behaviour and to minimise or prevent material brittleness, a disadvantage of FRP bars [24-26]. Hybrid FRP bars, consist of multiple fibres, one or more with a high elasticity modulus and one with a low modulus [24-26]. The rigid fibre is destroyed upon application of a tensile load, while the flexible fibre shows ductile behaviour with some plastic deformation before rupture [24-26]. Carbon fibre has a high elasticity modulus, and glass or aramid fibre has a low elasticity modulus. Under an applied load to a hybrid of these two fibres, the carbon fibre is destroyed, and the glass fibre or aramid fibre secures the plastic deformation region until rupture [24-26]. In addition, plastic deformation may be promoted by reducing the fibre volume fraction of the high elasticity modulus fibre and increasing that of the low elasticity modulus fibre. FRP bars show linear elastic, or brittle, behaviour [24-26]. In the existing research on FRP composites, many research groups have achieved ductile behaviour using 30% rigid and 70% elastic fibre (having a high deformation rate). In this study, glass fibre and natural jute fibre were used in the FRP composite. The elasticity modulus of glass fibre (71 GPa) is 29% higher than that of natural jute fibre (55 GPa). Similarly, the tensile strength of glass fibre is 8.65 times that of natural jute fibre (3400 MPa vs. 393 MPa), and the deformation rate of glass fibre is 6.75 times that of natural jute fibre (4.79% vs. 0.72%). The difference in deformation rate makes sufficient plastic deformation after destruction of the glass fibre almost impossible; thus, for fibre volume fractions of 50%, 70%, and 100%, the natural jute fibre (low elasticity modulus) was destroyed earlier than the glass fibre (high elasticity modulus), regardless of the mix proportion, as indicated by the brittle behaviour. At a fibre volume fraction of 30% glass fibre, the initial elastic behaviour of the FRP composite was followed by plastic deformation and finally, rupture. The addition of natural jute fibre to the FRP composite appeared to delay the onset of plastic deformation, which was not prominent because the tensile strength and deformation rate of natural jute fibre are 1/8.65 and 1/6.75 those of the glass fibre, respectively. FRP bars containing 100% natural jute fibre displayed some plastic deformation when the load was reduced following linear elastic behaviour, but the tensile load required to induce plastic deformation was small. This plasticity results from the process used to manufacture the glass/natural jute FRP bars; they were braided with PVA fibre to form ribs for improved surface adhesion to concrete. As PVA fibre has little resistance after destruction of the natural jute fibre, it has negligible impact on the behaviour of any of the FRP bars. Thus, upon rupture, the reduction in load was so large that the PVA fibre was simultaneously destroyed. This was expected; PVA fibre was used for adhesive purposes only.

3.2. Tensile strength

Figure 4 shows that tensile strength decreased in proportion to the volume fraction of natural jute fibre. These results confirm previous observations; the tensile strength of hybrid FRP bars containing two or

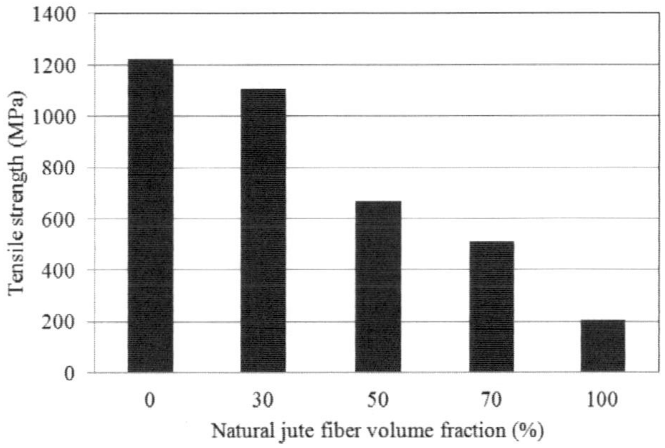

Figure 4. Tensile strength of the glass/natural jute fiber reinforced polymer bars.

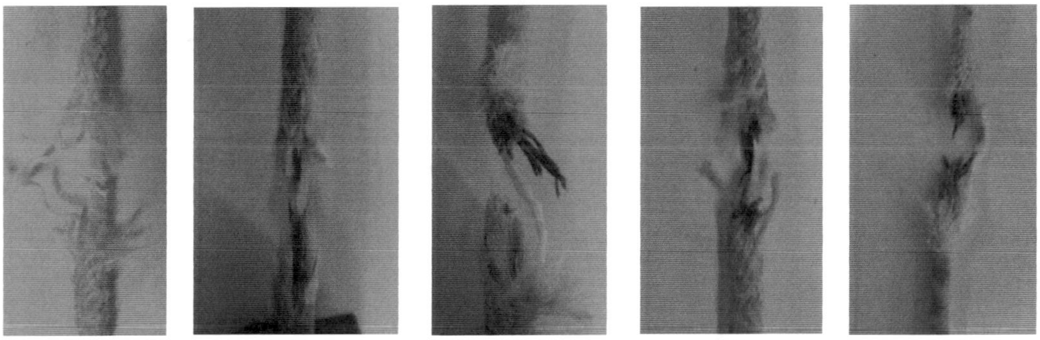

Figure 5. Fracture surface investigation of glass/natural jute fiber reinforced polymer bar after tensile strength tests: (a) Natural jute-0%, (b) Natural jute-30%, (c) Natural jute-50%, (d) Natural jute-70%, (e) Natural jute-100%

more fibres is known to decrease as the fibre volume fraction of the low elasticity modulus fibre (in this case, natural jute) increases. The tensile strength of the natural jute fibre used here is only 1/8.65 that of the glass fibre, and thus should significantly decrease the tensile strength of glass/natural jute FRP bars. For natural jute fibre volume fractions of 30%, 50%, 70%, and 100%, the corresponding tensile strength was 89.9%, 54.3%, 41.4%, and 16.5%, respectively, that of 100% glass fibre (1129 MPa). These results reveal a smaller impact of jute fibre than would be expected based on the difference in tensile strength; the glass/natural jute FRP bar containing 50% each glass fibre and natural jute fibre should show about 23.12% (1/4.325) of the tensile strength as that using 100% glass fibre. This is attributable to the difference in density between the two fibres; natural jute fibre is ½ as dense as glass fibre, so the number of fibre molecules doubles for the same volume. This larger number of fibres increases the tensile strength. FRP bars ruptured as a result of interfacial separation between the braided PVA surface-treated area and the inner portion containing the mixed glass/natural jute FRP composite (figure 5). This separation was more pronounced at high volume fractions of natural jute fibre. For example, there was almost no interfacial separation for FRP bars containing 0% and 30% natural jute fibre, but significant separation was observed for those containing 50%, 70%, and 100% natural jute fibre.

4. Conclusions

Glass/natural jute FRP bars were evaluated for use as a tensile reinforcement material in concrete structures. The tensile properties of the glass/natural jute FRP bars can be summarized as follows:

- The tensile load–displacement behaviour of the composite bars showed almost perfect linear elasticity up to a fibre volume fraction of 50% natural jute fibre. For 70% natural jute fibre, the linear elastic behaviour of the glass fibre was evident to the point of rupture, followed by plastic deformation, which was minimal as glass fibre has high elasticity modulus, tensile strength, and deformation rate. For a fibre volume fraction of 100% natural jute fibre, linear elastic behaviour was observed.

- The tensile strength of glass/natural jute FRP bars decreased as the fibre volume fraction of natural jute fibre increased. This was attributed to the tensile strength of natural jute fibre, which is about 1/8.65 that of glass fibre. However, the tensile strength did not decrease linearly with the mix proportion of natural jute fibre is about 1/2 as dense as glass fibre, meaning the same volume contains twice the number of fibres.
- Glass/natural jute FRP bars ruptured because of interfacial separation between the PVA fibre surface-treated portion and the inner portion containing the mix of glass fibre and natural jute fibre. This trend was more prominent for large fibre volume fractions of natural jute fibre.

Acknowledgment
This research was supported by the basic science research program through the national research foundation of Korea (NRF) founded by ministry of education (NRF-2013R1A1A4A01011776).

References

[1] Vivek M and Sandhyarani B 2013 Physical and mechanical properties of bi-directional jute fiber epoxy composites *Procedia Engineering* **51** 561-6
[2] Mohanty A K, Misra M and Drzal L T 2002 Sustainable bio-composites from renewable resources: Opportunities and challenges in the green materials world *Journal of Polymers and the Environment* **10** 19-26
[3] Nabi Saheb D and Jog J P 1999 Natural fiber polymer composites: A review *Advanced in Polymer Technology* **18** 351-63
[4] Shim J H, Cho D H and Yoon J S 2008 Natural fibers and biocomposites *Polymer Science and Technology* **19** 299-306
[5] Holbery J and Houston D 2006 Natural-fiber-reinforced polymer composites in automotive applications *The Journal of The Minerals, Metals & Materials Society* **58** 80-6
[6] Bledzki A K, Reihmane S and Gassan J 1996 Properties and modification methods for vegetable fibers for natural fiber composites *Jornal of Applied Polymer Science* **59** 1329-36
[7] Joshi S V, Drzal L T, Mohanty A K and Arora S 2004 Are natural fiber composites environmentally superior to glass fiber reinforced composites *Composites: Part A* **35** 371-6
[8] Dipa Ray B K, Sarkar S D and Rana A K 2002 Dynamic mechanical and thermal analysis of vinylester-resin-matrix composites reinforced with untreated and alkali-treated jute fibres *Compos. Sci. Techno.* **62** 911-7
[9] Igor M De Rosa, Santulli C, Sarasini F and Valente M 2009 Post-impact damage characterization of hybrid configurations of jute/glass polyester laminates using acoustic emission and IR thermography *Compos. Sci. Techno.* **69** 1142-50
[10] Codispoti R, Daniel V Oliveira, Renato S Olivito, Paulo B Lourenço and Fangueiro R 2015 Mechanical performance of natural fiber-reinforced composites for the strengthening of masonry *Compos. Part B: Engineering* **77** 74-83
[11] Shah A N and Lakkad S C 1981 Mechanical properties of jute-reinforced plastics *Fibre Sci. Techno.* **15** 41-6
[12] Won J P, Lee S J, Kim Y J, Jang C I and Lee S W 2008 The effect of exposure to alkaline solution and water on the strength-porosity relationship of GFRP rebar *Compos. Part B: Engineering* **39** 764-72
[13] Chin J W, Aouadi K and Nguyen T 1997 Effects of environmental exposure on fibre-reinforced plastic (FRP) materials used in construction *J. Compos. Techn. Res.* **19** 205-13
[14] Dutta P K and Hui D 1996 Low-temperature and freeze-thaw durability of thick composites

Compos. Part B: Engineering **27** 371-9

[15] Jones F R, Rock J W and Wheatley A R 1983 Stress corrosion cracking and its implications for the long-term durability of e-glass fibre composites *Composites* **14** 262-9

[16] Park C G, Won J P and Yoo J K 2003 Long-term effect of chemical environments on FRP reinforcing bar for concrete reinforcement *J. of the KCI* **15** 811-9

[17] ACI 440H 2000 Guide for the design and construction of concrete reinforced with FRP bars (American concrete institute committee 440)

[18] Lin X S and Zhang Y X 2014 Evaluation of bond stress-slip models for FRP reinforcing bars in concrete *Compo. Structur.* **107** 131-41

[19] Gravina R J and Smith S T 2008 Flexural behaviour of indeterminate concrete beams reinforced with FRP bars *Engineer. Structur.* **30** 2370-80

[20] Julio F Davalos, Chen Y and Indrajit R 2008 Effect of FRP bar degradation on interface bond with high strength concrete *Cem. Concr. Compos.* **30** 722-30

[21] Hayder A Rasheed, Nayal R and Melhem H 2004 Response prediction of concrete beams reinforced with FRP bars *Compos. Structur.* **65** 193-204

[22] Bos H L 2004 The potential of flax fibers as reinforcement for composite materials (Technische Universiteit Eindhoven, Nederland: Ph.D Thesis)

[23] Sparnins E 2006 Mechanical properties of flax fibers and their composites (Lulea University of Technology, Sweden: Licentiate Thesis)

[24] Harris H, Somboonsong W and Ko F 1998 New ductile hybrid FRP reinforcing bar for concrete structures *Journal of Composites Construction* **2** 28-37

[25] Won J P and Park C G 2006 Effect of environmental exposure on the mechanical and bonding properties of hybrid FRP reinforcing bars for concrete structures *Journal of Composite Materials* **40** 1063-76

[26] Won J P, Park C G and Jang C I 2007 Tensile fracture and bond properties of ductile hybrid FRP reinforcing bars *Polymers & Polymer Composit*es **15** 9-16

IOP Conf. Series: Materials Science and Engineering **103** (2015) 012010 doi:10.1088/1757-899X/103/1/012010

The enhancing effect of mesogen-jacketed liquid crystalline polymer PBPCS on epoxy resin

C P Chai[1], X J Li, Y Zhu, Y X Gao, G P Li and Y J Luo

School of Material Science and Engineering, Beijing Institute of Technology, Beijing 100081, China

E-mail: chaicp@bit.edu.cn

Abstract. The mixtures of mesogen-jacketed liquid crystalline polymers (MJLCP) and epoxy resin (E-51) have been prepared in a certain proportion. The category of the MJLCP is poly{2, 5-bis[(4-butoxyphenyl) oxycarbonyl] styrenes} (PBPCS). Methyl tetrahydro phthalic anhydride (MeTHPA) is served as a curing agent, and N, N-dimethylbenzylamine plays the role of catalyst. Then, based on the curing process, the modified materials have been acquired by casting molding. The effect of PBPCS's improvement has been researched through the mechanical properties test, dynamic thermal mechanical test and scanning electron microscope (SEM). The results indicated that PBPCS could apparently improve the mechanical properties of resin E-51. Compared with unmodified materials, the elongation at break and the tensile strength of PBPCS have been improved remarkably through modification. Both mechanical properties of the 4wt% PBPCS/E-51/MeTHPA were enhanced by 48% and 153% separately, and the characteristics of the 3wt% one were raised by 47% and 19% respectively. Also, the ductile fracture morphology of the resins was exhibited in SEM photograph clearly.

1. Introduction

Owing to greater mechanical, chemical resistance, and dielectric properties, epoxy resins are in multiple utility in the fields of aviation, aerospace, chemical engineering, transportation industry and electronic and electric technologies, etc [1, 2]. Nevertheless, as a thermosetting resin, brittleness becomes a major barrier of this kind of material. What's more, poor thermal stability also restricts further application in advanced industry [3, 4]. Some researches attribute the formation of self-reinforcing structures in epoxy resins to the blending with liquid crystalline polymers, which consist of rigid mesogenic groups and flexible chains. In addition, liquid crystalline polymers, which has been detected and witnessed in epoxy resins in the form of disperse particles or fibers, can also lead to microcracks and shear zones that absorb a large number of the fracture energy in a resin matrix. As a result, they can strikingly improve the toughness and strength of epoxy resins. Compared to epoxy resins toughened by other ordinary methods, incorporating liquid crystalline polymers into epoxy networks exhibits higher thermal stability, toughness and tensile strength.

The genre of MJLCPs [5-8] is a new type of liquid crystalline polymer. In the late 1980s, Zhou, et al. reported the MJLCPs, which are side-on side-chain liquid crystalline polymers, has a short spacer or a single covalent bond that connects the mesogen to the polymer backbone [9]. PBPCS, whose chemical structure is same as that of side-chain liquid crystalline polymer, is a typical kind of

[1] Address for correspondence: C Pchai, School of Material Science and Engineering, Beijing Institute of Technology, Beijing 100081, China. E-mail: chaicp@bit.edu.cn.

medogen-jacketed liquid crystiline that based on vinly terephthalic. But some properties of PBPCS are similar to those of main-chain liquid crystalline polymer. This means that MJLCPs [10], a special category of SCLCPs, can be easily synthesized and possess stable liquid crystalline phases like main-chain liquid crystalline polymers do. Recently many studies on the morphology of liquid crystal phase transition and its applications to advanced functional materials have become of great importance. However, there has been no report on the method of resin's modification with MJLCP up to date.

In this work, stoichiometric amounts of PBPCS were dissolved in epoxy resins (E-51). By casting and molding, all samples were cured in accordance with curing process that was determined by curing kinetics. Static and dynamic mechanical properties of the epoxy resins modified by PBPCS would be studied in detail.

2. Experimental procedure

2.1. Materials

Epoxy resin E-51 (the content of epoxy was about 0.051mol•kg^{-1}) was supplied by Sinopec Group(Baling, China) and used without further purification; Methyl tetrahydro phthalic anhydride (MeTHPA), N, N-dimethylbenzylamine and dichloromethane with the purity of analytical level were provided by Aladdin corporation. The PBPCS (Mn=77000, Mw/Mn=1.7) was synthesized according to literature [11]. Its structure stated as figure 1.

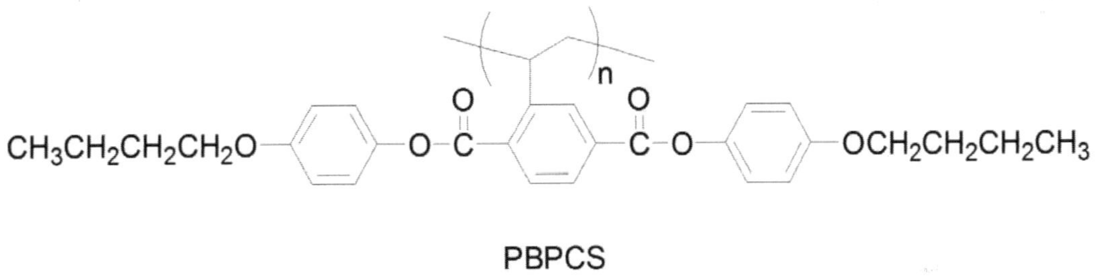

Figure 1. The structure of PBPCS.

2.2. Characterization

Tensile and flexural properties were menstruated on the basis of Chinese National Standard GB/T 2567-2008 on a universal testing machine of type Instron26022(Shimazu, Kyoto, Japan). And Impact strengths were tested by a TCJ LCD type pendulum impact testing machine (Jilin, China), also according to Chinese National Standard GB/T 2567-2008. Morphology and structure of the fracture surfaces was observed with a SEM(S-4800, Hitachi Construction Machinery Co., Ltd. Japan) at the acceleration voltage of 20kV. Dynamic mechanical analysis was carried out with the assistance of METTLER TOLEDO DMA(USA) by using three-point bending mode at a frequency of 1Hz with the temperature limits from 50℃ to 200℃ at the heating rate of 3℃per minute. Ultrasonic cleaner, the type of which was SB 5200DTDN, was provided by Xinzhi Biological Technology Co., Ltd. Ningbo (China).

2.3. Preparation of PBPCS/E-51/MeTHPA materials

Five samples were prepared according to corresponding mass fraction of PBPCS in epoxy resin: 1%, 2%, 3%, 4% and 5%. The PBPCS and epoxy resin E-51 were mixed homogeneously in dichloromethane, and the solvent was removed at 90℃. After the temperature of mixture cooling down to the normal temperature (25℃), the stoichiometric amounts of MeTHPA and N, N-dimethylbenzylamine were added to the mixture under continuous stirring. The blends were then poured into a preheated mold coated with silicone resin. After degassing in the vacuum for about half an hour, all the samples were cured at 90℃ and 120℃ for 2 hours respectively and at 140℃ for 1.5

hours [12].

3. Results and discussion

3.1. Mechanical properties of PBPCS/E-51/MeTHPA materials

3.1.1. Tensile properties. Variations of elongation at break and tensile strength of E-51/MeTHPA with varied mass ratio of additive PBPCS are illustrated in the table 1.

Table 1. Tesile properties of 0-5wt%PBPCS/E-51/MeTHPA.

Materials / Properties	Tensile strength/MPa	Elongation at break/%
E-51/MeTHPA	51.09±1.36	3.60±0.91
1%PBPCS/E-51/MeTHPA	66.31±2.21	5.22±1.35
2%PBPCS/E-51/MeTHPA	66.92±1.88	6.61±1.27
3%PBPCS/E-51/MeTHPA	68.83±1.72	7.08±1.22
4%PBPCS/E-51/MeTHPA	75.48±2.79	9.10±1.73
5%PBPCS/E-51/MeTHPA	73.46±1.35	8.79±0.94

Obviously, the additive PBPCS has improved the tensile strength and elongation at break of pure E-51/MeTHPA. The statistics in the table 1 showed the trend of increasing of tensile strength and elongation at break from the mass ratio of additive PBPCS of 1wt % to that of 4wt%, followed by a tendency of decreasing, with the growth of 1wt% of the content of additive PBPCS, at that of 5%. The maximum tensile strength and elongation at break both corresponded at the point of 4wt% of the amount of additive PBPCS added into E-51/MeTHPA (4wt% PBPCS/E-51/MeTHPA) materials, where both were respectively 48% and 153% higher than value of those of pure E-51/MeTHPA materials. As a result of the alignment of the rigid mesogenic groups of PBPCS during the solidity and the enclosure of orientation in epoxy networks, the materials were to be reinforced. Simultaneously, flexible chains of PBPCS and epoxy resin matrix, interpenetrating to form micro rubber phase that can absorb part of the energy, accumulated inside the materials. So elongation at break of the modified materials has been greatly improved. When the mass ratio of the additive PBPCS reached 5wt%, the tensile strength and elongation at break of modified materials appeared to have a slight decrease, but were still higher than those of the pure E-51/MeTHPA, which possibly attributed to the uneven dispersion of excessive PBPCS in epoxy resin matrix.

3.1.2. Impact properties. The result of impact measurement of PBPCS/E-51/MeTHPA materials is illustrated in figure 2.

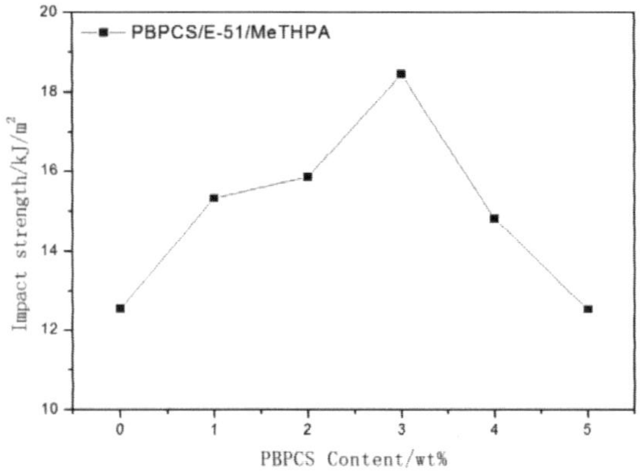

Figure 2. Impact properties of PBPCS/E-51/MeTHPA materials.

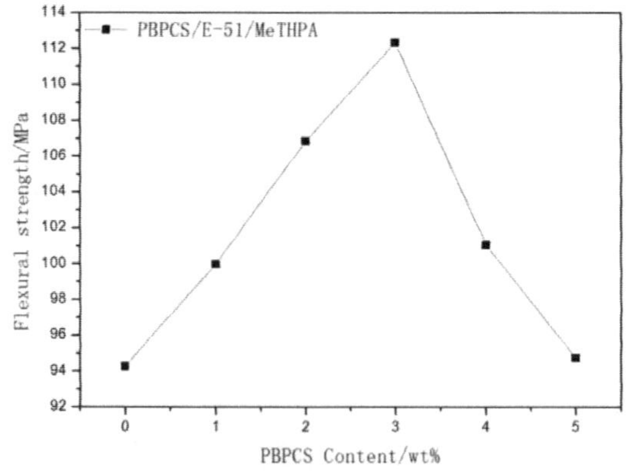

Figure 3. Bending strength of of PBPCS/E-51/MeTHPA materials.

The curve indicated that the impact strength value of PBPCS/E-51/MeTHPA materials was higher than that of pure E-51/MeTHPA materials. In particular, the impact strength value of 3wt% PBPCS/E-51/MeTHPA was 1.47 times as high as that value of pure E-51/MeTHPA. Since the statistics of impact strength showed the energy absorption before fracture, the results in figure 1 could be considered as that in the existence of additive PBPCS, the epoxy resin E-51 could result in impacting crack deformation of the surrounding matrix networks. However, the impact strength of the materials was to decrease when exceeding the certain mass ratio of the additive PBPCS because the compatibility between the epoxy resin matrix and the PBPCS became less effective. So the results of impact measurement indicated that PBPCS was able to toughen epoxy resin E-51 effectively.

3.1.3. Flexural properties. The result of flexural strength measurement derived from the static three-point winding tests was shown in figure 3.

Apparently, the trend of variation of varied mass ratio of the additive PBPCS in the materials of E-51/MeTHPA resembled tensile and impact strength. The flexural strength initially exalted with the mass ratio of the additive PBPCS up to 3 wt %. It reached the maximum when the content was about 3 wt %. After that, it suffered a slight decline with the increase of the mass ratio of the additive PBPCS. The maximum flexural strength value of PBPCS/E-51/MeTHPA was 19% more than that of the pure E-51/MeTHPA. When the mass ratio of PBPCS that added into E-51 was less than 3 wt %, PBPCS presented island structure and had a uniform dispersion in epoxy resin matrix. While the mass ratio of PBPCS was greater, phase separation was observed in the modified epoxy resins.

3.1.4. SEM. To provide a direct comparison, the morphology of the tensile fracture surfaces of all the materials were investigated by SEM, as shown in figure 4(a). It showed the smooth fracture surface of pure epoxy resin E-51. These fissures were formed initially as river pattern and spread swiftly by the method of parallel lines. It was a sign of brittle rupture. But with the incorporation of PBPCS to epoxy resin E-51, as shown in figures 4(b)-4(f), a wide range of ductile sunken areas were shown at the fracture surfaces, indicating a typical tough feature [13]. It proved that PBPCS has played a part in toughing and reinforcing the materials.

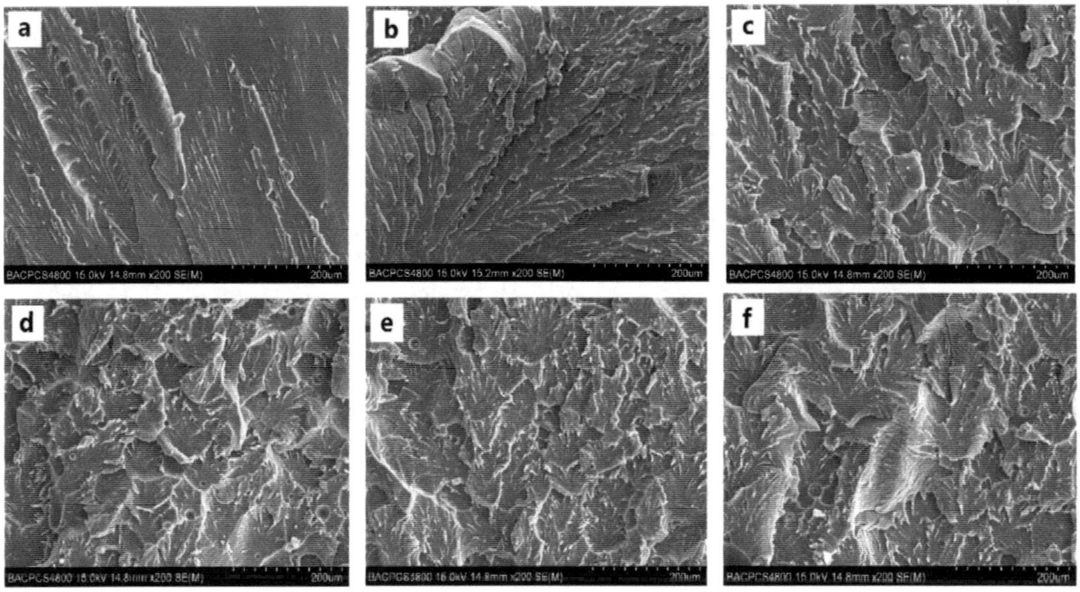

Figure 4. SEM graphs of PBPCS/E-51/MeTHPA materials (a:0wt% PBPCS; b:1wt% PBPCS; c:2wt% PBPCS; d:3wt% PBPCS; e:4wt% PBPCS; f:5wt% PBPCS).

3.2. Dynamic mechanical behaviors (DMA)

DMA is an efficient tool to acquire the information about the microstructure of thermoset networks and the mechanical damping of materials as the function of thermometer [14, 15]. The temperature dependence of dynamic storage modulus (E') and loss angle (tan delta) were displayed in figures 5 and 6.

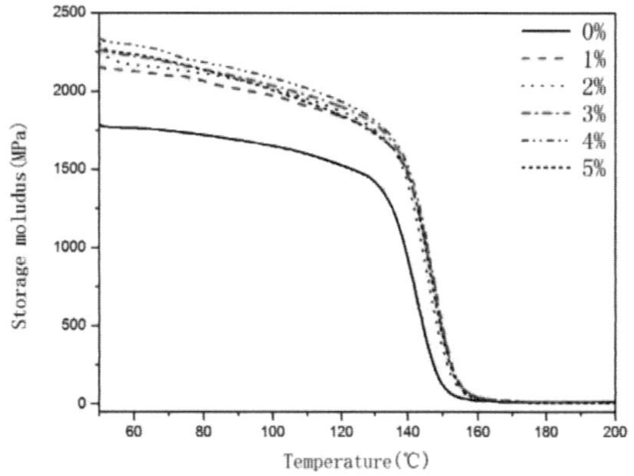

Figure 5. Storage modulus versus temperature of PBPCS/E-51/MeTHPA materials.

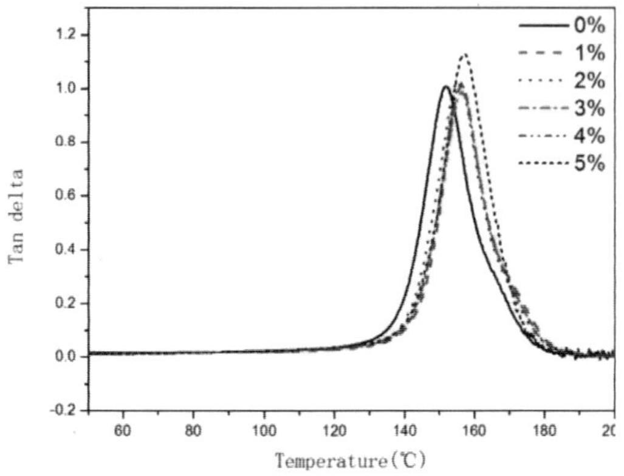

Figure 6. Tan delta versus temperature of PBPCS/E-51/MeTHPA materials.

Table 2 summarized the value of E' and the value of glass transition temperature (Tg) at 60℃. The value of storage modulus of all PBPCS/E-51/MeTHPA materials was higher than the value of that of pure E-51/MeTHPA at 60℃, suggesting that the former was more rigid than the latter. What's more, PBPCS/E-51/MeTHPA materials with greater mass ratios of PBPCS tended to possess greater E' values, proving the dominated role of the semi-rigid structure of the MJLCP.

Table 2. DMA data of 0-5wt%PBPCS/E-51/MeTHPA.

Materials \ Properties	E'/GPa	Tg /℃
E-51/MeTHPA	1.763	151.4
1%PBPCS/E-51/MeTHPA	2.126	156.2
2%PBPCS/E-51/MeTHPA	2.166	155.5
3%PBPCS/E-51/MeTHPA	2.222	156.3
4%PBPCS/E-51/MeTHPA	2.293	156.1
5%PBPCS/E-51/MeTHPA	2.235	157.2

From the curves of tan delta versus temperature (figure 6), there was no existence of other α-transition peak in the high-temperature range, potentially owing to the dispersion of PBPCS in epoxy resin E-51. On the tanδ-temperature chart, the temperature that corresponded to the maximum tan δ was delimited as Tg, indicating rotational freedom in the segments between crosslinks [16]. In table 2, the Tg of the modified epoxy resins was slightly higher than pure epoxy resin E-51, hence the rigid chains of PBPCS were aligned during the process of curing and the enclosure of location in epoxy networks. In order to keep orientation in the networks, rigid units needed more energy, so Tg of PBPCS/E-51/MeTHPA materials went higher than that of pure E-51/MeTHPA.

4. Conclusions

Based on the analysis above, the introduction of PBPCS into epoxy resin E-51 could increase both toughness and strength of the epoxy resin matrix. Compared to unmodified materials, the results showed that PBPCS could distinctly harden the epoxy resin E-51. The 4wt% of mass ratio of modifier enhanced the elongation at break and tensile strength by 153% and 48% separately. The bending and impact strength were raised by 19% and 47% apart with 3wt% of mass ratio PBPCS added. In addition, the value of T_g was increased by 4-6℃, and the SEM image showed obvious ductile fracture morphology. Consequently, the MJLCP PBPCS could act as an effective and successful reinforcing and toughening modifier of epoxy resin.

Acknowledgment
This material is based upon work supported by the National Science Foundation of China (grant no. 51343008)

References

[1] Landry B, Laplante G and Leblanc L R 2012 Environmental effects on mode II fatigue delamination growth in an aerospace grade carbon/epoxy composite *Compos. Part A-Appl. S* **43** 475-85
[2] Zhao Y, Chen Z K, Liu Y, Xiao H M, Feng Q P and Fu S Y 2013 Simultaneously enhanced cryogenic tensile strength and fracture toughness of epoxy resins by carboxylic nitrile-butadiene nano-rubber *Compos. Part A-Appl. S.* **55** 178-87
[3] Yi X F, Mishra A K, Kim N H, Ku B C and Lee J H 2013 Synergistic effects of oxidized CNTs and reactive oligomer on the fracture toughness and mechanical properties of epoxy *Compos. Part A-Appl. S.* **49** 58-67
[4] Vyas A and Iroh J O 2013 In situ growth of multilayered crystals in amorphous matrix: Thermal, dynamic mechanical and morphological analysis of nylon-6/epoxy composites *J. Appl. Ploym. Sci.* **130** 3319-27
[5] Zhou Q F, Li H M and Feng X D 1987 Synthesis of liquid-crystalline polyacrylates with laterally substituted mesogens *Macromolecules* **20** 233-4
[6] Zhou Q F, Zhu X L and Wen Z Q 1989 Liquid-crystalline side-chain polymers without flexible spacer *Macromolecules* **22** 491-3
[7] Chen X F, Shen Z H, Wan X H, Fan X H, Chen E Q, Ma Y G and Zhou Q F 2010 Mesogen-jacketed liquid crystalline polymers *Chem. Soc. Rev.* **39** 3072-101
[8] Chen X F, Shen Z H, Chen E Q, Wan X H, Fan X H and Zhou Q F 2012 Side-chain jacketing effect and mesogen-jacketed liquid crystalline polymers *Science China Chemistry* **42** 606-21
[9] Finkelmann H, Ringsdorf H and Wendorff J H 1978 Model considerations and examples of enantiotropic liquid crystalline polymers, polyreactions in ordered systems *Die. Makromol. Chem.* **179** 273-6
[10] Zhao Y F, Fan X H, Wan X H, Chen X F, Yi Y, Wang L S, Dong X and Zhou Q F 2006 Unusual phase behavior of a mesogen-jacketed liquid crystalline polymer synthesized by atom transfer radical polymerization *Macromolecules* **39** 948-56
[11] Zhang D, Liu Y X, Wan X H and Zhou Q F 1999 Synthesis of a new side-chain type liquid crystal polymer poly[dicyclohexyl vinylterephthalate] *Macromolecules* **32** 4494-6
[12] Li X J, Chai C P, Li C F, Li G P, Luo Y J, Fan X H, Shen Z H and Zhou Q F 2013 Non-isothermal cure kinetics of mesogen-jacketed liquid crystal polymer PBPCS modified epoxy resin *Acta Polymerica Sinica* **9** 1190-6
[13] Le Q H, Kuan H C, Dai J B, Zaman I, Luong L and Ma J 2010 Structure-property relations of 55 nm particle-toughened epoxy *Polymer* **51** 4867-79
[14] Lu S R, Yang X W and Wei C 2010 Preparation and characterization of liquid crystalline epoxy/epoxy resin composites *Plastics, Rubber and Composites* **39** 67-73
[15] Zhang X Y, Gu A J, Liang G Z, Zhuo D X and Yuan L 2011 Liquid crystalline epoxy resin modified cyanate ester for high performance electronic packaging *J. Polym. Res.* **18** 1441-50
[16] Konecny P, Cerny M, Voldanova J, Malac J and Simonik J 2007 Dynamic mechanical properties of filled styrene butadiene rubber compounds: Comparison of tensile and shear data *Polym. Adv. Technol.* **18** 122-7

Revealing the effect of electrode materials on the performance of low-voltage pentacene thin film transistor by in-situ surface potential measurement

Q L Chen[1], M Ouyang[2], Q C Liu[1], H Y Jin[1,3] and W G Xie[2]

[1] Department of Medical Devices, Guangdong Food and Drug Vocational College, Guangdong 510520, PR China
[2] Siyuan Laboratory, Department of Physics, Jinan University, Guangzhou, Guangdong 510632, PR China

E-mail: jinhy@gdyzy.edu.cn

Abstract. We investigate the effect of contact potential between Au, CuO_x and Au-CuO_x, and pentacene thin films by Kevin Probe microscopy Compared to the Au contact, although the injection barrier is lower for CuO_x contact, the extraction barrier is higher, and the total contact resistance is only slightly smaller. The better performance of CuO_x contact device results from the doping effect from the p-type CuO_x. The introduction of additional Au top layer deposited on the thin CuO_x layer lowersthe extraction barrier and facilitates better carrier transfer, which further reduces the contact resistance. Therefore, both the apparent mobility and channel mobility are the best in the device with Au-CuO_x double layer contact.

1. Introduction

Organic thin film transistors (OTFTs) have received intensive research in recent years because of their potential application in flexible electronics [1, 2]. Gold is the most commonly used electrode material in OTFTs. Owing to the high work function, good conductivity, and environmental stability, high-performance pentacene thin film transistors (TFTs) have been developed in recent years [3, 4]. However, large-scale application of Au electrode is limited by the high cost and scarcity of gold. There have been continuous efforts to replace gold with low-cost materials, for example, Cu-based materials [5-7]. The energy offset at the interface of Cu/pentacene is expected to be higher than that of Au/pentacene. However, Wang et al. [5] showed that the contact resistance of pentacene TFT with Cu electrodes was smaller than that of Au contact device, which was attributed to the narrower trap distribution in Cu contact. Di et al. [6] found that a CuO_x interfacial layer would form between the Cu and pentacene thin film, which improved the hole injection because of better energy level alignment. Modification of the Cu electrode with 7,7,8,8-tetracyanoquinodimethane (TCNQ) could further reduce the contact resistance. In our recent study, we also demonstrated that a partially oxidized CuO_x electrode greatly raised the mobility of the pentacene TFT. Furthermore, the introduction of an additional Au top layer deposited on the thin CuO_x layer results in the increase of apparent mobility from 1.0 cm^2/Vs to 2.0 cm^2/Vs. Transfer line method has shown that the optimization of the electrode leads to a dramatic decrease of contact resistance [7]. The above findings show that the performance

[3] Address for correspondence: H Y Jin, Department of Medical Devices, Guangdong Food and Drug Vocational College, Guangdong 510520, PR China. E-mail: jinhy@gdyzy.edu.cn.

of OTFT has been successfully improved by the use of Cu-based materials, however, the underlying mechanism is not systematically revealed. In this study, we investigate the surface potential distribution in the channel of the working pentacene TFTs using Kelvin probe force microscopy (KPFM). The effects of the electrode materials and structural optimization on the performance of the devices are discussed.

2. Experimental methods

An Al_2O_y/TiO_x (ATO) dielectric layer modified by octadecylphosphonic acid (ODPA) was used as gate dielectric for pentacene TFTs [8]. A 30-nm-thick pentacene film was deposited onto the ODPA/ATO substrates at a rate of 0.1 Å/s and pressure of 3×10^{-4} Pa with the substrate temperature at 60°C. A 40-nm-thick S/D electrode layer was then deposited onto the pentacene layer through a shadow mask at a rate of 0.3 Å/s. For Au electrodes, the deposition pressure was about 5×10^{-4} Pa. For CuO_x electrodes, Cu source is deposited onto the pentacene at a deposition pressure of about 5×10^{-3} Pa. X-ray photoelectron emission spectroscopy (XPS) has confirmed that Cu can be oxidized to CuO_x (mainly Cu_2O) at low vacuum condition [7]. For the Au-CuO_x complex electrodes, a CuO_x layer was deposited first. An Au top layer was then deposited at high vacuum condition. The device performance was measured in ambient conditions using Keithley 4200SCS. The in-situ surface potential profile was measured by integrating a Bruker DI 3100 atomic force microscopy with KPFM mode to the device measurement setup, which is shown in figure 1(a) [9].

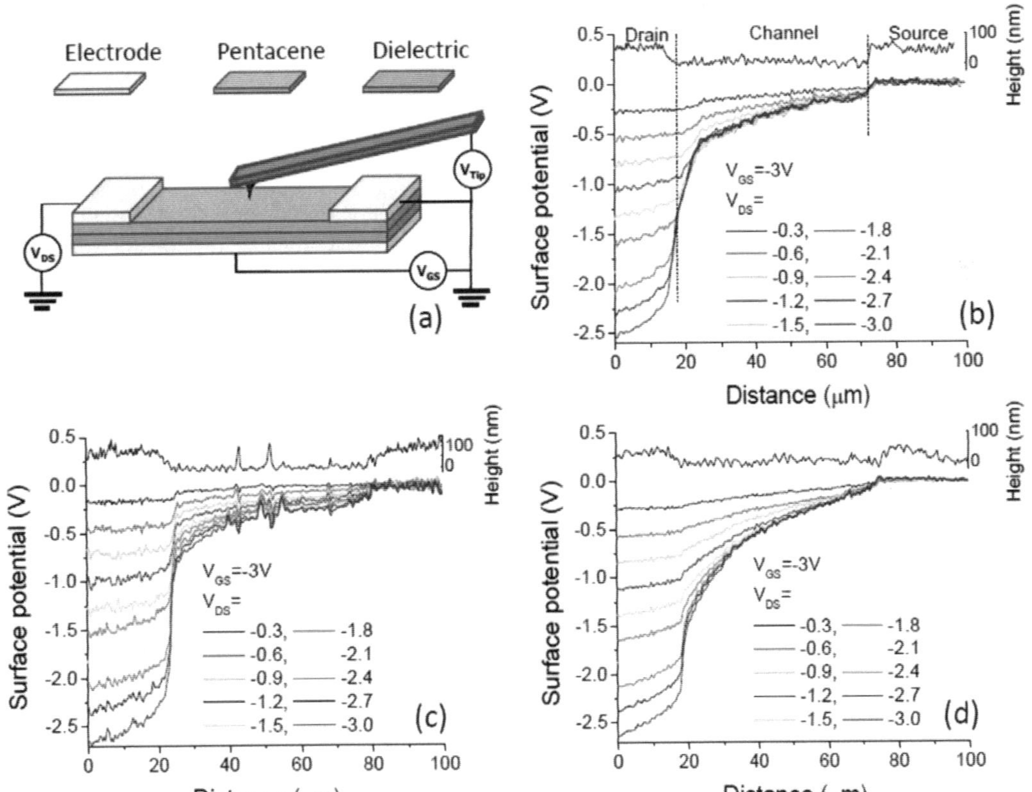

Figure 1. (a) Schematic setup of in-situ surface potential measurement; Normalized surface potential profile along the channel of Au contact (b), CuO_x contact (c) and Au-CuO_x contact (d) devices.

3. Results & discussions

To clearly observe the potential distribution induced by the applied V_{DS}, the surface potential profile at

V_{GS}=V_{DS}=0 V was subtracted in advance, and the voltage drop along the channel of Au contact device is shown in figure 1(b). At small V_{DS}, the linear variation of surface potential across the channel is observed, which identifies that the device is working in the linear region. At higher V_{DS}, the surface potential near the drain electrode changes more rapidly. It means that a depletion layer forms and the device is working in the saturation region.

The total resistance of a TFT device is the sum of resistances of the source (R_S), drain (R_D) and channel (R_{Ch}), which can be directly calculated from the potential profiles as shown in figure 2. We find that in the linear operation regime, the contribution from the contact resistance R_C=R_S+R_D is the smallest in the Au-CuO_x contact device (R_{Ch}/R_C=2.33, V_{DS}= -0.3 V). In the other two devices, the contribution from the contact resistance is similar (R_{Ch}/R_C ~ 0.3-0.4, V_{DS}= -0.3 V). It's also expected that in the linear operation regime, the contact resistance R_C is the sum of R_S and R_D. The R_S approximately equals to R_D [5]. However, we find that this is correct only for Au-CuO_x. In the Au contact device, the ratio of R_D/R_S is 2 – 3. For CuO_x contact, this ratio is as high as 10.

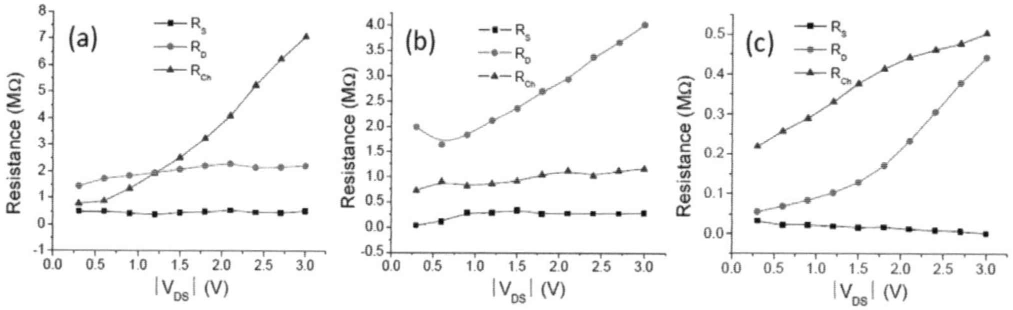

Figure 2. The calculated resistance of R_S, R_D, R_{Ch} in the devices with contact of Au (a), CuO_x (b) and Au-CuO_x (c).

To understand the effect of electrode materials on the device performance, we consider the vacuum level alignment of the device at V_{DS}=V_{GS}=0 V (figure 3(a)). When a gate bias is applied, the vacuum level will slightly increase initially, but then slowly decrease to its original value, implying that the energy level is pinned. It can be seen from figure 3(a) that there is an energy level offset of -0.27 eV between the Au electrode and pentacene. When the contact is changed to CuO_x (mainly Cu_2O, IP=5.3 eV [10]), a clear upward shifting of the energy band is observed. The energy diagrams are constructed in figures 3(b)-3(d) according to the energy level alignment in figure 3(a).

As CuO_x is a p-type semiconductor, the shifting of the energy band can be understood as the hole doping from the electrode. The doping from the electrode is easier to observe from the Au-CuO_x contact. In this case, the CuO_x layer is thinner than the full CuO_x electrode so that the doping effect is smaller. The potential near the electrode drops down, whereas at the center of the channel, the potential increases again. Due to the doping effect, the hole density will increase, and hole traps at the interface of pentacene/dielectric should be passivated, which facilitates better hole transport. As a consequence, the depletion of the channel, which indicated by the rapid increase of R_{Ch} with V_{DS}, is more profound in the Au contact device (figure 2). In the other two contacts, especially in the case of CuO_x contact, the depletion region is restricted near the vicinity of the drain contact (figures 1(c) & 1(d)).

We then consider R_S in the linear operation region at -0.3 V, Where R_S is not affected by the depletion of the carriers. We find that $R_{S\text{-}Au}$ (0.5M) > $R_{S\text{-}CuOx}$ (0.04 M) > $R_{S\text{-}Au/CuOx}$ (0.03 M). Considering the energy level alignment in figure 3(b), the large contact resistance in the case of Au contact should be partially caused by the large injection barrier of 0.27eV. In addition, studies have revealed that the Au atoms will diffuse into the pentacene layer, which will cause interfacial defects and raise the contact resistance [11]. In figure 3(c), holes at CuO_x/pentacene interface do not feel the barrier when injected from the electrode to the pentacene thin film. The resistance of the source is thus caused by the high bulk resistance of CuO_x.

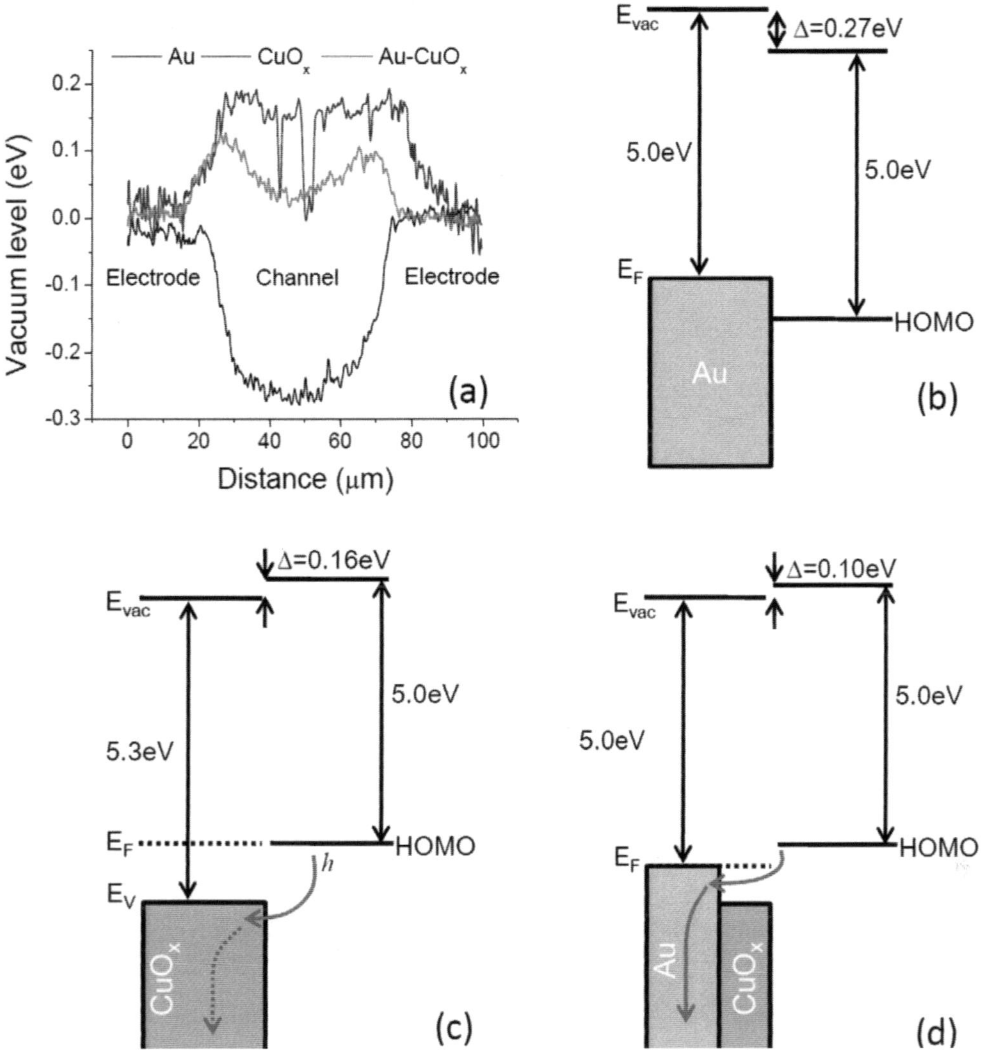

Figure 3. (a) Surface potential profile along the channel at V_{DS}=V_{GS}=0 V; (b)-(d) Energy diagram at the contact of Au, CuO_x and Au-CuO_x respectively;

The resistance of the drain contact is determined by the extraction barrier and the depletion extent. For Au contact in figure 3(b), the extraction of the holes from the channel should be easier than the injection of the holes, so it cannot account for the large R_S/R_D ratio. Considering the trend of R_D and R_{Ch} with the V_{DS} in figure 2(a), we find that the R_D comes to a saturated value at low bias. At the same time, the channel depletion becomes obvious. This indicates that the drain contact is depleted, which leads to a relative high R_D/R_S ratio. The easy depletion of the contact should be caused by the traps induced by the pentacene/dielectric interface and the diffusion of Au atoms into the pentacene thin film. In the case of CuO_x/pentacene contact, the large R_D/R_S ratio at low V_{DS} should come from the large barrier of 0.46 eV for hole when it injects from the pentacene into the CuO_x (figure 3(c)). Such a high barrier seriously limits the carrier extraction, leading to high R_D. The R_D will decrease dramatically when an increasing V_{DS} bend down the barrier. It comes to a minimum resistance at V_{DS} = 0.6 V, after which, the depletion effect leads to increase of resistance again (figure 2(b)). Thinning of the CuO_x layer may lead to increase of tunneling possibility of the carriers; however, the relative poor resistance of CuO_x itself will also limit the transfer of holes as well (figure 3(b)). When the CuO_x

layer was thinned, and a top Au layer was used to encapsulate the electrode, the problem solved. The holes can tunnel easier to the Au layer and be transferred out (figure 3(d)). This is why we observed a dramatic decrease of R_D/R_S ratio in the contact of Au-CuO_x.

From the above discussion, we can see that the contact effect in the three types of device is different. With the Au-CuO_x contact, the contact resistance is reduced to a minimum value. To ensure that the device still keeps the best performance after the removal of the contact effect, we evaluate the real device performance by calculating the mobility of the channel from the potential profiles near the source electrode. The channel mobility at a V_{DS} can be extracted from the potential profile where it varied linearly using the following equation:

$$I_{DS} = eWn(x)\mu(V_g', E)E(x) \tag{1}$$

where W is channel length, and e is the elemental charge. $E(x)$ is the local electrical field that can be determined by $-\partial V / \partial x$ from the measured surface potential profile. The interfacial carrier density $n(x)$ is given by $C_{ox}(V_{GS}-V_t)/e$.

Figure 4(a) shows that the saturated electrical field in the channel of CuO_x contact device is the smallest. The small electrical field is due to the localization of the potential near the drain contact, which leads to the low voltage drop in the channel. In the Au-CuO_x device, the contact effect is the smallest. Most of the applied voltage will drop at the channel, which strongly raises the channel electrical field.

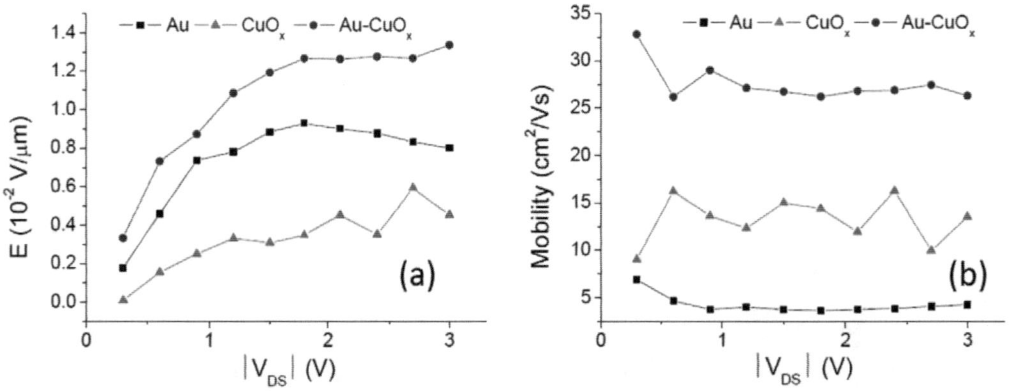

Figure 4. Electrical field (a) and mobility (b) of the channel calculated from the linear region in the surface potential profile

As shown in figure 4(b), the carrier mobility is almost the same under different V_{DS}. The mobility of the Au-CuO_x device is still the highest. It is partially caused by the high electrical field in the channel, which enhances hoping transport of carriers. Moreover, although the electrical field of the CuO_x contact device is smaller than that of the Au contact device, its mobility is higher. This indicates that doping from the electrodes, which pre-passivates the interface traps, also contributes to the high carrier mobility by reducing the trapping and detrapping probability of carriers. The results confirm that the optimization of the contact materials improves the device performance.

4. Conclusions

In this work, the effect of the electrode materials on the performance of pentacene TFT devices is revealed. Although the CuO_x electrode is not so efficient for holes extraction, it reduces the injection barrier and induces hole doping to the pentacene channel, which favor the hole transport easier. With the additional Au top layer covered, the CuO_x interlayer reduces the extraction barrier. As a consequence, the contact resistance of pentacene TFT is dramatically reduced by the use of Au-CuO_x double layer electrode. Both the apparent mobility and channel mobility of the OTFT are greatly improved over that of the Au or CuO_x contact devices. The findings show that the OTFT performance can be optimized by carefully choosing the materials types and electrode structures.

Acknowledgments
The research work was financially supported by National Natural Science Foundation of China under Grant No. 61106093, the Guangdong Natural Science Foundation (Grants No. 2014A030313381), the Fundamental Research Funds for the Central Universities (Grants No. 21614368), the Research Grant of Guangdong Food and Drug Vocational College (No. 2014YZ001), and Medical Research Grants Council of Guangdong, via Grant No. B2014071.

References

[1] Sekitani T, Yokota T, Zschieschang U, Klauk H, Bauer S, Takeuchi K, Takamiya M, Sakurai T, Someya T 2009 Organic nonvolatile memory transistors for flexible sensor arrays *Science* **326** 1516-9
[2] Subramanian V, Frechet J M J, Volkman S K, et al. 2005 Progress towards development of all-printed RFID tags: materials, processes, and devices *Proceedings of the IEEE* **93** 1330-8
[3] Sun X, Zhang L, Di C, Wen Y, Guo Y, Zhao Y, Yu G and Liu Y 2011 Morphology optimization for the fabrication of high mobility thin-film transistors *Adv. Mater.* **23** 3128-33
[4] Kumaki D, Yahiro M, Inoue Y and Tokito S 2007 Air stable, high performance pentacene thin-film transistor fabricated on SiO_2 gate insulator treated with β-phenethyltrichlorosilane *Appl. Phys. Lett.* **90** 133511
[5] Wang S D, Minari T, Miyadera T, Tsukagoshi K and Aoyagi Y 2007 Contact-metal dependent current injection in pentacene thin film transistors *Appl. Phys. Lett.* **91** 203508
[6] Di C A, Liu Y Q, Yu G and Zhu D B 2009 Interface engineering: An effective approach toward high-performance organic field-effect transistors *Acc. Chem. Res.* **42** 1573-83
[7] Su Y R, Wang M D, Xie F Y, Chen J, Xie W G, Zhao N and Xu J B 2013 In situmodification of low-cost Cu electrodes for high-performance low-voltage pentacene thin film transistors (TFTs) *Org. Electron.* **14** 775-81
[8] Su Y R, Xie W G, Li Y, Shi Y, Zhao N and Xu J B 2013 A low-temperature, solution-processed high-kdielectric for low-voltage, high-performance organic field-effect transistors (OFETs) *Journal of Physics D: Applied Physics* **46** 095105
[9] Su Y R, Ouyang M, Liu P Y, Luo Z, Xie W G and Xu J B 2013 Insights into the interfacial properties of low-voltage CuPc field-effect transistor *ACS Appl. Mater. & Interfaces* **5** 4960-5
[10] Wei H M, Gong H B, Chen L, Zi M and Cao B Q 2012 Photovoltaic efficiency enhancement of Cu_2O solar cells achieved by controlling homojunction orientation and surface microstructure *J. Phys. Chem. C* **116** 10510-5
[11] Kim Y and Jeon D 2010 Effect of deposition temperature on the morphology and contact resistance of Au on pentacene *J. Appl. Phys.* **108** 016101

IOP Conf. Series: Materials Science and Engineering **103** (2015) 012012 doi:10.1088/1757-899X/103/1/012012

Application of alcohol based spraying coating on green sand mould for steel casting

Z L Xu[1,3], J Wang[2], S S Yang[2], Q L He[2] and H Sh Xiong[2]

[1]Wuhan Textile University, Wuhan, 430073, China
[2]Tongling Branch, CSR Yangtze Co., Ltd., Tongling, Anhui, 244000, China

E-mail: xzl9675@163.com

Abstract. A kind of coating suitable for green sand steel casting was developed. Thepractical application showed that the strength of the coating was high enough with no crack and no peeling under room temperature after drying the spraying coating, the performance of the coating for anti-cracking was good under high temperature, and the gas evolution of the coating was low. Using the coating, the casting surfaces finish appearedvery good.

1. Introduction

Sodium silicate or resin sand molding is most used for steel casting production. Due to high cost of sodium silicate sand and resin sand molding, also difficulty for shakeout, deoxidization and recycling, many foundries began to use green sand molding instead them. Green sand molding which is low cost, high efficiency, occupies an important position in the casting production [1-3]. However, directly using green sand mould to cast steels is liable to bringing about burnt-on sand, sand inclusion, sand holes or/and other casting defects, therefore, the appropriate coating matching is necessary, and the development of fireproof coating for steel casting by green sand molding is of great significance.

2. The characteristics of green sand casting steel coating

Pouring temperature in steel casting is high, up to 1550~1600℃, it requires coating being with resistance to high temperature and burnt-on sand for steel casting mould. The strength of green sand is poor, brushing coating on the surface of green sand mould will be easy to bring up the sand, so it is not suitable for brushing, and spraying is preferred. Water-based paint will further increase the sand surface moisture, causing sand strength lower, at the same time increase the gas evolution, and drying process influence the production efficiency, so alcohol based spraying coating is more appropriate. Since the permeability of green sand mould is low, and gas evolution tendency of it is larger, the gas evolution of alcohol-based coating should be as small as possible. Spraying process requires that the refractory particle should be fine, requires that the coating should be in good adhesion strength on mould sand, be good permeability, and be more resistant to cracking and peeling.

In general, after spraying, the coating could dried by ignition, also can be natural dried. But light dry may be easy to cause coating peeling and cracking, because the volatile solvent burn fast, the coating dry too quickly, shrink sharply, and the strength of sand mould surface is basically low and get to be lower due to the solvent infiltration. Therefore, air drying after spraying coating would be

[3] Address for correspondence: Z L Xu, Wuhan Textile University, Wuhan, 430073, China. E-mail: xzl9675@163.com.

advisable.

3. The component of spraying coating for green sand steel casting

3.1. Refractory aggregate

It is generally believed that zircon powder is the most suitable refractory aggregate for cast steel coatings [4]. However, the production practice shows that the single zircon powder coating may not completely resist burnt-on sand, and zircon powder coating is expensive [5]. Besides zircon powder, other high fire resistance of the aggregate such as white corundum, brown corundum, bauxite, is relatively economy, with which the burnt-on sand resistance would be better than that of the single zircon powder coatings as long as the ingredient is appropriate[6, 7].

For a long time, the ingredient of coatings only focus on composition of additives such as binder, suspending agent and accessory ingredient, ignore the collocation of refractory aggregate. Really for one kind of good quality coating, the kinds of refractory aggregate, in which the collocation about particle size are also important to be considered, in addition to the appropriate additives proportion. If the aggregate particles are too coarse, the spraying nozzle would be blocked, too fine, the coating will be not conducive to reducing crack tendency [4]. Taken together, a certain proportion of white corundum and brown corundum blended in with zircon powder as refractory aggregate was adopted in our scheme. Aggregate particles mesh distribution was between 220 ~ 325 mesh.

3.2. Agglomerant

Coating must have sufficient strength both at room temperature and at high temperature. Agglomerant of alcohol-based coating for room temperature involve resin, rosin, polyvinyl butyral (PVB), and for high temperature involve bentonite clay, refractory clay, phosphate, ethyl silicate and so on. Considering coating strength, crack resistance and fast drying characteristics, an appropriate constituent of organic and inorganic agglomerant blended as a multiple adhesion agent was proposed to satisfy the necessary strength for both room temperature and high temperature [8].

3.3. Suspending agent

Coating must be of good suspension stability, so as to facilitate transportation, storage and stirring to be uniform seriflux before spraying. Due to the compatibility problems to be blended into alcohol solvent, the suspending agent for alcohol-based coating is not a lot, and the suspensibility of alcohol-based coating is always less than that of water-based coating. The suspending agent for alcohol-based coating involves PVB, sodium-based bentonite, lithium-based bentonite, organic bentonite, and some commercial suspending agent such as SN suspending agent, etc. PVB and resin would form a dense protective film which does not facilitate the resin combustion gas to go out of coating and prone to blistering while burning coating, should not be added much. Sodium-based or lithium-based bentonite need water to be swelled, water will be increased in coating, which reduce the strength of sand mould surface, and is not conducive to quick drying, meanwhile, water precipitation would decrease coating stability. It is clear that neither sodium-based bentonite nor lithium-based bentonite is suitable for spraying coating. No matter what kind of coating, bentonite can't be added much, otherwise cracking will be likely to occur on coating.[4, 8, 9] As comprehensive consideration, the organic bentonite, a small amount of PVB, were added as the suspending agent.

3.4. Solvent

The solvent of coating was a kind of mixture based on anhydrous ethanol with other alcohol solvent, in which compatibility of mixture within coating would be improved, leading to quick drying.

3.5. Additives

Adding additives can improve wettability, dispersion, permeability, adhesion strength of coating, reduce or eliminate the tendency of cracking and peeling. Additives commonly used that could enhance permeability and dispersion are JFC and OP-10[3, 4, 9].

In order to improve the agglomeration of the coating, and to reduce the tendency to burnt-on sand, an oxide sintering agent was appended.

4. Testing of coating performance

4.1. Viscosity, Baume degree and suspension property

Test of viscosity was by flow cup with a tapping hole of diameter Φ6 mm on the bottom, Baume degree test was by Baume scale, and the test of suspension property was by quiet placing in graduated flask.[4]

4.2. Coating strength at room temperature

Sand particles falling down from a viscosity cup to hit a coated glass surface until the coating was chafed, weighing the weight of the sand dropped out once the coating was abraded, the total weight of the sand was as the evaluation index of surface coating strength, seeing the reference for specific operation. [4]

4.3. Crack resistance at high temperature

A Φ60mm × 60mm cylinder specimen was made from green sand with washing silica sand 50/100 mesh (old and new sand ratio of 1:9), which contain 3.0% bentonite and 3.5% water. The sample was sprayed coating with thickness of 0.5-0.8 mm, air dried, then was put inside a furnace within which high temperature atmosphere was to 1200℃, staying 2 minutes, the coating crack could be observed if it was likely to crack. [4, 5]

4.4. Gas evolution

Scraping coating powder and crushing down, weighing accurately for 1.0 g, the gas evolution of coating at 850℃ could be measured by SFL type recording gas evolution test apparatus. [4, 5]

4.5. Cracking and peeling resistance at room temperature

Evaluation of cracking situation of coating while drying after spraying on green sand moulds at room temperature was by actual observation.

4.6. Sintering and stripping property

Casting steels actually, sintering and strippingperformance of the coating was evaluated by actual observation.

5. Ingredient of coating and the main performance

Referring to the domestic and foreign literature [1-11], by some tests in laboratory and production trial in workshop, an appropriate ingredient of coating was determined, which is shown in table 1. Coating performance is shown in table 2.

Table 1. Ingredient of coating for spraying alcohol-based green sand mould (%).

Zircon powder + corundum	Organic bentonite	Organic binder	Inorganic binder	PVB	Additives	Mixture solvent
66	2.0-3.0	1.0-2.0	1.5-2.5	0.2-0.5	2.0-5.0	remainder

Table 2. Performance of coating for spraying alcohol-based green sand mould.

Viscosity (Φ 6 flow cups)/s	Baume degree/ Be	Suspension property (24 h)/%	Density/ $g\cdot cm^{-3}$	Coating strength at R.T./g	Resistance to cracking at H.T. (1200℃,120s)	Gas evolution $/ml\cdot g^{-1}$

14	63	>95	1.7~1.9	459	grade I	<18.5

6. Application of the coating

6.1. Production trial and effect show

Alcohol-based spraying coating for green sand was used in steel foundry, in Tongling Branch, CSR Yangtze Co., Ltd.. Surfaces of the green sand moulds for casting bolsters and frames were sprayed with coating. After drying in the air, no crack or peeling was observed on the coating film, and the casting surfaces finish were good, as shown in figures 1 and 2.

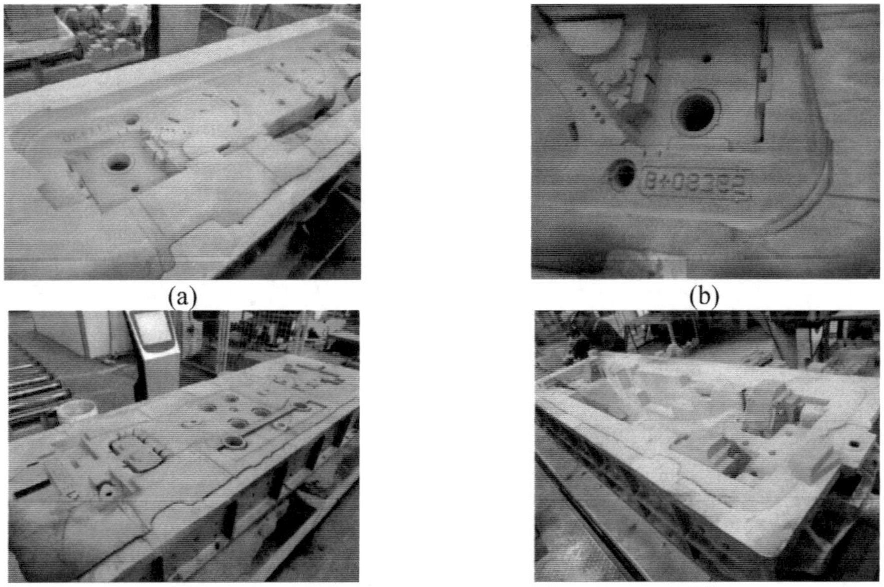

(a) (b) (c) (d)

Figure 1. Spraying coating on (a) half mould of a side frame (b)an end of mould of side frame (c) upper mould of a swing bolster and (d) lower mould of a swing bolster in production.

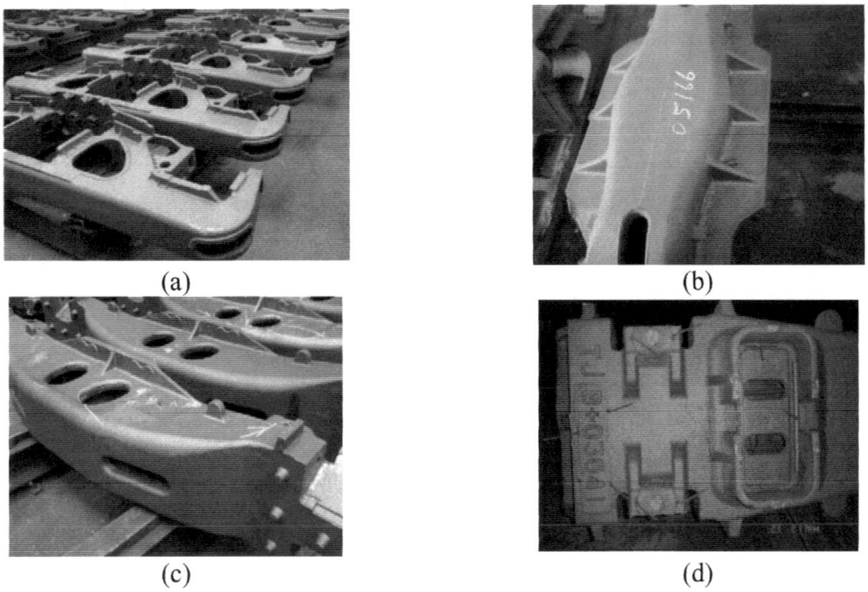

(a) (b) (c) (d)

Figure 2. Surface finish of (a) side frames (b) low beam surface of a side frame (c) swing bolsters and (d) local area of a swing bolster marked part number before shot blasting by spraying coating on green sand moulds.

6.2. Discussion

6.2.1. Influence of bentonite and agglomerant on coating. Appropriate amount of bentonite and agglomerant, is the guarantee for good suspending and rheological properties of the coating, also the guarantee for wet strength after spraying, for dry strength at room temperature after drying, for high temperature strength while casting, of the coating. If bentonite in coating were not enough, or too much, cracking or peeling is likely to occur on the coating after spraying. This is because insufficient bentonite could lead to low wet strength, and too much of bentonite could lead to big shrinkage and cracking. If agglomerant were not enough, dry and wet strength of the coating would be low, that is liable to cause crack. Once crack occur at wet coating in the process of drying, the crack will be easy to extend, become a long crack and cause peeling on the coating after drying, as a result, the coating would have to be out of effect. Too much of organic agglomerant, maybe beneficial to the improvement of the wet and dry strength, but would cause a large gas evolution, thus pinholes might appear on casting surfaces.

6.2.2. Influence of additives on coating. Two additives are worth noting. First, wetting agent, which promote wetting between coating and mould, should be in moderation, too much would lead to poor film-forming, even to local crack, and too little, would lead to poor permeability and poor adhesion on mould surface in some local surfaces, as a result, it would be difficulty for coating to "take root", bring about uneven shrinkage strain on the coating in the process of drying, thus would lead to peeling and cracking. Sometimes, even though no cracking and peeling at coating, the whole coating is going to fall off due to inadhesion to moulds, which immediately lose shielding effect because of cracking or being washed away at high temperature while casting pouring. The second, flake sintering additives, involve several oxides added, are to counter the oxide consumption in reaction with such as SiO_2 and Al_2O_3 existing in sand mould surfaces. Redundant oxide would accumulate on the metal surface, form a kind of molten sintering layer that would be easy to peel off from casting due to the difference of contraction coefficient, avoiding burnt-on sand on casting, and casting surfaces would be clean. Too little oxide added would lead to poor/no resistance to burnt-on sand, and too much, would lead to deterioration of the coating strength.

6.2.3. Influence of the solvent. A kind of mixture solvent based on anhydrous ethanol added other alcohol solvent such as isopropyl alcohol, methanol and n-butyl alcohol under the appropriate proportion, was used, which would be the guarantee of wet strength of coating. Single use of ethanol or industrial alcohol, coating is prone to crack. This may be the consequence that the single anhydrous ethanol solvent would evaporate too fast, causing the coating drying fast before it penetrate into sand mould.

6.2.4. Influence of coating spraying process. Spraying process is important besides the quality of coating must be qualified, and the adjustment of process would be necessary to adapt depending on the change of seasons. In 20-25℃ environment, the baume degree suggestion of coating 58-63 would be more appropriate. In hot summer above 30℃, the baume degree suggestion 56-60, and in cold winter below 10℃, the baume degree could be adjusted to 60-63, but not more than 65. If baume degree were too high, it is difficult for the coating to penetrate into the sand mould before air dried, and if baume degree were too low, solvent permeability is too much to deteriorate sand strength, that would be easy to cause the coating crack and peeling. Spraying pressure suggestion is 0.2-0.35 Mpa. If the pressure were too high, coating grain will rebound, make against form coating. The spraying nozzle height to the mould surfaces should be appropriate with 500-700 mm, too low for the height is similar to the consequences of too high for the pressure. Conversely, too high for the height is similar to the consequences of too high baume degree because the solvent would be lost fast in the air.

7. Summery

7.1. The developed coating is good and feasible

The spraying coating developed for green sand casting steel, is of high wet strength, of good cracking resistance at room temperature and high temperature. No burnt-sand occurrence on steel castings and the surface finish of castings were good, after spraying on green sand moulds before casting.

7.2. Matters needing attention about preparing coating

To prepare coating for green sand casting steel as spraying coating, attention should be paid to the amount of bentonite, appropriate addition of additives and the use of mixed alcohol solvent, so as to avoid crack and peeling, to ensure the coating sintering layer formed under high temperature reaction, then the surface of casting to be easily peeled off and to avoid burnt-sand.

7.3. The correct spraying process is necessary

The correct spraying process could ensure to achieve the best effect of the coating.

References

[1] Yin L L, Yang B and Zhao B Y 2010 The experiment and application of spray coating for wet mould casting *J. B.Voc.Tech. Coll.***11** 1-2
[2] Zhuo Y M 2006 Application of alcohol-based coating to green sand grey iron casting *J.Hot Work.Technol.* **35** 44-5
[3] Feng S S and Zhu S L 2005 The application of new alcohol based spraying coating for green sand mold *Foundry* **54** 65-9
[4] Li Y C 2008 *Foundry Coating and Applications* (Beijing: Mechanical)
[5] Wang G H 2000 Study of zircon and emery powder alcohol based compound coating *Journal of Shenyang University and Technology* **22** 196-9
[6] Wang G H 2000 Investigation of alcohol based compound coatingfor steel castings *Foundry* **49** 279-81
[7] Liu F L 2002 Development and application of alcohol-based coating *Modern Cast Iron* **2002(1)** 54-6
[8] Qu Y H, Gao Y M, Xing J D, Jiang X and Fu P 2002 Preparation of zirconite ointment coating and its usage in thick section steel castings *Hot Working Technology* **2002(4)** 51-2
[9] Zhang X D 2007 Study of coating for heavy steel castings Master Dissertation (China: Kunming University of Science and Technology) pp 30-67
[10] Hong Y 2009 Choice of burning-on defect preventing coating for heavy castings *Modern Cast Iron* **2009(1)** 81-3
[11] Xiao Q G and Jin Y 2010 Development of efficient coating with environmental protection for casting *Ningxia Machinery* **32** 21-3

The effect of adding SiO_2 on the pore structure and the color fastness to washing of PVA sponge

Y I Chang[1,4], T Y Lin[1], W Y Cheng[2] and L Jang[3]

[1] Department of Chemical Engineering, Tunghai University, Taichung, 40704, Taiwan
[2] Kung-Jim Chemical Ind. Co., Ltd. Taichung Industry Park, 40755, Taiwan
[3] Department of Chemical Engineering, California State University, Long Beach, CA 90840, USA

E-mail: yichang@thu.edu.tw

Abstract. The present study investigates the role of pore-forming SiO_2 particles on the pore structure, and color fastness following washing of a porous PVA sponge. We found that the SiO_2 micron particle consideration plays a decisive role on the pore-structural type and the color fastness. Moreover, the particles also influence the mechanical modulus and the water adsorption capacity.

1. Introduction

It is well known that porous Poly (vinyl formal) foams can be made by using hydrophilic chemicals of different grades of Poly (vinyl alcohol) (PVA) [1, 2]. PVA sponges are usually prepared by the condensation reaction (i.e. acetalization) between PVA and formaldehyde by using suitable pore-making agent under acidic condition. Different types of the pore-making agents, such as potato and wheat starches, and their content used in the acetalization process is the most important role in affecting the morphology and the resilience of the PVA sponge. Because of its availability from the food industry since the 1960s, wheat starch has been commonly used as the pore-making agent in the making of PVA sponges [3-5]. However, when those wheat starches of high COD value were washed out in the final step of manufacturing PVA sponges, the discharged wastewater always caused a serious environmental impact problem.

In order to minimize the above pollution problem, in recent years SiO_2 micron scale particles have been used as the pore-making agent to replace the wheat starch in making PVA sponge [6]. Since fine SiO_2 particles can be synthesized when sulfuric acid is added to the sodium silicate (i.e. water glass) solution [7, 8], so we had adopted sodium silicate solution derived SiO_2 micron particles as the pore-making agent when PVA sponges were manufactured under acidic condition (named as PVA/SiO_2 sponge below) [6]. When different contents of sodium silicate solution serving as the pore-making agent were mixed with 12 wt% PVA aqueous solution, a closed-cell type PVA/SiO_2 sponge was found in our previous work [9]. Because the water molecules cannot easily to go through the pore structure of close-cell type, we found that the adsorption rate of PVA/SiO_2 sponge for water molecules is not as good as those of open-cell type PVA/starch sponge, which adopted wheat starches as the pore-making agent as always (named as PVA/starch sponge below). However, we did find that the wastewater COD

---

[4] Address for correspondence: Y I Chang, Department of Chemical Engineering, Tunghai University, Taichung, 40704, Taiwan. E-mail: yichang@thu.edu.tw.

(chemical oxygen demand) value for making PVA/SiO_2 sponge can be reduced efficiently [6].

Since we already proved that the content of PVA does affect the pore-structural type and the pore size when PVA/starch sponges were made, in order to improve the super-absorbent ability of those PVA/SiO_2 sponges made by using SiO_2 micron particles as the pore-making agent, the main purpose of the present paper is to see if it is possible to change the pore morphology of PVA/ SiO_2 sponges from the closed-cell type to the open-cell type when the raw material content of PVA is adjusted. Also, because those SiO_2 particles residue in the PVA polymer matrix perform a good ink receptivity, so their role on improving the sponge's color fastness is investigated, too.

2. Experimental

2.1. Materials
The pure white PVA powders (BF-05; 99.3 wt%) was obtained from Chang-Chun Petrochemical Co., Ltd., Taiwan, whose averaged polymerization degree was 500 and the hydrolysis degree was 98.5-99.2 mol%. The sodium silicate solution was purchased from Rong Xiang Ind. Co., Ltd., Taiwan (EG, 50 wt%, Na_2O:SiO_2=1:1, CAS No. 6834-92-0). All reagents adopted in the present experiments were analytical grade.

2.2. Preparation of PVA/ SiO_2 sponge
In making the present PVA/SiO_2 porous sponges, the solutions of 12, 13 and 14 wt% of PVA and 0.1, 0.3 and 0.5 wt% of sodium silicate solution were adopted. The detailed experimental procedure can be found in our previous paper [6, 9].

2.3. Scanning electron microscope imaging
The porous structure of the present sponge was obtained using a scanning electron microscope (SEM, JSM-7000F, JEOL, Japan) at the operating voltage 15 KV. After coating with a thin layer of gold at 5.0 Pa and 20 mA for 80 seconds, the samples were then monitored by.

In addition to taking those SEM images, we also determined the glass transition temperature ($T_{g.}$, Perkin-Elmer DSC Pyris-1), the Young's modulus of tensile test (Instron, Model 4467) and the water adsorption capacity of PVA/ SiO_2 samples. The detailed analyze methods can be found in our previous papers, too [6, 9].

2.4. Color fastness to washing
Color fastness to washing is one of the most important parameters for testing the quality of PVA sponges from the viewpoint of consumers. In order to prove the ability of the present PVA/SiO_2 sponges to fasten various color dyes, at this point we adopt the AATCC 61-2010 test method to determine the loss of sponge and any change in the sponge's color in the washing process and the possible staining of other garments (or lighter portion) that may be washed with it. In this test method, a six-multifibre swatch (acetate, cotton, nylon, polyester, acrylic and wool) is mechanically washed with the sample sponges for 45 minutes at 40±2°C in 0.37% WOB solution (without optical brighteners, 1993 AATCC standard reference detergent), along with #10 steel balls and 200 ml liquor in volume. The conventional PVA/starch sponges made by using the wheat starches as the pore-making agent are also adopted as the contrast samples in this test (i.e. its manufacture method can be found in our previous paper [6]).

3. Results and discussion

3.1. The pore morphology
The SEM pictures of the pore structures of PVA/SiO_2 sponges made at 12 wt% PVA and different content of 0.1, 0.3 and 0.5 wt% sodium silicate solution are shown in figures 1(a)-1(d). We can find that the PVA/SiO_2 sponges shown in these four pictures were all belong to a closed-cell type with

many micron voids of 2 μm in diameter appear on the pore walls. The SEM example of those SiO_2 micron particles resulted from the acidification of the sodium silicate solution is shown in figure 2, which illustrates that the average diameter of the present SiO_2 is about 2 μm, too, hence we believe that those micron voids were made by those non-aggregated SiO_2 micron particles. As shown in figures 1(a)-1(d), the average diameter of those closed-cell type pores is 300 μm (which is the same as in our previous paper [6, 9]), and the pore size of PVA/SiO_2 sponges was found only slightly increased when sodium silicate content is increased from 0.1 wt% to either 0.3 or 0.5 wt%. In addition to those silica particles dispersed in the PVA matrix which will reinforce the mechanic strength of sponges (see the tensile test results below), the aggregation of those SiO_2 particles reside in the aqueous solution associate with their surface adsorbed H_2O molecules and those water molecules condensed from acetalization is the main reason to cause this close-cell type pore structure [10].

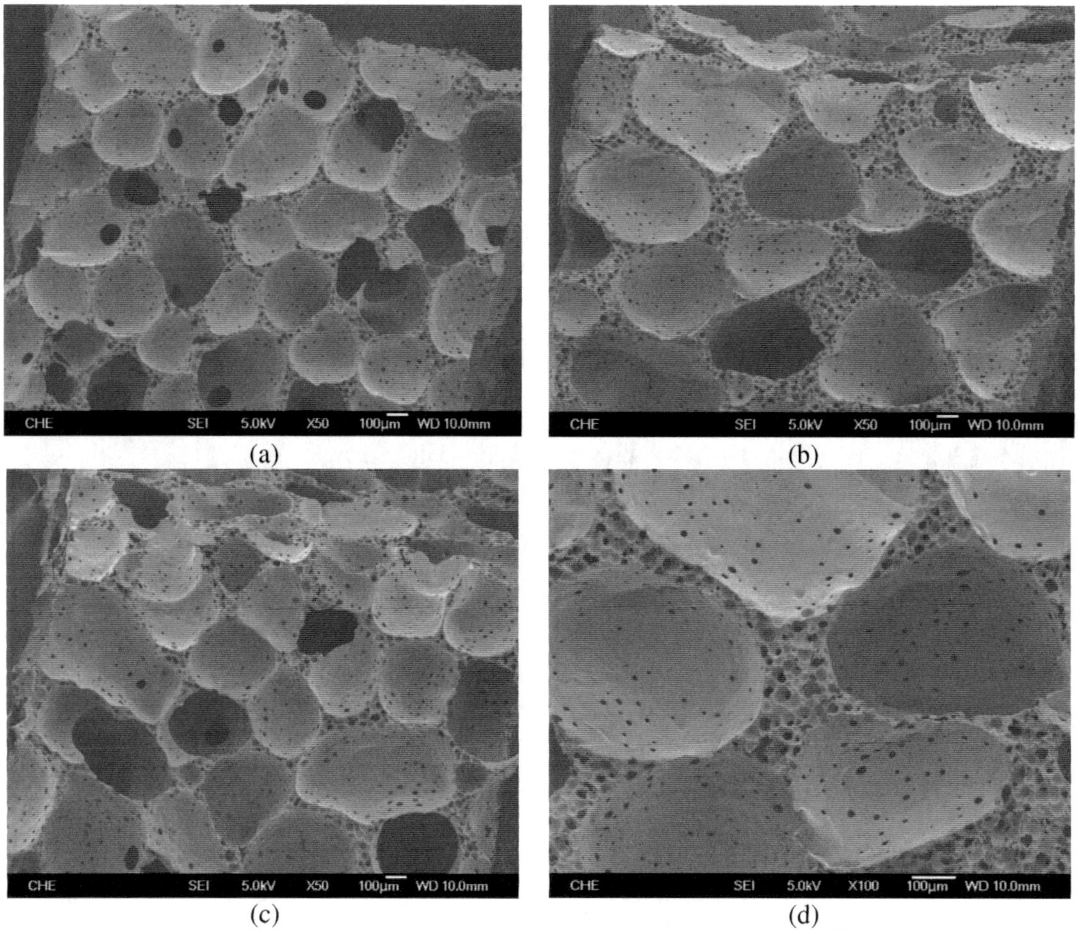

Figure 1. SEM pictures of PVA/SiO_2 sponges when different amounts of 0.1 (a), 0.3 (b) and 0.5 wt% (c) sodium silicate solutions are added in a 12 wt% PVA aqueous solution, respectively. (d) is the enlarged part of (c).

As shown in figures 3(a)-3(d), due to the increased PVA content, the average pore size of PVA sponge decreased from 300 μm to 40 μm when PVA content is increased from 12 wt% to 13 wt%. Fewer micron voids are observed on the pore walls under the contents of 0.1, 0.3 and 0.5 wt% sodium silicate solution as observed in these four pictures. When comparing with the pore type between

Figure 2. SEM picture of the SiO_2 micron particles obtained in the present experiments.

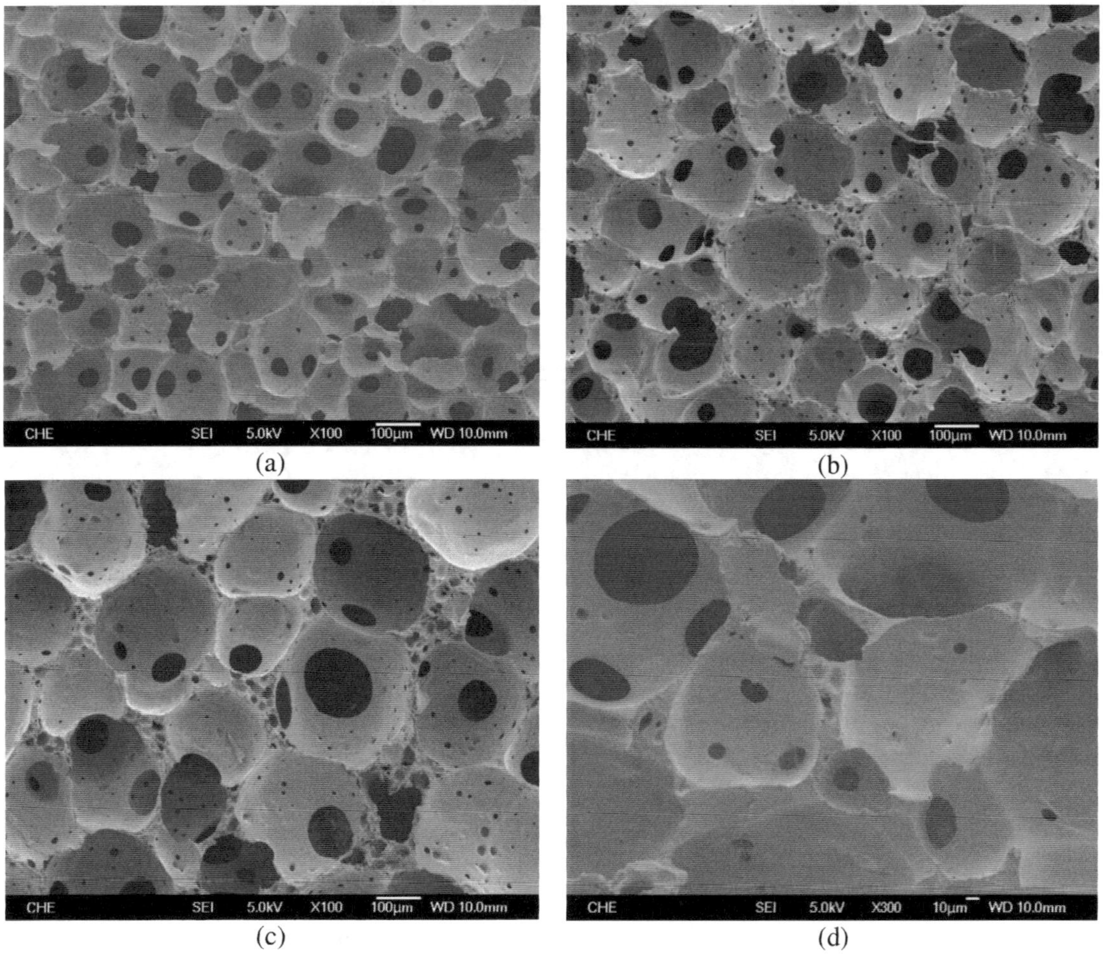

Figure 3. The pore structure SEM pictures of PVA/SiO_2 sponges when 0.1 (a), 0.3 (b) and 0.5 wt% (c) sodium silicate solutions are added in a 13 wt% PVA aqueous solution. (d) is the enlarged part of (c).

figures 1(a)-1(d) and figures 3(a)-3(d), it is interesting to note that more pores in figures 3(a)-3(d) become a more open type than those pores in figures 1(a)-1(d). Also, because of the formation of a larger silica aggregate, the pore size of the open part of those pores made by 13 wt% PVA increases with the increase of the content of sodium silicate solution. The same results are observed in figures 4(a)-4(d) when the PVA content is further increased to 14 wt%; i.e. the pore size of sponge can be further decreased but the open part of those pores increases with the increase of the sodium silicate solution content. Note that there is no micrometer void observed in figures 4(a)-4(d). By using the Image-Pro Plus software to measure the pore size distribution [11], the results for those sponges made by 13 and 14 wt% PVA are summarized in figure 5. It can be found that, instead of the proportions of those 10 μm voids appear on the pore walls of PVA/SiO_2 sponges made by 13 wt% PVA, the pore sizes of those PVA sponges made by 14 wt% PVA all become increased and locate at 30, 40 and 50 μm, respectively. This result proves that the pore structure of PVA/ SiO_2 sponges can be changed from the close-cell type to the open-cell type when PVA content is increased from 13 to 14 wt%.

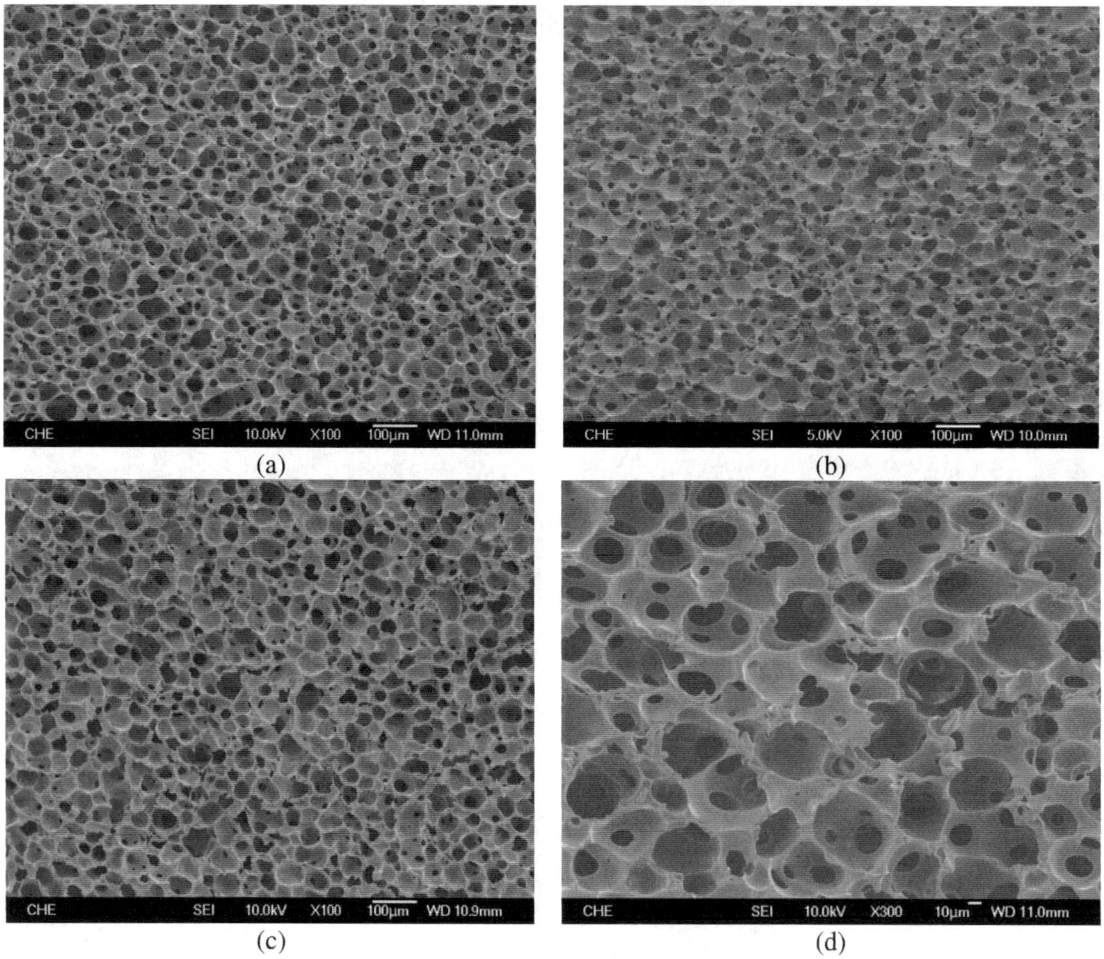

Figure 4. The pore structure SEM pictures of PVA/SiO_2 sponges when 0.1 (a), 0.3 (b) and 0.5 wt% (c) sodium silicate solutions are added in a 14 wt% PVA aqueous solution. (d) is the enlarged part of (c).

The above results can be explained by the fact that, as illustrated by the schematic diagram of figure 6, since the drainage velocity of PVA polymer between two water bubbles in 13 wt% PVA solution is faster than that in 14 wt% PVA solution due to its increased viscosity, therefore a larger pore size but thinner pore wall can be induced in PVA sponge made by 13 wt% than that made 14 wt% PVA. This result is evidenced by the examples of the enlarged SEM pictures shown in figures 3(d) and 4(d) where PVA content is fixed at 13 and 14 wt%, respectively, for the case of adding 0.5 wt% sodium silicate solution.

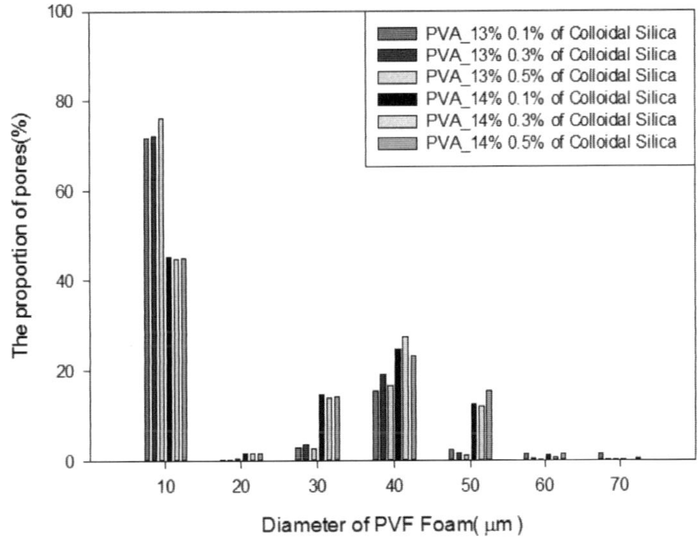

Figure 5. The pore-size distribution of PVA/SiO_2 sponges analyzed by using the Image-Pro Plus software [11], when 0.1, 0.3 and 0.5 wt% sodium silicate solutions are added in the 13 wt% and 14 wt% PVA aqueous solution, respectively.

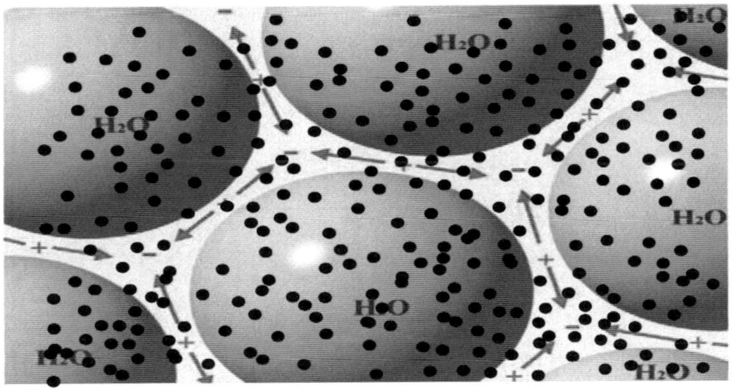

Figure 6. The schematic diagram illustrates the drainage phenomenon of PVA polymer occurred in the pore walls of PVA/SiO_2 sponges. In this figure, the SiO_2 micron particles are represented by the dots as shown by the black color. The capillary pressure difference between PVA polymer and plateau borders (i. e. at positions of inter-film contact) leads to rapid PVA wall thinning (over pressure is indicated by +, low pressure by −).

3.2. DSC analysis

From the DSC analysis results, we found that T_g of the present PVA/SiO_2 sponges made by different PVA and sodium silicate contents are all fixed at 70°C. This result indicates that the present SiO_2 particles do not affect the hybrid bond formation between PVA and formaldehyde molecules as acetalization proceeds.

3.3. Mechanical property

From the slope of the strain-stress curves obtained from the tensile test, the Young's modulus of present PVA/SiO_2 sponges can be determined (i.e. based on the five times of measurements) and the results are shown in figure 7. It can be found that the Young's modulus of PVA/SiO_2 sponges increases as PVA content increases. Also, under the same content of PVA, the sponge's modulus increases with the increase of the sodium silicate solution content. The reinforcement role played by those silica particles dispersed in the PVA matrix is believed to be the primary cause of this result [12, 13].

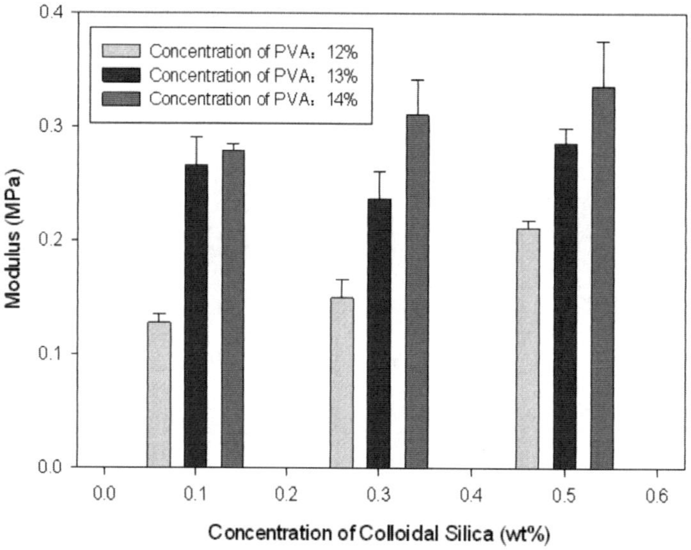

Figure 7. The mechanical modulus of PVA/SiO_2 sponges made by 12, 13 and 14 wt% PVA, respectively, when 0.1, 0.3 and 0.5 wt% of sodium silicate solution are added.

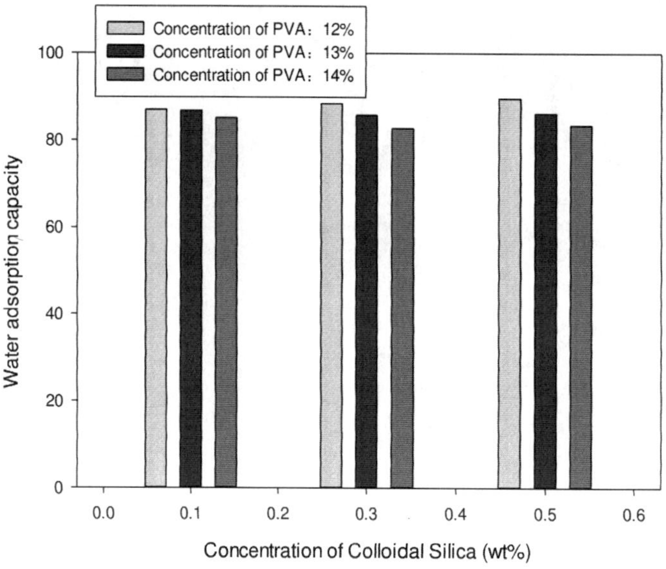

Figure 8. The water adsorption capacity of PVA/SiO_2 sponges made by 12, 13 and 14 wt% PVA, respectively, when 0.1, 0.3 and 0.5 wt% of sodium silicate solutions are added.

3.4. The water adsorption capacity and the water desorption rate

The water adsorption capacity of the present PVA/SiO_2 sponges was determined from the loss of the sample weight at 100°C by using TGA, when 20 mg of dry sample was inserted into 100 mL water for one minute. As shown in figure 8, there is an 84% weight loss of water (therefore the water adsorption capacity) obtained for the 12 wt% PVA made sponge when 0.1 wt% sodium silicate solution was added. The same value of 84% was also obtained for other PVA/ SiO_2 samples when 0.3 and 0.5 wt% sodium silicate solution was added, respectively. Hence, the water adsorption capacity of the present sponge is determined by its PVA content and that it is not affected by the added sodium silicate solution [9]. However, because of the less amount of water can be adsorbed in those smaller but become more open type pores than those close type but bigger pores of 12 wt% PVA made sponge, an average 80% water adsorption capacity was obtained for those 13 and 14 wt% PVA made PVA/SiO_2 sponges. These water adsorption capacities of 80 and 84% water weight meet the requirement of superabsorbent [1]. This conclusion can be evidenced by the temporal water retention curves as shown in figures 9(a)-9(c) when the present sponge samples were heated at 50°C in TGA for 50 minutes (i.e. since the value of 16 wt% water retention is always reached after heating 30 minutes at 50°C for all of curves shown in figures 9(a)-9(c). The water desorption rates of the sponges can be therefore determined by the slopes of those temporal water retention curves. Since more open-type pores can be obtained for those sponges made by increased PVA as shown in figures 4(a)-4(d), which will drive those water molecules to go through pores easier, hence the water desorption rates of 14 wt% PVA made sponges are always higher than those sponges made by 12 and 13 wt% PVA.

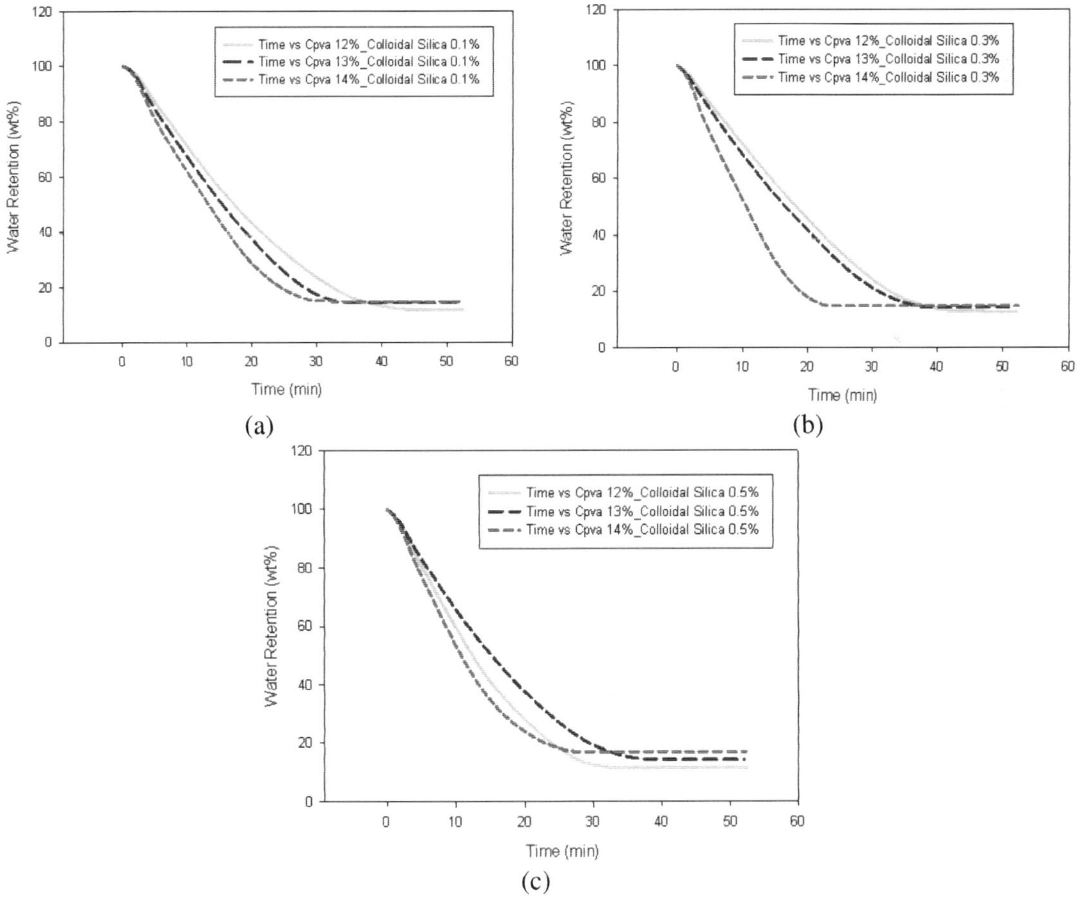

Figure 9. The temporal water retentions of PVA/SiO_2 sponges made by 12 (a), 13 (b) and 14 wt% (c) PVA, respectively, when 0.1, 0.3 and 0.5 wt% of sodium silicate solutions are added.

3.5. Test results of Color fastness to washing

One of the test results are summarized in table 1, where the PVA/SiO_2 sponges were manufactured by 13 wt% PVA and 0.3 wt% sodium silicate solutions and the PVA/starch sponges were manufactured by 10 wt% PVA and 3.0 wt% wheat starch. The concentrations of dyes employed in those sample sponges were all fixed at 0.05 wt%. It can be found that the PVA/SiO_2 sponges own a better fastness on navy-blue and crimson dyes (marked by the red color), and therefore on the darker colored dyes than that of the PVA/starch sponges. The good ink receptivity of those SiO_2 particles residue in the PVA/SiO_2 polymer matrix is the major reason to improve sponge's color fastness.

Table 1. The test results of color fastness to washing for the present PVA/starch and PVA/SiO_2 sponges.

	Sample (1)	Sample (2)	Sample (3)	Sample (4)	Sample (5)
Change in shade	Grade 4.5	Grade 4.5	Grade 4.5	Grade 4.5	Grade 4.5
Staining of Acetate	Grade 4.0	Grade 4.5	Grade 4.5	Grade 4.5	Grade 4.5
Staining of Cotton	Grade 4.0	Grade 3.0	Grade 4.5	Grade 3.0	Grade 4.5
Staining of Nylon	Grade 4.0	Grade 4.5	Grade 4.5	Grade 4.5	Grade 4.5
Staining of Polyester	Grade 4.5	Grade 4.5	Grade 4.5	Grade 4.5	Grade 4.5
Staining of Acrylic	Grade 4.5	Grade 4.5	Grade 4.5	Grade 4.5	Grade 4.5
Staining of Wool	Grade 4.5	Grade 4.5	Grade 4.5	Grade 4.5	Grade 4.5
	Sample (6)	Sample (7)	Sample (8)	Sample (9)	Sample (10)
Change in shade	Grade 4.5	Grade 4.5	Grade 4.5	Grade 4.5	Grade 4.5
Staining of Acetate	Grade 4.0	Grade 4.5	Grade 4.5	Grade 4.5	Grade 4.0
Staining of Cotton	Grade 4.0	Grade 3.0	Grade 4.5	Grade 3.0	Grade 3.5
Staining of Nylon	Grade 3.5	Grade 4.5	Grade 4.5	Grade 4.5	Grade 3.5
Staining of Polyester	Grade 4.5	Grade 4.5	Grade 4.5	Grade 4.5	Grade 4.0
Staining of Acrylic	Grade 4.5	Grade 4.5	Grade 4.5	Grade 4.5	Grade 4.0
Staining of Wool	Grade 3.5	Grade 4.5	Grade 4.5	Grade 4.5	Grade 3.0

[a] Sample (1): PVA/SiO_2 sponge+Navy blue dye;
[b] Sample (2): PVA/SiO_2 sponge+Purple dye ;
[c] Sample (3): PVA/SiO_2 sponge+Green dye;
[d] Sample (4): PVA/SiO_2 sponge+Yellow dye;
[e] Sample (5): PVA/SiO_2 sponge+Crimsom dye;
[f] Sample (6): PVA/starch sponge+Navy blue dye;
[g] Sample (7): PVA/starch sponge+Purple dye;
[h] Sample (8): PVA/starch sponge+Green dye;
[i] Sample (9): PVA/starch sponge+Yellow dye;
[j] Sample (10): PVA/starch sponge+Crimsom dye.

4. Conclusion

Accompanying with the method of preparing 12, 13 and 14 wt% PVA made porous PVA/SiO_2 sponge through the simultaneous acidifications of 0.1, 0.3 and 0.5 wt% sodium silicate solutions, respectively, the effect of their contents on the sponge's pore-structural type are investigated in the present paper. With the help of the analyses of SEM, the tensile test and the water adsorption capacity, there are five conclusions can be obtained: (1) The pore-structural type of sponges can be changed from the closed-cell type to the open-cell type when PVA content is increased from 12 to 14 wt%, especially when 0.5

wt% sodium silicate solution was served as the pore-making agent. (2) The pore size of PVA decreases with the increase of PVA content, but the the open part of those pores increases with the content of sodium silicate solution. (3) The unchanged T_g indicates that the present sodium silicate solution does not involve in the formation of the hybrid bonds through the acetalization between PVA and formaldehyde. (4) The sponge's modulus increases with the increase of PVA content. Also, under the same PVA content, the sponge's modulus can be reinforced with the increase of the sodium silicate content. (5) Even with the same water adsorption capacity, but because more pores become opened, the water desorption rates of 14 wt% PVA made sponges are always higher than those sponges made by 12 and 13 wt% PVA, when different contents of sodium silicate solution were added in the present study. (6) Finally, but not at least, from the test results of color fastness to washing, we find that the present PVA/SiO_2 sponges own a better fastness on the darker colored dyes than that of the conventional PVA/starch sponges.

Acknowledgment
The financial support received from the Ministry of Science and Technology of Taiwan R.O.C. research grant no. MOST 104-2221-E-029-022, is greatly appreciated.

References

[1] Buchhlz F L and Graham T 1997 *Modern Superabsorbent Polymer Technology* (New York: Wiley-VCH Pub)
[2] Hallenslenben M L 2011 *Polyvinyl Compounds* (Others' Ullmann's Encyclopedia of Industrial Chemistry, 7th edition, vol 29) (Weinheim, Germany: Wiley-VCH) pp 605-622
[3] Wilson C L 1952 Method of making expanded polyvinyl alcohol-formaldehyde reaction product and product resulting therefrom (US Patent 2609347)
[4] Sueoka A, Okamoto T, Ohmori A, Kawai S and Ueda M 1981 Kuraray Co Ltd: 'Polyvinyl alcohol semi-permeable membrane and method for producing same (US Patent 4279752)
[5] Rosenblatt S 1996 Injection molded PVA sponge (US Patent 5554658A)
[6] Chang Y I, Cheng W Y and Jang L 2014 A new method of preparing superabsorbent PVF porous foam through the simultaneous acidification of water glass solution – Aspect of environmental protection *J. of Applied Polymer Sci.* **131** 1001-7
[7] Stöber W and Fink A 1968 Controlled growth of monodisperse silica spheres in the micron size range *J. Colloid Interface Sci.* **26** 62-9
[8] Olphen H. van 1977 *An Introduction to Clay Colloid Chemistry* 2nd edn. (New York: John Wiley & Sons, Inc.)
[9] Lin T Y 2013 The effect of adding silica gel on the physical properties of PVA sponge (Chem. Eng. Dept., Tunghai Univ.: Master thesis)
[10] Plieva F M, Karlsson M, Aguilar M R, Gomez D, Mikhalovsky S, Galaev I Y and Mattiasson B 2006 Pore structure of macroporous monolithic cryogels prepared from poly (vinyl alcohol) *J. of Applied Polymer Sci.* **100** 1057-66
[11] Image-Pro Plus, version 4.5, 2001 (Bethesda, Maryland: Media Cybernetics Inc.)
[12] Pagliaro M 2009 *Silica Based Materials for Advanced Chemical Applications* (London, UK: Royal Society of Chemistry) chapter 7 pp 160-175.
[13] Theng B K G 2012 *Formation and Properties of Clay Polymer Complexes* (Oxford, UK: Elsevier) 2nd edn Chapter 6.3 pp 166-185.

IOP Conf. Series: Materials Science and Engineering **103** (2015) 012014 doi:10.1088/1757-899X/103/1/012014

Economical evaluation of damaged vacuum insulation panels in buildings

Y M Kim[1], H Y Lee, G S Choi and J S Kang

Building and Urban Research Institute, Korea Institute of Civil Engineering and Building Technology, South Korea

E-mail: ymkim@kict.re.kr

Abstract. In Korea, thermal insulation standard of buildings have been tightened annually to satisfy the passive house standard from the year 2009. The current domestic policies about disseminating green buildings are progressively conducted. All buildings should be the zero energy building in the year 2025, obligatorily. The method is applied to one of the key technologies for high-performance insulation for zero energy building. The vacuum insulation panel is an excellent high performance insulation. But thermal performance of damaged vacuum insulation panels is reduced significantly. In this paper, the thermal performance of damaged vacuum insulation panels was compared and analyzed. The measurement result of thermal performance depends on the core material type. The insulation of building envelope is usually selected by economic feasibility. To evaluate the economic feasibility of VIPs, the operation cost was analyzed by simulation according to the types and damaged ratio of VIPs

1. Introduction

1.1. Background of study

The Korean government has pursued intensive building energy saving policies in the area of building that takes about 21% of total energy consumptions [1]. In addition, various kinds of related technology have been developed and utilized for application of passive house and zero energy building. One of the core technologies for such zero energy building is the application of high-performance thermal insulation [2-5]. Recently, Vacuum insulation panels (VIPs) are installed in building envelopes since VIPs have considerably high thermal performance compared to existing thermal insulation [6-8]. There are various kinds of core materials for VIPs such as glass wool, fumed silica, polyurethane, and polystyrene. Core materials keep the form of an envelope and block the movement of internal gas molecule, minimizing heat transfer. Glass wool and fumed silica are mainly used for the core material [9].

1.2. Application of VIPs in building

It is necessary to evaluate the performance variation of VIPs by vacuum condition of the panel since vacuum condition can be destroyed at the time of construction. When the vacuum of VIPs is destroyed, convective heat transfer inside of the insulation panel greatly increases, the thermal performance decreases.

[1] Address for correspondence: Y M Kim, Building and Urban Research Institute, Korea Institute of Civil Engineering and Building Technology, South Korea. E-mail: ymkim@kict.re.kr.

The insulation of building envelope is usually selected by economic feasibility which is based on life cycle cost [10, 11]. In case of VIPs, thermal performance which significantly influences the operation cost depends on the condition of vacuum condition of VIPs. In this paper, to evaluate the economic feasibility of VIPs, the initial and operation cost was analyzed by simulation according to the types and the ratio VIPs whose vacuum condition was destroyed.

2. Experimental methods

2.1. Measurement of Thermal Conductivity of VIPs

This study measured thermal conductivity of normal VIPs and damaged VIPs which vacuum is destroyed by boring a hole. GW VIP and FS VIP were measured and the core materials of the panels were glass wool and fumed silica respectively. The thermal conductivity was measured heat flux method according to KS L 9016 [12] with HFM 436/3/1E Lambda of NETZSCH Instruments, Inc. as shown in figure 1.

Figure 1. Equipment for thermal conductivity measurement.

2.2. Energy simulation and economic feasibility

TRNSYS software was used to simulate heating energy consumption of apartment housing according to the types and ratio of damaged VIPs. 9 cases of damaged ratio from 0 to 100% were applied to the simulation and the damaged ratio was calculated by dividing the number of the VIPs whose vacuum condition was destroyed by the total number of VIPs which were installed.

VIPs for building envelope are usually installed as external insulation system and other heat bridge was not considered in this simulation. The condition of the simulation is listed in table 1 and the floor plan of the housing is shown in figure 2. Because the heating load is dominant in housing buildings and the cooling load depends on the lifestyle of occupants in Korea, the simulation was focused on the heating load.

Table 1. Conditions of simulation.

Contents	Condition	Remark
Weather	Seoul(TMY2)	
Gloss area	80 m^2	
Wall area which VIPs are installed	74.389 m^2	
Occupants	3 persons	2 adults and 1 child
Set temperature for heating	23℃	
Operating Schedule	Weekday : 19:00 – 08:00 Weekend : 17:00 – 11:00	
Ventilation rate	0.5 Air change rate per hour	
Infiltration rate	0.2 Air change rate per hour	Constant
Heating source	City gas	
Efficiency of the boiler	87%	Typical boiler

IOP Conf. Series: Materials Science and Engineering **103** (2015) 012014 doi:10.1088/1757-899X/103/1/012014

Figure 2. Floor plan of the typical apartment housing.

In life cycle cost analysis, the unit price of gas for heating was 72.02 \$/MWh [13]. To calculate the initial cost, the cost of VIPs and installation cost of VIPs were surveyed. The costs of GW VIP and FS VIP were 36.36 \$/m^2 and 32.73 \$/m^2 respectively and the cost of installation was 13.64 \$/m^2 for both types of VIPs.

The nominal interest rate was applied to the average of the interest on a loan for the last ten years of public and other parts. The inflation rate and gas price increase rate was applied to the average of the last ten years by Consumer Price Index and Korea Gas Safety Corporation, respectively. The discount rate is determined by the equation (1). Table 2 shows the nominal interest rate, the inflation rate, and gas price increase rate. The present value (P_v) is determined by the equation (2).

Table 2. Variables for economic analysis.

Period	The nominal interest rate, %	The inflation rate, %	The increase rate of gas price, %	The discount rate, %	
				Price	Gas
20 years	6.13	2.96	10.94	3.08	-4.34

$$i = \frac{(1+i_n)}{(1+f)} - 1 \tag{1}$$

$$P_v = \frac{N_v}{(1+i)^n} \tag{2}$$

3. Results and discussion

3.1. Measurement results

As a result of measurement, thermal conductivity of GW VIPs and FS VIPs before vacuum condition is destroyed was 0.0020, 0.0035 W/mK respectively. To examine thermal performance change of damaged VIPs, we artificially destroyed internal vacuum of VIPs and measured thermal conductivity. As a result of measuring thermal conductivity in 5 days after destroying the vacuum condition of VIPs, change rate depended on the type of core material of VIPs. Thermal performance of GW VIPs and FS VIPs decreased 18.3 times and 5.9 times respectively after the panels were damaged. Table 3 shows the results of thermal conductivity of VIPs before and after damage. To apply thermal conductivity of VIPs obtained from the measurement to building simulation, U-value was calculated for the walls which VIPs are installed. Table 4 shows the composition of the walls and thermal conductivity and thickness of each material. U-values of the walls which the VIPs are installed were calculated as shown in table 5.

Table 3. Results of measured thermal conductivity.

	GW VIPs (Normal)	GW VIPs (Damaged)	FS VIPs (Normal)	FS VIPs (Damaged)
Conductivity, W/mK	0.002	0.037	0.120	0.279

Table 4. Properties of materials applied in external insulation system.

Materials	Thickness, m	Thermal Conductivity, W/mK	Thermal Resistivity. mK/W
Internal surface	-	-	0.110
Concrete	0.180	1.600	0.113
Polyester	0.030	0.038	0.789
Glass wool 64K	0.025	0.034	0.735
VIPs	0.020	Values from the measurement	
Glass wool 64K	0.025	0.034	0.735
Air	0.100	1.163	0.086
Masonry	0.030	2.800	0.011
External surface	-	-	0.043

Table 5. U-value of external insulation system according to the types of VIP.

	GW VIPs (Normal)	GW VIPs (Damaged)	FS VIPs (Normal)	FS VIPs (Damaged)
U value, W/m^2K	0.079	0.316	0.120	0.279

3.2. Energy performance and economic feasibility

Table 6 shows the annual heating load variation according to the types and the ratio of damaged VIPS for the typical apartment housing. Because the thermal conductivity of GW VIP is better than the conductivity of FS VIP, the heating load can reduced by installation of GW VIP compared to FS VIP when the ratio of damaged VIPs is lower than 50%. However when the VIP is damaged, FS VIP has better performance than GW VIP, as the ratio of damaged VIPs is increased more than 50% , FS VIP is superior to GW VIP as shown in table 6.

Table 6. Heating load variation according to the types and damaged ratio of VIP.

Damaged ratio of VIPs, %	GW VIPs		FS VIPs	
	Heating load, kWh	Ratio of Increased heating load, %	Heating load, kWh	Ratio of Increased heating load, %
0	6,514	Base case	6,720	Base case
5	6,575	0.9	6,761	0.6
10	6,640	1.9	6,807	1.3
20	6,775	3.9	6,902	2.6
30	6,915	5.8	7,000	4.0
40	7,061	7.7	7,100	5.4
50	7,202	9.6	7,199	6.7
75	7,556	13.8	7,448	9.8
100	7,877	17.3	7,671	12.4

To evaluate the life cycle cost of typical apartment housing which VIPs are installed, the initial cost was calculated including the material cost and construction cost. The initial cost for installation of GW VIPs and FS VIPs were 3,719 $ and 3,449 $ respectively. Figures 3-5 shows the life cycle cost according to ratio of damaged VIPs by 3 cases of useful life respectively. In this analysis, the useful life is assumed as 10, 20 and 30 years and the cost of destruction was not considered as the cost is the same for both types of VIPS.

In case the useful life is 10 years, the life cycle cost of FS VIP is lower than the cost of GW VIP regardless the ratio of damaged VIPs as shown in figure 3 because the initial cost of GW VIP is higher. If the useful life is 20 years, GW VIP is economic when the ratio damaged VIPs is lower than 26% however when the ratio of damaged VIPs is higher than 26%, FS VIP is economic because operation cost of GW VIP increase as the ratio of damaged VIPs increased. In case the useful life is 30 years, the life cycle cost is the same for both VIPs when the ratio of damaged VIPs is 38% as shown in figure 5. For economic feasibility, the VIPs should be selected in consideration of the predicted ratio of damaged VIP in the construction field and the useful life of the VIPs.

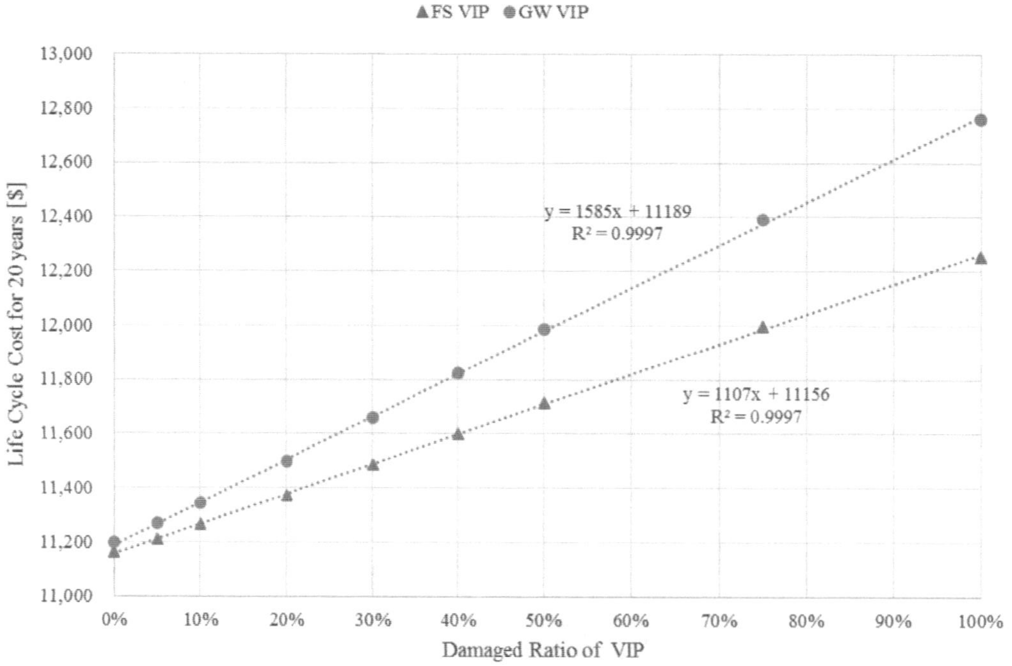

Figure 3. Life cycle cost according to damaged ratio of VIPs (useful life: 10 years).

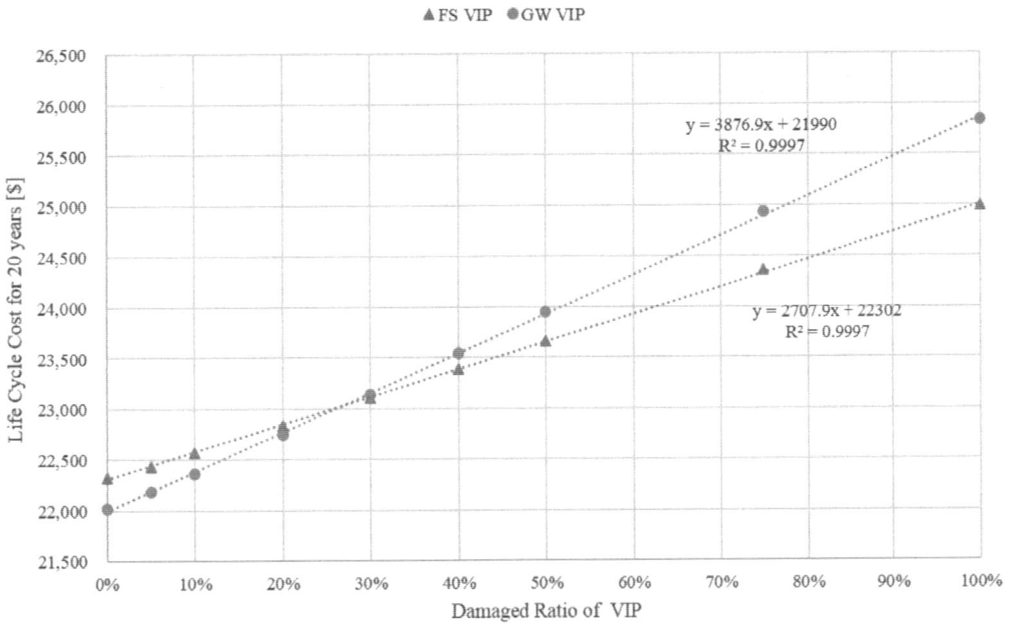

Figure 4. Life cycle cost according to damaged ratio of VIPs (useful life: 20 years).

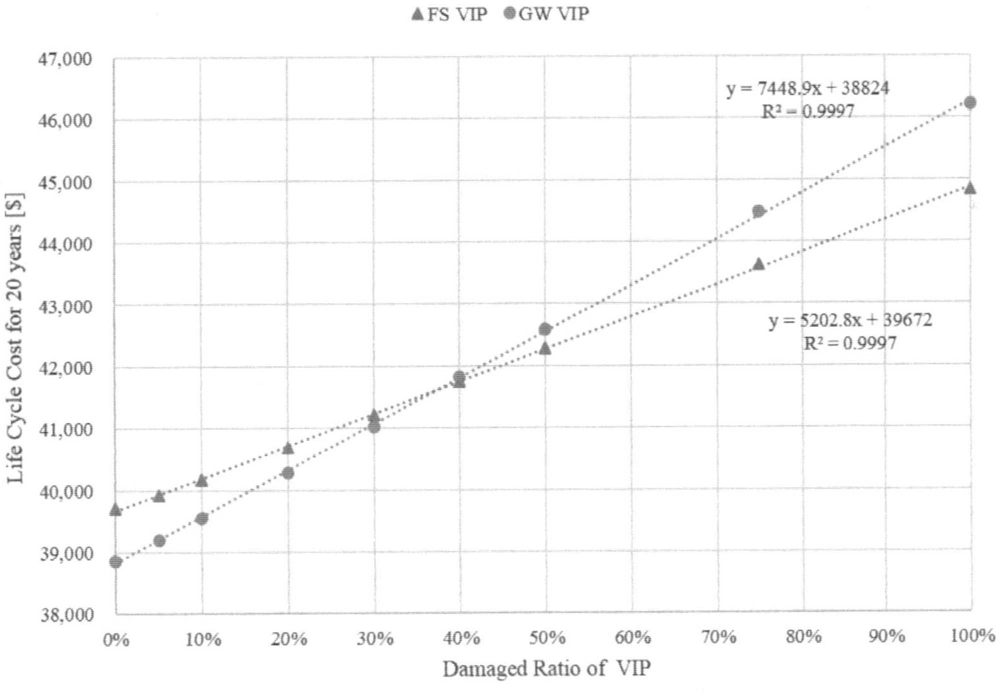

Figure 5. Life cycle cost according to damaged ratio of VIPs (useful life: 30 years).

4. Conclusions

Vacuum insulation panel (VIP) is one of most efficient insulations, however, the thermal performance of VIP is significantly influenced by the vacuum condition. This study evaluated the influence of thermal conductivity variation by destruction of vacuum condition to energy performance and economic feasibility in Korea.

- The thermal conductivity of GW VIP is better than the conductivity of FS VIP in the other

hand when the vacuum condition is destroyed, FS VIP has better thermal performance than GW VIP.
- For Korean typical apartment housing, the heating load can reduced by installation of GW VIP compared to FS VIP when the ratio damaged panels is lower than 50%.
- Economic feasibility differs according to the ratio of damaged panels and useful life. GW VIP is economic when useful life is longer and the ratio of damaged panels is lower in the other hand FS VIP is economic when useful life is shorter and the ratio of damaged panels is higher.

In the future work, the prediction of damaged ratio and the useful life will be studied to select proper VIP in the construction field

Acknowledgments
This research was supported by a grant (15AUDP-B079104-02) from Architecture & Urban Development Research Program funded by Ministry of Land, Infrastructure and Transport of Korean government.

Appendix: List of symbols

i : Discount rate, % i_n : Nominal interest rate, %

f : Inflation rate, % P_v : Present value, $

N_v : Net value, $ n : Elapsed years

References

[1] Lee S E 2010 Technology and policy trend toward zero-energy building *Architecture* **54** 52-8
[2] Airmin M et al. 2005 Vacuum insulation in the building sector *IEA/ECBCS annex* **39** 6-7
[3] Simmler H and Brunner S 2005 Vacuum insulation panels for building application basic properties, aging mechanisms and service life *Energy and Buildings* **37** 1122-31
[4] Kim J M and Song T H 2013 Vacuum insulation properties of glass wool and opacified fumed silica under variable pressing load and vacuum level *International Journal of Heat and Mass Transfer* **64** 783-91
[5] Lee H Y and Kim Y M 2014 A study on thermal performance for damaged vacuum insulation *Proceedings of Architectural Institute of Korea Autumn Annual Conference* **34** 319-20
[6] Lee S R, Jeung N G, Lee K K, Lee S H and Oh M D 2010 Life cycle cost analysis of energy saving performance for apartment *Proceedings of the SAREK Winter Annual Conference* **30** 254-9
[7] Sung U J, Cho S and Song K D 2012 An analysis on thermal performance and economic of heat recovery ventilation system integrated with window *Korean Journal of Air-Conditioning and Refrigeration Engineering* **24** 646-55
[8] Koo B K, Park S H, Lim J H, Song S Y and Ryu S R 2012 Energy performance analysis of EIFS using vacuum insulation panel *Proceedings of Architectural Institute of Korea Autumn Annual Conference* **34** 333-4
[9] ASTM c1484-10 2010 *Standard Specification for Vacuum Insulation Panels* (West Conshohocken, PA: ASTM International)
[10] Sim H, Kim J H and Kim J T 2011 A study on energy performance of standardized detached house with vacuum insulation panels *Proceedings of the KIEAE Autumn Conference* **21** 57-61
[11] Song S Y, Park S H, Koo B K, Lim J H and Ryu S R 2013 Performance evaluation of EIFS using vacuum insulation panels for passive and zero-energy apartment buildings *Journal of the AIK* **29** 219-28
[12] KS L 9016 2010 Thermal Conductivity Measurement Method of Insulation Materials (Korean Agency for Technology and Standards)
[13] Seoul City Gas 2015 Cost of city gas according to end use (www.seoulgas.co.kr)

IOP Conf. Series: Materials Science and Engineering **103** (2015) 012015 doi:10.1088/1757-899X/103/1/012015

Effect of hydrogen storage alloy on combustion properties of ammonium perchlorate /glycidylazide polymer -based propellant

G P Li, Y M Dou, C P Chai and Y J Luo[1]

School of Materials Science and Engineering, Beijing Institute of Technology, Beijing 100081, China

E-mail:yjluo@bit.edu.cn

Abstract. Hydrogen storage alloys can serve as good potential fuels for propellant design, by improving the energy and combustion properties. The influence of hydrogen storage alloy (A30) on the combustion properties of ammonium perchlorate/glycidylazide polymer (AP/GAP)-based on propellant were studied. The results showed that A30 could increase the burning rate of propellants by 29.75% and 74.78%, compared with B30 and Al. The combustion model of AP/GAP-based propellant containing different fuel was built. Firstly, A30 reduced the high decomposition temperature and promote condensed phase reaction heat of AP. Secondly, A30 deduced the burning surface temperature. Thirdly, A30 might prove the explosive heat of propellant. Therefore, A30 could greatly improve combustion properties of AP/GAP-based propellant.

1. Introduction

Raising the energy stored in a material has been the main line of the development for solid propellant technology. High formation enthalpy, low gas molecular weight and high combustion heat release of the propellant components all contribute to improvement in the propellants energy. Some important areas include the developments and applications of new energetic binders, high-energy oxidants and energetic plasticizers [1-3]. Introducing a certain amount of H into metal fuel is a feasible way to raising its energy performance. H_2 has a high combustion heat value up to 142900 kJ/kg [4]. Besides, H_2O generated from the combustion of H_2 has the smallest molecular weight among all the oxides. Therefore, propellant energy performance could be improved by using a combustion agent containing H in place of conventional metal fuel. In recent years, AlH_3[5, 6] has been received extensive attention because it can significantly improve specific impulse of the propellant, and yet it has not been applied extensively due to its poor stability and compatibility. Hydrogen storage alloy, by contrast, is a potential combustion agent because of its controlled alloy structure and adjusted performances in a wide range. A30 is a new kind of hydrogen storage alloy whose structure is MgNiB-Al and contains 70% Al and 0.98% H, wherein H exists as MgH_2 form and releases from it at 372.24℃ (DSC data, 0.1 MPa) and has a good stability [7, 8]. A30 (31525 kJ/kg) has a higher combustion heat than Al (25500 kJ/kg), so it can improve the energy level of the propellant and has a good application prospect when used instead of Al. In our work, A30 is applied into AP/GAP-based propellant and combustion

[1] Address for correspondence: Y J Luo, School of Materials Science and Engineering, Beijing Institute of Technology, Beijing 100081, China. E-mail: yjluo@bit.edu.cn.

properties are investigated, which will lay the foundation for further applications of hydrogen storage alloys in AP/GAP-based propellants.

2. Experiments

2.1. Materials and specimen
Al(13 μm), B30(B30 has the same structure with A30 but for the hydrogen storing characteristic, 16 μm) and A30 (16 μm) was used as fuels to prepare respectively AP/GAP-based propellants coded for GA0, GB30 and GA30 separately. Propellant formulation was as follows: (GAP + N100): 14%, AP: 63%, Fuel:17%, others:6%.

2.2. Equipment and experimentations

2.2.1. Burning rate. Strand burner method was adopted to test propellant burning rate at 20℃, N_2 atmosphere and 3 MPa, 5 MPa, 7 MPa, 9 MPa pressure. The burning rate at each pressure was the average value of 5 duplicate results. And pressure exponents are got according to Vieille burning rate equation $u= u_1p^n$ by line regression.

2.2.2. Condensed phase reaction performance. Condensed phase reaction performance was characterized by thermo gravimetry (TG)-differential scanning calorimetry (DSC) analyzer of TG-DSC1 type produced by METTLER company at the condition of 2.0 mg sample weight, N_2 atmosphere, 40 ml/min gas flow rate, Al_2O_3 crucible, 10 ℃/min heating rate and 0.1 MPa pressure.

2.2.3. Burning surface morphology. Steady state quenched surface sample was obtained employing thermal contact conductance method [5] in closed system under N_2 atmosphere, 5 MPa pressure and 22℃ ignition temperature. And then, quenched surface morphology is characterized by JSM-5800 type scanning electron microscope (SEM) produced by Japan.

3. Results and discussions

3.1. Burning rate
Burning rates of AP/GAP-based propellants with different fuel at 3~9 MPa and the corresponding pressure exponents (*n*) were all listed in Table 1. As could be seen in table 1, GA30 had a lower pressure exponent and the highest burning rates among three kinds of propellants. Burning rate of GA30 was increased by 29.75% and 74.78% respectively compared to GB30 and GA0 at 5 MPa. In order to find the reasons, the combustion models of AP/GAP-based propellants were studied as following.

Table 1. Burning rates and pressure exponents for three kinds of propellants.

Samples	Burning rates/(mm/s)				*n*
	3 MPa	5 MPa	7 MPa	9 MPa	
GA0	6.43(0.23)*	7.81(0.07)	8.38(0.06)	8.94(0.10)	0.2901
GB30	9.86(0.28)	10.34(0.14)	10.52(0.11)	11.56(0.32)	0.1330
GA30	12.90(0.02)	13.65(0.19)	14.91(0.13)	15.73(0.27)	0.1869

a"*" value in parentheses represents variance of the 5 sample results.

3.2. TG/DSC
Condensed phase reaction performance of propellant at steady state could be reflected by decomposition performance, so TG/DSC is a reliable mean for researching the effects of A30 on AP/GAP propellant condensed phase reactions. TG and DTG curves of 30, GB30 and GA0 were shown in figure 1 and DSC data was listed in table 2.

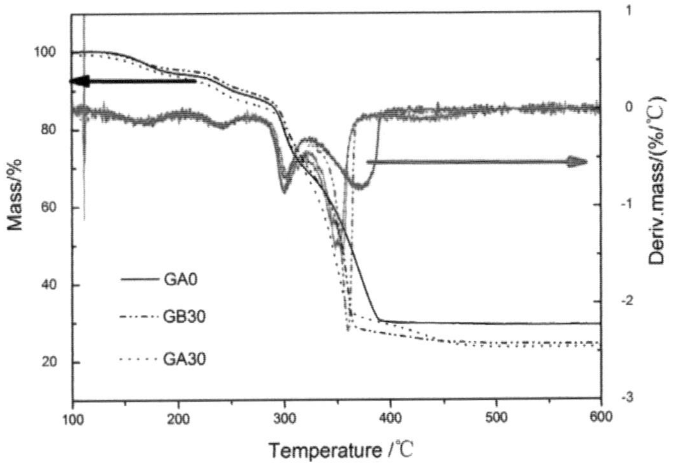

Figure 1.TG and DTG curves of GA30, GB30 and GA0.

Table 2.DSC data of GA30, GB30 and GA0.

Propellants	Heat generations/ (J/g)			Total heat generations/ (J/g)
	Secondary zone *	Tertiary and fourth zone	Fifth zone	
GA0	157.45	277.96	-	435.41
GB30	165.39	1140.62	30.05	1336.06
GA30	158.39	1369.25	28.63	1556.27

*Note: Heat generation of the secondary area is only the heat generation of GAP bind decomposition

As is seen from figure 1, the TG curve of GA0 was divided into 4 zones assigned into volatilization of the plasticizer, decomposition of GAP, decomposition of AP at lower and higher temperature respectively. Other from these, there was one another decomposition area belonging to the further decomposition of AP only in GB30 and GA30.So, B30 and A30 had little effect on decomposition of GAP. But, the effect of B30 and A30 on AP was obvious, for example, low-temperature decomposition peak (TL) of AP was the main exothermic peak for GA0, but higher-temperature decomposition peak (TH) of AP was the main exothermic peak for GB30 and GA30. Possible reasons for the additional decomposition process of AP in GB30 and GA30 was as follows: Compared to GA0, GB30 and GA30 produced a large amount of gas in the fourth area at a lower temperature as results of their intense reaction activities. The generated gases covered over AP surface to affect its decomposition performance, and when reaction active centers of AP are completely covered, low-temperature decomposition of AP terminated. At this situation, there was a part of undecomposed AP. This remaining AP would go on decomposing when the covered gases left again at a higher temperature and formed the fifth decomposition zone. If the interference is right, there will be the following conclusions: the more violent reactions of AP higher-temperature decomposition are, the more generated gases are, the stop temperature of AP higher-temperature decomposition is, and the higher of step AP decomposition temperature is. And this might be proved by GB30's lower fifth decomposition peak temperature compared to GA30.

It was shown from table 2 that B30 and A30 do not substantially affect GAP thermal decomposition heat possibly because of the low activities between fuel and GAP and the corresponding generated oxidizing gases, but could largely increase AP thermal decomposition heat generation, and total heat generations with respect to GA0 have the increase of 2.07% and 2.57% respectively, because of the high combustion heats of B30 and A30, which increase the condensed phase reaction heat generations of GB30 and GA30 at the same reaction rates between fuel and oxides.

3.3. Quenched surface morphology and residual elements

Steady state burning surfaces of GA30, GB30 and GA0 propellants at a certain pressure could be preserved by the quick flameout method shown in figure 2.

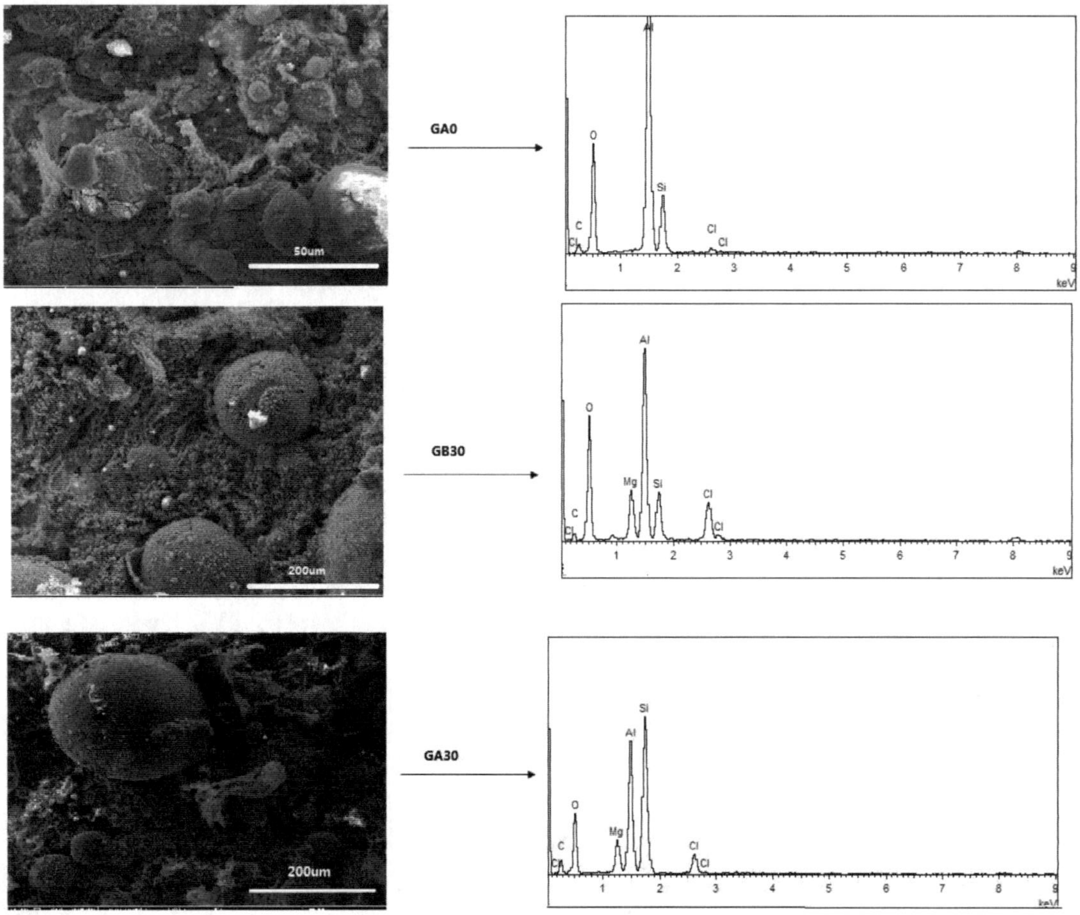

Figure 2. Quenched surface SEM images (100 times) and the corresponding EDS images of 3 propellants at 5 MPa.

GA0 quenched surface was flat and had spherical particles connecting with each other by coral-like materials. And with further amplified, it can be seen from the images (figure 2) that the spherical particles were surrounded with much smaller ones. EDS analysis showed that spherical particles are the melt and partially oxidized Al particles, and the coral-like materials are residual carbons generated from the decomposition of binder and other organic components. Similar spherical particles and coral-like materials which were alloy particles that were much larger and widely oxidized are also appear on GB30 quenched surface. At the center and upper left side of GB30 SEM image (figure 2) there were some irregular smooth large particles which were also alloy particles exist. These particles were at the transition state between small and large particles and were preserved due to rapid flameout. GA30 had the similar quenched surface with GB30, but for the very irregular surface and the less oxidized spherical particles rising above the surface. At the top left of GA30 SEM image (figure 2), there was an alloy group in maximum size, which had already been melted and got ready to leave the burning surface but was preserved by sudden flameout. Apart from this one, the sizes of other particles are all smaller than that of GB30 alloy particles. The cause of phenomena may be with the combustion process of fuel in propellant. For example, combustion of Al in composite propellants has to go through the following processes [9]: melted, accumulated, ignited and leave from burning surface. It can be known from Al ignition model [9] shown in figure 3(a) that main heats supporting the above processes of Al were supplied by diffusion flames generated from combustion of heterogeneous AP and premixed flame generated from combustion of binder. Al started to melt while the temperature is higher than its melting point (660℃), and then the neighboring Al accumulated together until ignitions happen. Diameter of Al particles in GA0 quenched surface is up to 0.3~0.4 mm which is bigger than that of raw materials, also proving the occurrence of the above processes. Furthermore, condensed phase reaction heat and gas phase reaction heat passing onto the burning surface with conduction and feedback ways would also accelerate occurrence of the whole above processes. Ignition model of B30 was established based on Al ignition model shown as figure 3(b). These models of B30 and Al were similar but for the following differences. Firstly, Mg/Al alloy in B30 made it easy to be ignited which makes GB30's burning rate higher than GA0. Secondly, B30 could form bigger alloy particles easily at a fixed temperature because of its lower melt point (652℃, DSC data). A30 ignition model was established based on B30 ignition model shown as figure 3(c). There were some new features compared with B30 ignition model. Firstly, A30 had a hydrogen release process before melted and accumulated. Secondly, A30 melting point 657℃ (DSC data) after hydrogen release was higher than B30 which is advantageous in forming little size alloy particles. Thirdly, more heat would be generated from H_2 combustion which could reduce ignition time. Finally, the size of A30 is smaller and specific surface area of A30 was bigger after hydrogen release and alloy powdered process which will be advantageous in shorting ignition time and reducing alloy sizes.

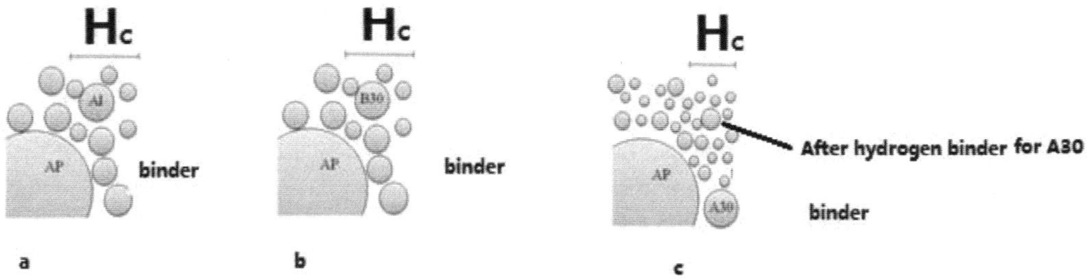

Figure 3. Ignition models of fuels in propellant. (a) GA0, (b) GB30, (c) GA30.

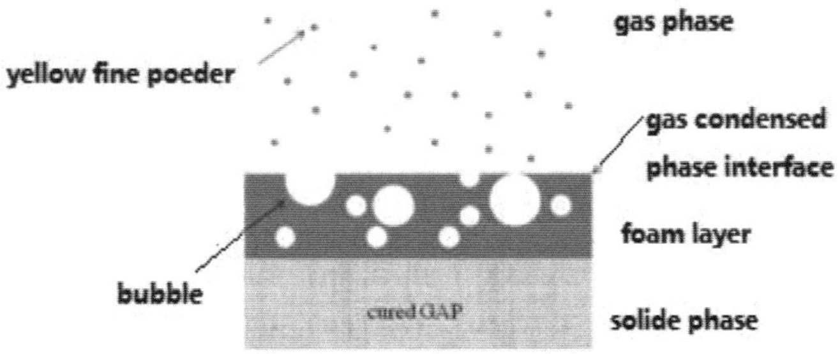

Figure 4. Combustion model of cured GAP binder in propellant.

In addition, the huge Al or alloy particles on burning surface of GA30, GB30 and GA0 were also affected by the combustion performances of GAP binder (figure 4). Cured GAP binder has a bubble zone according to its combustion model [10] shown in figure 4. GAP decomposed and generated initial gas and liquid products, then some evaporating liquid products together with gas products to form one after another bubbles in which gas-gas reactions occur. Therefore, coexistence of condensed phase and gas phase was the most important feature of this zone. And then gases were sharply produced at the burning surface of GAP binder. Further thermal decomposition and oxidation-reduction reactions were proceeding around nearer burning surface to produce huge heat. Burning surface temperature of GAP binder was about 427~527℃, lower than any burning surface temperature of GA30, GB30 and GA0 (table 3). Therefore, it can be deduced that combustion heat of GAP is largely absorbed by burning surface of the GA30, GB30 and GA0 which is positive to form huge Al or alloy particles.

Table 3. Explosion heats and burning surface temperatures of GA30, GB30 and GA0.

Propellants	Explosion heats/(kJ/kg)	Burning surface temperatures/℃
GA0	6135	738.53
GB30	6592	774.52
GA30	6604	668.67

So, the combustion model of propellant was different, when the fuel was A30, B30, Al. At the same, the fuel might affect the energy level of propellant. Seen from table 3, the explosion heats of GA30 was largest in the three kinds of propellant caused by easier ignition performance, higher combustion heat and higher combustion efficiency of A30.

4. Conclusions

It is found that alloy compositions and hydrogen storage/release properties of hydrogen storage alloy were all favored the combustion performances of AP/GAP-based propellant. Alloy compositions favored the improvement of its combustion heat and further the level of condensed phase reaction heat and explosion heat of the propellant. Catalysis of hydrogen storage alloy to AP was reflected on its influences to decomposition of AP, making high decomposition temperature of AP decreased and decomposition reactions of AP more violently. Ignition and energy performances of the propellant were improved due to hydrogen storage/release and H_2 combustion properties of A30. Ignition performance can be further improved owing to powder feature of A30, which also promotes both condensed phase and gas phase reaction efficiencies of the propellant. Easier ignition performance, higher combustion heat and higher combustion efficiency of hydrogen storage alloy were the main reasons to make AP/GAP-based propellant have much marked improvements. Besides, the released H_2 which has acceleration effect on the leave of the components at burning surface is one another reason to the improvements of combustion performance of the propellant.

Acknowledgment

The authors gratefully acknowledge the helps from Institute of Metal Research from Chinese Academy of Sciences and Xi'an Modern Chemistry Research Institute.

References

[1] Golofit Tand Zysk K 2015 Thermal decomposition properties and compatibility of CL-20 with binders HTPB, PBAN, GAP and polynimmo *J. Therm. Anal. Calorim.* **11** **9**1-9

[2] Zygmunt A, Cieslak Kand Golofit T 2014 Magnesium-an important component of high-energy composition *J. Elementol* **4** 526-37

[3] Gu Jand Wu J H 2011 Effect of lead-salt burning-rate catalyst LBC on thermal decomposition behaviors of key constituents of GAP propellan *Chinese J. Solid Rocket Tech.***34** 492-6

[4] Dou Y M, LuoY J, Li G P and Ge Z 2011 Basic performance of hydrogen storage alloy combustion agent *Chinese J. Solid Rocket Tech.***34** 760-3

[5] Weiser V, Eisenreich N, Koleczko A and Roth E 2007 On the oxidation and combustion of AlH_3 a potential fuel for rocket propellants and gasgenerators *Propel. Explos. Pyro.***32** 213-21

[6] Graetz Jand Hauback B C 2013 Recent developments in aluminum-based hydrides for hydrogen storage *MRS Bullet.***38** 473-9

[7] Dou Y M, LuoY J, Li G Pand Ge Z 2013 Investigation on combustion properties of hydrogen storage alloy/AP/HTPB propellant*Chinese J. Prop. Tech.***34**285-8

[8] Yan Q L, Li X J, Wang Y, Zhang W H and Zhao F Q 2009 Combustion mechanism of double-base propellant containing nitrogen heterocyclic nitroamines (I): The effect of heat and mass

transfer to the burning characteristics *Combust. Flame* **156** 633-41
[9] Deluca L, Maggi F, Dossi S, Weiser V, Franzin A, Gettwert V and Heintz T 2013 High-energy metal fuels for rocket propulsion: Characterization and performance *Chinese J. Explos. Propell.* **36** 1-14
[10] Makoto K 2013 Mechanical characteristics and thermal decomposition behavior of polytetrahydrofuran binder using glycerol propoxylate (Mn=260) as crosslinking agent*Propel. Explos. Pyro.***38** 366-71

IOP Conf. Series: Materials Science and Engineering **103** (2015) 012016 doi:10.1088/1757-899X/103/1/012016

Preparation of CNTs rope by electrostatic and airflow field carding with high speed rotor spinning

J F Dai[1,2,3], J F Liu[1], J T Zou[1] and Y L Dai[1]

[1] School of Science, Lanzhou University of Technology, Lanzhou 730050, China
[2] State Key Laboratory of Advanced Processing and Recycling of Non-ferrous, Lanzhou 730050, China

E-mail: Daijf@lut.cn

Abstract. The large-scale preparation of disorderly CNTs with a length larger than 3 mm using CVD method were aligned in polymer monomer airflow fields in a quartz tube with an internal diameter of 200 μm and a length of 1.5 m. The airflow aligned CNTs at the output end of the pipe connects to a copper nozzle with an electrostatic field of applied voltage 5×10^5 V/m and space length of 0.03 m, which were further realigned using via electrostatic spinning. End to end spray into the high speed rotor twisted single-stranded carbon nanotubes threads via rotor spinning technology. The essential component of this technique was the use of carbon nanotubes at a high rotory speed (200000 r/min) combined with the double twisting of filaments that were twisted together to increase the radial friction of the entire section. SEM micrography showed that carbon nanotube thread has a uniform diameter of approximately 200 μm. Its tensile strength was tested up to 2.7 Gpa, with a length of several meters.

1. Introduction

Single carbon nanotubes demonstrate high strength and excellent mechanical properties [1-3] but cannot form a macro-scale structural material. Due to its small specific density, high intensity, acid corrosion resistance and good low temperature and high performance, thermoset carbon nanotubes were twisted into an axially-aligned structure with excellent mechanical properties, forming a uniform diameter carbon nanotube fiber rope. It has promising applications as an excellent fiber material and could replace high-performance carbon fibers or military "armored Guardian" Kevlar (aramid) [4] The CNT thread has vital application prospects in the fields of advanced structural materials and functional materials; the application value is incalculable and widespread.

Researchers have made significant advancements in carbon nanotube fibers and composite materials [5, 6]. Currently, there are two primary methods to prepare CNT thread. CNTs prepared by A Lin H were achieved via a one-step floating CVD method [7], prepared by floating CVD spinning. The advantage of this method is that it began with entangled, disordered CNT powder as a raw material for processing of the fibrous material. However, with the addition of a catalyst, the impurity content of the CNTs increased, forming tiny amorphous carbon particles and other defects that may weaken its strength and contribute to the inhomogeneity of the CNT thread and decrease the tensile strength of the thread. Another method is preparing a directional growth of the CNT array, reported by

[3] Address for correspondence: J F Dai, School of Science, Lanzhou University of Technology, Lanzhou 730050, China. E-mail: Daijf@lut.cn.

Fan S S [8] Initially, this approach was modified so that CNT thread was pulled from the array and post-spin twisting was initiated by an ultra-fine probe to make a CNT thread with a high tensile strength of 1076 MPa [9]; this technique offers controllability of the spinning process, fewer impurities and stable mechanical properties of the resultant thread. The disadvantage of the method is that directional growth efficiency of CNT arrays becomes low. Prepared fiber length is affected by array size. Furthermore, low-speed twisting or small twists result in uneven diameter and low mechanical properties that limits the industrial applications. Solving uninterrupted and directional alignment problems are necessary for the preparation of long CNT threads. This paper presents a new method for preparation of CNT thread on a macro-scale with uniform diameter and excellent mechanical properties, in which carbon nanotubes are aligned along the velocity gradient field in the polymer monomer airflow and electric fields before the high speed rotor spinning and double twisting process.

2. Experiment

2.1. Preparation carbon nanotubes thread

Carbon nanotubes were prepared by enhanced chemical vapor deposition which inputs a three-channel entrance (nitrogen, hydrogen, hydrocarbon gas) at the bottom of the device. The reactor was evacuated with a vacuum pump and flushed with pure nitrogen. When the temperature of the reactor was raised to 750°C under a flowing nitrogen atmosphere, a hydrogen gas stream 150°C for heat preservation was continuously injected for 10 minutes ($Fe(C_5H_5)_2$) into the quartz tube with diameter of 50 mm and a length of 1.5 m by a precision metering pump at a rate of 0.1 mm/min. Meanwhile, thiophene was added to the catalyst precursor as a growth promoter. By operating the controller at a flow of 800 ml/min, a hydrogen: acetylene ration of 3:1 were mixed and reacted for 5 hours. Next, the system was cooled to room temperature in a nitrogen atmosphere to acquire CNTs with lengths exceeding 3 mm, as shown in figure 1. If the reaction time is too short, it will result in CNTs too short to match the rotor speed and form a thread. However, long reaction times will lead to serious problems such as wall thickness increases or the reunion phenomenon.

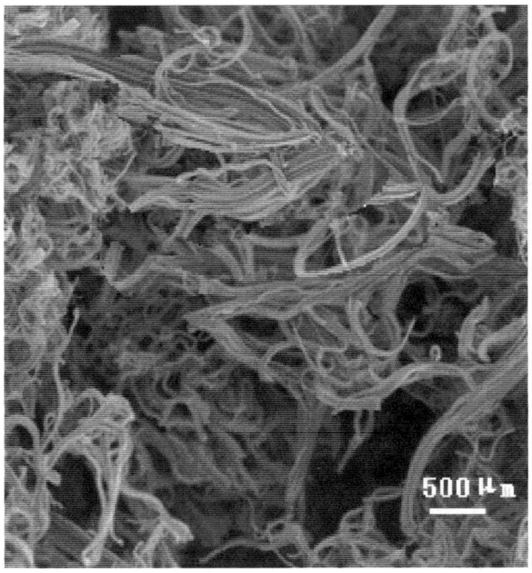

Figure 1. SEM image of CNTs prepared by CVD.

The traditional composite fabrication strict with layer and impurities of CNTs. To achieve a lot of carboxyl and carbonyl active group on the surface of CNTs and easily interface bonding with other material, CNTs was required further purified using acid mixture with V (96%H_2SO_4): V (65% HNO_3) 3:1 [10]. However, it was found that strong oxidizer will truncate CNTs and has a negative effect during the preparation of CNTs thread. It's very gratifying that there was residual chemical bond and dangling bond surface states itself during the synthesis of CNTs by CVD method. Due to the polymer monomer molecules together with the abundant surface chemical bond of it, the CNTs will obtain infrangible interface bonding bringing the CNT in closer contact to each other. Heated polymer monomer formed continuously and steadily airflow field in the quartz tube, then CNTs continuous through the quartz tube with diameter of 200 μm and a length of 1.5 m by automatic vacuum feeder in flow of 350 ml/min. Polymer flow field teased out disorderly arrangement of CNTs along the axial directional movement in capillary quartz tube, which improved the mechanical strength of CNTs. At the output quartz tube's end connects copper nozzle(150 μm of nozzle diameter) applied voltage electrostatic field of 5×10^5 V/m and space length of 3 cm. CNTs are further realigned end to end spray into the high speed rotor and twisted single stranded thread. This process solved the low mechanical strength and fiber interrupted caused by CNT arranged clutter. In comparison with the normally monofilaments, the mechanical properties of the thread were obviously improved that tensile strength increased to 150% and process equipment as shown in figure 2.

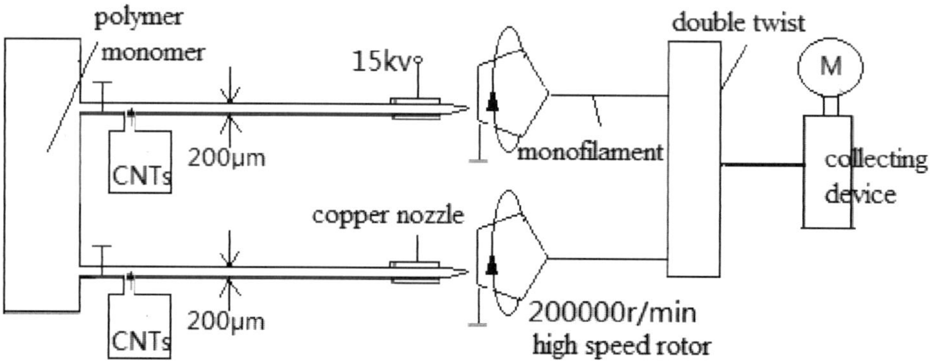

Figure 2. The flow chart of CNT thread prepared by airflow and electrostatic field combing high speed rotor spinning.

2.2. Electrostatic and airflow field carding CNTs

The monomer 4, 4 '- dichloro diphenyl ketone of Poly ether ketone (PEEK) replaces acetone and ethanol adopted by other research group [11]. Heating steam was made through high heat, the addition of polymer monomer improves the connectivity of CNTs and plays an auto-action for the preparation of super long nanotubes. Using high pressure blower (the pressure of jet cylinder piston is 0.5×10^6 Pa), the pipe were filled with high polymer airflow. Injecting CNTs into capillary quartz tube. Steam flow not only makes the directional arrangement of CNT (2 mm in length , less than 100 nm in diameter) , but also affects with carboxyl groups of ionization of CNT after being processed, which form stable chemical bonds in order to improve the cohesiveness. Then, applying high voltage (15 kv) of positive power on quartz capillary outlet, rotor grounding. Direction controlled single nanowire is obtained by high voltage electric field. The appropriate diameter ratio guarantees accelerated motion of airflow in quartz tube and reduce the fiber bending in the capillary. To further make the orientational alignment of CNT, the top of the nozzle is tangent to condensation of rotor slot. Injecting spinning into an improved German Autocoro360 suction type rotor at a speed of 5 m/s, rotor speed is 200000 r/min, the diameter of rotor is 18 mm as shown in figure 3. The increase of rotor speed makes short fiber higher tension and arranged tightly in a row. Along with the negative pressure air flow increases of the high speed forming, the continuous filament is strengthened. Since high negative pressure (7.5 kpa) can make the air draft, enhance coagulation closely, improve the quality of the silk and guarantee the stability of spinning.

2.3. The preparation of CNT thread with the double twisting

The thread is made up of fiber bundle along the whole length of the fiber or filament direction. Friction will make not loose or slippage between monofilament and monofilament after holding closely each other. The group adopted the Oerlikon Alma TCS - ML experimental machine. Spindle number is 4, spindle pitch is 600 mm. With drawing speed (8 cm/min), the two CNT fiber bundles were twisted into a diameter of less than 200 um by air electrostatic method, length up to several meters of CNT thread. The single yarn strength of traditional cotton fiber airflow spinning is 8.4 CN/tex. Under suitable twist modulus and rotor speed, twisting strength is up to 130% [12].

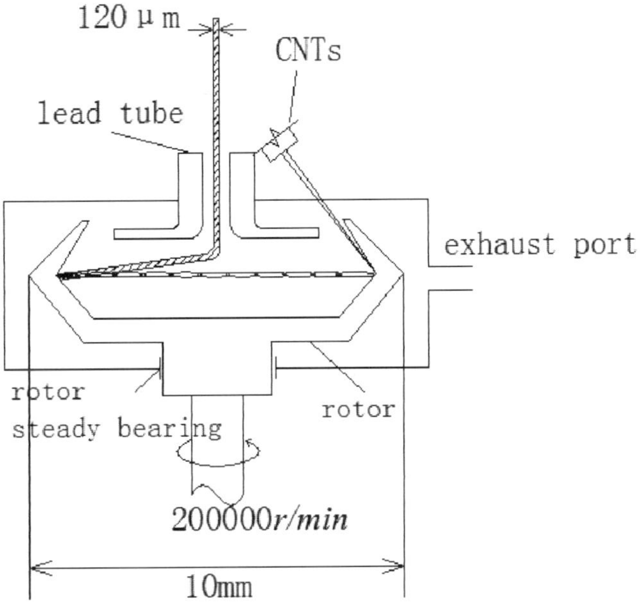

Figure 3. The schematic diagram of high speed rotor.

3. Results and discussion

3.1. The microstructure and mechanical strength

Scanning electron microscopy (SEM) analysis adopted the field emission of JSM-6700-F, which is produced by the Japanese electronics company. The preparation samples were detected by SEM, as it is shown in figure 4. It indicated that the CNT thread monofilament in diameter is about 120 μm and has good continuity, and quality line density is 30 g/km. It differs by selecting different process parameters. The mechanical properties of CNT thread are measured by INSTRON 1122 universal material testing machine, the results show that a tensile strength is up to 2.7 Gpa.

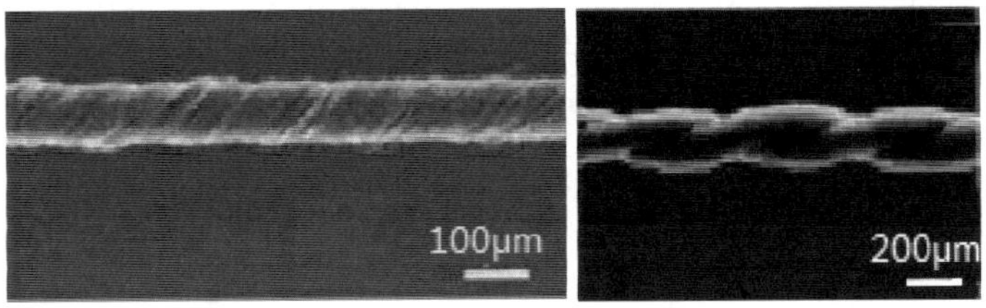

Figure 4. SEM photos of CNTs thread & double twisted CNTs rope.

3.2. CNTs orientation function of the airflow field

Based on the mass conservation law, the fluid flow steadily in the tube, liquid of per unit time flowing through a pipe of arbitrary across section are equal in quality. The continuum equation for fluid, V_1A_1 and V_2A_2 is constant. Thus the fluid flow steadily in the same pipeline, the flow rate is a constant. The smaller the cross pipe section is, the larger the flow velocity is. The author designed a new type of air ventilation with polymer monomer capillary quartz tube, the air does not only have directional effect of CNTs in the quartz tube, but also form the steady chemical bond, which improve the cohesiveness between monofilament. In conclusion, CNT has arranged the trend along the flow direction. The longer the process is, the more obvious directional trend has [13].

3.3. CNTs orientation function of the electric field

The orientation degree of CNT is in high demand of high speed rotor spinning. Single flow has a significant orientation function in carding CNTs process, what is more, adopting the electrostatic field will promote further directional of CNTs. Because the CNTs is polarized in the static electric field, bound charge is formed at both ends, single root CNT become a dipole electric in a static electric field. Electric dipole suffered the electric field force generated dipole moment M ($M = Fl\sin\alpha$), as shown in figure 5. Dipole moment makes CNTs rotated into the direction of the electric field and it also suppress that directional CNTs deviates from the direction of the electric field in the process of spinning. Experimental results show that electric field with intensity of 5×10^5 V/m can make combing CNTs oriented arrangement higher degree in the electric field. If the field voltage is too high, it can easily breakdown CNTs. If the electric field voltage is too low, then the dipole moment is not sufficient to overcome the resistance moment.

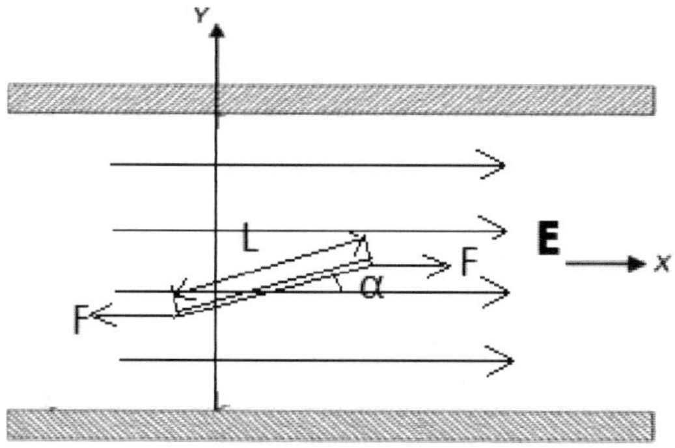

Figure 5. Illustration of CNTs in electrostatic field.

4. Conclusion

Airflow field and electric field play a crucial role in carding CNTs, along with high speed rotor spinning. The orientation, orderly and high mechanical properties of the long CNTs thread has been prepared. The technology of polymer monomer assisted that using the air induction principle make CNT oriented in the direction of the airflow in the pipe and increased the property of the bonding. Preparation of monofilament is about 120 μm, the quality density of line is 30 g/km, Two share twisting diameter of monofilament is 200 μm in diameter. Compared with monofilamentits, the mechanical strength is increased by 150% at 2.7 Gpa. A continuous supply of CNTs avoided the situation that the preparation of continuous thread cannot reel off in array due to a small array. Double twisting function as outer layers of the fiber turned into the inner layer formed inner pressure, which make the thread obtained friction along fiber radial obtain the friction and accumulate micromechanics strength of filaments, and further improve the mechanical strength of CNTs.

Acknowledgments

This material is based upon work supported by the National Natural Science Foundation of China (grant no. 50873047).

References

[1] Baughman R H, Zakhidov A and Heer W A 2002 Carbon nanotubes-The route toward applications *Science* **297** 787-92

[2] Wei F and Zhang Q 2007 Progress on aligned carbon nanotube arrays *New Carbon Materials* **22** 271-9
[3] Zhang X F and Li Q W 2007 Strong carbon-nanotube fibers spun from long carbon-nanotube arrays *Small* **3** 244-8
[4] Dai J F, Wang Q and Li W X 2007 Properties of well aligned SWNT modified poly (methyl methacrylate) nanocomposites *Materials Letters* **61** 27-9
[5] Pham G T, Park Y B and Kramer L 2008 Mechanical and electrical properties of polycarbonate nanotube buckypaper composite sheets *Nanotechnology* **19** 325705
[6] Bekyarova E, Thostenson E T and Haddon R C 2007 Multiscale carbon nanotube-carbon fiber reinforcement for advanced epoxy composites *Langmuir* **23** 3970-4
[7] Motta M, Moisala A and Windle A H 2007 High performance fibers from dog bone carbon nanotubes *Adv Mater.* **19** 3721-6
[8] Jiang K L, Li Q Q and Fan S S 2002 Nanotechnology: Spinning continuous carbon nanotube yarns *Nature* **419**(6909) 801
[9] Li S and Zhao J N 2012 Studies on preparation and properties of continuous carbon nanotube fibers spun from CNT array *Fiber Reinforced Plastics/Composites* **6** 8-13
[10] Dai J F, Zhang C and Wang Q 2008 Preparation and characterization of polymethylmethacrylate/aligned SWCNTS composites with by repeated stretching *New Carbon Materials* **23** 201-3
[11] Wang Z, Liang Z Y and Wang B 2004 Processing and property investigation of single-walled carbon nanotube (SWNT) buckypaper/epoxy resin matrix nanocomposites *Composites Part A* **35** 1225-32
[12] Lord P R and Shangguan X B 1988 Strength of friction, airflow and ring spinning twine *Textile Technology Overseas* **10** 19-20
[13] Dai Y L and Dai J F 2013 Alignment of carbon nanotubes in ultra-long carbon nanotube polymethylmethacrylate composite nanofibers by electrospinning *New Carbon Materials* **28** 101-7

IOP Conf. Series: Materials Science and Engineering **103** (2015) 012017 doi:10.1088/1757-899X/103/1/012017

A parametric design of ceramic faced composite armor subject to air weapon threats

Y N Guo[1] and Q Sun

School of Aeronautics, Northwestern Polytechnical University, Xi'an, China

E-mail: gyn@nwpu.edu.cn

Abstract. By taking into consideration the two categories of military projectile threats to aircraft structures, an optimal layer configuration of ceramic faced composite armor was designed in this paper. Using numerical simulations and the same layer arrangement of ceramic, UHMWPE, and carbon fiber laminates, a parametric finite element model using LS-DYNA code was built. Several thickness combinations were analyzed in order to determine the final lightest configuration that is capable of supporting a high-speed impact load and HEI blast wave load, which implements a high anti-penetration design for aircraft armor. This configuration can be used to improve the anti-impact ability of aircraft structures as well as achieve a structure/function integration design that considers a lighter weight.

1. Introduction

Air weapons are playing an increasingly important role in modern warfare. However, as ground air defense forces continue to develop, many countries are becoming increasing concerned with the improvement of military aircraft survivability when subjected to ground fire. To ensure safety during flights, structural integrity should be maintained, and the main airframe should have sufficient impact resistance. Lightweight armor systems can efficiently improve the maneuverability, attack ability, and survivability of aircrafts, and they have become a trend for bulletproof armor systems.

For evaluating aircraft structure degradation resulting from a projectile impact, military projectiles of concern in aircraft design can be loosely cataloged into two generic types [1]: (a) non-exploding projectiles and (b) exploding projectiles. For the non-exploding type, typical encounter condition criteria for a low-speed aircraft might be a normal impact at a velocity of 2000ft/s. However, for HEI (High Explosive Incendiary) projectiles, the aircraft damage is a combined effect of fragment impacts and blast pressures. Under fire from HEI, the aircraft should be able to survive the impact and return safely to base.

The armor structure that equips aircrafts needs to possess a good anti-impact performance as well as be able to meet certain special requests, such as aerodynamics and load transfer. Additionally, it must also satisfy weight constraints. Since the 1960s, in which it was first demonstrated that multi-layered armors present advanced strength efficiency when subjected to a high velocity impact, extensive progress has been made in this field. For instance, Wilkins [2] studied the ballistic performance of ceramic tiles backed by aluminum plates against a 7.62 mm projectile. The impact response of ceramic-composite hybrid armor was studied by Woodward using a one-dimensional

[1] Address for correspondence: Y N Guo, School of Aeronautics, Northwestern Polytechnical University, Xi'an, China. E-mail: gyn@nwpu.edu.cn.

approach [3]. Additionally, Wang and others optimized multi-layered armors in order to generate the best performance [4-6]. To improve the ballistic performance, a UHMWPE layer was incorporated with the traditional ceramic armor backed by fiber reinforcement composites, so a three layer configuration was designed in which a ceramic layer works as an energy absorption and dissipation layer, a carbon fiber layer works as a support layer and also a load transfer layer, and a UHMWPE layer works as a flexible layer to absorb the residual energy of the projectile. This configuration can effectively improve the anti-impact ability as well as achieve a structure/function integration design that considers a lesser weight.

In order to improve the anti-impact ability of aircraft structures, research into the anti-impact behaviors of composite armors subject to high-speed impact loads and blast wave loads is implemented based on the theory of impact dynamics and numerical simulation. Through kinematics analysis and ballistic characteristics analysis, an optimal layer configuration is obtained for better ballistic performance.

2. Numerical model

A parametric finite element model was built by using the finite element program LS-DYNA™ [7]. Three layer models were meshed separately and then assembled with an index file. Using the parametric tool built into LS-DYNA, the thickness of every layer could be defined, and the contact relationship could be changed accordingly when the index file calls every layer model as subroutines.

2.1. Index file

The information contained in the index file includes the following:

- Solver control options, such as solve termination time, time step control, and contact parameter options;
- Material mechanical properties of every layer and projectile;
- Contact definitions between the different layers and projectile;
- Initial velocity of the projectile. According to reference 1, vi= 609.6m/s(2000ft/s);
- Amplify factor definitions of every layer using keyword * define transform and simultaneously the location of every layer;
- Call FEA sub-model of every layer using keyword * include transform.

2.2. Sub-model

The index file defined nearly every detail, except for the model geometry information. In the sub-model, the node coordinates and element constructions were built to describe the geometric information of every layer.

The first layer, the ceramic layer, was meshed using solid elements. The central zone mesh was refined to achieve accuracy and to avoid negative volume during impact analysis. The second layer, the carbon fiber layer, was meshed using shell elements, and each element was created by mapping the above ceramic layer. The layup information, such as the thickness and angle of every lamina, was defined by applying the keyword * INTEGRATION_SHELL. Finally, the third layer, the UHMWPE layer, was also built with solid elements.

The projectile (weight of 46 g) was modeled by solid elements with a diameter of 12.7 mm and length of 64.5 mm.

The finite element models of each layer and the projectile are shown in figure 1.

The final complete finite element model is given in figure 1(d), and it indicates the three composite armor layers (ceramic 6 mm, carbon fiber 3 mm, and UHMWPE 5 mm) and the projectile.

2.3. Material model

For the ceramic, carbon fiber, and UHMWPE, three constitutive model were respectively utilized: Johnson Holmquist Ceramics, Enhanced Composite Damage, and User Defined Material Models. Additionally, the Johnson Cook constitutive model was used for the steel projectile. Each material's mechanical properties are detailed in table 1.

Table 1. Mechanical properties of materials.

Steel(Johnson Cook Model) [8]					
ρ (Kg/m^3)	G(Gpa)	K_1(Gpa)	K_2(Gpa)	K_3(Gpa)	Tm(℃)
7800	78	164	294	500	1520
Cp(J/Kg℃)	A(Gpa)	B(Gpa)	N	C	M
477	0.792	0.510	0.26	0.014	1.03
SiC(Johnson Holmquist Ceramics Model) [9]					
ρ (Kg/m^3)	G(Gpa)	A	B	C	M
3163	183	0.96	0.35	0.0	1.0
N	EPSI	T(Gpa)	SFMAX	HEL(Gpa)	PHEL(Gpa)
0.65	1.0	0.37	0.8	14.567	5.9
D1	D2	K1	K2	K3	BETA
0.48	0.48	204.785	0	0	1.0
CCF-1/5228 Carbon Fiber (Enhanced Composite Damage Model)					
ρ (Kg/m^3)	E_1^{0t} (Gpa)	E_2^0 (Gpa)	G_{12}^0 (Gpa)	G_{23}^0 (Gpa)	ν_{12}^0
1700	146	8.53	4.22	9	0.3
Y_C	Y_0	Y_C'	Y_0'	Y_S'	Y_R
4.29	1.0	1.294	1.0	5.0	1.0
$d_{\max}$	ε_i^{ft}	ε_u^{ft}	d_u^{ft}	E^{0C}	
0.97	0.001	0.005	0.97	23.4	
UHMWPE (User Defined Material Models Hybrid model) [10, 11]					
E(Gpa)	ν	μ_A (Gpa)	λ_A^{lock}	s_{Bi}	s_{Bf}
2.452	0.46	0.00921	2.83	40.0	10.0
α_B	τ_B^{base} (Gpa)	m_B	$\hat{p}$	τ_p^{base} (Gpa)	m_p (Gpa)
27.0	0.0246	9.50	200	0.008	0.0033

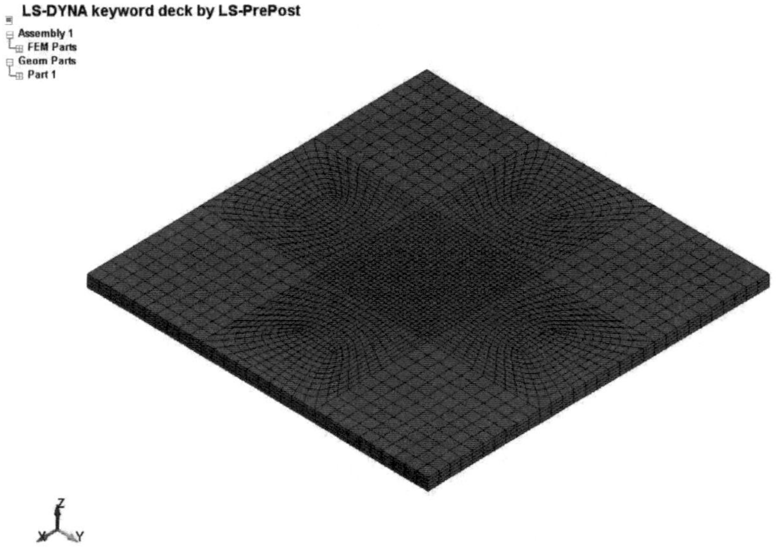

(a)

IOP Conf. Series: Materials Science and Engineering **103** (2015) 012017 doi:10.1088/1757-899X/103/1/012017

(b)

(c)

(d)

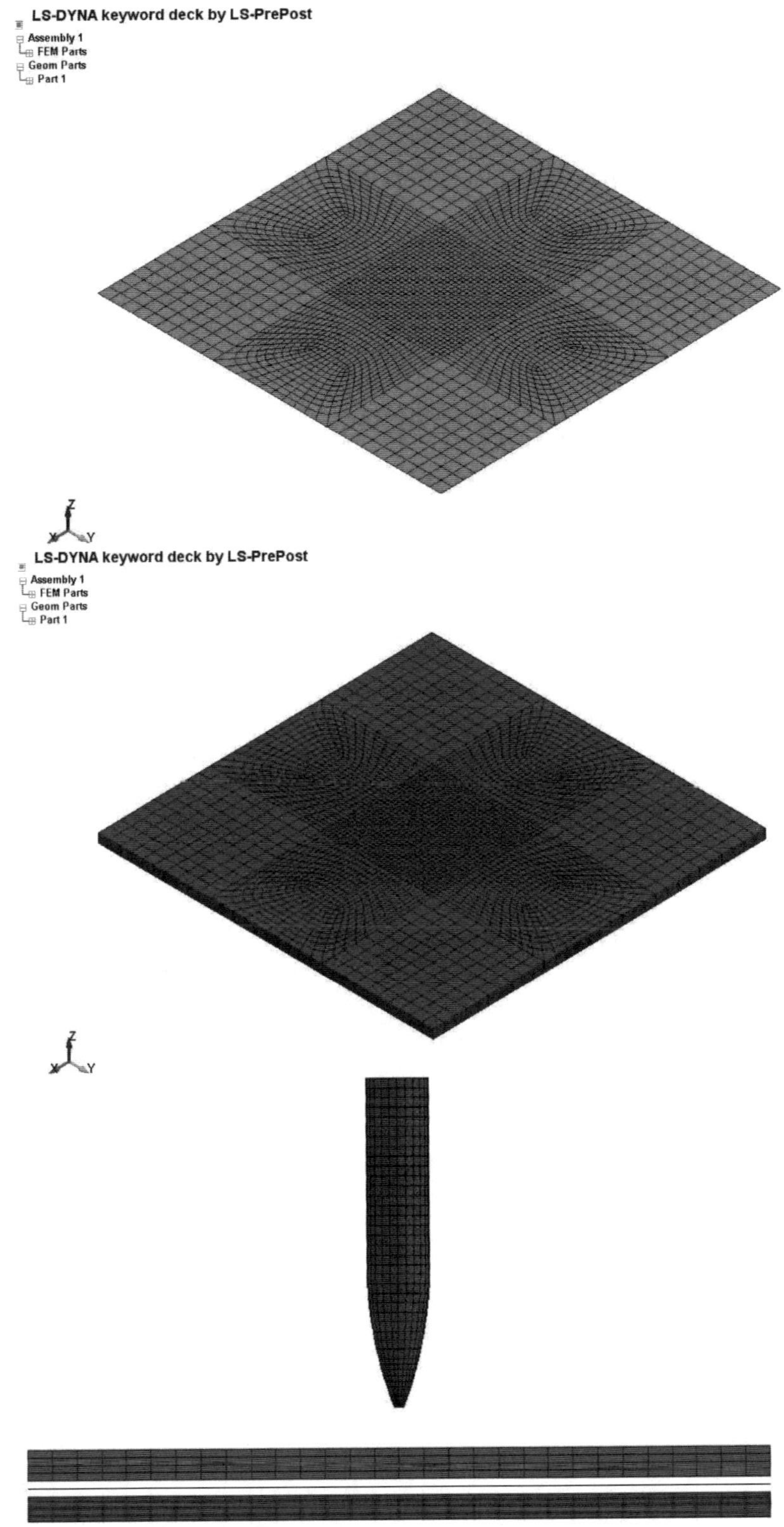

Figure 1. Finite element model. (a) Ceramic layer; (b) Carbon Fiber Layer; (c) UHMWPE Layer; (d) projectile.

3. Selective preference design for high anti-penetration armor

In designing aircrafts, the optimization method is usually the optimal means of achieving a structure that performs well but is also relatively light weight. However, armor impact analysis is very time-consuming, and the process includes some indispensable manual intervention to solve certain issues, such as negative volume, abnormal penetration, and hourglass control. Thus, it is difficult to automatically obtain the best thickness combination for different layers by using the optimization method.

A selective preference design was used, which employed the parametric model so as to intentionally change the amplification factor of every layer. After analyzing every thickness combination, the areal density and the residual kinetic energy are evaluated in order to determine the best armor construction.

Under the initial speed of 609.7 m/s (initial kinetic energy 8.55 KJ), 25 combinations of armor construction were analyzed, and the results are shown in table 2.

Table 2. Armor impact results of various thickness combination.

Armor Arrangement (mm)			Total thickness (mm)	Areal density (g/cm^2)	Residual speed (m/s)	Residual kinetic energy (KJ)	Impact results
SiC	Carbon fiber	UHMWPE					
9.24	0	0	9.24	29.20	-90.3	0.187544	rebound
3	1	11.9	15.9	22.12	413.4	3.93069	penetration
3.5	1	10.2	14.7	22.12	391.2	3.519861	penetration
4	1	8.4	13.4	22.12	388.15	3.46519	penetration
4.5	1	6.8	12.3	22.12	441.3	4.479151	penetration
5	1	5.1	11.1	22.12	436.25	4.377223	penetration
5.5	1	3.4	9.9	22.12	438.9	4.430564	penetration
6	1	1.7	8.7	22.12	439.35	4.439654	penetration
3	2	10.2	15.2	22.12	441.6	4.485243	penetration
3.5	2	8.50	14.00	22.12	430.2	4.256657	penetration
4	2	6.83	12.83	22.12	427.4	4.201427	penetration
5	2	3.44	10.44	22.12	467.0	5.016047	penetration
5.5	2	1.75	9.25	22.12	478.8	5.272737	penetration
6	2.5	4	10.25	26.60	137.9	0.437377	penetration
6.5	2.5	4.28	13.28	28.44	16.4	0.006186	penetration
7	2.5	2.58	12.08	28.44	32.3	0.023996	penetration
3	3	8.52	14.52	22.12	442.1	4.495405	penetration
3.5	3	6.83	13.33	22.12	427.4	4.201427	penetration
4	3	5.13	12.13	22.12	415.2	3.964994	penetration
4.5	3	3.44	10.94	22.12	402.9	3.733553	penetration
5	3	5	13	25.16	49.1	0.055449	penetration
6	3	3	12	26.45	-11.7	0.003148	inlay
3	4	6.84	14.84	22.12	427.3	4.199462	penetration
3.5	4	5.15	12.65	22.12	392.5	3.543294	penetration

From table 2, it can be seen that all the combinations with an areal density of 22.12 were penetrated. With increased total thickness and areal density, two armor constructions were capable of meeting the non-exploding projectile impact design request: the ceramic 6 mm/carbon fiber 3 mm/UHMWPE 3 mm and the single layer ceramic. However, the hybrid construction has a lower areal density than the does the single layer ceramic. The residual velocity of the bullet is 11.7 m/s in the negative direction, which denotes that most of the kinetic energy was absorbed by the armor. The deformation of every layer and the projectile of the preference construction are given in figure 2, and the speed-time curve of the projectile is shown in figure 3.

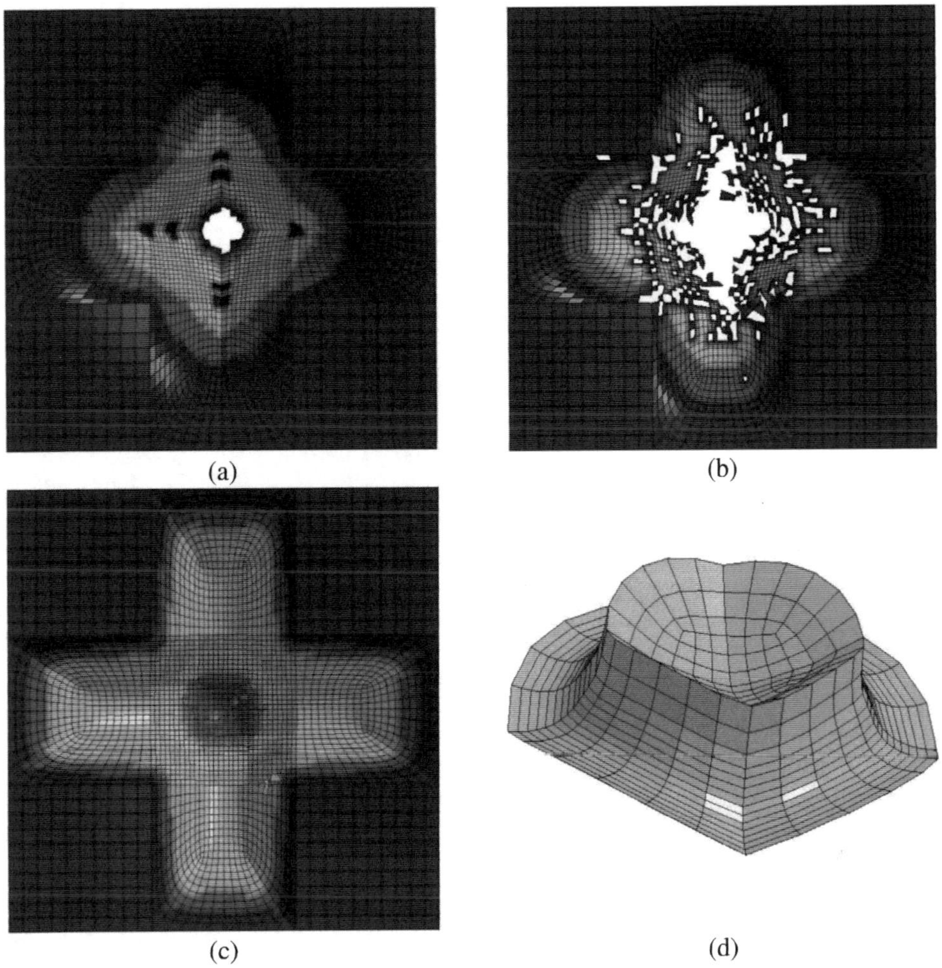

Figure 2. The deformation of every layer and projectile. (a) Deformation of ceramic layer; (b) Deformation of carbon fiber layer; (c) Deformation of UHMWPE layer; (d) Deformation of projectile.

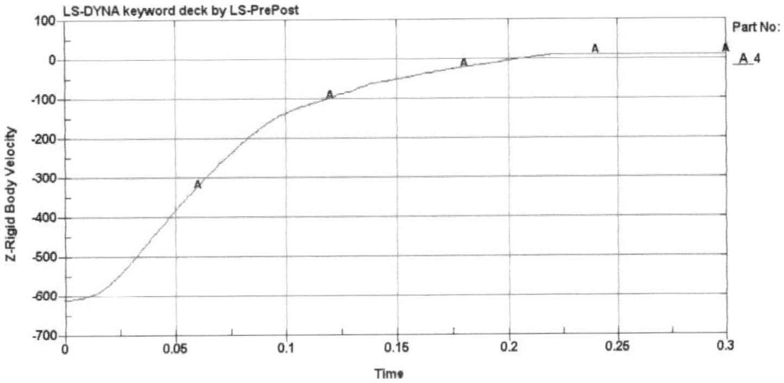

Figure 3. The speed-time curve of the projectile.

4. HEI blast wave load simulation

The armor's response when subjected to another threat, the HEI blast wave load, was analyzed based on the CJ theory of the blast wave. According to the construction built in reference [12], a finite element model was built to simulate the HEI projectile, as is shown in figure 4. The model contains four parts including the explosive, shell case, infilling, and fragments. For 23 mm HEI, 13.28 g of RDX was used as the ammunition, and the explosive process was calculated with the JWL equation of state. The EOS parameters for RDX are described in table 3.

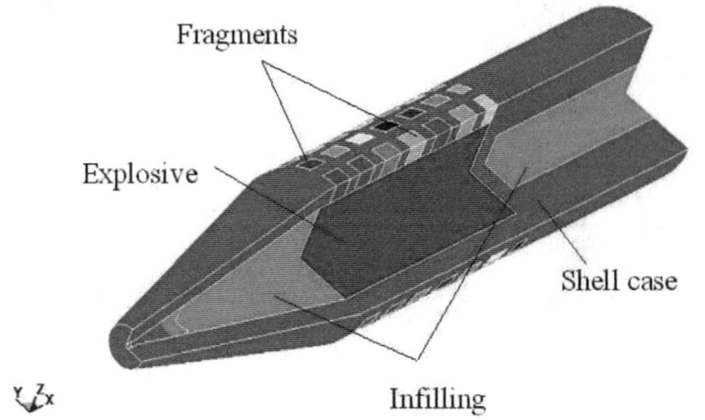

Figure 4. HEI projectile model.

Table 3. RDX explosive parameters.

component		γ	Density ρ $(g/cm)^2$	Explosive speed (m/s)	Chemical energy release (cal/g)
RDX	(CH2NNO2)3	2.70	1.65	8180	1280

Inside of the HEI projectile, a cylindrical charge (34 mm long and 17 mm in diameter) was meshed. The two-part infillings were used to simulate other concentrated masses other than the explosive, and the total weight of the infillings could be adjusted by changing the density. On the surface of the shell case, 112 rigid parts were distributed so as to simulate the fragments created during an explosion. The mass of each fragment was calculated based on the detonation parameters and the shell case dimensions detailed in reference [12]. The finite element model for each part is shown in figure 5.

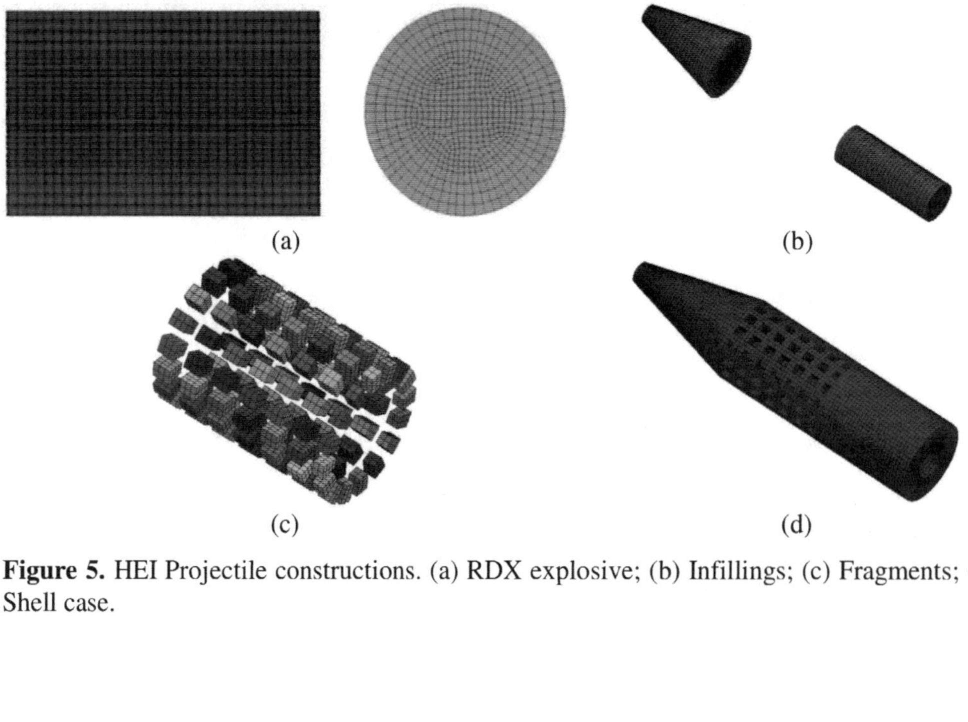

Figure 5. HEI Projectile constructions. (a) RDX explosive; (b) Infillings; (c) Fragments; (d) Shell case.

Following impact with an HEI, the armor experiences a double effect resulting from both the blast wave and the fragment impact. The projectile case would be destroyed because it reached the damage criteria after the explosion. In addition, the rigid fragments flew off at approximately 1500 m/s. The HEI explosion is shown in figure 6, and the speed of some of the fragments are given in figure 7.

Figure 6. The explosive of HEI (0.012ms).

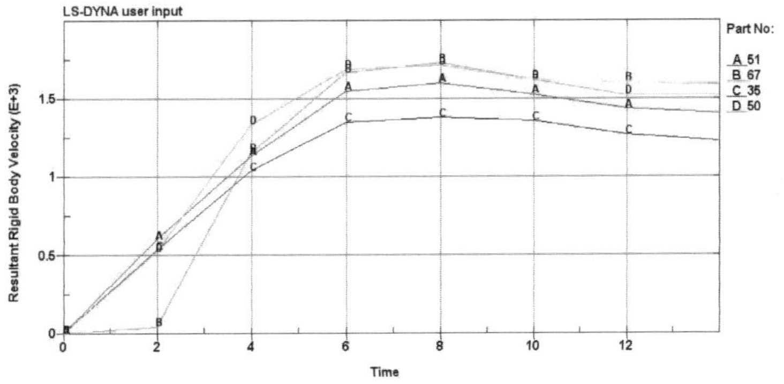

Figure 7. Speed of some fragments.

To test the construction (ceramic 6 mm/carbon fiber 3 mm/UHMWPE 3 mm) obtained in the previous section, an impact response analysis was implemented. The distance between the HEI projectile and the armor was 100 mm. The normal deformation and the maximum stress of the different layers are shown in figure 8.

Figure 8(a) provides a visual representation of the entire impact process. It can be seen that the load takes effect in 0.028 ms, and due to the high speed of the fragment as well as its short distance, no peak deformation appears due to the blast wave until the fragment impacts the armor. At 0.126 ms, under the coupling effect of the blast wave and the fragment, the UHMWPE layer has the first peak displacement, which is 0.286 mm. Then, all three layers return to the equilibrium position due to the recovery elastic force and then continue to deform backwards until 0.24 ms, which is when the ceramic layer reaches its negative peak deformation of 0.126 mm.

Figures 8(b)-8(d) give the stress contour of every layer. The data reveals that no stress exceeded the failure criteria. From the stress contour in figure 8(b), it can be seen that the maximum stress appears in the same pattern as the fragment impact, which indicates that the ceramic layer's function is high-velocity impact resistance. Meanwhile, the patterns in figures 8(c) and 8(d) give a relatively symmetrical distribution, thus indicating that blast wave absorption is the main function of both the carbon fiber layer and UHMWPE layer.

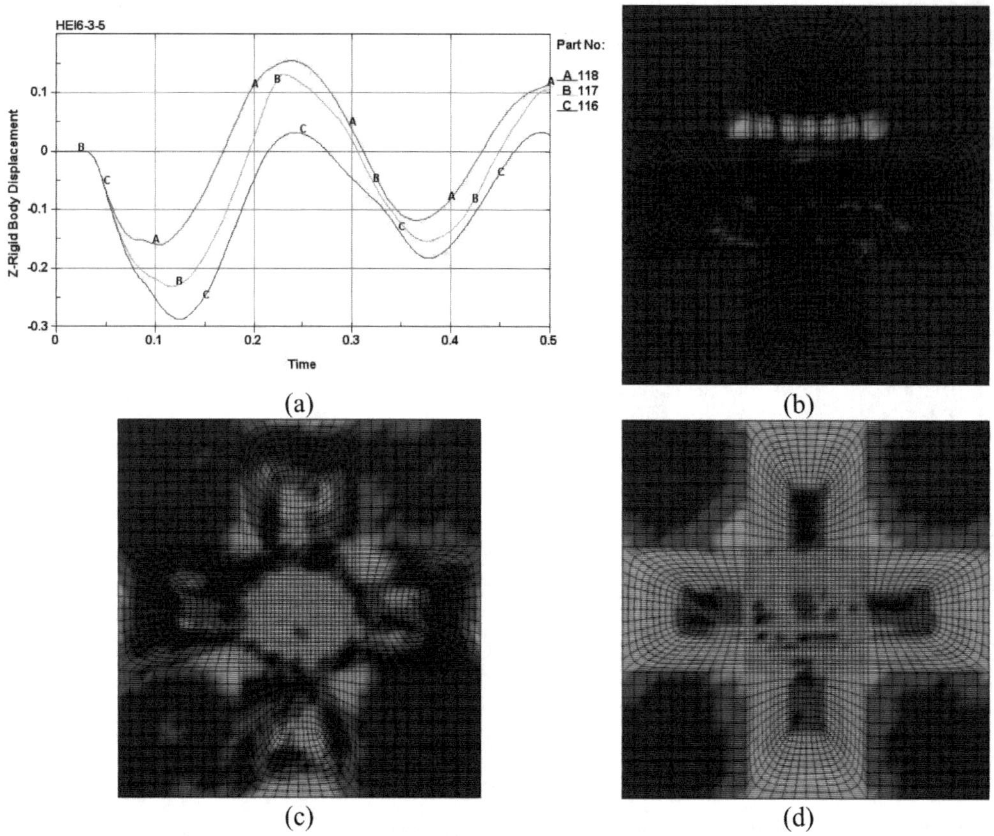

Figure 8. The normal deformation and the maximum stress of different layer. (a) The deformation of each layer(ceramic 118, carbon fiber 117, UHMWPE 116); (b) Maximum stress contour of ceramic layer (maximum stress 6.17 Gpa at 0.09 ms) ; (c) Maximum stress contour of carbon fiber layer(maximum stress 0.672 Gpa at 0.114 ms); (d) Maximum stress contour of UHMWPE layer(maximum stress 0.026 Gpa at 0.092 ms).

5. Conclusions

A parametric finite element analysis was performed in order to obtain an armor arrangement that could meet the criteria for both a non-exploding projectile and an exploding projectile. A 12.7 mm caliber bullet was modeled for a numerical test that evaluated different thickness combinations of composite armor. The best performing armor was attained by a selective preference design, and then, a numerical test with a 23 mm HEI was utilized in order to observe the armor's response to an exploding projectile. The results demonstrated that the armor arrangement is very efficient for both of the two projectile loadings. The armor arrangements and the engineering method could both be important references for improving armor structures.

Acknowledgments

This work is supported by the Fund for basic research from Northwestern Polytechnical University of China (grant no. 0100-GCKY1005).

References

[1] Avery J G 1982 Design manual for impact damage tolerant aircraft structure (AD-A 109290)

[2] Wilkins M L, Landingham R L and Honodel C A 1971 *Fifth Progress Report of Light Armor Program* (Lawrence Radiation Lab. Rept. UCRL-50980)
[3] Woodward R L 1990 A simple one-dimensional approach to modelling ceramic composite armour defeat *Int J Impact Eng.* **9** 455-74
[4] Wang B and Lu G 1996 On the optimisation of two-component plates against ballistic impact *Journal of Materials Processing Technology* **57** 141-5
[5] Park M, Yoo J and Chung D T 2005 An optimization of a multi-layered plate under ballistic impact *International Journal of Solids and Structures* **42** 123-37
[6] Ben-Dor G, Dubinsky A and Elperin T 2005 Optimization of two-component composite armor against ballistic impact *Composite Structures* **69** 89-94
[7] Livermore Software Technology Co. 2007 *ANSYS/LS-dyna Keyword User'S Manual* Vol 1
[8] Johnson G R and Cook W H 983 A constitutive model and data for metals subjected to large strains, high strain rates and high temperatures *Proceedings of the 7th International Symposium on Ballistics* (The Netherlands: Den Haag) vol 1 pp 541-543
[9] Johnson G R and Holmquist T J 1999 Response of boron carbide subjected to large strains, high strain rates, and high pressure *J. Appl. Phys* **85** 8060-73
[10] Bergstrom J S, Rimnac C M and Kurtz S M 2004 An augmented hybrid constitutive model for simulation of unloading and cyclic loading behavior of conventional and highly crosslinked UHMWPE *Biomaterials* **25** 2171-8
[11] Bergstrom J S 2006 Development and implementation of an advanced user material model for UHMWPE *9th International LS-DYNA Users Conference* (Dearborn, MI)
[12] Anderson Jr C E, Sharron T R , Walker J D and Freitas C J 1999 Simulation and analysis of a 23-mm HEI projectile hydrodynamic ram experiment *International Journal of Impact Engineering* **22** 981-97

IOP Conf. Series: Materials Science and Engineering **103** (2015) 012018 doi:10.1088/1757-899X/103/1/012018

Effect of turbine materials on power generation efficiency from free water vortex hydro power plant

P Sritram, W Treedet and R Suntivarakorn[1]

Department of Mechanical Engineering, Faculty of Engineering, Khon Kaen University, Khon Kaen, 40002, Thailand

E-mail: ratchaphon@kku.ac.th

Abstract. The objective of this research was to study the effect of turbine materials on power generation efficiency from the water free vortex hydro power plant made of steel and aluminium. These turbines consisted of five blades and were twisted with angles along the height of water. These blades were the maximum width of 45 cm. and height of 32 cm. These turbines were made and experimented for the water free vortex hydro power plant in the laboratory with the water flow rate of 0.68, 1.33, 1.61, 2.31, 2.96 and 3.63 m^3/min and an electrical load of 20, 40, 60, 80 and 100 W respectively. The experimental results were calculated to find out the torque, electric power, and electricity production efficiency. From the experiment, the results showed that the maximum power generation efficiency of steel and aluminium turbine were 33.56% and 34.79% respectively. From the result at the maximum water flow rate of 3.63 m^3/min, it was found that the torque value and electricity production efficiency of aluminium turbine was higher than that of steel turbine at the average of 8.4% and 8.14%, respectively. This result showed that light weight of water turbine can increase the torque and power generation efficiency.

1. Introduction

At the present the turbines have been developed with the different sizes and figures and with high effectives. These turbines are most important components for hydro power plant because they can transform the kinetic energy into mechanical energy of water to make turbines rotation and it results in rotation of the axis for connected generator. This process leads to electricity production. The turbines can be classified into two types consisted of Impulse turbine and Reaction turbine [1].

The water free vortex is a power seriously studied and brought for electricity production [2-4]. There is using the process of increasing the kinetic energy in the natural water flow for being higher power with throwing water into the rim of whirlpool. And the effect of water flows into pipe with the holes parched in the bottom causing the flow of free water vertex. This makes water has a higher speed from the influence of the Earth's gravity having the maximum speed of water tide on the radius of free water vertex [2-10]. According to the rule of angular momentum, more there are many radiuses of whirlpool and rate of water flow, more there are angular momentums also. When water flows into the turbine, the angular momentum of water will be transformed as torque values to rotate generator and to product the electricity [11].

According to Sujate and Ratchaphon [2], they studied Design and Development of Water Free Vortex Micro Hydro Power Plant [2-10] and found that turbines could produce the electricity using five blades made of steel twisted with angle along the height of water [2, 11]. However, this research was found the problem about the weight of steel because it must spend much power for rotation turbines. So, this research was aimed to adapt the materials for making turbines to have lighter weight and to study the effect of turbine materials on power generation efficiency from the water free vortex hydro power plant.

[1] Address for correspondence: R Suntivarakorn, Department of Mechanical Engineering, Faculty of Engineering, Khon Kaen University, Khon Kaen, 40002, Thailand. E-mail: ratchaphon@kku.ac.th

2. Theory for experimentation

2.1. Force and velocity of water

According to figure 1, the power of water could affect blades when water tides flowed against with rotating turbine. If the rapidity of water was more than the rapidity of turbine, the water power could flow against with blades affects acceleration. Then water turbines would rotate and there was relative movement between liquid and moving materials affecting to have power. The rapidity of liquid related to moving objects could be written as this following equation [12]:

$$F = \sqrt{F_x^2 + F_y^2} \quad (1)$$

$$F_x = \rho Q_r V_r (\cos\beta + 1) \quad (2)$$

$$F_y = \rho Q_r V_r (\sin\beta) \quad (3)$$

$$V_r = V_1 - V_0 \quad (4)$$

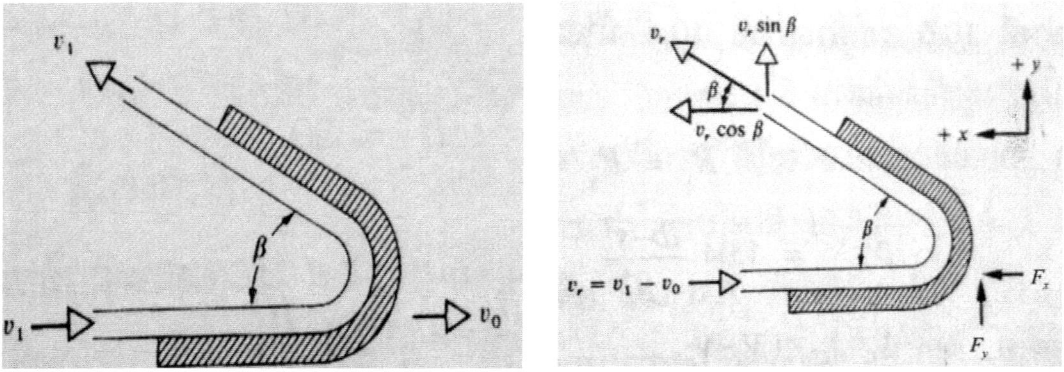

(a) Moving of Venn Plate (b) Unmoving of Venn Plate

Figure 1. Showing power of water affecting turbine [12].

When F = Force turbine
F_x = Force of flowing water affecting turbine in axis X
F_y = Force of flowing water affecting turbine in axis Y
V_r = Relative Velocity
V_1 = Velocity of flowing water against with moving materials
V_0 = Turbine velocity motion
β = Angle of water turbine

2.2. Torque and power

Torque values and power received from this energy could be calculated from this following equation [11]:

$$P_m = \rho g Q H \quad (5)$$

$$T = \frac{60 P_m}{2\pi N} \quad (6)$$

When ρ = density of water 1,000 kg/m^3
g = acceleration from earth's gravity 9.81 m^2/s
Q = water flow rate (m^3/min)
P_m = Power from hydro power (W)

T = Torque of water turbine (Nm)
N = rotation a speed (rpm)
H = Height of water (w)

2.3. Power generation efficiency (η)
The efficiency of electric system could be calculated from the rate between real powers received from electricity with mechanical power. It could be found from this equation [13]:

$$\eta = \frac{P_e}{P_m} \tag{7}$$

According to

$$P_e = EI \tag{8}$$

When E = Voltage (V)
I = Electric current (A)
P_e = Electrical power (W)

3. Instruments for experimentation and experimental methodology

3.1. Instruments for experimentation
Small generator from water free vortex hydro power plant in the laboratory according to the figure 2 consisted of four parts:

(a) The real working photo in the laboratory

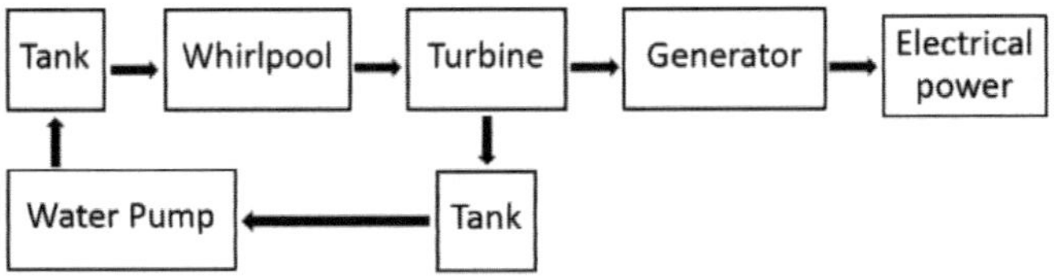

(b) Working diagram

Figure 2. System of small electricity production from the free vortex hydro power plant [3].

- A water tank size on the top with of 1.5 × 1.5 × 1.2 m. was connected to gutter.
- Vortex pond consisted of diameter 1 m. and height 1 m. The bottom of pond had water hole of diameter 0.2 m.
- Lower reservoir.
- Five water pumps could control the value of water flow rate during 0.65-3.63 m^3/min.
- The steps of experimentation consisted of using turbines made of steel and aluminium. These turbines consisted of 5 blades and were twisted angle along the height of vertex water [2]. The figures of water turbines made of steel and aluminium was shown in table 1. These turbines were the maximum width of 45 cm. and height of 32 cm. A figure of single turbine could be seen in figure 3 and construction of water turbines was in figure 4.

Table 1. A comparative property of steel (Fe) and aluminum (Al) materials.

Material	Density (kg/m^3)	Young's modulus (Pa)	Shear modulus (Pa)	Poisson Ratio	Max. stress × 10^7 Pa	Weigh of turbine (kg)
Fe	7,830	196.5	76	0.3	35	14.894
Al	2,700	68.95	26	0.33	14	6.184

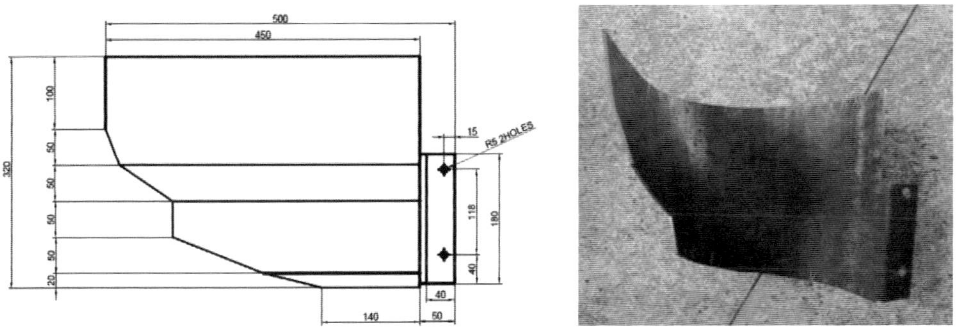

(a) A drawing form of turbine (b) Real turbine

Figure 3. Figures of water turbine for the test [2].

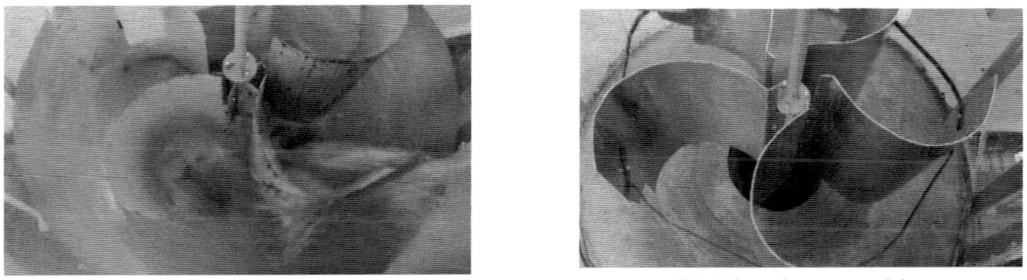

(a) Steel turbine (b) Aluminum turbine

Figure 4. Figures of turbines were set in the small generator from the free water vortex hydro power plant.

3.2. Experimental methodology

The experimental methodology consisted of using steel turbine was set in the vertex water of small generator from the free water vortex hydro power plant in the laboratory. (According to figure 2) For this experimentation, there was setting the water flow rate flowed to turbines with the water flow rate of 0.98, 1.33, 1.61, 2.31, 2.96 and 3.63 m^3/min respectively using electric load size of 5 incandescent of 20 W.

The experimentation was set the water flow rate of 0.98 m^3/min and connected to the electricity production system at size load of 20 W. After that there was measurement of water height, number of axel rotation round, generator, voltage, and electric current received from the generator system. Later there was a number of load as 40, 60, 80 and 100 W respectively and recorded the values and there was repeated experimentation changed flowing rate as they were set. After that changing blade of aluminum turbine and set it in the small generator from the water free vortex hydro power plant according to figure 4. Then there was experimentation as the case of steel turbine and recorded the values. At last brought the values received from experimentation to calculate the value of electric power, power of electric flowing and torque value of turbine and efficiency of electricity production.

4. Result of experimentation

4.1. The speed of turbines

According to figure 5, it was found that the low flow rate affected a round speed of aluminum turbine rotation much more than steel turbine. However, when there was increasing of the much load and flow rate, the round speed of turbine rotation was equal values. According to result analysis, it made us know that if there was increasing much flow rate, there was higher power of water affecting turbine. So, the different results of round speed happened from weight of turbine would be less and if the power of water affected turbines which had very high value. It could make the equal round speed of turbine rotation. The result was related with the equation of flow power affected material moving if it was set as $Q_r = Q_1 = Q_2$ and $\beta = \beta_1 = \beta_2$ when Q_1 = the water flow rate was used with steel turbine, Q_2 = the water flow rate was used with aluminum turbine, β_1 = angle in steel turbine and β_2 = angel in aluminum turbine. This equation could be written as (1), (2) and (3) as these followings:

$$\sqrt{[\rho Q_1 V_{r1}(\cos\beta_1 + 1)]^2 + [\rho Q_1 V_{r1}(\sin\beta_1)]^2} = \sqrt{[\rho Q_2 V_{r2}(\cos\beta_2 + 1)]^2 + [\rho Q_2 V_{r2}(\sin\beta_2)]^2}$$
$$(\rho Q_r)^2 (V_1 - V_{0Fe})^2[(\cos\beta + 1)^2 + (\sin\beta)^2] = (\rho Q_r)^2 (V_1 - V_{0Al})^2[(\cos\beta + 1)^2 + (\sin\beta)^2]$$
$$V_{0Fe} = V_{0Al} \quad (9)$$

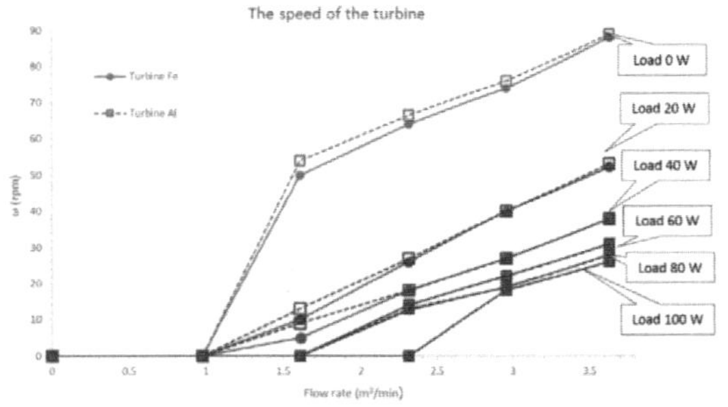

Figure 5. A comparative round speed of both turbine made of steel and aluminum.

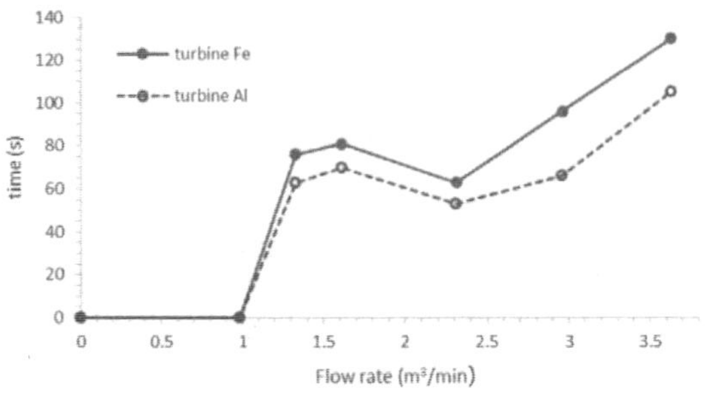

Figure 6. Time for moving of turbines from stopping points to highest speed of other flow rates.

According to the equation, it could explain that the highest speed of steel and aluminum rotation was the equal rapidity which was not depending on the materials for making turbines.

According to figure 6, the result was found that the steel turbine spent time for first moving to the highest speed much more than all flow rates of aluminum turbine. This result could analyze that the same flow rate and water power affected both turbines which were equal values. The weight of steel turbine was much more than aluminum turbine affected time which was spent for moving of turbine to the highest speed. It could spend much time than aluminum turbine. This resulted in long time for electricity production. This experimental result was related with theory of Newton's motion said that, "an object will only accelerate if there is a net or unbalanced force acting upon it". It got from this relation of $F = mg = m\Delta v\Delta t = \dot{m}\Delta v$ [12]. It was set X as fixed rate and $X \geq 1$ so, $m_1 = Xm_2$ by m_1 was mass of steel turbine and m_2 was mass of aluminum turbine and it was analyzed as the same flow rate became $\dot{m}_1 = \dot{m}_2$

$$\frac{m_1}{\Delta t_1} = \frac{m2}{\Delta t2}$$
$$\Delta t_1 = X\Delta t_2 \quad (10)$$

It was found that when Δt_1 was time for moving steel turbine which was higher speed than spent time of aluminum, so Δt_2 was related with experimentation on figure 6.

4.2. Torque values and power

According to the experimental table 2, it could be calculated to find out the torque values of turbine from equation (6) and electric power from equation (8). The water flow rate of 0.98 m^3/min could not have electric power and torque values because the water power was not enough for turbine rotation with the same load value. When the value of water flow rate increased, it would have electric power and torque values got from the both turbines could increase. This result showed the value of electric power and torque changed as the value of water flow rate. The value of electric power received from aluminum turbine was higher value than steel turbine. At the flow rate at 3.63 m^3/min, the power using aluminum turbine was between 21.14–51.62 W, while the power using steel turbine was 19.98–49.8 W. This result came from the different weight of aluminum for making turbine which had less weigh than steel turbine.

Table 2. The result of experimentation to find out the torque power and efficiency.

Flow rate	Load	H		Turbine Torque (Nm)		Water power (W)		Electrical power (W)		Efficiency	
(m^3/min)	(W)	Turbine Fe	Turbine Al	Turbine Fe	Turbine Al	Turbine Fe	Turbine Al	Turbine Fe	Turbine Al	Turbine Fe	Turbine Al
1.61	20	0.29	0.33	2.03	6.15	76.34	86.87	2.13	3.31	2.8%	3.81%
	40	0.30	0.34	1.26	5.30	78.97	89.50	0.66	1.19	0.8%	1.33%
	60	0	0	0	0	0	0	0	0	0%	0%
2.31	20	0.33	0.31	6.02	7.31	124.64	117.08	16.39	20.83	13.15%	17.79%
	40	0.34	0.34	5.17	5.41	128.41	128.41	9.20	10.20	7.16%	7.94%
	60	0.36	0.34	3.58	4.04	135.97	128.41	5.24	5.92	3.85%	4.61%
	80	0.36	0.34	2.83	3.20	135.97	128.41	3.74	4.36	2.75%	3.40%
	100	0	0	0	0	0	0	0	0	0%	0%
2.96	20	0.25	0.24	6.16	8.04	120.99	116.15	25.80	33.66	21.32%	28.98%
	40	0.29	0.27	7.19	7.87	140.35	130.67	20.49	22.26	14.60%	17.04%
	60	0.30	0.29	5.64	7.07	145.19	140.35	13.00	16.29	8.95%	11.61%
	80	0.31	0.30	4.89	6.19	150.03	145.19	9.73	12.32	6.49%	8.49%
	100	0.31	0.30	4.11	4.95	150.03	145.19	7.74	9.33	5.16%	6.47%
3.63	20	0.25	0.25	9.11	9.30	148.38	148.38	49.80	51.62	33.56%	34.79%
	40	0.30	0.31	9.83	11.45	178.05	183.99	39.12	45.57	21.97%	24.77%
	60	0.32	0.32	9.47	9.98	189.92	189.92	30.73	32.38	16.18%	17.05%
	80	0.33	0.32	8.26	9.19	195.86	189.92	24.22	26.94	12.36%	14.18%
	100	0.34	0.33	7.34	7.76	201.79	195.86	19.98	21.14	9.90%	10.79%

4.3. The total efficiency

According to figure 7, it showed the total efficiency of electricity production received from steel turbine and aluminum turbine with the highest flow rate of 3.63 m^3/min and electric power 20 W which was valuable as 33.56% and 34.79% respectively. This result showed that the total efficiency was received from the test of both turbines was very close value. The total efficiency of aluminum turbine was higher than steel turbine. Moreover, the trends of total efficiency increased when the flow rate increased as figure 8 and it was found that the electric efficiency increased when there were fewer loads.

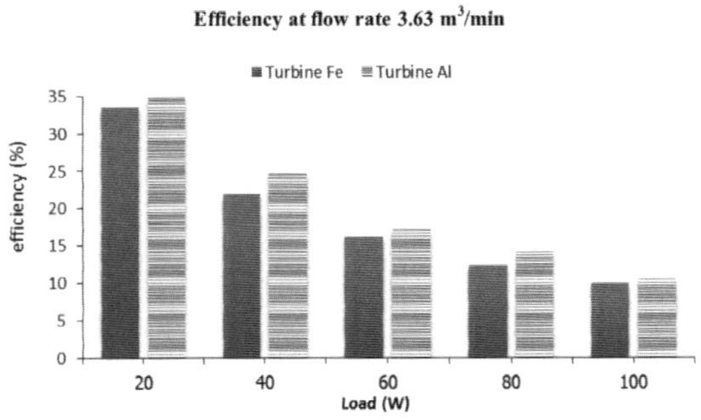

Figure 7. The total efficiency of the both turbines which had water flow rate of 3.63 m^3/min.

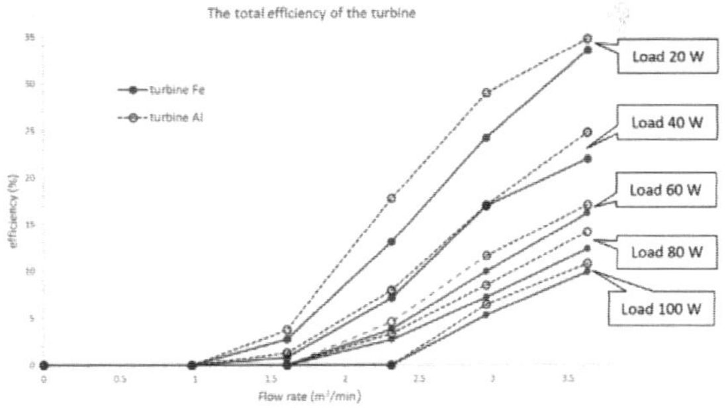

Figure 8. The total efficiency of both turbines which was different load values.

5. Conclusion

Changing materials to make light weight of water turbines with the unchanged size and figures could increase the efficiency of electricity production. According to the experimentation at the highest flow rate of 3.63 m^3/min, the maximum efficiencies of the steel and aluminum turbine were 33.56% and 34.79% respectively. It was found that the torque value and electricity production efficiency of aluminum turbine was higher than that of steel turbine at the average of 8.4% and 8.14% respectively. Moreover, aluminum turbine could spend the less time for moving of the highest speed than the equal round speed because the weight of aluminum turbine was lighter than steel turbine. So, we could conclude that the materials for making the turbines were effective for electricity production.

Acknowledgment

The authors are grateful to thank Department of Mechanical Engineering and Farm Engineering and Automation Technology Research Group, Khon Kaen University, Thailand for supporting the tools and equipments in this research.

References

[1] Natthawut J and Chirakarn S 2010 Performance testing of a vertical axis water TURBINE with friction reduction: Blade performance comparison *The 9th National Kasetsart University Kamphaena Sean Conference* (Chonburi, Thailand: University of Kasetsart) pp 121-128

[2] Sujate W 2014 Design and development of water free vortex micro hydro power plant (University of Khon Kaen, Khon Kaen, Thailand: PhD thesis)

[3] Sujate W, Ratchaphon S and Sujin W 2013 A parametric study of a gravitation vortex power plant *Advanced Materials Research* **805-806** 811-817

[4] Sujate W and Ratchaphon S 2012 *Preliminary Design of a Vortex Pool for Electrical Generation Advanced Science Letters* **13** 173-177

[5] Zhao Y Z, Gu Z L and Yu Y Z 2003 Numerical analysis of structure and evolution of free water vortex *Journal of Xi'an Jiaotong University* **37** 85-88

[6] Granger R 1966 Steady three-dimensional vortex flow *Journal Fluid Mechanics* **25** 557-576

[7] Anwar H O 1965 Flow in a free vortex *Journal of Water Power* **17** 153-161

[8] Gordon J L 1970 Vortices at vertical intakes *Water Power* **4** 137-8

[9] Reddy Y R and Pickford J A 1972 Vortices at intakes in conventional sumps *Water Power* **3** 108-109

[10] Chen Y L, Wu C and Ye M 2007 Hydraulic characteristics of vertical vortex at hydraulic intakes *Journal of Hydrodynamics* **19** 143-149

[11] Nut P and Ratchaphon S 2014 A study of Torque and hydro turbine efficiency from free vortex energy *The 15th Graduate Research Conferences* (Khon Kaen, Thailand, University of Khon Kaen) pp 273-278

[12] Komsun W 2004 *Fluid Mechanics* (Thailand: Odean stone publisher) pp 215-223

[13] Giorgio R 2011 *Principles and Applications of Electrical Engineering* 5th edn (Mc Graw Hill , New York, International Enterprise Inc.) 566

IOP Conf. Series: Materials Science and Engineering **103** (2015) 012019 doi:10.1088/1757-899X/103/1/012019

3D X-rays application for precision measurement of the cell structure of extruded polystyrene

J Y Lim[1], K Y Kim[2], H S Shin[2], S Yeom[2] and S E Lee[1,3]

[1] Building and Urban Research Institute, Korea Institute of Civil Engineering and Building Technology, South Korea
[2] Geo-technical Engineering Research Institute, Korea Institute of Civil Engineering and Building Technology, South Korea

E-mail: Selee2@kict.re.kr

Abstract. While the thermal performance of existing insulation materials have been determined by blister gases, the thermal performance of future insulation materials will be dependent on the cell size and independent foam content as we use eco-friendly blister gases with a higher thermal conductivity. However, with the current technology we are only able to guess the whole cell size and independent foam content through SEM applied 2D fragmentary scanning but are still far from the level of accurate cell structure data extraction. Under this situation, we utilized X-ray CT scanned 3D images to identify and shape the cell structure and proposed a method of inferring the whole distribution and independent foam content as accurately as possible. According to X-ray CT scanning images and SEM images, the shape was similar but according to tracer applied CT scanning images, the cell size distribution was 380~400 μm within the range of the general insulation diameter distribution which had the highest reliability. As for extrusion foaming polystyrene, we need additional image processing to identify the independent foam content as its density is too low. So, it is recommended to raise the 3D cell structure completeness of XPS by improving the scanning accuracy.

1. Introduction

In 1992, Korea officially signed the Montreal Protocol as a developing country and the ozone layer depleting substance reduction schedule was fixed separately for advanced countries and developing countries, starting from the substances having the ODP index of 1.0. In case of CFC compound, according to the total abolition schedule by the end of 2009, the extruded polystyrene industry of Korea began CFC-12 reduction and total abolition from 1996 for the first time in the domestic industry but started to use HCFC foaming agents, a potential substitute under the Protocol [1].

As for HCFC foaming agents, the average amount of production and consumption in Korea were fixed in 2009 and in 2010 respectively, and actual reduction of them is scheduled to commence from the beginning of 2015 [2]. Previously, Korea adopted the quota system in 2014 and has executed the reduction schedule according to the agreement of staged HCFC compound reduction, by 35% until 2020, by 67.5% until 2025, and total abolition until 2030.

Under the Montreal Protocol reduction schedule, Korea is obliged to complete total abolition by 2030, starting from 10% reduction in 2015, so we are in urgent need of new XPS insulation

[3] Address for correspondence: S E Lee, Building and Urban Research Institute, Korea Institute of Civil Engineering and Building Technology, South Korea. E-mail: Selee2@kict.re.kr.

development based on eco-friendly foaming agent while we are maintaining the existing thermal conductivity of XPS or securing better thermal performance [3].

To use blister gases having a higher thermal conductivity as an alternative, we must not only have a discussion on the method of upgrading the insulation performance to enter the existing insulation market but also study the cell size and independent foam (Closed Cell) whose thermal conductivity is calculated by the Knudsen Effect theory. Accordingly, in this paper we observed the insulation cell structure using X-ray applied 3D and compared the result with the existing 2D scanning images in order to study the insulation X-ray scanning method which is used for the observation of cell size and independent foam.

2. Relation between cell structure and thermal conductivity

Knudsen Effect, the basic foundation concept of Aerogel which is now taking the center stage as the next generation insulation says the gas based thermal conductivity will significantly decrease due to lower adiabatic gas liquidity in a foam if the foam size is below 70 μm [4]. In other words, if we reduce the foam size to the nanometer level, we are able to reduce the thermal conductivity innovatively thanks to radiant heat, gas convection, and conduction. As a result, total thermal conductivity of the polymeric foam (λ_{total}) can be expressed in equation (1) as the sum of the three contributions: thermal conductivity through gas phase (λ_{gas}) and solid phase (λ_{solid}), and radiation (λ_{rad}):

$$\lambda_{total} = \lambda_{gas} + \lambda_{solid} + \lambda_{rad} \tag{1}$$

The plastic foam insulation has a thermal conductivity of about 0.03 W/m·K. According to the Almanza study, the thermal conductivity of low-density Polyolefin foam whose cell size exceeds 100μm is known to be determined by heat radiation (20~40%), the thermal conductivity of plastic (10~15%), and the adiabaticity of blister gases injected in the foamed plastic (40~65%) [5, 6]. However, as shown on Fig 1, we can see its insulation performance is lower than that of R134a in HCFC series, the existing foaming agent because it has a higher temperature based thermal conductivity of CO_2 [3].

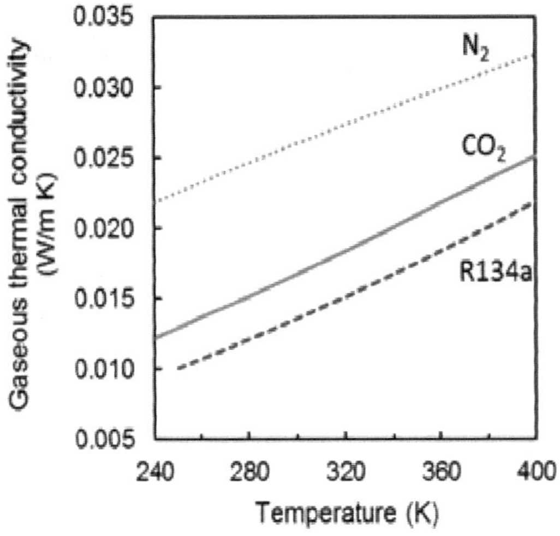

Figure 1. Gaseous thermal conductivity [1].

As thermo-plastic resin foaming is greatly dependent on the thermal conductivity of foaming agent (blister gas), we need the technology of lowering the ultimate thermal conductivity by reducing the cell size for commercialization because the thermal conductivity tends to increase when we replace the existing HCFC blister gas with blister gases having a higher thermal conductivity, for example, CO_2 gas.

3. 3D X-ray application for organic insulation cell structure measurement

While the thermal performance of existing insulations have been determined by blister gases, the thermal performance of future insulations will be dependent on the cell size and independent foam content as we use eco-friendly blister gases with a higher thermal conductivity. However, with the current technology we are only able to guess the whole cell size and independent foam content through SEM applied 2D fragmentary scanning as showing Fig2 but still distant from the level of accurate cell structure data extraction. To keep up with the tendency of NIMs development for insulation, we utilized X-ray scanned 3D images to identify and shape the cell structure to propose the method of calculating the whole cell size distribution and independent foam content as accurately as possible.

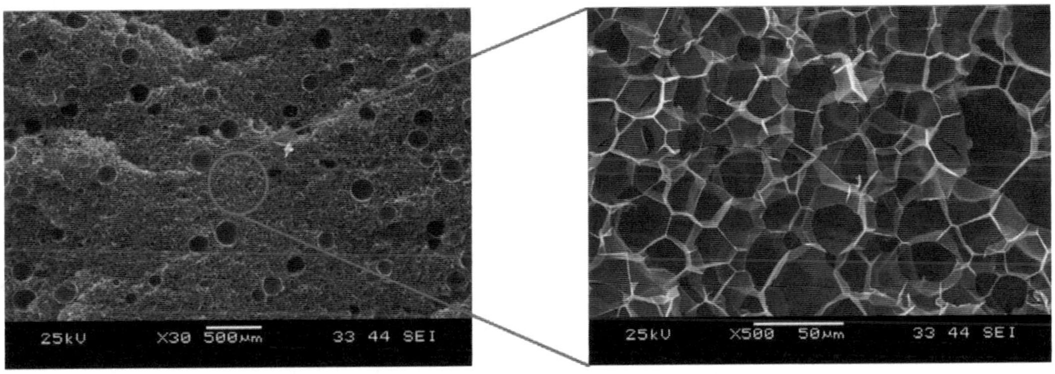

Figure 2. Foam morphology of PS/CNT 0.25 wt%.

The X-ray CT equipment located at Korea Institute of Building & Civil Engineering Institute was implemented to analyse the cell structure of samples in terms of cell size distribution. It is capable of attaining high resolution up to 3.5 μm 3 and the maximum voltage and current are 225 kV and 3.0 mA, respectively. Although the cell size of sample is expected to be large enough for the CT system can identify, it is very restricted to separate each cell structure from the CT image due to relatively thin walled structure of polystyrene which is hard to be identifiable in CT image. We contrived new method to shape and size the cell structure by infiltrating a tracer, i.e., potassium iodide, into the cell structure, because the tracer can be well attenuated as the X-rays pass through it.

Each 2D sliced image was binarized to separate the cells filled with the tracer followed by consequential stacking along the height. However, it does not necessarily ensure the identification of individual cells because the boundaries between cells were relatively fuzzy. Therefore, we deployed the relevant image processing techniques to separate them. The reconstructed cells image is illustrated in Fig 3 and Fig 4 shows 3D patch and isosurface of cell structure extracted from CT image analysis.

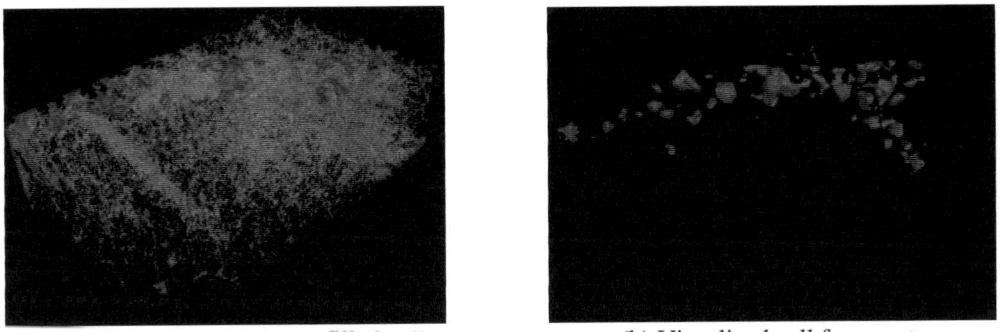

(a) 3D CT image with tracer filled cells (b) Visualized cell fragments

Figure 3. 3D X-ray CT images acquired from tracer infiltrated specimen.

(a) Isosurface (b) patch

Figure 4. Illustration of identified cell example from CT image.

The volume of individual cell fragments is calculated by voxel counting and the cell size could be obtained by converting it into a diameter of equivalent volume of sphere (Fig 5). The average cell size is estimated to be approximately 384 μm which corresponds to the range (380 ~ 400 μm) of the cell size calculated from SEM image.

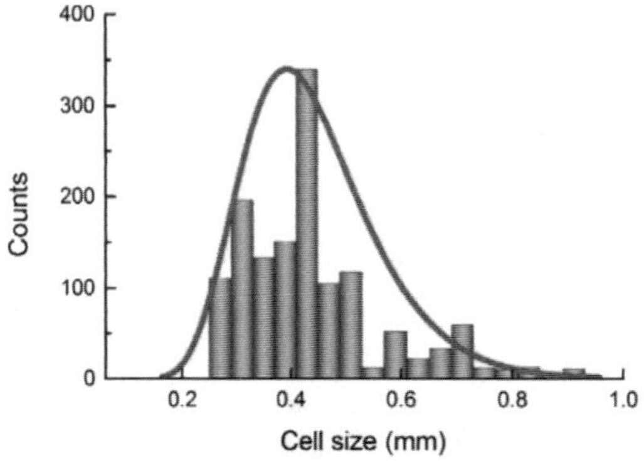

Figure 5. Cell size distribution calculated as a diameter of equivalent volume of sphere.

4. Conclusion

This study investigated the acquisition of size for 3D cell fragments in extruded polystyrene using X-ray computed tomography combined with a newly contrived method of using a tracer and image processing techniques.

The size distribution of cell fragments follows a log-normal distribution and corresponds well to measured SEM data. According to X-ray CT scanning images and SEM images, the shape was similar but according to tracer applied CT scanning images, the cell size distribution was 380~400 μm within the range of the general insulation diameter distribution which had the highest reliability. As for extrusion foaming polystyrene, we need additional image processing to identify the independent foam content as its density is too low. So, it is recommended to raise the 3D cell structure completeness of XPS by improving the scanning accuracy. This stereological analysis of particulate objects has substantial potentials in estimating size or shape dependent parameters for building materials.

Acknowledgments

This study was conducted with the support of the "Development of 0.02 W/m·K micro cellular organic insulation materials through CO_2 foaming" project, the energy and fusion resource, energy technology development project of Korea Institute of Energy Technology Evaluation and Planning. This research team acknowledges their financial support for our project.

References

[1] Lim J Y, Choi G S, Kang J S and Lee S E 2014 Construction insulation materials domestic and foreign trends and developmnet on CO_2 foaming plastic insulation materials *Conference of*

SAREK **2014** 656-9

[2] Korea Institute of Construction Technology 2013 GHG reduction method pool development on building and transportation fields (KEMCO)

[3] Lim J Y, Lee S E, Choi G S, Buahom P and Park C B 2015 Trends of non-CFC insulation materials and thermal characteristics of CO_2 foamed plastic insulation materials Materials Research Innovations **19** 1041-3

[4] Tenpierik M J, Cauberg J J M and Thorsell T 2006 Integrating vacuum insulation panels in building constructions: An integral perspective *Construction Innovation: Information, Process, Management* **7** 38-53

[5] O. Almanza, Rodriguez-Perez M A and De Saja J A 2000 Applicability of the transient plane source method to measure the thermal conductivity of low-density polyethylene foams *J Polymer Science: Part B: Polymer Physics* **38** 993-1004

[6] Yoon S H, Lee J C and Lee S H 2013 Study on the physical properties of CO_2 as a blowing agent utilized insulation Journal of Architectural Institute of Korea **29** 59-66

IOP Conf. Series: Materials Science and Engineering **103** (2015) 012020 doi:10.1088/1757-899X/103/1/012020

Effect of tool eccentricity on surface periodic banded structures in friction stir welding

N Guo[1,4], M R Wang[2], Q Meng[3], L Zhou[1] and D Y Tang[2]

[1] Shandong Provincial Key Laboratory of Special Welding Technology, Harbin Institute of Technology at Weihai, Weihai 264209, China
[2] Department of Chemistry, School of Science, Harbin Institute of Technology, Harbin 150001, China
[3] Beijing Aeronautical Manufacturing Technology Research Institute, Beijing 100024, China

E-mail: gn21c@126.com

Abstract. This paper describes the relationship between tool eccentricity and surface formation of periodic banded structures in friction stir welding. Motion characteristics of welding tool are calculated to explore the forming mechanism of banded structures. The results reveal that the welding tool motion differences on advancing side and retreating side caused by eccentricity are crucial for the formation of banded structures. The crests and troughs of banded structures form during tool motion on retreating side and advancing side, respectively.

1. Introduction

As known, the surface structures and internal microstructures of friction stir welding (FSW) joints show a spatial periodic distribution characteristic, and the space of periodic structures always equals to the distance traveled by the welding tool in one revolution [1, 2]. The surface periodic structures display as semi-circle strips packed closely along welding line, and there is a very obvious sense of crests and troughs. The internal periodic microstructure presents as dark and light strips, which commonly calls “onion ring”. Yang, et al. [3] and Reynolds [4] reported that the grain size of light strips was larger than that of dark strips. Furthermore, texture characteristic and second phase particles distribution are also different in those areas [3, 5], and all of these can affect the fracture behavior of FSW joints [6-8].

Although the periodic phenomena in FSW are widespread and easy to observe, the causes and mechanism are not known very clearly yet. At present, studies suggest some completely different views about the formation mechanism of periodic phenomena in FSW: banded structures are formed by (1) the effect of pin thread on metal flow [9, 10]; (2) a repeating occurrence of the tool stop traveling for a very short time to make material plastify [2]; (3) a sucking-extruding mechanism [11]; (4) a cooler walls mechanism [12]. However, theories above cannot clearly explain the common phenomenon that the periods of banded structures always correspond with tool rotation and same shapes of banded structures are formed on the surface of joints whatever the tool has pin or not [13, 14]. Moreover, for common FSW structures, welding tools have an axisymmetric geometrical shape,

[4] Address for correspondence: N Guo, Shandong Provincial Key Laboratory of Special Welding Technology, Harbin Institute of Technology at Weihai, Weihai 264209, China. E-mail: gn21c@126.com.

and welding tracks are straight lines. Theoretically, banded structures are not able to be formed in the welding process.

In recent years, some researchers pointed out that the tool eccentricity may play a key role in periodic metal flow of FSW. Gratecap, et al. [15] proposed that the welding tool could present a periodic up-and-down motion in the vertical direction under the combination effect of tilt angle and eccentricity. Recent results contributed to understanding FSW, but some conclusions lacked enough evidence, and theoretical defects also existed. Therefore, in this study, microstructures observation and mathematical calculation have been carried out, and the forming process of banded structures on the surface of joints has been discussed systematically.

2. Experimental procedure

The base material used in present work is 7 mm thick commercially pure lead rolled plate with lead amount more than 99.9% (wt.%). The rotational speed of 60 revolutions per minute (rpm), traveling speed of 60 mm s^{-1}, the tilt angle of 2.5° and counterclockwise of the sense of rotation are used during the welding process. An H13 steel pinless welding tool with double ring concave and shoulder with the diameter of 11.5 mm is used. Two eccentricity values are designed: (1) eccentricity of 60 μm made by the welding system spontaneously when the welding tool is installed directly; (2) eccentricity of 400 μm, by using a welding tool which has been grinded the handle appropriately.

FSW equipment applied in this study is modified from an XK5032 computer numerical control milling machine. 15 mm × 15 mm samples crossing the joints are cut by electro-discharge cutting machine (EDCM), and well cleaned by using anhydrous alcohol. The glue is applied evenly over the banded structures for several times in order to prevent further deformation. The stir zone is cut along the longitudinal section and carefully polished using 2000-grit SiC paper, the micro appearance of banded structures is observed by Axio Imager A1m optical microscope.

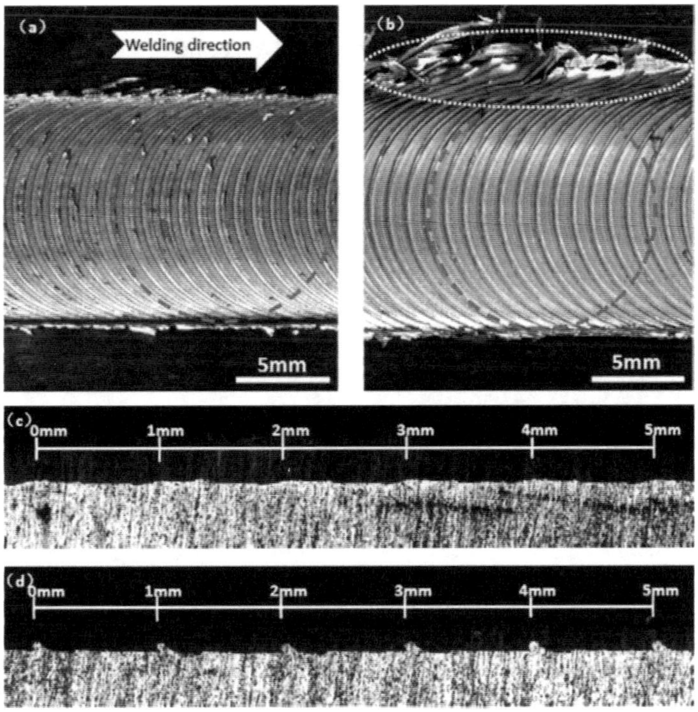

Figure 1. Surface banded structures on welds made with eccentricity of (a) 60 μm and (b) 400 μm; longitudinal cross-section with eccentricity of (c) 60 μm and (d) 400 μm.

3. Results and discussion

Figures 1(a) and 1(b) present the images of banded structures on joints surface with tool eccentricity of 60 μm and 400 μm, respectively. By comparing the banded structures with dash circles in figures 1(a) and 1(b), it can be observed that the banded structures are more obvious with the eccentricity of 400 μm. Moreover, in the dotted oval area shown in figure 1(b), a string of close packing bands appear on the surface of the flash on retreating side (RS), which connect with the surface bands one by one. It indicates that the forming mechanisms of the two kinds of bands are same.

Longitudinal section images of banded structures are shown in figures 1(c) and 1(d). When the eccentricity is 60 μm, the longitudinal cross-section profile of banded structures appears as a gradual periodic curve, with the average height of 34 μm between crests and troughs. But when the eccentricity is 400 μm, the average height is increased to 82 μm. Furthermore, spaces of each wave in figures 1(c) and 1(d) are both equal to 1 mm accurately. The remarkable different morphologies of banded structures only can be caused by variation of tool eccentricity.

The eccentric rotating and traveling process of the tool in FSW is illustrated in figure 2. The outer circle and inner circle with dash line are the motion trail of tool's edge and center, respectively. The velocity of tool movement along welding direction is defined as v that is the sum of v_1 and v_2, where v_1 is the feeding speed, v_2 is velocity component along welding direction only caused by eccentric rotation. One revolution of the tool can be divided into two stages: tool center O near advancing side (AS) and near RS. In the former case, the directions of v_1 and v_2 are same, but in the latter case, directions of v_1 and v_2 are opposite.

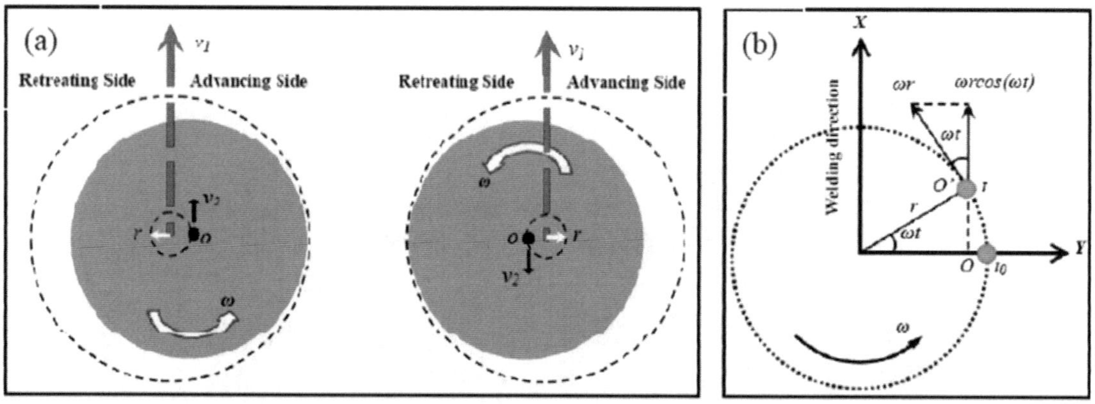

Figure 2. Eccentric rotating and traveling process of tool in FSW: (a) schematic of traveling process; (b) relationship between the angular velocity, eccentricity and tool velocity components in welding direction.

According to figure 2(b), v_2 can be expressed as equation (1), where ω is the angular velocity of tool, r is eccentricity, and α is tilt angle of the tool.

$$v_2 = \omega r \cos\alpha \cos(\omega t) \quad (1)$$

Considering that α is small, and then v can be defined as

$$v = v_1 + \omega r \cos(\omega t) \quad (2)$$

According to equation (2), the curves of tool velocity along welding direction changing with time under different eccentricities are plotted in figure 3. It can be clearly seen that tool velocity shows significant cyclical variation. However, the frequency of fluctuation is 1 s, which well coincides with the period of banded structures in figure 1. In figure 3(a), the maximum speed reaches 1.38 mm s^{-1}, and the minimum speed is 0.62 mm s^{-1}. The changing amplitude is 0.76 mm s^{-1}, which is a relatively high value compared with preset velocity. The eccentricity of 60 μm is fairly small for current FSW equipment, but it has already made a significant effect on tool motion. With the eccentricity of 400 μm, the maximum speed is 3.51 mm s^{-1}, and the minimum speed is -1.51 mm s^{-1}. The changing amplitude reaches 5.02 mm s^{-1}, which is almost five times of preset velocity. Meanwhile, it indicates that if the eccentricity and angular velocity of tool reach an enough high value, the velocity will be negative. It means tool moves towards the reverse direction of welding for a period.

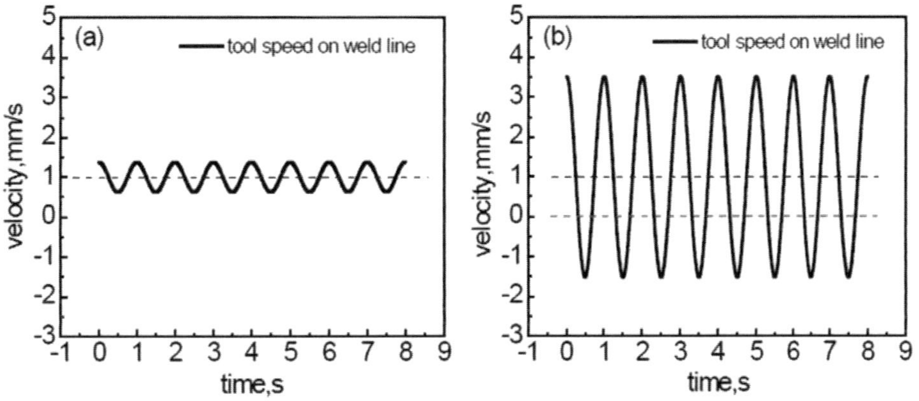

Figure 3. Velocity varied with time under different eccentricities (a) 60 μm and (b) 400 μm.

Under the condition of FSW tool rotating with eccentricity, feeding length of tool center on AS is longer than that on RS. In another word, on unit welding length, the friction time of AS is shorter than that of RS. In addition, during the welding process, FSW equipment provides down force continuously to avoid the upwards motion of welding tool. Therefore, under the effect of down force and rotating motion, the tool will be plunged into parent metal deeper on RS than that on AS. So the forming process of banded structure can be described as follows. When tool center travels on RS, rotating with eccentricity and tilt angle can make the welding tool moving downwards and plunging into parent metal much deeper. It causes more metal extruded out and accumulated at the rear edge of shoulder. Sequentially, the crest of the banded structure is formed. Conversely, the tool tends to move upward on AS, which induces less metal extruded. Thus the trough is formed. Under the condition of tool eccentricity 400 μm, L_R is negative, which indicates that shoulder will push the surface metal backward and form horns.

It is impossible to make FSW equipment without eccentricity. Therefore, it can't be authenticated that banded structures cannot be formed without eccentricity. However, it had been proved in many numerical simulation papers of FSW [16, 17]. Even though the eccentricity is very small, it is bound to affect the metal flow, macrostructure and microstructures of FSW joints.

4. Conclusions

From the present study, it can be concluded that the formation of surface periodic banded structures in friction stir weld is caused by the different motion characteristic of welding tool on AS and RS induced by tool eccentricity. The trough and crest is formed on AS and RS, respectively, due to the different flow behavior and extruded metal volume. So tool eccentricity considered or not has great effect on the accuracy of the flow models, because the flow laws of metal is totally different, which is one of the core issues of FSW.

Acknowledgments

We are grateful to the Weihai Science and Technology Development Planning (grant no.2013DXGJ07) for the financial support for this study.

References

[1] Xu S W and Deng X M 2008 A study of texture patterns in friction stir welds *Acta Mater* **56** 1326-41

[2] Krishnan K N 2002 On the formation of onion rings in friction stir welds *Mater. Sci. Eng A* **327** 246-51

[3] Yang B C, Yan J H, Sutton M A and Reynolds A P 2004 Banded microstructure in AA2024-T351 and AA2524-T351 aluminum friction stir welds Part I. Metallurgical studies *Mater. Sci. Eng. A* **364** 55-65

[4] Reynolds A P 2008 Flow visualization and simulation in FSW *Scr. Mater.* 58 338-42

[5] Schneider J A and Nunes A C 2004 Characterization of plastic flow and resulting microtextures in a friction stir weld *Metall. Mater. Trans. B* **35** 777-83

[6] Sutton M A, Yang B C, Reynolds A P and Yan J H 2004 Banded microstructure in 2024-T351 and 2524-T351 aluminum friction stir welds Part II. Mechanical characterization *Mater. Sci. Eng. A* **364** 66-74

[7] Sutton M A, Reynolds A P, Yang B C and Taylor R 2003 Mode I fracture and microstructure for 2024-T3 friction stir welds *Mater. Sci. Eng. A* **354** 6-16

[8] Sutton M A, Reynolds A P, Yang B C and Taylor R 2003 Mixed mode I/II fracture of 2024-T3 friction stir welds *Eng. Fract. Mech.* **70**, 2215-34

[9] Mahoney M W, Rhodes C G, Flintoff J, Spurling R A and Bingel W H 1998 Properties of friction-stir-welded 7075 T651 aluminum *Metall. Mater. Trans. A* **29** 1955-64

[10] Colligan K 1999 Material flow behaviour during friction stir welding of aluminium *Weld. J.* **75** 229s-37s

[11] Ke L M, Pan J L, Xing L and Wang S L 2009 Sucking-extruding theory for the material flow in friction stir welds *Chin. J. Mech. Eng* **45** 89-94

[12] Biallas G, Braun R and Donne C D 1999 Mechanical properties and corrosion behavior of friction stir welded 2024-T6 *First International Conference on Friction Stir Welds* (Thousand Oaks, CA) paper no. S2-P3

[13] Simoncini M and Forcellese A 2012 Effect of the welding parameters and tool configuration on micro- and macro-mechanical properties of similar and dissimilar FSWed joints in AA5754 and AZ31 thin sheets *Mater. and Des* **41** 50-60

[14] Zhang G F, Su W, Zhang J, Wei Z X and Zhang J X 2010 Effects of shoulder on interfacial bonding during friction stir lap welding of aluminum thin sheets using tool without pin *T. Nonferr. Metal. Soc.* **20** 2223-8

[15] Gratecap F, Girard M, Marya S and Racineux G 2012 Exploring material flow in friction stir welding: Tool eccentricity and formation of banded structures *Int. J. Mater. Form.* **5** 99-107

[16] Hamilton R, MacKenzie D, Li H J 2010 Multi-physics simulation of friction stir welding process *Eng. Computation* **27** 967-85

[17] Kim D, Badarinarayan H and Kim J 2010 Numerical simulation of friction stir butt welding process for AA5083-H18 sheets *Eur. J. Mech. A-Solid* **29** 204-15

IOP Conf. Series: Materials Science and Engineering **103** (2015) 012021 doi:10.1088/1757-899X/103/1/012021

Effect of ageing on phase evolution and precipitation behaviour of duplex steel

P Podany[1], M Kover and J Dlouhy

COMTES FHT a.s., Prumyslova 995, 334 41 Dobrany, Czech Republic

E-mail: pavel.podany@comtesfht.cz

Abstract. The isothermal formation of secondary phases in duplex stainless steel was studied. Samples were isothermally heat treated (aged) at temperatures of 700, 800 and 850°C in a quenching dilatometer. Microstructured evolution of secondary phases was analysed by means of optical and scanning electron microscopes. Both common phases, Chi and Sigma, were observed. The resulting shrinkage curves from dilatometric measurements show the potential for research into the formation of minor phases and the estimation of transformation kinetics in this kind of steel.

1. Introduction

The duplex family of stainless steels (DSS) is composed of a mixture of ferrite and austenite and combines some typical properties of each of these phases. The duplex microstructure and chemical composition lead to high strength and good resistance to corrosion, abrasion and wear [1]. Different grades of DSS are increasingly used to store and transport highly corrosive fluids and gases in a wide range of industry sectors. A further enhancement to the mechanical properties and corrosion resistance of DSS can be achieved by the addition of various alloying elements, such as chromium, nickel, molybdenum, tungsten, nitrogen etc [2].

However, this alloying could also lead to microstructure or phase instability and thus to matrix depletion of the main alloying elements (e.g., chromium, molybdenum, and niobium) connected with the formation of secondary phases at high temperatures associated with processing forging, heat treatment or during the service itself. Several precipitation reactions usually occur when this steel is exposed in the specific temperature range. Usually it is between 600 and 1300°C. It especially leads to the precipitation of secondary phases: χ - Chi, σ - Sigma, carbides and nitrides [2]. Of all these phases, the Sigma phase is the most important, as its formation is accompanied by a decrease in ductility and toughness, especially if these are measured at normal temperatures. It is an intermetallic compound with a tetragonal crystal structure. In the mixture with the alpha phase, the Sigma phase contains between 20 and 70 wt. % chromium at relatively long thermal exposure times in the interval between 500 and 900°C. It is possible to put the resulting Fe-Cr Sigma phase into solid solution using short-time heating (for 1 hour or more). Fe-Cr alloys with manganese, nickel and molybdenum require longer times or higher temperatures for the Sigma phase to dissolve [3].

In recent years, great effort has been devoted to research into the precipitation of the Sigma phase and Chi phase in duplex steels. In general, the Sigma phase is considered undesirable as it depletes the

[1] Address for correspondence: P Podany, COMTES FHT a.s., Prumyslova 995, 334 41 Dobrany, Czech Republic. E-mail: pavel.podany@comtesfht.cz.

matrix of chromium. This depletion impairs the material's resistance to pitting corrosion. On the other hand, presence of the Sigma phase increases the creep resistance of the material [4].

The research into precipitation of the above phases focused on how some alloying elements suppress this precipitation during annealing. For instance, Kang, et al. showed that substituting part of the molybdenum content with tungsten suppresses precipitation of the Sigma phase. Like molybdenum, tungsten is known to improve resistance to pitting corrosion [5, 6]. Akisanya, et al. studied super duplex steels with increased tungsten content. They explored the effect of age hardening on mechanical properties [2]. Other studies indicate that adding a small amount of cerium helps to suppress the precipitation of secondary phases. Kim S, et al. has proven that adding 50-100 ppm cerium inhibits precipitation of the Sigma and Chi phases. If the cerium content is increased to 450 ppm, it merely leads to the formation of cerium-rich particles and the depletion of cerium in the matrix. In such case, cerium cannot suppress the precipitation [7]. Similarly, Yoo Y H, et al. studied the effects of alloying with cerium and other rare earth elements, lanthanum and barium [8]. According to Jeon S H, et al., another element that suppresses precipitation of the Sigma phase is copper. According to their results, adding approx. 1.30% copper practically eliminates precipitation of the Sigma phase and causes a slight increase in precipitation of the Chi phase. Despite that, the resulting total amount of secondary phases is smaller than without the addition of copper [9, 10]. As the above studies show, the precipitation of undesirable phases can be effectively controlled by adding some elements to the material. Despite that, designers are frequently required to use a particular commercially available grade of material specified by clients. It is, therefore, important to know the temperature ranges and heat treating sequences which lead to the precipitation of unwanted phases that can substantially alter the corrosion resistance, as well as the mechanical properties of the material. In the presented paper, the authors used a quenching dilatometer to study the precipitation process and to carry out the heat treatment. The purpose was to map primarily the initial stages of Sigma phase precipitation within three ranges of age hardening temperatures for a particular heat.

2. Experimental methods

2.1. Experimental material and sampling

The chemical composition of the experimental material is shown in table 1. The material was supplied in the form of forged pieces of 120 × 120 mm cross-section. Specimens were taken from the forged pieces for dilatometric measurement. The diameter and length of these specimens were 4 mm and 10 mm, respectively. After dilatometric measurement, these specimens were used for making longitudinal metallographic sections. As the other authors of research institute showed earlier [11], these specimens can also be used for making miniature mechanical testing specimens.

Table 1. Chemical composition (wt %) of experimental steel.

C	Si	Mn	P	S	Cu	Cr	Ni	Mo	Ti	W	N	Fe
0.032	0.316	1.449	0.017	0.007	0.084	21.771	5.660	3.138	0.002	0.023	0.135	bal

2.2. Dilatometric experiments

Samples with a length of 10 mm and a diameter of 4 mm were heat treated in the quenching dilatometer Linseis L78 RITA. The temperature was measured and controlled by a type K thermocouple, welded to the sample centre. The length change was measured by an LVDT sensor through a quartz push-rod. The dilatometric measurements were performed in a protective environment (He). Heating was performed using an inductive coil. The sample was cooled with protective gas. Prior to heat treatment, the sample chamber was refilled (evacuated and filled with He gas) three times. The samples were solution annealed at 1100°C for 1 hour, quenched at the rate of 180°C/s and aged at 700°C, 800°C and 850°C.

2.3. Characterisation of microstructure and phases

The microstructure was studied on the metallographic sections through the dilatometer specimens. The microstructure was revealed using two-phase electrolytic etching. The first stage involved a voltage of 5.5 V applied for 20s in combination with 10% oxalic acid. In the second stage, a voltage of 2.3 V was used for 8 s with 20% NaOH. In the first stage, the precipitates are attacked. In the second stage, the ferrite becomes coloured. The microstructure was documented using a Zeiss Axio Observer optical microscope. Specimens for scanning electron microscopy were etched with Beraha's reagent. Quantitative image analysis was carried out using NIS-Elements software. EBSD analysis was performed and scanning electron micrographs taken by means of JEOL 6380 and JEOL 7400F microscopes and an HKL Nordlys EBSD (Electron backscatter diffraction) camera from Oxford Instruments.

3. Results and discussion

3.1. Evolution of microstructure during aging

The as-received microstructure of the forged stock is shown in Fig. 1. In the micrograph, ferrite appears dark, whereas austenite appears bright. The quantitative analysis of phases by means of the NIS Elements software showed a ferrite fraction of 49.6%. No intermediate phases were found in the as-received microstructure. The same applies to the microstructure of solution-annealed specimens (Figs. 1(a) and 1(b)).

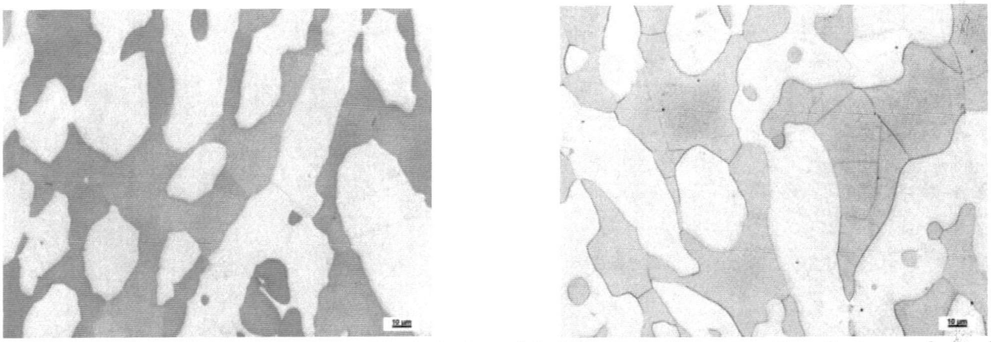

(a) As-received state (forged)-2-phase electrolytic etching

(b) After solution annealing-2-phase electrolytic etching

Figure 1. Microstructure of analysed duplex steel.

The microstructure evolution with the increasing holding time at 800°C is illustrated in Figs. 2(a)-2(d). As these figures show, the amount of the Chi and Sigma phase increases with increasing annealing time (see Figs. 2(a)-2(c)) up to 16,000s. When the Sigma phase is developed, the Chi phase starts to transforms into the Sigma phase and the amount of the Chi phase decreases (compare Figs. 2(c) and 2(d)). This assumption has been confirmed by other studies, as well [12].

Several basic shapes of intermediate phase particles were found in the microstructures. The Chi phase occurs first on grain boundaries at the ferrite-austenite interface. Later, it can be found on grain boundaries within the ferrite. Its particles are of irregular shape and sometimes elongated. They appear light grey in backscattered electron micrographs. This is due to the higher level of molybdenum than in the Sigma phase. The Sigma phase appears dark grey (2). According to other authors, chromium carbonitrides often precipitate together with the Chi phase [13-14]. This seems to be confirmed by figure 3. It shows very small precipitates on grain boundaries in a specimen annealed at 850°C for 107 seconds. Due to their size, their chemical composition could not be measured using EDX.

The Sigma phase occurs both in the interior of the original ferrite grains and as precipitates within the Chi phase particles which are found on the grain boundaries [12]. For the purpose of monitoring the precipitation of various phases and carbonitrides in steel, the use of electrolytic etching with oxalic acid has proven very useful. It is clear that even those particles which are too small to be resolved by

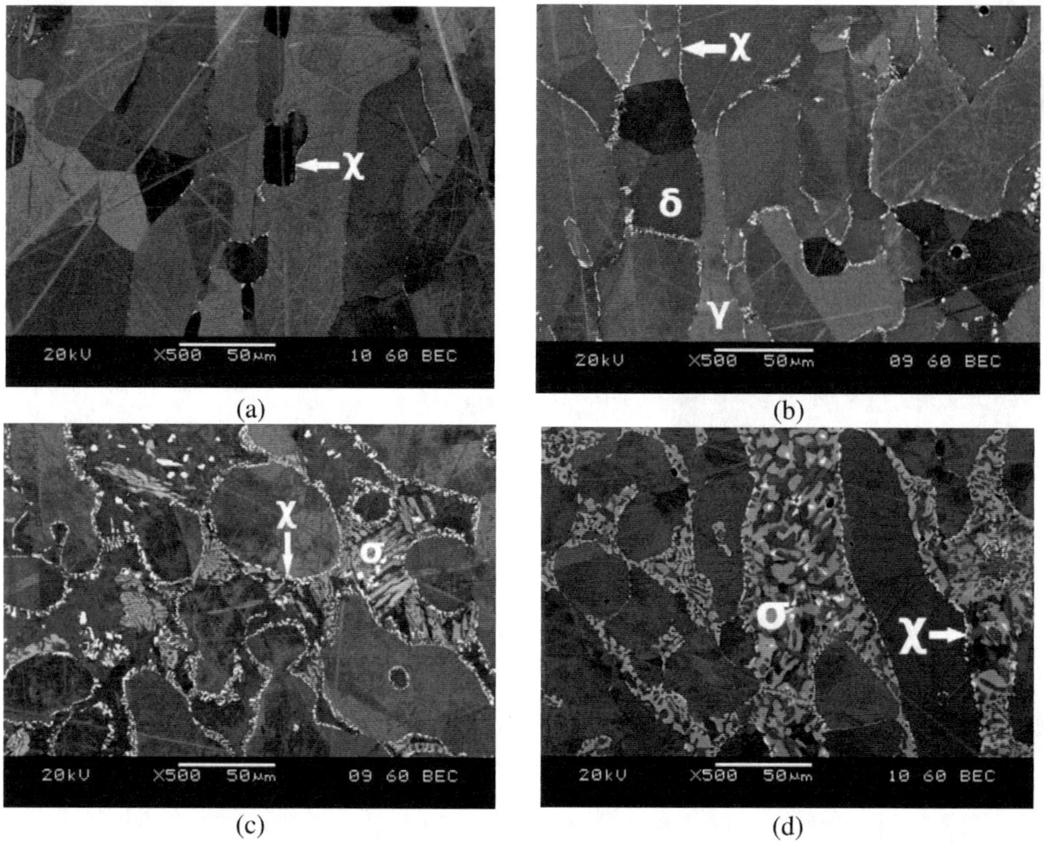

Figure 2. The microstructure of the experimental steel after ageing at 800°C for (a) 1000s, (b) 3,600s, (c) 16,000s, and (d) 230,000s.

Figure 3. Sample aged at 850°C for 107s – precipitation of Chi or CrC(N).

scanning electron microscopy can initiate a very vigorous reaction between the etchant and the specimen surface, leading to an intensive attack on the particle and its immediate vicinity. For illustration, optical micrographs of specimens etched with oxalic acid and backscattered scanning electron micrographs of identical polished specimens are shown in Figs. 4(a) and 4(b). Using specimens etched with oxalic acid to document microstructure by means of scanning electron microscopy is not useful, as the Chi phase, carbonitrides and other phases are completely etched away by the electrolyte.

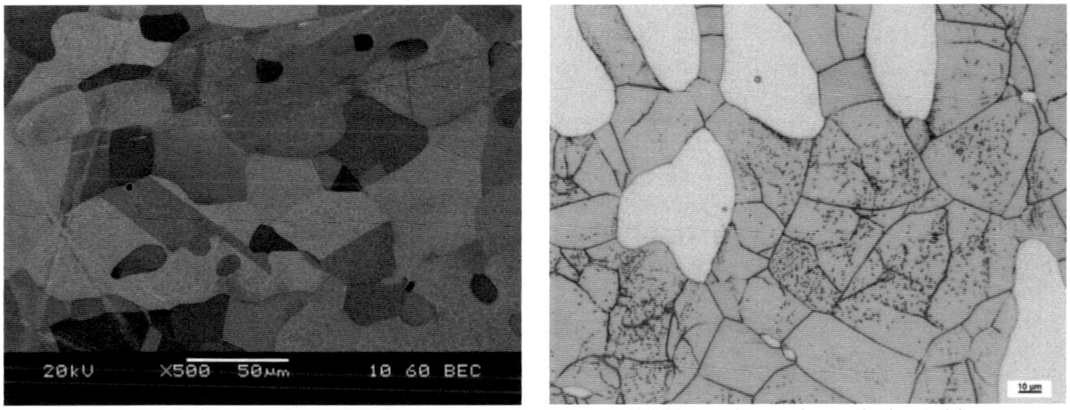

(a) BSE signal on SEM (b) Two-phase electrolytic etching

Figure 4. Microstructure of sample aged at 700°C for 156s.

3.2. EBSD analyses

Fig. 5 shows an EBSD (Electron backscatter diffraction) micrograph of the steel sample aged at 800°C for 16,000 s. The Sigma phase (coloured yellow in Fig. 5(b)) was clearly detected. The Chi phase is difficult to distinguish (by EBSD) between the ferrite and austenite because it has a cubic lattice [14].

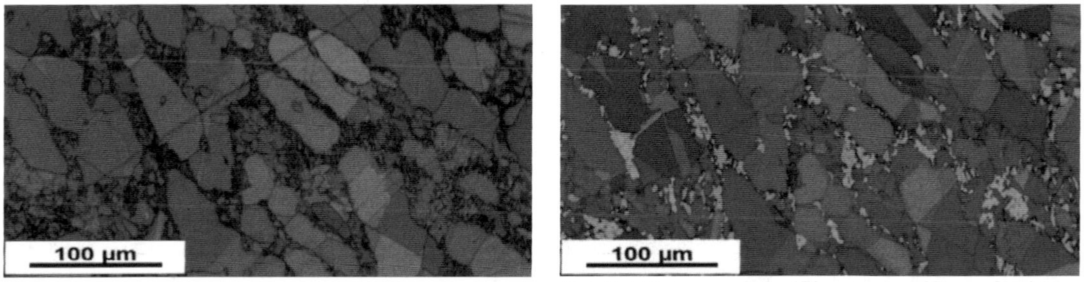

(a) Band contrast (b) EBSD mapping (Sigma - yellow, ferrite – red, austenite – orientfigation)

Figure 5. EBSD microstructure of sample aged at 800°C for 16,000s.

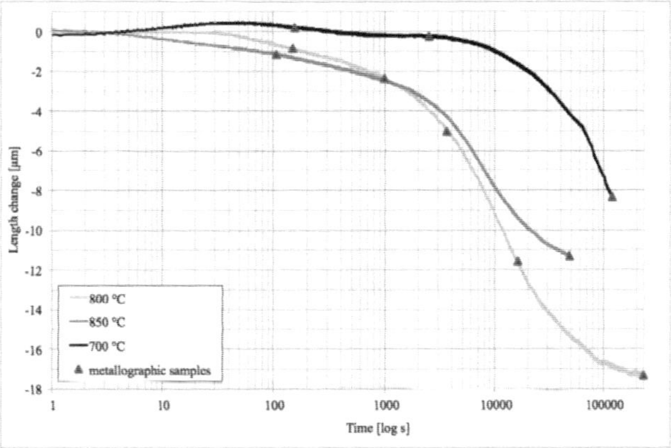

Figure 6. Shrinkage of the samples during ageing.

3.3. Dilatometric curves

During isothermal ageing in the dilatometer, shrinkage of all samples was observed. However, the length change at 700°C differs from the length change at 800°C and 850°C (figure 6). This behaviour might be attributed to different equilibrium phase ratios and ageing kinetics. Similar conclusions were stated in [13-16]. The initial length increase at 700°C might be attributed to the transport of the alloying elements to future nucleation sites. According to SEM observation, the first rapid decrease in length is a result of the formation of austenite, carbides and nitrides. Nucleation of the Chi and Sigma phases is expected to follow, leading to less shrinkage because the contraction during the phase formation at the interface is compensated by transport of the alloying elements. As the Sigma phase grows into ferrite grains, the shrinkage decreases even more. Formation of the Chi and Sigma phases cannot be distinguished from each other, although their formation kinetics is obvious from the curves. The most visible differences between the kinetics relate to nitride and Sigma/Chi formation at 700°C. The nucleation phase was estimated for all temperatures. The final phase of the ageing process was estimated for 800°C and 850°C. The estimated Chi and Sigma phase formation start and end times are listed in table 2.

Table 2. Estimated start and end of Chi and Sigma formation.

Temperature [°C]	Chi/Sigma formation start [s]	Chi/Sigma formation end [s]
700	2500	-
800	151	120,000
850	107	48,000

4. Conclusion

According to the results presented here, a quenching dilatometer can be used not only for precise heat treatment, but also for research into the formation of minor phases and estimation of transformation kinetics in duplex steels. The formation of precipitates led to contraction. The length change, which reflects the transformation kinetics, has a typical "S shape" in the logarithmic time graph. Although the early stage of Chi and Sigma precipitation could not be determined by SEM, a future TEM investigation might confirm the nucleation start identified by means of a dilatometer.

Very early stages of precipitation of undesirable phases can be observed using optical microscopy and specimens following electrolytic etching with oxalic acid. For instance, the Chi phase becomes fully discernible using the backscattered electron mode in scanning electron microscopy only after the specimens have been isothermally held for an extensive period of time, although evidence from both electrolytically etched specimens and dilatometric curves shows that precipitation occurs earlier.

Acknowledgment

These results were created under the project entitled Development of West-Bohemian Centre of Materials and Metallurgy No.: LO1412, financed by the Ministry of Education, Youth and Sports of the Czech Republic.

References

[1] Nilsson J O, et al 1992 Super duplex stainless steel *Mater. Sci. Technol.* **8** 685-700
[2] Akisanya A R, Obi U and Renton N C 2012 Effect of ageing on phase evolution and mechanical properties of a high tungsten super-duplex stainless steel *Materials Science and Engineering A* **535** 281-9
[3] Chvatalova K 2007 First-principles calculations of the total energy of systems of ordered structures of transition metals (Faculty of Science, Masaryk University: Dissertation thesis)
[4] Shek C H, Li D J, Wong K W and Lai J K L 1999 Creep properties of aged duplex stainless steels containing σ phase *Materials Science and Engineering A* **266** 30-6
[5] Kang T H, Li D M, Lee Y D and Lee C S 1998 Alloying and aging effects on the fatigue crack growth of duplex stainless steels *Materials Science and Engineering A* **251** 192-9

[6] Jeon S H, Kim S T, Lee I S, Kim J S, Kim K T and Park Y S 2012 Effects of W substitution on the precipitation of secondary phases and the associated pitting corrosion in hyper duplex stainless steels *Journal of Alloys and Compounds* **544** 166-72

[7] Sun M K, Ji S K, Kim K T, Park K T and Chong S L 2013 Effect of Ce addition on secondary phase transformation and mechanical properties of 27Cr–7Ni hyper duplex stainless steels *Materials Science and Engineering A* **573** 27-36

[8] Yoo Y H, Choi Y S, Kim J G and Park Y S 2010 Effects of Ce, La and Ba addition on the electrochemical behavior of super duplex stainless steels *Corrosion Science* **52** 1123-9

[9] Jeon S H, Kim S T, Lee I S, Kim J S, Kim K T and Park Y S 2013 Effects of Cu on the precipitation of intermetallic compounds and the intergranular corrosion of hyper duplex stainless steels *Corrosion Science* **66** 217-24

[10] Ran Q, Xu Y, Li J, Wan J, Xiao X and Yu H 2013 Effect of heat treatment on transformation-induced plasticity of economical Cr19 duplex stainless steel *Materials and Design* **46** 758-65

[11] Dzugan J and Konopik P 2012 Small punch test application to material properties evolution determination *Journal of Solid Mechanics and Materials Engineering* **6** 782-91

[12] Calliari I, Pellizzari M and Ramous E 2011 Precipitation of secondary phases in super duplex stainless steel ZERON100 isothermally aged *Materials Science and Technology* **27** 928-32

[13] Fritz J D 2003 Effects of metallurgical variables on the corrosion of stainless steels in ASM *Handbook Volume 13A Corrosion: Fundamentals, Testing, and Protection* pp 266-74

[14] Kim Y J, Ugurlu O, Jiang C, Gleeson B and Chumbley L S 2007 Microstructural evolution of secondary phases in the cast duplex stainless steels CD3MN and CD3MWCuN *Metallurgical and Materials Transactions A* **38** 203-11

[15] Elmer J W, Palmer T A and Specht E D 2003 Direct observations of sigma phase formation in duplex stainless steels using In situ synchrotron X-ray diffraction *Metallurgical Transactions A* **38** 464-75

[16] Magnabosco R 2009 Kinetics of sigma phase formation in a duplex stainless steel *Mat. Res.* **12** 321-7

Effects of cryogenic treatment on mechanical property and microstructure of JIS SKH51 high-speed steel

C C Wu[1,3], W C Chen[2] and L H Chen[1]

[1]Department of Mechanical Engineering, Southern Taiwan University of Science and Technology, Tainan 71005, Taiwan, ROC
[2]Institute of Mechatronic Technology, Southern Taiwan University of Science and Technology, Tainan 71005, Taiwan, ROC

E-mail:wcc@mail.stust.edu.tw

Abstract. When SKH 51 high-speed steel was first cryogenically treated and then tempered at 540℃, the surface hardness of the SKH 51 steel increased from HRC 64 to HRC 65. This result indicates the reduction of retained austenite as well as the secondary precipitation of tiny carbides. TEM analyses indicated that cryogenic treatment can induce high-density dislocations and stacking faults within the martensite matrix. In the subsequent tempering process, conducted at 540 ℃, tiny secondary carbides precipitated nearby the existing dislocations and stacking faults, resulting in the hardening of SKH 51 high-speed steel. These secondary carbides mainly consisted of V-rich MC carbides and (Mo, W)-rich M_2Ccarbides, which were identified by STEM/EDS techniques.After being cryogenically treated and then tempered at 590℃, both the surface hardness and wear resistance of SKH 51 high-speed steel were weakened below acceptable ranges. This result indicates the promotion of rapid decomposition and growth of tempered martensite. Therefore, to achieve the best surface hardness, SKH 51 high-speed steel should first be cryogenically treated, and then tempered at a temperature below 590℃.

1. Introduction

Cryogenic treatments, namely the treatment processes involving continuous cooling of materials to -196℃, have been widely used in the manufacturing of various high-precision components of precision machinery. Cryogenic treatment has been acknowledged as an effective method of microstructure refinement, residual stress relief, and dimensional stability, and shown to induce improvements in the mechanical properties of machine components, thus enhancing the service life of tools [1-9]. As a result, several companies in Taiwan have shown great interests in cryogenic treatment technology and instrumentations.

Cryogenic treatment provides some processing benefits in wear performance, thermal stability, homogenization of hardness distribution, and enhancement of tool life, among other mechanical properties. However, the effects of cryogenic treatments on the microstructure and mechanical properties of high-precision steels and components of precision machinery are not well-known. The licensed patents and previous literature that have already provided experimental data regarding the

[3] Address for correspondence:CC Wu, Department of Mechanical Engineering, Southern Taiwan University of Science and Technology, Tainan 71005, Taiwan, ROC. E-mail: wcc@mail.stust.edu.tw.

mechanical properties of tool steels are very few. Therefore, it is benefit to the precision machinery industry to determine the actual benefits of cryogenic treatments, and verify the qualitative conclusions with experimental results. However, no previous research has provided relative experimental data to prove the advantages of cryogenic treatment, which has limited the applications of cryogenic treatment technology.

Much recent interest has focused on the effects of cryogenic treatment on the mechanical properties of SKD 11 and SKD 61 tool steels. However, little information exists concerning the effects of the sequence procedures of cryogenic treatment on SKH 51 high-speed steel. KelkarP [8] and HuangJ Y [9] have investigated the effects of cryogenic treatment on the strengthening mechanism and microstructural changes in M2 tool steels. However, further studies concerning the relationship between the mechanical properties and the phenomenon of secondary precipitation hardening are needed. Therefore, the present study examines the effects of cryogenic treatment on microstructural developments and the mechanical properties of SKH 51 high-speed steel by means of hardness testing, x-ray diffraction, scanning transmission electron microscopy and energy dispersive spectrometry.

2. Experimental procedures

JIS SKH 51 (corresponding to ASTM M2) high-speed steel was used as the raw material, the chemical composition of the steel is shown in table 1. The SKH 51 high-speed steel was preheated to 750℃ for 1 hour, and then solution heat-treated for 1 hour at 1190℃ before being rapidly quenched to room-temperature by liquid nitrogen gas. The tempering processes were carefully performed at temperatures ranging from 540℃ to 590℃. SKH 51 steel can achieve excellent surface hardness at 540℃, and displays excellent toughness at 590℃. Therefore, the tempering processes in the present study were performed at 540℃ and 590℃, respectively. The cryogenic treatment was performed at temperatures ranging from room temperature to -196℃, then cooled by liquid nitrogen at a rate of 30℃/hr. The samples were then kept at -196℃ for 8 hours.

Table 1. Chemical compositions of the studied JIS SKH 51 steel by GDS analysis (wt%).

	C	Cr	Mo	W	V	Co	Fe
SKH 51	0.86	3.95	4.75	5.83	1.82	0.56	Bal.

Specimens for transmission electron microscopy were prepared by using a double-jet electropolisher with an electrolyte consisting of 10% perchloric acid, 30% acetic acid and 60% ethanol. The polishing temperature ranged from -5℃ to -15℃, and the current density ranged from 3.5 to $4.0\times10^4 A/m^2$. Electron microscopy was performed on a Philips FEG-20 scanning transmission electron microscope (STEM) operating at 200 kV.

3. Results and discussion

3.1. Effects of cryogenic treatment on the mechanical properties of SKH 51 high-speed steel

Figures 1(a) and 1(b) depict the hardness and wear mass loss properties of SKH 51 high-speed steel after being quenched, cryogenic treated and tempered at 540℃, respectively. The hardness of the quenched and then cryogenically treated SKH 51 high-speed steel was HRC66, which is similar to that of the quenched specimen that did not undergo cryogenic treatment. It indicates that the influence of cryogenic treatment on the as-quenched SKH 51 high-speed steel is not apparent in relation to surface hardness. However, the hardness of the SKH 51 high-speed steel after being quenched, cryogenically treated and then tempered at 540℃ is approximately HRC65, which is excellent compared to the specimen tempered at 540℃ without cryogenic treatment. Our experimental results indicates that the SKH 51 high-speed steel which was cryogenically before being tempered at 540℃ demonstrates excellent surface hardness and wear resistance. This necessitates the investigation of the microstructural changes in SKH 51 high-speed steel caused by cryogenic treatment described in the following section.

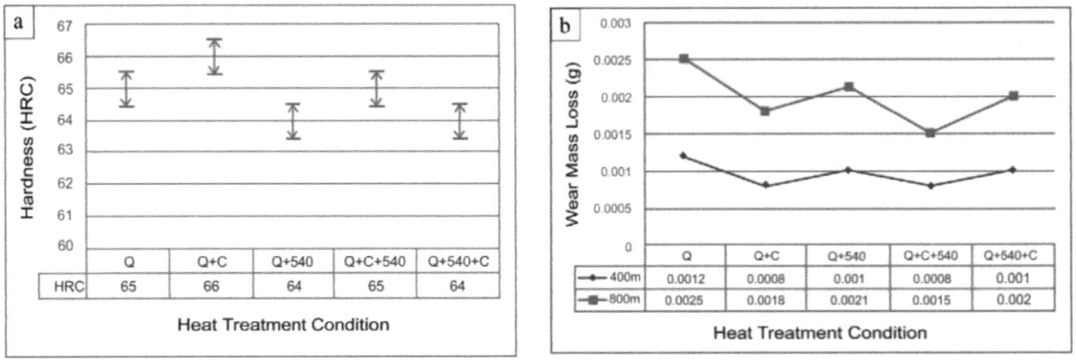

Figure 1. Effects of cryogenic and tempering treatments at 540℃ on the mechanical properties of the studied SKH 51 steel. (a) the surface hardness; (b) the wear resistance. (Q: quenched; C: cryogenic).

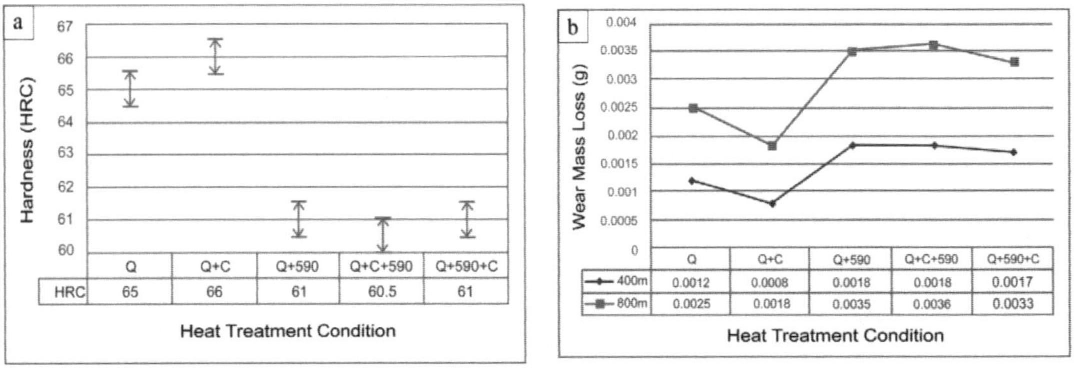

Figure 2. Effects of cryogenic and tempering treatments at 590℃ on the mechanical properties of the studied SKH 51 steel. (a) the surface hardness; (b) the wear resistance.

Figures 2(a) and 2(b) demonstrate the surface hardness and wear mass loss properties of SKH 51 high-speed steel under as-quenched condition, after cryogenic treatment and after being tempered at 590℃, respectively. After being cryogenically treated and then tempered at 590℃, both the surface hardness and wear resistance of the SKH 51 high-speed steel are compromised. The surface hardness of SKH 51 high-speed steel after being quenched and then tempered at 590℃ is HRC 60.5, which is relatively similar to that measured from the specimen that was cryogenically treated and then tempered at 590℃. The influence of cryogenic treatment was not obvious for SKH 51 high-speed steel after being tempered at 590℃. Therefore, to achieve better surface hardness and wear resistance at 590℃, the SKH 51 high-speed steel should be tempered at 590℃ before undergoing cryogenic treatment. It was not necessary to perform the cryogenic treatment before the tempering processes. The above heat treatment procedure was also convenient and suitable for the industrial production process.

3.2. Effects of cryogenic treatment on the microstructure of SKH 51 high-speed steel

Figure 3(a) represents an optical micrograph of SKH 51 high-speed steel in the as-quenched and then cryogenically treated condition, clearly showing the finer prior austenite grains as well as particle-like carbides. The microstructure is similar to that observed in the as-quenched specimen without cryogenic treatment, as shown in figure 3(b). It was difficult to visually distinguish the martensite and retained austenite phases in the optical micrograph. Figures 4(a) and 4(b) show x-ray diffractions of the SKH 51 high-speed steel under different heat treatment conditions. Figure 4(a) displays an x-ray diffraction (XRD) profile of the as-quenched specimen without cryogenic treatment, demonstrating the presence of martensite, carbides and retained austenite. Figure 4(b) shows an XRD profile of the specimen after being quenched and then cryogenically treated, showing the presence of martensite and carbides. This result indicated that the amount of retained austenite decreased significantly with cryogenic treatment.

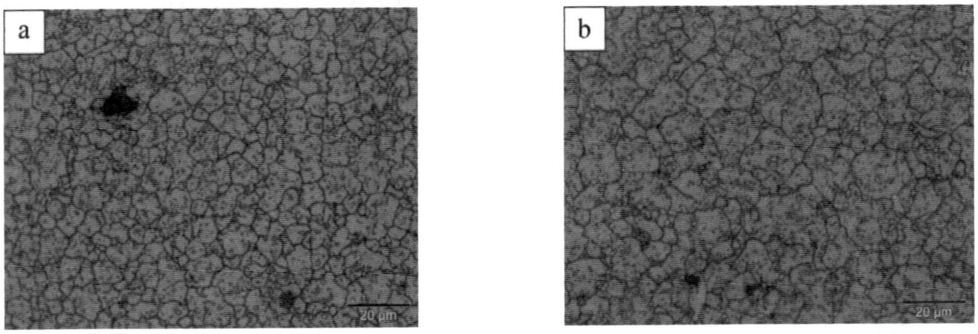

Figure 3. Optical micrographs of the SKH 51 high-speed steel. (a) solution-treated at 1190℃ and then cryogenically treated; (b) solution-treated at 1190℃ without cryogenic treatment.

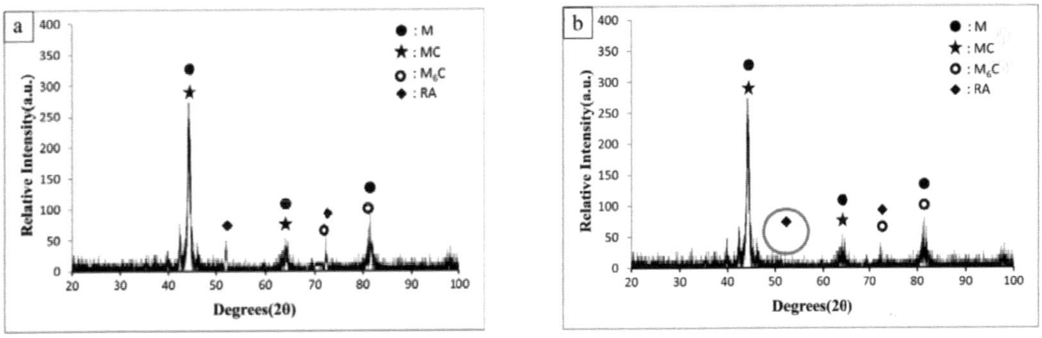

Figure 4. XRD patterns of the SKH 51 steel under different heat treatment conditions. (a) The as-quenched SKH 51 steel without cryogenic treatment; (b) The as-quenched and then cryogenically treated steel. (M: martensite; RA: retained austenite).

Figure 5(a) shows a bright-field (BF) electron micrograph of the as-quenched SKH 51 steel, demonstrating the presence of dispersed coarse carbides within the martensite matrix. Figures 5(b) and 5(c) show two selected area diffraction patterns (SADPs) of the dispersed carbides. Diffraction analysis demonstrates that the carbides consisted primarily of MC and M6C carbides within the martensite matrix. Figures 5(d) and 5(e) show two EDS spectra of the particle-like M6C carbides and the surrounding martensite matrix, respectively, from the as-quenched SKH 51 high-speed steel. The primary differences between these two EDS spectra are the tungsten and molybdenum contents. It is clear that the particle-like carbides within the martensite matrix are tungsten-rich and molybdenum-rich M6C carbides, as confirmed by electron diffraction analysis. Figures 6(a) to 6(c) depict electron micrographs of the SKH 51 high-speed steel after being quenched and then cryogenically treated. Besides the particle-like carbides and the martensite matrix, the high density of plate-like twin faults and dislocations can be observed within the martensite matrix.

Figure 7(b) displays an optical micrograph of the SKH 51 high-speed steel being solution-treated, cryogenically treated, and then tempered at 540℃, clearly displaying the finer tempered martensitegrain as well as the presence of particle-like carbides. The microstructure was similar to that observed in the as-quenched specimen without cryogenic treatment, as shown in figure 7(a). However, thecryogenically-treated SKH 51 steel displays better surface hardness and wear resistance. Based on

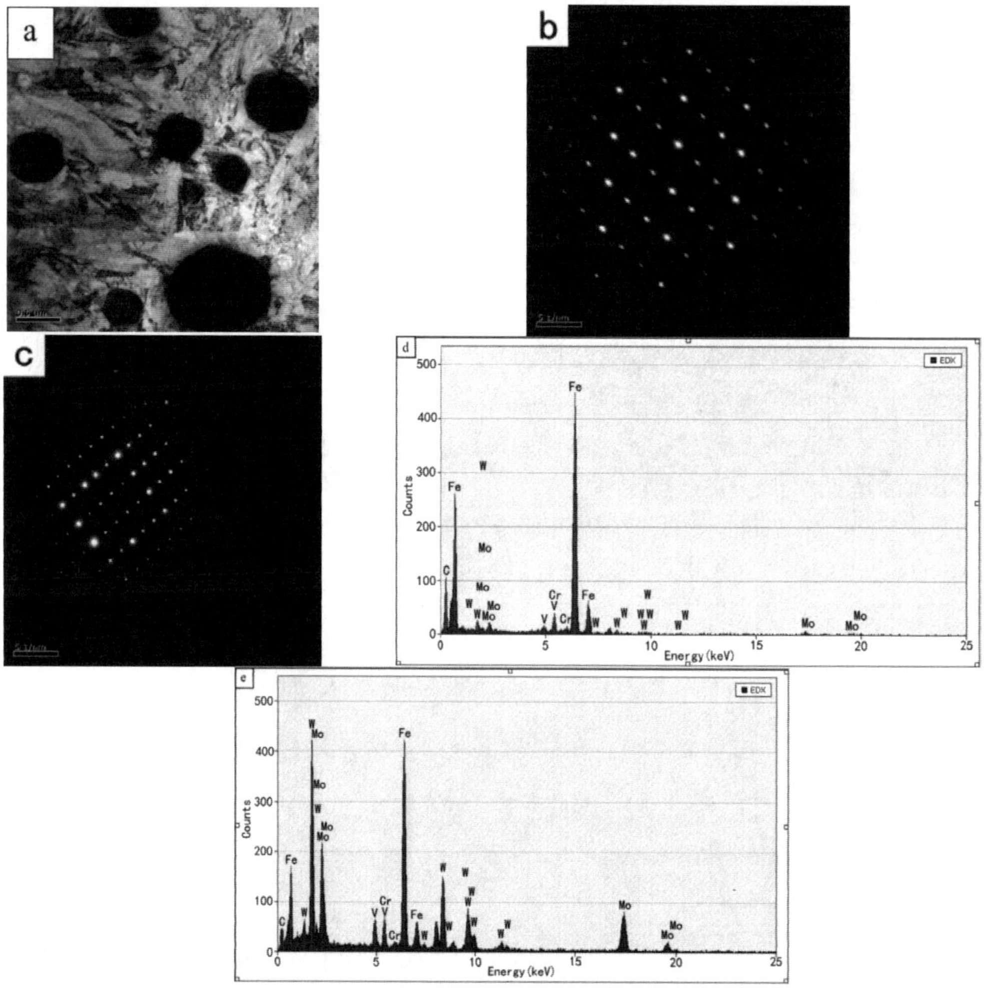

Figure 5. TEM electron micrographs of the SKH 51 steel under as-quenched conditions. (a) BF electron micrograph showing the carbides dispersed within the martensite matrix. (b) and (c) show SADPs of the dispersed carbides in (a); (d) and (e) show EDS spectra obtained from the M_6C carbide and its surrounding martensite matrix, respectively.

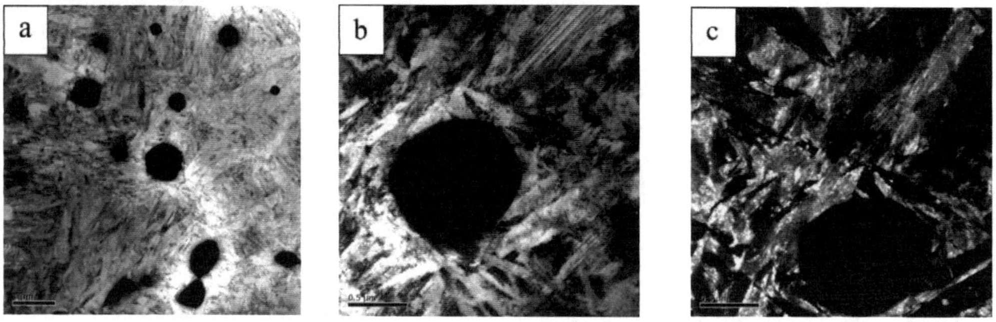

Figure 6. TEM electron micrographs of the SKH 51 high-speed steel after undergoing solution treatment at 1190℃ and then cryogenic treatment. (a) BF electron micrograph showing the presence of carbides dispersed within tiny laths structure of the martensite matrix; (b) BF electron micrograph, showing twins and dislocations within the martensite matrix; (c) DF electron micrograph taken from the same martensite region shown in (b).

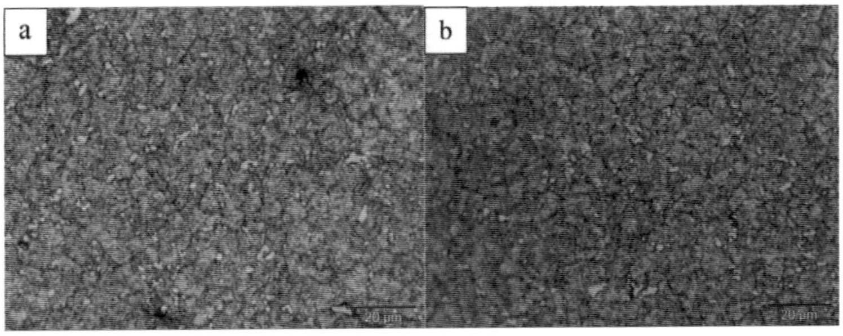

Figure 7. Optical micrographs of the SKH 51 steel under the following conditions: (a) solution treated at 1190℃ and then tempered at 540℃ without cryogenic treatment; (b) solution-treated at 1190℃, then cryogenically treated and then tempered at 540℃.

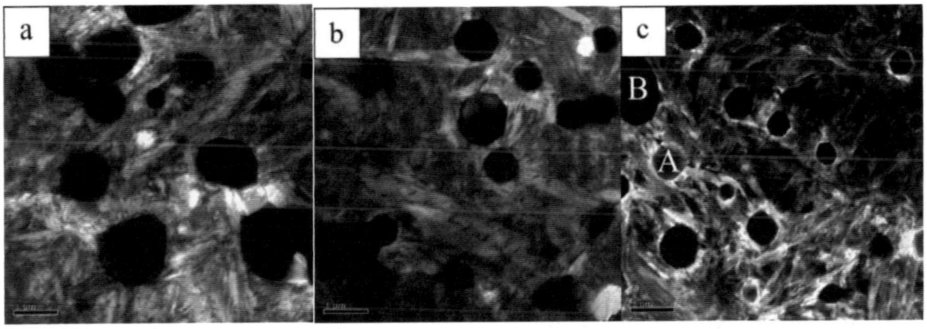

Figure 8.TEM electron micrographs of the SKH 51 high-speed steel. (a) BF electron micrograph of the specimen which was solution treated and then tempered at 540℃ without cryogenic treatment; (b) and (c) show BF electron micrographs of the specimen which was solution treated, before being cryogenically treated and then tempered at 540℃, demonstrating the high density of the tiny secondary precipitated carbides.

STEM/EDS examinations, coarse carbides were observed within the tempered martensite matrix of the SKH 51 steel tempered at 540℃ without cryogenic treatment, as shown in figure 8(a). However, these results also displayed a higher density of tiny secondary carbides which precipitated near dislocations and stacking faults within the tempered martensite matrix, as shown in figures 8(b) and 8(c). Two types of tiny secondary carbides were observed within the martensite matrix of the SKH 51 steel after being cryogenically treated and then tempered at 540℃. The first type carbide, the MC carbide, contained a much higher Vanadium content, as determined by EDS analysis and shown in figure 9(a). However the other type of carbide, the M2C type, contained higher tungsten and molybdenum contents, as shown in figure 9(b). STEM/EDS examinations indicated that the cryogenic treatment promoted the precipitation of Vanadium-rich MC carbides and (Mo, W)-rich M2C carbides within the martensite matrix of the SKH 51 steel after being cryogenically treated and then tempered at 540℃.

After being solution-treated at 1190℃, then cryogenically treated and finally tempered at 590℃, both the surface hardness and wear resistance of the SKH 51 high-speed steel became compromised. In addition to the growth of coarse carbides, TEM examinations indicated that the decomposition of martensite to ferrite and Fe3C was promoted by cryogenic treatment, as shown in figures 10(a) and 10(b). Therefore, to achieve better mechanical properties, the SKH 51 high-speed steel should be tempered first and then cryogenically treated at temperatures higher than 590℃. The above procedure was also suitable for industrial production.

Figures 11(a) and 11(b) shows two x-ray diffractions of SKH 51 high-speed steel under different heat treatment conditions. Figure 11(a) shows an XRD profile taken from the as-quenched and tempered at 590℃ specimen without cryogenic treatment, showing the presence of martensite and

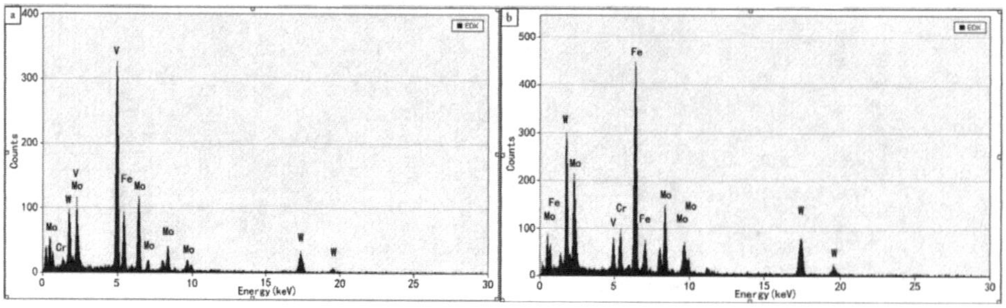

Figure 9. EDS spectra obtained from the SKH 51 steel which was cryogenically treated and then tempered at 540℃. (a) EDS spectrum taken from the particle-like MC carbide marked as "A" in figure8(c); (b) EDS spectrum taken from the particle-like M_2C carbide marked as "B" in figure 8(c).

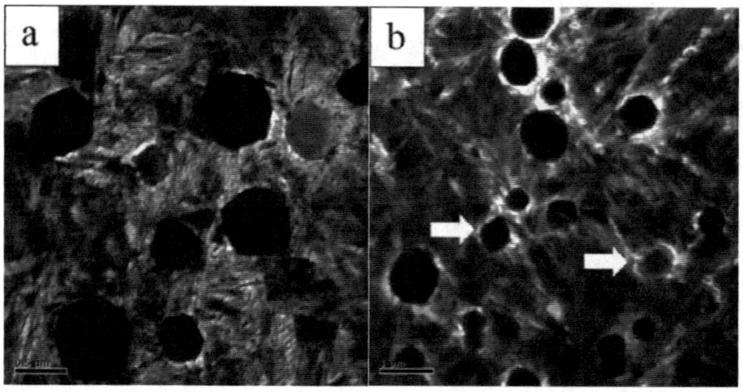

Figure 10.TEM BF micrographs of the SKH 51 steel. (a) solution-treated and then tempered at 590℃ without cryogenic treatment; (b) solution-treated, then cryogenically treated and finally tempered at 590℃.

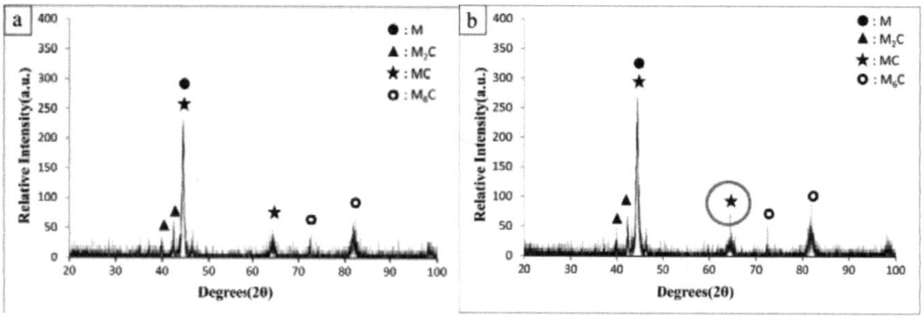

Figure 11. XRD patterns of the SKH 51 steel under different heat treatment conditions. (a) solution treated and then tempered at 590 ℃ without cryogenic treatment; (b) solution treated, then cryogenically treated and finally tempered at 590℃.

carbides. Figure 11(b) shows an XRD profile of the specimen after undergoing cryogenic treatment and then being tempered at 590℃, which is similar to that observed from the specimen which was quenched and then tempered at 590℃. However, the higher peak relating to MC carbides was derived from the cryogenically treated specimen. The amount of MC carbides was greater than that observed in SKH 51 steel that was only tempered at 590℃ without cryogenic treatment. This result indicated that cryogenic treatment enhanced the secondary precipitation of tiny MC carbides during the tempering process, which was consistent with the visual observation of carbides as shown in figure 10(b).

Based on the above observations, the surface hardness of the as-quenched SKH 51 steel generally decreased during the tempering processes. However, the SKH 51 steel which was cryogenically treated and then tempered at 540 ℃ demonstrating excellent surface hardness as well as wear resistance. It was observed that cryogenic treatment can reduce the amount of retained austenite, as well as induce a high density of dislocations and stacking faults within the martensite matrix. During the subsequent tempering at 540℃, tiny secondary MC and M2C carbides precipitated near the existing dislocations and stacking faults, which resulted in the secondary precipitation hardening of the SKH 51 high-speed steel. When the tempering processes was performed at 590℃, the secondary precipitation of MC carbides was observed. However, both the surface hardness and wear resistance of the SKH 51 high-speed steel became compromised. This result may have been caused by the promotion of rapid decomposition and growth of tempered martensite. Further investigation of the effects of cryogenic treatment on the toughness of SKH 51 high-speed steel is necessary to further understand the industrial applications of cryogenic treatment.

4. Conclusions

When the SKH 51 high-speed steel was cryogenically treated and then tempered at 540℃, the surface hardness of the SKH 51 steel increased from HRC 64 to HRC 65. This result indicates the reduction of retained austenite as well as the secondary precipitation hardening of tiny carbides.

TEM analysis indicated that cryogenic treatment can induce high density dislocations and stacking faults within the martensite matrix after cryogenic treatment. During the subsequent tempering at 540℃, tiny secondary carbides precipitated near the existing dislocations and stacking faults, which resulted in the secondary precipitation hardening of the SKH 51 high-speed steel. These secondary carbides consisted primarily of Vanadium-rich MC carbides as well as (Mo, W)-rich M_2C carbides, as identified by STEM/EDS techniques.

After being cryogenically treated and then tempered at 590℃, both the surface hardness and wear resistance of the SKH 51 high-speed steel became compromised. This result indicates the promotion of rapid decomposition and growth of tempered martensite. The influence of cryogenic treatment was not obvious for SKH 51 high-speed steel after being tempered at 590℃.

Therefore, to achieve the best surface hardness, the SKH 51 high-speed steel should be cryogenically-treated first and then tempered at 540℃.

References

[1] BarronR F 1982 Cryogenic treatment of metal to improve wear resistance *Cryogenics* **22** 409-13
[2] de SilvaF J, FrancoS D, MachadoA R, EzugwuE O and Souza Jr.A M 2006 Performance of cryogenically treated HSS tools*Wear* **261** 674-85
[3] Das D, DuttaA K and RayK K2009Correlation of microstructure with wear behavior of deep cryogenically treated AISI D2 steel*Wear* **267** 1371-80
[4] ZhirafarS, Rezaeian Aand PughM2007Effect of cryogenic treatment on the mechanical properties of 4340 steel*J. Mater. Proc. Tech.* **186** 298-303
[5] LeskovsekV, Kalin Mand VizintinJ2006Influence of deep-cryogenic treatment on wear resistance of vacuum heat-treated HSS *Vacuum* **80** 507-18
[6] LalD M, Renganarayanan Sand KalanidhiA2001Cryogenic treatment to argument wear resistance of tool and die steels*Cryogenics* **41** 149-55
[7] DasD, DuttaA K, Toppo Vand RayK K2007The effect of cryogenic treatment on the carbide precipitation and tribological behavior of D2 steel*Materials and Manufacturing Processes* **22** 474-80
[8] KelkarP, Nash Pand ZhuY T2003The mechanism of property enhancement in M2 tool steel by cryogenic treatmentthe 45th MWSP Conference Proceedings Chicago, Illinois, November 9 - 12, 2003 pp 13 9
[9] HuangJ Y, ZhuY T, LiaoX Z, BeyerleinI J, BourkeM A and MitchellT E 2003Microstructure of cryogenic treated M2 tool steel *Materials Science and Engineering* **A339** 541-4

IOP Conf. Series: Materials Science and Engineering **103** (2015) 012023 doi:10.1088/1757-899X/103/1/012023

Improvement, characterization and use of waste corn cob ash in cement-based materials

P Suwanmaneechot[1], T Nochaiya[2,3] and P Julphunthong[3,4,5,6]

[1] Department of Civil Engineering, School of Engineering, University of Phayao, Phayao, Thailand
[2] Department of Physics, Faculty of Science, Naresuan University, Phitsanulok, Thailand
[3] Research Center for Academic Excellence in Applied Physics, Faculty of Science, Naresuan University, Phitsanulok, Thailand
[4] Research Unit for Innovative Construction Materials, Faculty of Engineering, Naresuan University, Phitsanulok, Thailand
[5] Department of Civil Engineering, Faculty of Engineering, Naresuan University, Phitsanulok, Thailand

E-mail: pop_civil@hotmail.com

Abstract. This work investigates the development of waste corn cob ash as supplementary cement replacement materials. The study focused on the effects of heat treatment on chemical composition, physical properties and engineering properties of corn cob ash. The results suggest corn cob ash that was heat treated at 600°C for 4 h shows percentage of $SiO_2 + Al_2O_3 + Fe_2O_3$ around 72%, which can be classified as Class N calcined natural pozzolan, as prescribed by ASTM C618. The X-ray diffraction patterns indicated that the amorphous silica phase increased with increasing calcining temperatures. The water requirement, initial setting time and final setting time of specimens increased with increasing replacement percentage of raw or treated corn cob ash. The morta cubes which used 20% of treated corn cob ash replaced cement showed 103% of the 28 days compressive strength as compared to reference samples. The corn cob ash that was treated at 600°C for 4 h samples shows slightly higher effectiveness for improving the splitting tensile strength and compressive strength of concrete when compared to the untreated corn cob ash.

1. Introduction

Ordinary Portland Cement (OPC) is the most common type of cement used in construction applications. The current cement production rate worldwide is approximately 1.2 billion tons/year. This consumption is expected to grow to about 3.5 billion tons/year by 2015 [1]. However, it is an expensive binder due to the high cost of production associated with the high energy requirements of the manufacturing process itself [2, 3]. Nowadays, industrial wastes such as blast furnace slag, fly ash and silica fume are being used as supplementary cement replacement materials [4-7]. These materials consist of silica oxide or alumina oxide, which react with calcium hydroxide in concrete, forming the cementitious materials. The agricultural waste or industrial waste which is used for partial cement

[6] Address for correspondence: P Julphunthong, Research Center for Academic Excellence in Applied Physics, Faculty of Science, Naresuan University, Phitsanulok, Thailand. E-mail: pop_civil@hotmail.com.

replacement is called pozzolanic materials [8]. The waste products which possess pozzolanic properties and which have been studied for use in blended cement are fly ash [9, 10], rice husk ash [11-13], palm oil ash [14-16] and sugarcane bagasse ash [17-19]. Corn cob is an agricultural waste product obtained from maize or corn, which are important for the food industry. Corn cob ash (CCA) waste is produced in the boilers of the animal feed industry. Adesanya, et al. [1] extensively studied the pozzolanic properties of CCA from South Africa. The sum of SiO_2, Al_2O_3 and Fe_2O_3 in the chemical composition of CCA satisfies the requirements of ASTM C618, suggesting its pozzolanic nature.Beside strength properties, the thermal conductivity of concrete is decreased and the insulation properties are improved by blending the CCA in concrete [20].

The quality of pulverized fuel ash or biomass fuel ash, used as a supplementary cementing materials in cement and concrete, varies on account of the variations in the firing processes, plant design, engineering parameters and dust collection process. It is well known that firing conditions strongly influence the physical characteristics and the chemical compositions of burned ash. For instant, the content of amorphous silica phase is decreased and carbon fraction is eliminated from the system with an increase firing temperature. The optimum firing temperature also provides the highest amount of amorphous silica and highest specific surface area, which results in satisfactory engineering properties when the burned ash is applied in concrete. To the best of our knowledge the improvement of physical and chemical properties of CCA using the heat treatment has not been yet investigated. Therefore, the aim of this study is to investigate the chemical and physical properties of CCA produced in Thailand. The effect of firing temperatures on the chemical properties, physical properties and engineering properties of CCA-cement replacement in concrete was investigated.

2. Materials and experiments

2.1. Materials

The CCA was obtained from the thermal power plant of Charoen Pokphand Foods Public Company Limited Plant in Phitsanulok, Thailand. The CCA was weighed and ground by the ball-milling machine for 12 h and then sieved into a fine powder. The chemical reaction of CCA at high temperature was investigated by thermogravimetric analysis and differential thermal analysis (TGA–DTA) using a heating rate of 10°C/min from room temperature to 700°C. The CCA was heated from 200 to 600°C for 4 h and the changed physical and chemical characteristics of the treated ash were observed.

2.2. Physical and chemical properties characterization

The chemical composition of raw CCA and treated CCA were analyzed using X-ray fluorescence (XRF) analyzer. Perkin-Elmer FTIR instrument with resolution 3 cm^{-1} was used to monitor the vibrational stretching and bending bands in silica of the samples. Mineralogical analysis was studied using the X-ray diffraction with monochromatic Cu-Kα radiation at a scanning speed of 1.5° (2θ) min^{-1} in a conventional diffractometer. The crystalline phases were identified by comparing the intensities and positions of the Bragg peaks with those listed in the JCPDS databases. The microstructure of raw CCA and treated CCA were also studied by low vacuum scanning electron microscope or SEM (Model JOEL JSM-5910LV) with the accelerating voltage of 15 kV. The specific surface area and particle size distribution of raw CCA and treated CCA were measured by Brunauer, Emmett and Teller (BET) and particle size analyzer (PSA), respectively.

2.3. Engineering properties investigations

The raw CCA and treated CCA were prepared by replacing cement over the range of 0-20% by weight (step up 5%). The 0% replacement serves as the reference specimen. The amount of water required for normal consistency of samples was carried out in accordance with the ASTM C187 standard [21]. The ASTM C191 [22] was applied to determine the initial and final setting time of the cement-CCA paste. A mortar cube of 50×50×50 mm. was produced in accordance with ASTM C109 [23] subsequently with the pozzolanic activity of the mortar determined, based on its compressive strength according to ASTM C311 [24]. For a concrete application, fresh concrete was tested using slump cone test followed ASTM C1611 [25] to find the workability of concrete. The compressive strength test and the splitting tensile strength test were operated. For compressive strength test, cube specimens of dimensions 150 × 150 × 150 mm were cast with different proportions of cement replacement fraction similar to the compressive strength test of mortar cube. The compressive strength was tested for 7, 14, 28, and 56 days curing. To investigate the splitting tensile strength, the 150 mm × 300 mm cylinders were also cast from each mix. Splitting tensile strength test was conducted after 28 days of moist curing.

3. Results and discussion

3.1. Chemical reaction of CCA during heat treatment

A typical graphical illustration of the TGA–DTA curves for CCA is shown in figure 1. The TGA result showed three distinct weight losses. The first weight loss occurred between 30 and 137°C corresponding to the endothermic peak at 106°C in the DTA curve. These observations related to the vaporization of light volatile matters and water. The second weight loss occurred between 137 and 382°C corresponding to the endothermic peak at 170°C in the DTA curve. This observation can be attributed to the depolymerizations of the polysaccharides [26]. The third weight loss was continuously occurred at the temperature higher than 500°C and corresponded to the endothermic peak at 520°C in the DTA curve. This may be attributed to the carbon char was burned and evaporated [27].

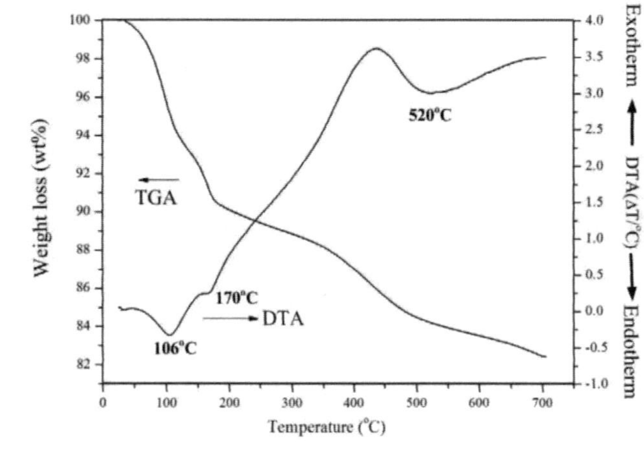

Figure 1. TGA–DTA curves for the CCA.

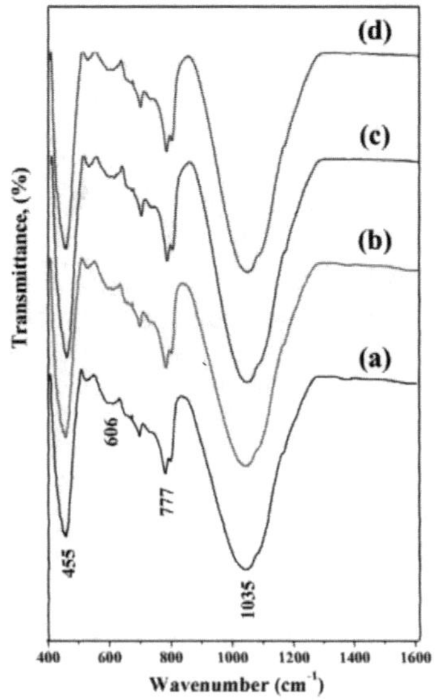

Figure 2. FTIR spectra of the corn cob ash and treated corn cob ash at various temperatures: (a) corn cob ash, (b) corn cob ash calcined at 200°C for 4 h, (c) corn cob ash calcined at 400°C for 4 h and (d) corn cob ash calcined at 600°C for 4 h.

3.2. Physical and chemical analysis of CCA and treated CCA

The FTIR spectrum of CCA, calcined corn cob ash at 200°C, 400°C and 600°C are presented in figure 2. The infrared spectrum of CCA is shown in figure 2(a) and three main signals are observed. The first one is a strong broad band at 1035 cm^{-1} is assigned to the asymmetric stretching vibration of tetrahedral SiO_4. The next signal is a sharp medium band at 777 cm^{-1}. This band is due to symmetric stretching vibrations of the same units. The third signal is a strong band at 455 cm^{-1}. This point is assigned to the Si-O bending mode. In addition, a very weak band at 606 cm^{-1} is observed due to the characteristic of the crystalline cristobalite [28, 29]. The spectrum patterns are insignificantly changed when the CCA was treated at various temperatures.

Figure 3 illustrates the X-ray diffraction patterns of CCA, calcined corn cob ash at 200°C, 400°C and 600°C for 4 h. In the case of CCA, the diffraction patterns correspond to quartz (JCPDS No: 33-1161). After heat treatments at 200, 400 and 600°C, the intensity of the observed peaks tends to decrease. This indicates that silica in the ash is transformed from crystalline to amorphous phase. This result corresponds to the thermal analysis investigation and the suggesting by Jal, et al. [30]. According to their report, naturally occurring silica is crystalline whereas synthetically obtained silica is amorphous in nature. Moreover, a broad hump at $2\theta = 22°$ to $36°$ observed in the CCA calcined at 600°C sample indicates the Si-O band structure is disordered [27].

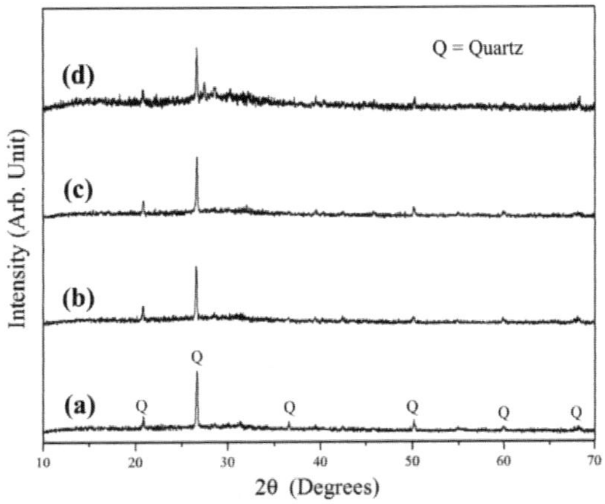

Figure 3. XRD patterns of: (a) corn cob ash, (b) corn cob ash calcined at 200°C for 4 h, (c) corn cob ash calcined at 400°C for 4 h and (d) corn cob ash calcined at 600°C for 4 h.

Table 1. Chemical composition, mean grain size and specific surface area of OPC, CCA and CCA600.

Chemical constituents (%)	OPC	CCA	CCA600
Silicon dioxide (SiO_2)	20.71	60.09	63.91
Aluminum oxide (Al_2O_3)	5.16	3.99	4.01
Iron oxide (Fe_2O_3)	3.22	4.03	3.95
Calcium oxide (CaO)	66.23	5.18	4.13
Magnesium oxide (MgO)	0.95	3.12	2.91
Potassium oxide (K_2O)	0.48	14.73	12.12
Sodium oxide (Na_2O)	0.19	-	-
Sulfur trioxide (SO_3)	2.14	0.82	0.88
Phosphorus pentoxide (P_2O_5)	-	6.72	6.49
SiO_2+ Al_2O_3+ Fe_2O_3	-	68.11	71.87
Mean grain size (μm)	15.37	23.56	28.45
BET specific surface area (m^2/kg)	3,200	9,050	7,441

The chemical composition and physical properties of OPC, CCA and corn cob ash calcined at 600°C for 4 h (CCA600) are compared in table 1. Chemical composition analysis of CCA indicates that CCA have $SiO_2 + Al_2O_3 + Fe_2O_3$ equal to 68.11%, which is lower than chemical requirements of the standard specification for natural pozzolan Class N according to ASTM C618 (summary of SiO_2,

Al_2O_3 and Fe_2O_3 higher than 70%) [31]. However, after heat treatment at 600°C for 4 h (CCA600), the percentage of SiO_2 + Al_2O_3 + Fe_2O_3 was increased to 71.87% with less than 4% of SO_3, which can be classified as Class N calcined natural pozzolan as prescribed by ASTM C618. An increase of SiO_2 + Al_2O_3 + Fe_2O_3 fraction in CCA600 ash is attributed to the vaporization of typical elements such as carbon, calcium or magnesium during heat treatment.

In terms of physical characteristics, heat treatment of CCA at 600°C for 4 h slightly increased mean grain size from 23.56 to 28.45 µm compared with the CCA (table 1). The specific surface area was determined by the Brunauer, Emmett and Teller (BET) theory as listed in table 1. The study results revealed CCA exhibits higher specific surface area compared with OPC. After heat treatment at 600°C for 4 h, specific surface area slightly decreased. This result corresponds with an increase in particle size. However, the higher means particle size with higher specific surface suggested the CCA and calcined CCA at 600°C samples show higher porosity than OPC.

The surface microstructural photographs of CCA and CCA calcining at different temperature are shown in figure 4. The CCA exhibits a polygonal shape and a wide range of particle size distribution (figure 4(a)). A higher narrow range of particle size distribution and formation of larger ash agglomerates appeared when the CCA was heat treated at 200°C as shown in figure 4(b). Increasing the calcining temperature to 400°C brought about an increase in the particle size due to smaller particles melting and merging into larger grains (figure 4(c)). An increase of the particle size and an agglomerate effect were obviously observed when the calcining temperature was increased to 600°C (figure 4(d)). The increase in mean particle size that accompanied the heat treatment process, observed from the SEM results, corresponds to the particle size analysis which was observed from the particle size analyzer.

Figure 4. SEM photographs of corn cob ash and calcined corn cob ash; (a) corn cob ash, (b) corn cob ash calcined at 200°C for 4 h, (c) corn cob ash calcined at 400°C for 4 h and (d) corn cob ash calcined at 600°C for 4 h.

From the XRF and XRD analyzes, the CCA600 showed higher amounts of SiO_2 + Al_2O_3 + Fe_2O_3 fraction and amorphous silica phase compared with raw CCA. These factors suggest the CCA600 probably exhibits more appropriate behaviors of pozzolanic material than raw CCA. Therefore, the CCA600 was selected to investigate the engineering properties for applied as supplementary cementing material when compared with the raw CCA material.

3.3. Consistency and setting time of CCA and CCA600 blended cement

The standard test method following ASTM C187 [21] is used to determine the amount of water required in preparing hydraulic cement pastes and cement-ash replacement with normal consistency. The results are listed in table 2. Replacing a fraction of OPC with CCA or CCA600 caused an increase in water requirement in a paste. This observation can be attributed to the internal porosity and specific surface area of CCA and CCA600 particles being higher than OPC particles. The water requirements of cement pastes with CCA and CCA600 replacement are slightly different. However, the standard specification for raw or calcined natural pozzolan for use in concrete (ASTM C618) [31] limits water requirements at 115% of reference specimen (water/binder ~30.87%). Hence, this is a problem in the application of CCA or CCA600 as a supplementary cementing material.

Table 2. Water/binder ratio, initial setting time and final setting time of cement paste, CCA-blended cement at various percentage and CCA600-blended cement at various percentage.

Samples	% replacement	W/B ratio at normal consistency (%)	Initial setting time (min.)	Final setting time (min.)
Reference	0	26.84	115	223
CCA	5	28.80	183	314
	10	30.98	222	349
	15	34.05	252	387
	20	35.77	271	405
CCA600	5	28.44	175	306
	10	32.48	229	370
	15	34.26	261	377
	20	36.43	270	414

The water/binder ratios are giving the paste normal consistency were selected to prepare specimens for initial and final setting time testing as shown in table 2. In the case of CCA, the initial and final setting times increased from 115 to 271 min. and 223 to 405 min., respectively when the replaced CCA percentage was enhanced from 0 to 20%. An increment of setting times with the increasing of replaced CCA ratio can be explained by two reasons. The first one is an increase of CCA content reduces the surface area of cement. As a result of this, the hydration process is slowed down, thus causing the setting times to increase. Another one reason is replacing of CCA caused the rate of the hydration process to slow down. Hence, the heat released from the hydration reaction is decreased when the evaporation of water slows which delays the setting times. The variation of setting times of the samples with CCA600 replacement shows a similar trend to CCA replacement. It can be observed that all the sample satisfy the NIS 439:2000 requirement of 45 minutes minimum initial setting time and maximum of 10 h final setting time [1].

3.4. Compressive strength of mortar cube

The compressive strengths of mortar cube at various percentage replacements of CCA or CCA600 are shown in figures 5 and 6, respectively. In the case of CCA replacement, an increasing of replacement percentage led the short term compressive strength, such as 24 h or 3 days, decreased. This is due to an increase of CCA decreased the OPC content, which induces lower hydration reaction than those of the samples with low CCA fraction. However, a differential in compressive strength between the CCA replacement and the 0% CCA replacement samples tended to decrease by increasing a soaked time to 28 days. An increment of compressive strength in the samples with CCA replacement can be attributed to the cement gel was produced by pozzolanic reaction between the amorphous silica from CCA and the calcium hydroxide ($Ca(OH)_2$) in paste. The mechanisms of cement hydration with CCA are shown in equations (1) and (2).

$$C_3S\ (OPC) + C_2S\ (OPC) + H_2O \rightarrow C\text{-}S\text{-}H + Ca(OH)_2 \quad (1)$$

$$SiO_2(CCA) + Ca(OH)_2 + nH_2O \rightarrow C\text{-}S\text{-}H + SiO_2(\text{unreacted}) \quad (2)$$

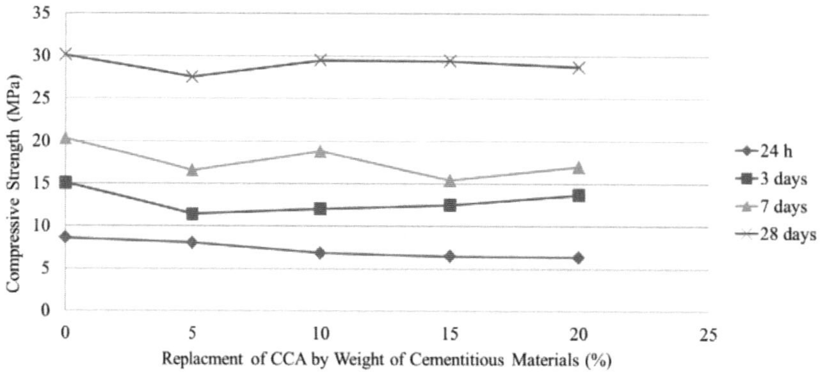

Figure 5. Effect of CCA replacement on the compressive strength of mortar cubes.

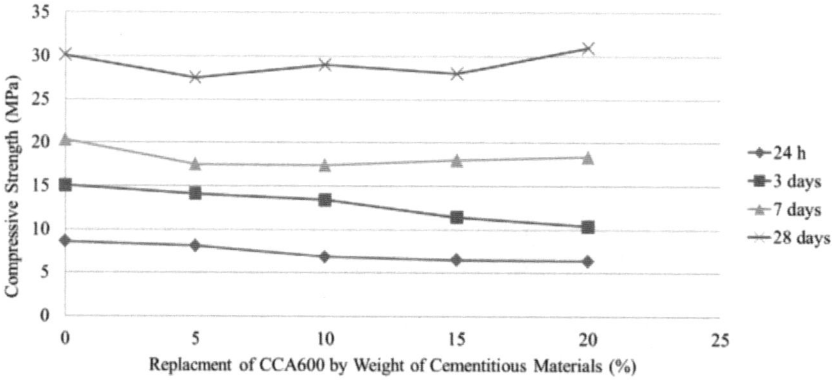

Figure 6. Effect of CCA600 replacement on the compressive strength of mortar cubes.

The relationship between the compressive strength of mortar and replacement of CCA600 showed a similar trend with the CCA replacement. Moreover, at the age of 28 days, the compressive strength of CCA600 with 20% replacement was higher than the sample with 0% replacement. This may be due to strong pozzolanic reaction between amorphous silica and $Ca(OH)_2$ and results on the higher compressive strength. The compressive strengths at 28 days were used to calculate the pozzolanic activity index. The samples of 20% replacing by CCA and the 20% replacing by CCA600 samples showed 95% and 103% of the pozzolanic activity index, respectively. These values are higher than minimum requirements of 75% as referred by ASTM C618 [31], which indicate that CCA and CCA600 are suitable pozzolanic materials to be used for replacing OPC for concrete.

3.5. Compressive strength and splitting tensile strength of concrete

Five different replacement percentages of concrete mixes (replacement ranging from 5% to 20% by weight of cement) including the control mix were prepared with a water binder ratio of 0.53 for a design cube compressive strength of 25 MPa 17. The slump test results indicate that the concrete slump decreases as the CCA or CCA600 content increases (table 3). These results indicate that concrete becomes less of their workable as the CCA or CCA600 replacement percentage increases and meaning that more water is required to make the mixes more workable. The high demand for water as the CCA or CCA600 content increases shows similar trend when compared with the W/B ratio at a normal consistency of cement paste results. These can be attributed to the silica (from CCA or CCA600) – lime (from hydrated cement) reaction requires more water in addition to a water required during hydration of cement [32-34].

Table 3. Compressive strength and splitting tensile strength at 28 days of CCA and CCA600 mixed concrete.

Sample	% replacement	W/B ratio	Slump (mm)	28 days compressive strength (MPa)	28 days splitting tensile strength (MPa)
Reference	0	0.53	102	38	4.2
CCA	5	0.53	104	34	4.2
	10	0.53	98	42	4.1
	15	0.53	90	41	4.3
	20	0.53	87	37	3.8
CCA600	5	0.53	97	41	4.0
	10	0.53	97	45	4.6
	15	0.53	94	39	4.1
	20	0.53	90	39	3.7

The compressive strength and the splitting tensile strength values of CCA and CCA600 blended concretes after 28 days of curing are listed in table 3. It can be clearly observed that the compressive strength reached highest value for the 10 replacement percentage of CCA or CCA600 samples. The 28 days compressive strength of the sample with 10% replacement of CCA and CCA600 are increased by 11% and 18%, respectively as compared to the control sample. The 28 days splitting tensile strength of the CCA or CCA600 replacement samples slightly changed with the increase of replacement fraction. The highest values of splitting tensile strength can be observed from 15% replacement percentage of CCA and 10% replacement percentage of CCA600. However, the splitting tensile strength tended to decrease when the CCA or CCA600 replacement fraction was increased to 20%.

The compressive strength of CCA and CCA600 blended concretes specimens for 7, 14, 28 and 50 days of curing times are shown in figures 7 and 8, respectively. It can be observed that the short-term strength (7 days and 14 days curing times) of the CCA and CCA600 replacement slightly changed with increasing replacement percentage. However, it was found that as CCA or CCA600 fraction increased, long-term strength (28 days and 50 days curing times) tended to increase and reached maximum value in the samples with 10% replacement percentage thereafter, it decreased for the samples with 15 and 20% replacement fraction. A decrease of the 28 days compressive strength of concrete with 15 and 20% of CCA replacement is not a similar trend when compare to a mortar cube, which shows constant strength with increase CCA fraction. This can be attributed to the water added affect. In the previous analysis, an increase of CCA caused increase in water requirement in a paste. The water added in mortar mixed was adjusted for a normal consistency. On the other hand, the W/B ratio of concrete mixed was fixed to 0.53 and may be caused an ineffectiveness of cement hydration reaction due to insufficient water.

An increase of long-term strength with increasing in CCA or CCA600 replacement ratio may be attributed to the pozzolanic reaction between the amorphous silica from CCA or CCA600 and the calcium hydroxide in a paste as shown in equation (2). As the CCA or CCA600 content increases, more silica is available to react with the calcium hydroxide produced during the hydration of cement, thus producing more cementitious materials, which contribute to the higher compressive strength. A distinct increase of long-term strength suggests that CCA or CCA600 blended cement gain strength slowly at early curing age. This result corresponds with the findings from previous studies [35, 36], which state that pozzolanic reaction is a slow rate, thus a longer curing time is needed to observe its positive effects [37]. A decrease in compressive strength of concrete is initially observed when the CCA or CCA600 replaced cement higher than 10% by weight. This result suggests that the presence of CCA or CCA600 above 10% is not advantageous in improving the compressive strength of concrete. Moreover, the compressive strength of the CCA600 replacement samples is slightly higher than the samples with CCA replacement. This result corresponds with the chemical and mineralogical analyzes. The compressive strength of concrete at 50 days curing period of CCA and CCA600 replaced cement shows higher values when compared with the control sample. This demonstrates the potential for the CCA and CCA600 for used as a cement replacement materials.

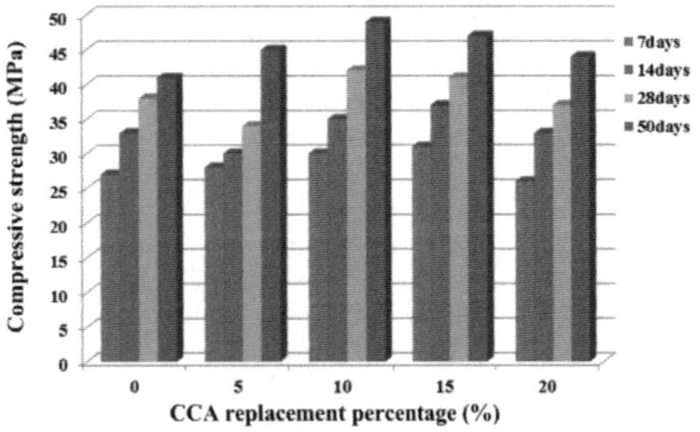

Figure 7. Compressive strength of CCA blended concretes.

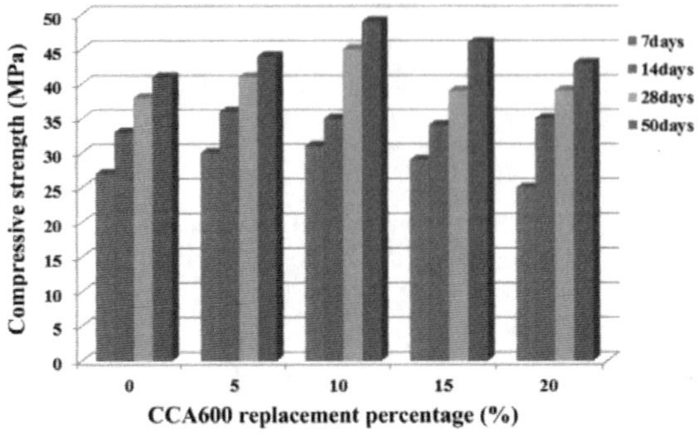

Figure 8. Compressive strength of CCA600 blended concretes.

4. Conclusion

From the results of various tests performed, the following conclusions can be drawn:

- The summary of SiO_2, Al_2O_3 and Fe_2O_3 and the amorphous silica phase tended to increase with increasing the calcined temperatures.
- The corn cob ash which treated at 600°C for 4 h exhibits satisfied chemical and physical properties for applied as pozzolanic material.
- An increment of the replacement fraction of CCA or CCA600 caused the water requirements, initial setting time and final setting time of the cement paste increased.
- The pozzolanic activity index, which calculated from the compressive strengths of mortar at 28 days of the 20% replacing by CCA and the 20% replacing by CCA600 samples are 95% and 103%, respectively.
- The CCA600 samples show slightly higher effectiveness for improvement the splitting tensile strength and compressive strength of concrete when compared with CCA.
- CCA and CCA600 can be used as a cement replacement material in concrete to improve the compressive strength at later ages. Due to the long-term strength development.

Finally, the test results showed that the CCA or CCA which treated at 600°C for 4 h have a potential to be used as alternative pozzolanic materials.

Acknowledgments

This work was financially supported by the Naresuan University. Thanks also to Department of Civil Engineering, Faculty of Engineering, Naresuan University for supporting facilities. Acknowledgments are also to Mr. Don Hindle, for helpful comments and corrections of the manuscript.

References

[1] Adesanya D A and Raheem A A 2009 Development of corn cob ash blended cement *Constr. Build. Mater.* **23** 347-52
[2] Hardjito D, Wallah S E, Sumajouw D M J and Rangan B V 2004 On the development of fly ash-based geopolymer concrete *ACI Mater. J.* **101** 467-72
[3] Badogiannis E, Kakali G, Dimopoulou G, Chaniotakis E and Tsivilis S 2005 Metakaolin as a main cement exploitation of poor greek kaolins *Cem. Concr. Compos.* **27** 197-203

[4] Arezoumandi M and Volz J S 2013 Effect of fly ash replacement level on the shear strength of high-volume fly ash concrete beams *J. Clean. Prod.* **59** 120-30

[5] Bentz D P 2010 Powder additions to mitigate retardation in high-volume fly ash mixtures *ACI Mater. J.* **107** 508-14

[6] Bhanja S and Sengupta B 2005 Influence of silica fume on the tensile strength of concrete *Cem. Concr. Res.* **35** 743-7

[7] Behnood A and Ziari H 2008 Effects of silica fume addition and water to cement ratio on the properties of high-strength concrete after exposure to high temperatures *Cem. Concr. Compos.* **30** 106-12

[8] Morales E V, Villar-cocina E, Frias M, Santos S F and Savastano H J 2009 Effects of calcining conditions on the microstructure of sugar cane waste ashes (SCWA): Influence in the pozzolanic activation *Cem. Concr. Compos.* **31** 22-8

[9] Subramaniam K V, Gromotka R, Shah S P, Olba K and Hill R 2005 Influence of ultrafine fly ash on the early age response and the shrinkage cracking potential of concrete *ASCE J. Mater. Civil Eng.* **17** 45-53

[10] Lam L, Wong Y L and Poon C S 2000 Degree of hydration and gel/space ratio of high-volume fly ash/cement systems *Cem. Concr. Res.* **30** 747-56

[11] Bui D D, Hu J and Stroeven P 2005 Particle size effect on the strength of rice husk ash blended gap-graded Portland cement concrete *Cem. Concr. Compos.* **27** 357-66

[12] Yu Q, Sawayama K, Sugita S, Shoya M and Isojima Y 1999 The reaction between rice husk ash and $Ca(OH)_2$ solution and the nature of its product *Cem. Concr. Res.* **29** 37-43

[13] Nair D G, Fraaij A, Klaassen A A K and Kentgens A P M 2008 A structural investigation relating to the pozzolanic activity of rice husk ashes *Cem. Concr. Res.* **38** 861-9

[14] Sata V, Jaturapitakkul C and Kiattikomol K 2007 Influence of pozzolan from various by-product materials on mechanical properties of high-strength concrete *Constr. Build. Mater.* **21** 1589-98

[15] Yin C Y, Ali W S W and Lim Y P 2008 Oil palm ash as partial replacement of cement for solidification/stabilization of nickel hydroxide sludge *J. Hazard Mater.* **150** 413-8

[16] Tangchirapat W, Jaturapitakkul C and Chindaprasirt C 2009 Use of palm oil fuel ash as a supplementary cementitious material for producing high-strength concrete *Constr. Build. Mater* **23** 2641-6

[17] Ganesan K, Rajagopal K and Thangavel K 2007 Evaluation of bagasse ash as supplementary cementitious material *Cem. Concr. Compos.* **29** 515-24

[18] Cordeiro G C, Filho R D T, Tavarse L M and Fairbairn E M 2008 Pozzolanic activity and filler effect of sugar cane bagasse ash in Portland cement and lime mortars *Cem. Concr. Compos.* **30** 410-8

[19] Cordeiro G C, Filho R D T, Tavarse L M and Fairbairn E M 2009 Ultrafine grinding of sugar cane bagasse ash for application as pozzolanic admixture in concrete *Cem. Concr. Res.* **39** 110-5

[20] Raheem A A and Adesanya D A 2011 A study of thermal conductivity of corn cob ash blended cement mortar *PJST* **12** 106-11

[21] American Society for Testing and Materials (ASTM) 2012 Standard test method for amount of water required for normal consistency of hydraulic cement paste (C187-98)

[22] American Society for Testing and Materials (ASTM) 2012 Standard test method for time of setting of hydraulic cement by vicat needle (C191-13)

[23] American Society for Testing and Materials (ASTM) 2012 Standard test method for compressive strength of hydraulic cement mortars using 2-in. or 50-mm cube specimens (C109/109M-02)

[24] American Society for Testing and Materials (ASTM) 2012 Standard test method for sampling and testing fly ash or natural pozzolans for use in Portland-cement concrete (C311/C311M-13)

[25] American Society for Testing and Materials (ASTM) 2012 Standard test method for slump flow of self-consolidating concrete (C1611/161M-14)

[26] Ramadhansyah P J, Mahyun A W, Salwa M Z M, Bakar B H, Johari M A and Ibrahim M H 2012 Thermal analysis and pozzolanic index of rice husk ash at different grinding time *Proceedings of the ICASCE 2012* (Jakarta, Indonesia) pp 24-25

[27] Mohanraj K, Kannan S, Barathan S and Sivakumar S 2012 Preparation and characterization of nano SiO_2 from corn cob ash by precipitation method *Optoelectron. Adv. Mat.* **4** 394-7

[28] Prasetyoko D, Ramli Z, Endud S, Hamdan H and Sulikowski B 2006 Conversion of rice husk ash to zeolite beta *Waste Manage.* **26** 1173-9

[29] Govindarajan D and Jayalakshmi G 2011 XRD, FTIR and microstructure studies of calcined sugarcane bagasse ash *Adv. Appl. Sci. Res.* **2** 544-9

[30] Jal P K, Sudarshan M, Saha A, Patel S and Mishra B K 2004 Synthesis and characterization of nanosilica prepared by precipitation method *Colloids Surfaces* **240** 173-8

[31] American Society for Testing and Materials (ASTM) 2012 Standard specification for coal fly ash and raw or calcined natural pozzolan for use in concrete (C618-12a)

[32] Syagga P M, Kamau S N, Sabuni B W and Dulo S O 2001 Potentials of using waste burnt clay as a pozzolanic material in Kenya *J. Discover. Innov.* **13** 114-8

[33] Sabuni B W, Syagga P M, Dulo S O and Kamau G N 2002 Rice husk ash cement-an alternative pozzolana cement for Kenyan building industry *J. Civil Eng.* **8** 13–26

[34] Shihembetsa L U and Sabuni B W 2002 Burnt clay waste as a pozzolanic material in Kenya *J. Civil Eng.* **7** 21-8

[35] Balendran R V and Buades W H M 2000 The influence of high temperature curing on the compressive, tensile and flexural strength of pulverized fuel ash concrete *Build. Environ.* **35** 415-23

[36] Antiohos S, Maganari K and Tsimas S 2005 Evaluation of blends of high and low calcium fly ashes for use as supplementary cementing materials *Cem. Concr. Compos.* **2** 349-56

[37] Adesanya D A and Raheem A A 2009 A study of the workability and compressive strength characteristics of corn cob ash blended cement concrete *Constr. Build. Mater.* **23** 311-7

Validity and reliability of estimated modulus of elasticity of cementitious materials

N U Kockal[1]

Akdeniz University, Department of Civil Engineering, 07058, Antalya, Turkey

E-mail: ukockal@yahoo.com

Abstract. Modulus of elasticity is a vital property in cementitious material (CM) design and analysis. A better understanding of the relationship between density, strength and stiffness is needed to construct proper structures especially with reinforced cementitious materials. For this purpose, modulus of elasticity of CMs with different waste aggregates was analyzed and the reliability of estimated values of various equations proposed by standards and sources in literature, suggested in this investigation and derived by a software chosen was discussed. The results demonstrated that as compared to experimental results, the model developed by software was the most accurate as the percentages of error in prediction were in a good agreement.

1. Introduction

The estimation of the hardened characteristics of CM by evaluating its properties has been of deep interest. Among these hardened characteristics, to design plain, reinforced, and prestressed structures, the elastic modulus E is a fundamental parameter that needs to be defined [1]. The elastic modulus of CM is very difficult to predict because the properties of CM is affected by the features and amounts of ingredients [2]. In addition to all these different ingredients, the differences in features of the different ingredients at the various scales further enhance the inhomogeneity of cementitious materials [3]. Differences in rigidity and strength of the ingredients possess the effect on the overall rigidity and failure behavior of the material. Particularly, E-modulus and shape of aggregate and cement paste structure have an important effect on the results [4]. Phase and its volumetric ratio considerably affect the CM strain due to the anisotropic structure. The transition zone between aggregate and cement paste also has a vital significance on elastic behavior as well as the phases. The non-homogeneity and anisotropy have made CM nearly impossible to determine its modulus of elasticity by either a theoretical approach or a directly experimental procedure [5]. The modulus of elasticity of CM is commonly defined in terms of compressive strength [6]. The equations derived from experimental investigations are present to predict the modulus of elasticity from the compressive strength, which is a conventional evaluation for determining the properties of CM [7]. Although most empirical expressions for estimating the modulus of elasticity have been suggested by some researchers, very few expressions are corresponded to all the data because the mechanical features of CM are remarkably related to the features and amounts of binders and aggregates [6]. Aggregates play an important role on both fresh and hardened features of CM [8]. The selection of the suitable fine

[1] Address for correspondence: N U Kockal, Akdeniz University, Department of Civil Engineering, 07058, Antalya, Turkey. E-mail: ukockal@yahoo.com.

aggregate for a specified implementation is of major significance as far as features of CM are taken into consideration.

Different formulas are proposed by many researchers to compute the modulus of elasticity of CM. Most of them based on the compressive strength are not suitable for both normal and high strength CMs. Many models suggested in the literature could not precisely predict the modulus of elasticity of CM specimens made with different types of aggregates.

The aim was to study the relationship between main parameters namely the density and the compressive strength that had an important influence on the modulus of elasticity and to find the most reliable formula which can be applicable to CMs containing different types of aggregates. Therefore, the experimentally obtained compressive strength, modulus of elasticity and density values were analyzed, and the coefficients of the parameters were arrived at and given by the regression equations that give the predicted values.

2. Experimental data and details

The models presented in this study are validated on experimental data from a previous research [9]. In the pronounced investigation, the materials used, and procedure applied on specimens for determination of E-modulus, density and compressive strength are as in the following.

CEM II/A-M (P-L) 42,5N was used as cement according to TS EN 197-1 Brick, marble and ceramic were collected from construction sites and ground until a similar grading as of natural crushed aggregate was obtained. Natural crushed fine aggregate utilized in the study was a calcareous aggregate. Super-plasticizer was used for providing flow diameter of CM in a constant range in order to prevent consistency difference effect on specimen properties.

The calculation of density of hardened CM is given in equation (1).

$$density = \frac{A}{B-C} \tag{1}$$

Where A is the dry mass in the air (g), B is the saturated surface-dry mass in the air (g), and C is the saturated surface-dry mass in the water.

The modulus of elasticity of the ø50x100 mm cylinder specimens was measured according to ASTM C 469-02 at the age of 28 days and calculated by using the equation (2):

$$E = \frac{\sigma_2 - \sigma_1}{\varepsilon_2 \text{ - } 0.000050} \tag{2}$$

Where: E = chord modulus of elasticity, MPa, σ_2= stress at 40% of ultimate load, σ_1 = stress at a longitudinal strain, ε_1, of 50 millionths, MPa, ε_2= longitudinal strain produced by stress σ_2.

3. Tools used for modulus of elasticity estimation

As also stated in another study [6], CM with a modulus of elasticity of 0 MPa possesses no compressive strength, therefore, the equations except for software is expressed as in equation (3).

$$E = a * \sigma^b * \gamma^c \tag{3}$$

3.1. Models specified by codes

ACI 318 and BS 8110 propose the equations (4) and (5), respectively to calculate the modulus of elasticity values of specimens.

$$E_c = \gamma^{1,5} * 0,043 * \sqrt{\sigma} \tag{4}$$

$$E_c = \gamma^2 * 0.0017 * \sigma^{0.33} \tag{5}$$

IOP Conf. Series: Materials Science and Engineering **103** (2015) 012024 doi:10.1088/1757-899X/103/1/012024

Where: E_c= modulus of elasticity, γ = air dry density of the specimen, σ= compressive strength.

3.2. Model suggested by software

Analysis of variance (ANOVA) is a method to analyze outputs. The variance of the observations is the sum of the variances of the independent sources when several sources of variation are acting simultaneously on a set of observations [10]. Therefore, regression and graphical analysis of the obtained data were made by a software TableCurve 3D. In order to find the ideal equation to describe three-dimensional empirical data, TableCurve 3D combines a powerful surface fitter with the ability. In TableCurve 3D, the dependent variable is Z. All equations express Z as a function of the two independent variables X and Y.

The equation (6) is an empirical relationship between the modulus of elasticity (*Z*) and density (*X*) and compressive strength (*Y*) and was yielded by the application of response surface methodology.

$$E = a+b*\sigma+c*\gamma \tag{6}$$

3.3. Model derived in current study

The best regression yielded with an expression of the power type stated by equation (7) gave the relationship between the modulus of elasticity and the compressive strength.

$$E = 318{,}8*\sigma^{0{,}863},\ R^2 = 0{,}86 \tag{7}$$

In addition, regression analysis was experienced on the modulus of elasticity and density in oven dry condition (OD) using power regression model which presented the best fit against the other models with the equation (8):

$$E = 6985*\gamma^{1{,}936},\ R^2 = 0{,}79 \tag{8}$$

By multiplying equations (7) and (8), modulus of elasticity was calculated in equation (9).

$$E = 1492{,}25*\sigma^{0.929}*\gamma^{1.391} \tag{9}$$

Where E is the modulus of elasticity (MPa), σ is the compressive strength (MPa) and γ is the density in OD condition (kg/dm^3).

4. Results and discussion

The observed results and predicted modulus of elasticity values are shown in table 1 while figure 1 exhibits the relationship between observed and estimated modulus of elasticity and compressive strength of CMs. All models indicated that the E value can be predicted by considering the compressive strength and density. Although elastic properties are not sufficient to explain mechanical behavior, they can help us understand the effects of microstructural properties more easily, because their prediction does not involve many unknown parameters [11].

Table 1. Experimental and estimated results.

Modulus of Elasticity/GPa				
Actual	Software	ACI 318	BS 8110	Calculated
28,5	29,1	27,2	25,8	29,0
27,6	26,9	25,3	24,2	27,1
27,9	27,6	25,5	24,1	27,6
28,2	28,0	25,6	24,0	27,8
28,1	27,8	25,4	23,7	27,6
24,3	25,3	23,3	21,9	25,5
23,9	23,7	21,4	19,9	23,8

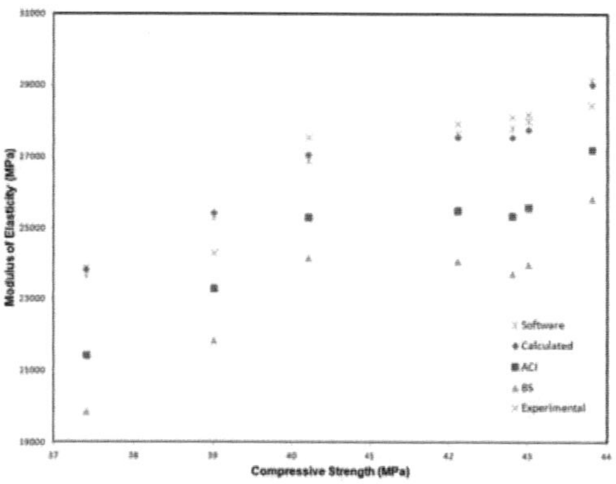

Figure 1. Relationship between modulus of elasticity and compressive strength.

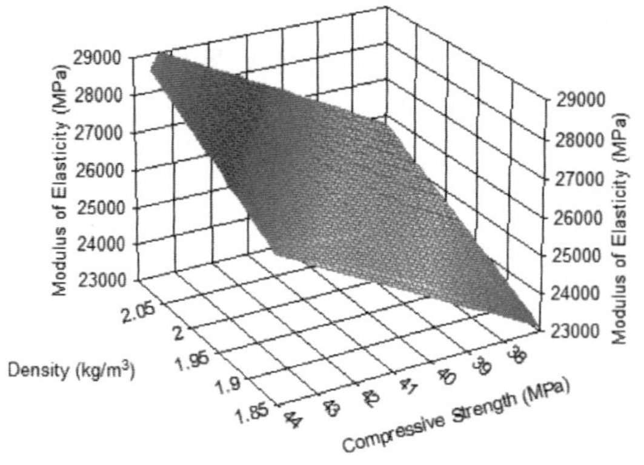

Figure 2. 3D view-response surface plot showing the effect of density and compressive strength on modulus of elasticity.

Power regression models were used to correlate the parameters except software. The modulus of elasticity obtained by all models mostly underestimated the results. However, equations proposed by software and developed in the current study presented close predicted values to the experimental results. In a study [12], the modulus of elasticity values of CMs were overestimated by the models given by standards whereas in another study [13], the obtained modulus of elasticity using code formulae were less than the experimental result. In the determination of modulus of elasticity, γ used was the density (OD) instead of air dry density and σ was the cube compressive strength instead of cylinder strength specified in standards.

When the design of the construction is based on elasticity considerations, the modulus of elasticity of cement-based material is a parameter essential in the structural analysis for the determination of the strain distributions and displacements [7]. The features of cementitious materials remarkably are related to the multi-scale distribution of fine aggregates and voids inside the matrix as well as on the shape and size of aggregates [14]. Hence, manufacturing process of such materials commonly results in a great diversity in the microstructure.

In CM with high strength, the modulus of elasticity of the hardened cement matrix is high and the variation between the modulus of elasticity of the aggregates and the hardened matrix becomes small enough to lead to larger bond strength and monolithic behavior [15]. This can be attributed to the low modulus of elasticity of the aggregate, leading to a smaller variation between its value and that of the hardened matrix, and to the increased aggregate–paste bond due to the rough and porous surfaces of the aggregates. The modulus of elasticity of the aggregate depends not only on the density, on the pore structure as well and the surface texture of aggregate [16]. Therefore, an aggregate with a compact structure and evenly distributed pores will give a higher modulus of elasticity and more CM rigidity than a more porous aggregate. As compared to normal weight CM, a much lower modulus of elasticity would be expected for lighter CM [17]. The lower E-value absorbs energy from impact and cyclic loading, decreasing the occurence of micro-cracking at the transition zone between cement and aggregate. A lower E-value provides more flexibility because rigidity is the product of the modulus of elasticity and the moment of inertia, EI [18]. Reduced rigidity could be beneficial at times, and utilization of CM with low density should be considered in these cases rather than normal-weight CM. In cases requiring improved impact or dynamic response, where differential settlement of foundation may take place, and in configurations of shell roofs, reduced rigity may be necessary.

Figure 2 illustrated the three dimensional view of the estimated values obtained by the software. The actual data of the compressive strength, density and modulus elasticity of CMs were analyzed by applied statistical method. X represented density, Y represented compressive strength and Z represented modulus of elasticity. The value of the determination coefficient (R^2 = 0.90) confirms the proper fit of the model, thus indicating a discrepancy of 0.10 % for total variation, which is a normally acceptable range of experimental error. The value of the adjusted determination coefficient (adjusted $R^2 = 0.80$) is also high which indicates a high significance for the model.

Relationships between experimental and estimated values of modulus of elasticity are shown in figure 3. The evidence for a good relationship is large R^2 value which indicated that there were negligible differences between the observed and predicted values of modulus of elasticity. The R^2 values were 0.901, 0.875, 0.845 and 0.895 for estimated values obtained by using the software, ACI 318, BS 8110 and model developed in the current investigation, respectively. All the predicted values were close to each other and exhibited small variations with the actual data. The software suggested the best fit model which possessed the highest R-square value.

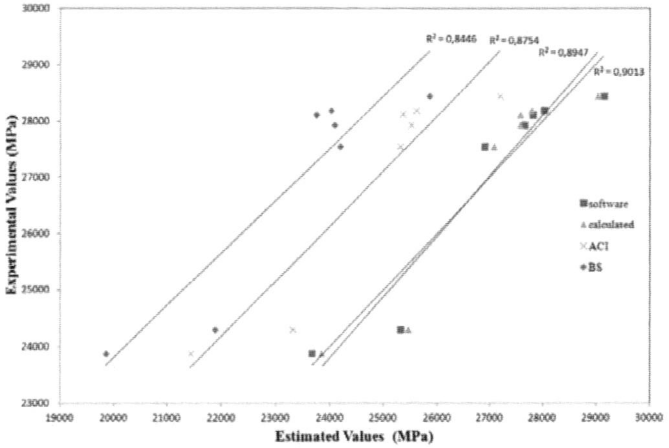

Figure 3. Relationships between experimental values and estimated values of modulus of elasticity.

5. Conclusions

The final models which were empirical relationships between density and compressive strength and modulus of elasticity derived in the current study, developed by software using response surface methodology and suggested by codes were compared with experimental results. According to the obtained results, all developed models were statistically accurate.

The modulus of elasticity of the cementitious materials was related to compressive strength as well as density, a strong dependency being observed. However, the relationship presented by ACI 318 and the suggested one by BS 8110 were similar, and their R-square values were close and relatively low. Besides, the largest R-square values were obtained from the models both given by software and calculated in the current study.

References

[1] Noguchi T, Tomosawa F, Nemati K M, Chiaia B M and Fantilli A P 2009 A practical equation for elastic modulus of concrete *ACI Struct J* **106** 690-6

[2] Yang C C, Lin Y Y and Huang R 1996 Elastic modulus of concrete affected by elastic moduli of mortar and artificial aggregate *J Mar Sci Technol* **4** 43-8

[3] Bernard F, Kamali-Bernard S and Prince W 2008 3D multi-scale modeling of mortar mechanical behavior and effects of microstructural changes *Cement Concrete Res* **38** 449-58

[4] Topçu İ B and Uğurlu A 2007 Elasticity theory of concrete and prediction of static e-modulus for dam concrete using composite models *Teknik Dergi* **18** 4055-67

[5] Nilsen A U, Monteiro P J M and Gjorv O E 1995 Estimation of the elastic moduli of lightweight aggregate *Cement Concrete Res* **25** 276-80

[6] Nemati K M 2007 Relationship between the compressive strength and modulus of elasticity of high-strength concrete *CBM-CI International Workshop* (Karachi, Pakistan) 511

[7] Vilardell J, Aguado A, Agullo L and Gettu R 1998 Estimation of the modulus of elasticity for

IOP Conf. Series: Materials Science and Engineering **103** (2015) 012024 doi:10.1088/1757-899X/103/1/012024

dam concrete *Cement & Concrete Research* **38** 93-101

[8] Gonçalves J, Tavares L M, Toledo Filho R, Fairbairn E R and Cunha E 2007 Comparison of natural and manufactured fine aggregates in cement mortars *Cement Concrete Res* 37 924-32

[9] Kockal N U 2015 Behavior of mortars produced with construction wastes exposed to different treatments *Indian J Eng Mater S* **22** 203-14

[10] Chi J M, Huang R, Yang C C and Chang J J 2003 Effect of aggregate properties on the strength and stiffness of lightweight concrete *Cement Concrete Comp* **25** 197-205

[11] Chamrova R 2010 Modelling and measurement of elastic properties of hydrating cement paste (École Polytechnıque Fédérale de Lausanne, Switzerland: PhD Thesis)

[12] Nassif H 2003 Development of high-performance concrete for transportation structures in New Jersey (In cooperation with New Jersey Department of Transportation Division of Research and Technology and U.S. Department of Transportation Federal Highway Administration) FHWA NJ 2003-016 (New Jersey)

[13] Al-Khaiat H and Haque M N 1998 Effect of initial curing on early strength and physical properties of a lightweight concrete *Cement Concrete Res* **28** 859-66

[14] Escoda J, Willot F, Jeulin D, Sanahuja J and Toulemonde C 2011 Estimation of local stresses and elastic properties of a mortar sample by FFT computation of fields on a 3D image *Cement Concrete Res* **41** 542-56

[15] Kayali O, Haque M N and Zhu B 2003 Some characteristics of high strength fiber reinforced lightweight aggregate concrete *Cement Concrete Comp* **25** 207-13

[16] Kockal N U and Ozturan T 2011 Strength and elastic properties of structural lightweight concretes *Mater Design* **32** 2396-403

[17] Lo T Y, Cui H Z and Li Z G 2004 Influence of aggregate pre-wetting and fly ash on mechanical properties of lightweight concrete *Waste Manage* **24** 333-8

[18] Committee ACI 213 2003 Guide for structural lightweight aggregate concrete (ACI 213R-03)

Accelerated carbide spheroidisation of 1.2343 tool steel by induction heating

J Dlouhý[1] and M Kövér

Comtes FHT a.s., Průmyslová 995, 334 41, Dobřany, Czech Republic

E-mail: jaromir.dlouhy@comtesfht.cz

Abstract. Tool steels undergo spheroidisation or soft annealing to enhance machinability and cold formability. Conventional soft annealing takes several hours. The final microstructure is composed of globular carbides in a ferritic matrix. We present an alternative process of carbide spheroidisation and steel softening. Accelerated carbide spheroidisation and refinement (ASR) was achieved by induction heating at temperatures close to the A_1 temperature. The spheroidised structure was obtained in less than 5 minutes. The carbide particles that formed during the ASR were significantly finer than for the conventional soft annealing. The hardness after ASR was higher than the hardness after soft annealing because of the dispersion strengthening by finer and more densely distributed carbide particles. On the other hand, the fine structure is favourable for hardening. It enables smaller austenite grains and martensite laths to be obtained.

1. Introduction

Manufacturing routes for products from tool steels often include soft annealing, typically before machining [1]. Upon metallurgical processing, alloyed tool steels have pearlitic, bainitic or martensitic microstructures, depending on their chemical composition and the cooling rate after casting or hot forming. As these microstructures are not favourable for machining or cold forming to the final shape, soft annealing is carried out. The steel is heated to approximately the A_1 temperature and held there for several hours. Carbon diffusion leads to carbide spheroidisation driven by the need to minimise the ferrite-carbide interface area. The final step in soft annealing is slow cooling in a furnace in order to produce a microstructure composed of a soft ferritic matrix and globular carbides.

1.2343 tool steel is widely used for hot work moulds, dies or punches [2]. Its conventional thermal treatment route comprises soft annealing at approximately 800°C, heating and subsequent quenching from 1050°C, and tempering. Soft annealing is crucial to the material's machinability and has implications for the quenching schedule. Coarsened spheroidal carbides enriched with alloying elements need to be held at a relatively high austenitizing temperature for a considerable time (typically 1 hour) in order to dissolve. Austenite grain growth is controlled by the pinning effect of vanadium carbides, which remain stable up to 1100°C. The austenite grain size dictates the maximum size of the martensite crystals after quenching. Finer microstructure possesses better material performance [3-5].

The presented article is devoted to the process of accelerated carbide spheroidisation and

[1] Address for correspondence: J Dlouhý, Comtes FHT a.s., Průmyslová 995, 334 41, Dobřany, Czech Republic. E-mail: jaromir.dlouhy@comtesfht.cz.

refinement (ASR). The aim is to obtain a microstructure morphology identical to the soft-annealed microstructure with finer and more densely distributed carbides in the ferritic matrix. The ASR process used as a substitute for lengthy soft annealing should take no more than 10 minutes [6]. Two positive effects are sought: a reduction in the processing time and better performance characteristics of the steel. Bainitic and martensitic structures in the initial condition are favourable for obtaining fine globular carbides [7], especially in case of steel alloyed by Mo and V [8].

The ASR process consists in rapid heating to a temperature around the A_1 and may involve cycling around that temperature. The cycling causes repeated austenitization and divorced pearlitic transformation [9]. It is therefore particularly effective for the induction heat treatment of small products, in which rapid temperature changes can be induced. It also permits switching from batch processing (soft annealing in a furnace) to continuous processing, which is much more suitable for small series. Finer globular carbides lead to finer structures after quenching. Densely dispersed vanadium carbides retard the austenite grain growth more effectively than the coarse conventionally soft-annealed microstructure. On the other hand, smaller Cr and Mo carbides dissolve more rapidly. To obtain the same level of carbide dissolution, the fine microstructure can be austenitized more rapidly or at a lower temperature than a coarse microstructure.

2. Experimental

2.1. Material
The experimental material was 1.2343 tool steel with the chemical composition shown in table 1.

Table 1. Chemical composition of experimental steel.

Element	C	Si	Mn	Cr	Mo	V	Ni	Cu	Al	S	P
wt. %	0.40	0.96	0.35	4.98	1.19	0.42	0.22	0.11	0.03	0.003	0.024

The steel was in the form of hot-forged bars 19 mm in diameter. The bars contained upper and lower bainite and a small amount of martensite (figure 1). The hardness of the material in the initial state was 538 HV10.

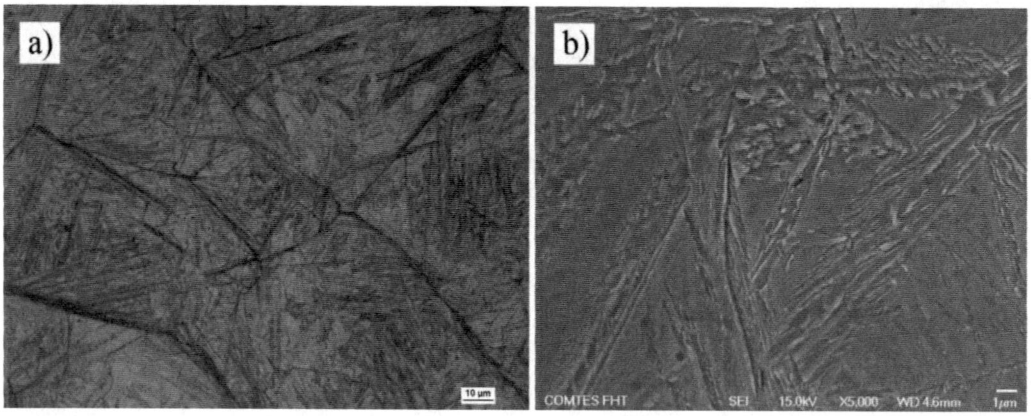

Figure 1. Microstructure of the material in its initial state: a) optical micrograph, b) scanning electron micrograph.

2.2. Heat treatment
The samples were treated in a quenching dilatometer Linseis L78 RITA, in which cylindrical samples with a diameter from 3 to 5 mm and a typical length of 10 mm can be tested. The present samples were heated by induction and cooled by gas flow. The available controlled heating and cooling rates are up to 200°C·s^{-1}. The temperature range is from -200°C to 1400°C. The chamber of the dilatometer

can be evacuated or filled with an inert atmosphere. The cylindrical samples used for heat treatment in the dilatometer were 4 mm in diameter and 10 mm in length. These dimensions permitted hardness measurement, microstructure evaluation, as well as micromechanical testing of the samples [10].

The heat treatment schedules comprised the following steps: heating at a rate of 35°C·s^{-1}, holding at the chosen temperature and subsequent gradual cooling at various cooling rates. Cooling from the holding temperature to 670°C took place at a rate of 1.97°C·s^{-1}. Between 670°C and 500°C, the cooling rate was 1.11°C·s^{-1}. Between 500°C and 400°C, it was 0.71°C·s^{-1}. Final cooling to ambient temperature took place at a rate of 10°C·s^{-1}. The heating and cooling rates were chosen to simulate induction heating and air cooling of a rod 15-20 mm in diameter. Specimens of this size are intended to be used in the next stage of the experimental programme. The samples in the first series were rapidly cooled from 700°C to ambient temperature. The purpose of this rapid cooling from 700°C was to preserve the microstructure characteristics present at that temperature, namely the austenite which had formed during the treatment and remained stable at temperatures below the A_1. The holding times at the annealing temperatures were chosen at 15, 45 and 150 seconds, based on previous experiments with 100CrMnSi6-4 chromium bearing steel.

The schedule designation codes contain information about holding times and temperatures. Where rapid cooling from 700°C was used, the schedule code ends with the suffix "q" (figure 2).

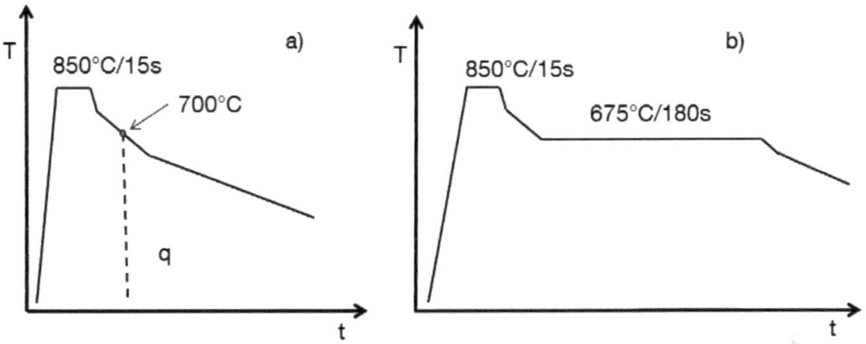

Figure 2. Time-temperature diagrams of heat treatment schedules. a) Schedules designated as "850C/15s" and "850C/15s-q" with rapid cooling from 700°C; b) Schedule "850C/15s 675C/180s".

2.3. Examination of samples

A dilatometric curve was recorded for each specimen in order to identify phase transformations during the ASR process. After the treatment, a longitudinal metallographic section through each sample was prepared. These sections were ground and polished. HV10 hardness was measured along the sample axis. The microstructure was revealed by etching with nital and observed under a Nikon Eclipse 200D optical microscope and JEOL JSM 7400F scanning electron microscope.

3. Results and discussion

3.1. Hardness and dilatometric measurement

The first stage of the experimental programme was focused on carbide coarsening in ferrite and austenite formation. The samples were heated to and held at temperatures between 730°C and 870°C for 15 seconds. Afterwards, they were gradually cooled to 700°C and then cooled rapidly at 200°C/s. The rapid cooling was employed to find the amount of austenite that had formed during the 15-second holding time. This austenite would transform to martensite during cooling. The hardness of the samples after these schedules is shown in table 2.

The final hardness decreases with increasing temperature of the 15-second holding period. This is due to the precipitation of carbides from the martensite and the spheroidisation and coarsening of the bainitic carbides. The significant increase in hardness after schedules with holding temperatures of

850°C and 870°C is caused by austenitization and martensitic transformation during rapid cooling.

Table 2. Hardness levels after schedules with 15-second and 50-second holding times and with quenching from 700°C. The letters IS stand for "initial state".

Schedule	Hardness HV10	Schedule	Hardness HV10
IS	538	…	…
730C/15s-q	497	…	…
750C/15s-q	438	…	…
770C/15s-q	391	…	…
790C/15s-q	362	790C/150s-q	303
810C/15s-q	332	810C/150s-q	286
830C/15s-q	319	830C/150s-q	292
850C/15s-q	343	850C/150s-q	353
870C/15s-q	473	…	…

The holding time was then extended to 150 seconds in those schedules which initially led to the lowest hardness values (table 2). The hardness behaviour was the same as with the 15-second holding time. The minimum hardness value of 286 HV10 was obtained with the 810C/150s-q schedule. The 830C/150s-q schedule led to a virtually equal hardness of 292 HV10.

As expected, higher temperatures and longer holding times at a particular temperature are more effective in softening the material. The temperature limit for softening is the temperature of austenite formation. The presence of austenite can be favourable for rapid spheroidisation of carbides, as it facilitates the dissolution of small carbide particles in bainite or those that precipitated from martensite during heating. However, austenite has to undergo divorced pearlitic transformation during cooling. Otherwise, lamellar pearlite, bainite or martensite form during cooling and softening of the material is reversed.

The purpose of the schedules without quenching, 850°C/15s, 850C/150s and 870C/15s, was to determine whether austenite decomposes during gradual cooling of the sample. 850C/15s and 830C/15s-q were the only schedules which led to lower hardness values without any austenite being formed. Metallographic analyses and dilatometric plots did not reveal any lamellar pearlite, bainite or martensite in the microstructure. Thus, it can be assumed that divorced pearlitic transformation took place during gradual cooling. Other samples listed in table 3 showed higher hardness and a small amount of martensite in the microstructure.

Table 3. Hardness of specimens upon schedules with gradual cooling to ambient temperature.

Schedule	Hardness HV10	Schedule	Hardness HV10
850C/15s	313	850C/150s	333
870C/15s	415	…	…

The final stage of the experimental programme was devoted to estimating the austenite stability. The schedules used at this stage involved austenitizing at temperatures of 850°C and 870°C and holding for 1800 s at 675°C (table 4). This holding time was chosen in order to ensure that the austenite fully decomposed. Dilatometric analyses showed the austenite decomposition times. The austenite that formed during the 15-second holding time at 850°C took 150 seconds at 685°C to decompose (counted from the holding start). The austenite that formed at 870°C was far more stable and decomposed for 1100 seconds. As this long decomposition time was not acceptable for the ASR process, the maximum temperature chosen for the final experiments was 850°C.

Table 4 shows hardness values after schedules involving partial austenitization and interrupted cooling. The austenitizing times at 850°C were 15, 45 and 150 seconds. The cooling was interrupted by 180-second holding at 675°C. The partial austenitization followed by divorced pearlitic transformation did not contribute to further carbide coarsening when compared to the 810C/150s-q schedule.

Table 4. Schedules with the hold at 675°C for austenite decomposition.

Schedule	Hardness HV10
850C/15s 675C/1800s	299
870C/15s 675C/1800s	294
850C/15s 675C/180s	308
850C/45s 675C/180s	298
850C/150s 675C/180s	312

3.2. Metallographic analysis

All samples contained carbides of similar morphology in the form of rounded or edged particles. Their size varied from approximately 20 nm to 300 nm. Carbides coarsened with increasing temperature of the induction treatment. The coarsening was visible in both the smallest and the largest carbides. The typical carbide size in the 750C/15s-q sample was between 20 nm and 150 nm (figure 3(a)). In the 810C/15s-q sample, it was between 30 nm and 200 nm (figure 3(b)). The lowest hardness was achieved in the 810C/150s-q sample: 286 HV10. The microstructure retained the differences between the bainitic and martensitic regions visible at lower magnifications (figure 4(a)). However, carbides were homogeneously distributed in the ferritic matrix. They were coarsened, with sizes up to 300 nm (figure 4(b)).

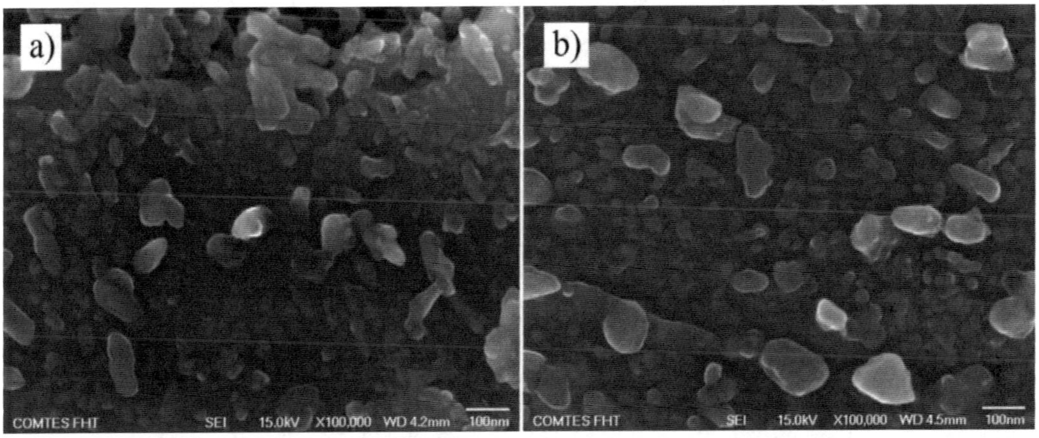

Figure 3. Microstructures upon schedules: a) 750C/15s-q, b) 810/15s-q.

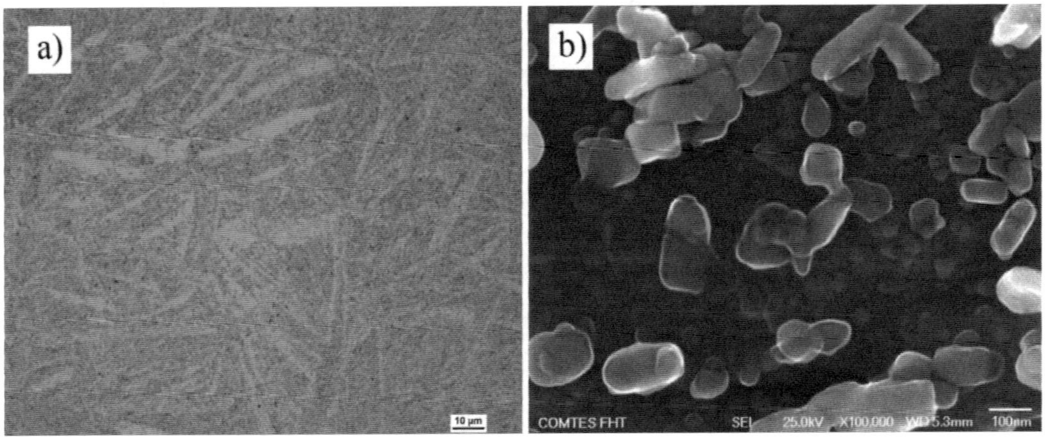

Figure 4. Microstructure upon schedule 810°C/150s-q: a) optical micrograph, b) scanning electron micrograph.

The lowest hardness achieved in the experimental programme was around 290 HV10. The hardness after conventional soft annealing is usually under 230 HV. From the morphological point of view, the ASR process produced an equivalent but much finer microstructure than soft annealing. In particular, the finest carbides below 100 nm in diameter are those which are probably responsible for the dispersion strengthening. There are two ways to ensure that coalescence of small carbides and the larger ones occurs. One mechanism is Ostwald ripening induced by annealing below the A_1 temperature. This is the conventional way used in soft annealing. The second approach involves austenitizing the ferritic matrix, dissolving the finest carbides and inducing divorced pearlitic transformation during cooling. If the ASR process is to be applied, the treatment should be as brief as possible. Rapid austenitization is therefore a suitable option.

Austenitization took place in schedules with heating at temperatures above 830°C. The divorced pearlitic transformation occurred during holding at 675°C. However, austenitization at 870°C for 15 seconds was not sufficient to dissolve even the finest carbides. The resulting size and distribution of carbides were the same as in schedules without austenitization. Future experimental work devoted to further softening of the steel should focus on fine carbide dissolution. This will entail finding the optimal temperature for austenite decomposition via divorced pearlitic transformation.

If a hardness of 290 HV is sufficiently low for a particular production process, then the schedules described in this paper could be used. Further work must be devoted to hardening of the fine-structured material obtained from the ASR process. The fine martensitic structure promises better performance of the final product.

4. Conclusion

Induction heat treatment of 1.2343 steel was performed. The initial state with a bainitic-martensitic structure underwent spheroidisation annealing for less than 5 minutes. A microstructure of globular carbides in a ferritic matrix was obtained. The most effective route for carbide spheroidisation and coarsening was annealing just below the A_1 temperature. After holding for 150 seconds at 810°C, the hardness dropped from the initial 538 HV10 to 286 HV10.

Rapid austenitization at temperatures of 850°C or higher, with subsequent divorced pearlitic transformation, led to similar results. Optimization of the austenitization and austenite decomposition cycle will be one of the aims for further research.

The process of accelerated carbide spheroidisation produces much finer globular carbides than conventional soft annealing. The influence of the microstructure fineness on the material's performance after hardening will be the subject of the next stage of the experimental programme.

Acknowledgment

This paper was prepared under the project entitled "Development of West-Bohemian Centre of Materials and Metallurgy" No.: LO1412, which is funded by the Ministry of Education of the Czech Republic.

References

[1] Roberts G, Krauss G and Kennedy R 1998 *Tool Steels* 5th Edn. (ASM International, Materials Park, OH 44073-0002, USA)

[2] Mennig G and Stoeckhert K 2013 *Mold-Making Handbook* 3rd edn. (Carl Hanser Verlag, Munich, Germany)

[3] Telasanga G, Majumdar J D, Padmanabham G and Manna I 2015 Wear and corrosion behavior of laser surface engineered AISI H13 hot working tool steel *Surf. Coat. Tech.* **261** 69-78

[4] Dobrzanski L A, Bonek M, Hajduczek E, Klimpel A and Lisiecki A 2005 Comparison of the structures of the hot-work tool steels laser modified surface layers *J. Mater. Process. Tech.* **164-165** 10014-1024

[5] Leskovsek V, Sustarsic B and Jutrisa G 2006 The influence of austenitizing and tempering temperature on the hardness and fracture toughness of hot-worked H11 tool steel *J. Mater.*

Process. Tech. **178** 328-34

[6] Hauserova D, Dlouhy J and Novy Z 2014 Effect of heating rate on accelerated carbide spheroidisation (ASR) in 100CrMnSi6-4 bearing steel *Arch. Metall. Mater.* Vol. **59** Issue **3** 1199-203

[7] Revilla C, Lopez B and Rodriguez-Ibabe J M 2014 Carbide size refinement by controlling the heating rate during induction tempering in a low-alloy steel *Mater. & Design* **62** 296-304

[8] Minwoo K, Gyujin P, Jae-Gli J, Byung-Hoon K and Young-Kook L 2015 The effects of annealing temperature and cooling rate on carbide precipitation behavior in H13 hot-work tool steel *J. Alloy. Compd.* **627** 359-66

[9] Verhoeven J D and Gibson E D 1998 The divorced eutectoid transformation in steel *Metall. and Mater. Trans. A* **29A** 1181-9

[10] Dzugan J and Konopik P 2012 Small punch test application to material properties evolution determination *J. of Solid Mech. and Mater. Eng.* **6** 782-91

IOP Conf. Series: Materials Science and Engineering **103** (2015) 012026 doi:10.1088/1757-899X/103/1/012026

Preparation of oxide powder by continuous oxidation process from recycled Fe-77Ni alloy scrap

J Y Yun[1], D H Park[1], G J Jung[2] and J P Wang[3,4]

[1] Powder Technology Research Group, Korea Institute of Materials Science (KIMS), Changwon 642-831, Korea
[2] Department of Architectural Engineering, Pukyong National University, Buean 608-737, Korea
[3] Department of Metallurgical Engineering, Pukyong National University, Busan 608-739, Korea

E-mail: jpwang@pknu.ac.kr

Abstract. The oxidation behavior of Fe-77Ni alloy scrap was studied under a 0.2 atm oxygen partial pressure at the temperature range of 400°C to 900°C. The oxidation rate was found to be increased with an increase of temperature and followed the parabolic rate law with linearly proportional to temperature. Microstructure and cross-sectional area of the oxide layer were examined by SEM, EDX, and XRD. It could be speculated that rate-limiting step was controlled by diffusion through either the spinel structure or the NiO layer, both of which were present in this alloy during oxidation at elevated temperatures. In the long run, oxide powder less than 10 μm from Fe-77Ni alloy scrap was obtained using ball-milling and sieving processes and recovery ratio approached up to 97% for 15 hours.

1. Introduction

In a field of industries, alloys containing high nickel content have been used due to their corrosion and high temperature resistance and these alloys, therefore, are adapted to sever environment in the power energy, chemical and petrochemical industries. Especially, nickel-based superalloys can be met to special requirements of high mechanical and chemical properties for strength and creep resistance as well as corrosion resistance at elevated temperatures. The aviation industries in the world are attempting to develop special structural materials that can resist high temperature to enhance the engine's fuel efficiency [1-5]. The nickel-based alloys are, therefore, representative metal that has been widely used in the hottest parts of gas turbines both for power generation and aircraft [6, 7].

A study on the oxidation behavior of iron-nickel alloys has been conducted by many researchers for decades. Many researches have already been performed, not only at low pressure and temperatures above room temperature, but also by air or during electrochemical oxidation and plasma formation [8-12] Liu et al. focused on the research of the corrosion phenomena for a single-crystal Ni-base superalloy at 900 and 1000°C in an air and oxidation was found to be occurred in two steps. It is noted that first oxidation is controlled by NiO growth and the second by Al_2O_3 growth until Al_2O_3 layer formed beneath NiO layer [13]. Li et al. investigated the oxidation layer for NiCrAlYSi deposited by

[4] Address for correspondence: J P Wang, Department of Metallurgical Engineering, Pukyong National University, Busan 608-739, Korea. E-mail: jpwang@pknu.ac.kr.

one-step arc ion plating and this coating layer was found to be thin and more effective protection of the substrate [14]. The pure and doped nickel alloys with different Co contents in air at 1000°C was studied for oxidation behavior by Zhou et al. It was found that coating efficiency was enhanced with an increase the Co content [15].

In this study, the specimen of Fe-77Ni alloy was studied for oxidation rate by thermal gravimetric analysis (T.G.A.) on the effect of temperature and oxygen partial pressure. The cross-sectional morphologies of oxide layer were examined and characterized by scanning electron microscopy provided by energy dispersive spectrometry (SEM-EDS) and X-ray diffraction (XRD). According to kinetic data and microstructure analysis, the high-temperature oxidation mechanism was discussed. Finally, the particle size distribution was conducted to find out the range of the oxide powder and fabrication of oxide powder less than 10 μm from Fe-77Ni alloy scrap was attempted by various physical processes.

2. Materials and experimentation

Scrap plate of Fe-77Ni alloy containing 77 wt.% of Ni was purchased from TV recycling center. Samples for oxidation were machined to be a size of 1.5 cm and 2 cm. In pretreatment process, polishing machine was used to treat on the surface of the specimens and subsequently diamond polishing suspension of 1 m grade was used to polish them accurately. In cleaning stage, an ultrasonic cleaning bath with acetone was used and samples were washed again with ethanol and water, and dried under the hood.

The oxidation behavior of Fe-77Ni alloy was continuously studied under a 0.2 atm using thermogravimetic apparatus (T.G.A.) at different temperatures. The gas mixtures with oxygen and argon were going through drierites for removal of moisture before their input into the reactor tube. Also, oxygen content as an impurity in argon gas was eliminated by copper turning furnace. The gas mixture was flowing to the quartz tube from the bottom to approach reaction zone while U.H.P. argon gas was flowing through the electrobalance to minimize contamination during the experiment. The specimen of the quartz tube was suspended with a platinum wire and then heated under ultra-purity argon atmosphere. When the temperatures were reached from 400°C to 900°C, oxidizing gases was used until oxidation was completed The oxygen partial pressure was controlled by each flowmeter of argon and oxygen gases. The weight gain of the specimen during oxidation was continuously measured and recorded for up to 24 hours. The cross-sectional area of the oxidized specimens and chemical composition of each layer were examined by SEM, EDX, and XRD.

3. Results and discussion

The change in the specimen weight of Fe-77Ni alloy with time during oxidation was observed for 24 hours in temperature range of 400 °C to 900 °C with gas mixture of an oxygen partial pressure of 0.2 atm using TGA shown in Fig. 1. It was found that the weight increased as the temperature rose, and it was shown to be the highest at 900°C.

The oxidation rate for iron-nickel alloys may be represented by the parabolic rate equation that is of great importance. As per this law, the oxide grows at a continually decreasing oxidation rate, and the diffusional and integral forms of the parabolic rate equations are given as follows:

$$dx/dt = k_p/x$$

$$x^2 = 2k_p t + C$$

Where kp indicates the parabolic rate constant and C the integration constant, plotting the variation of x with time. Parabolic oxidation at elevated temperature represents that the oxidation takes place by the diffusion of ions or electrons through the oxide scale [16]. The oxidation reaction dependence on the temperature is well described by an Arrhenius-type equation as:

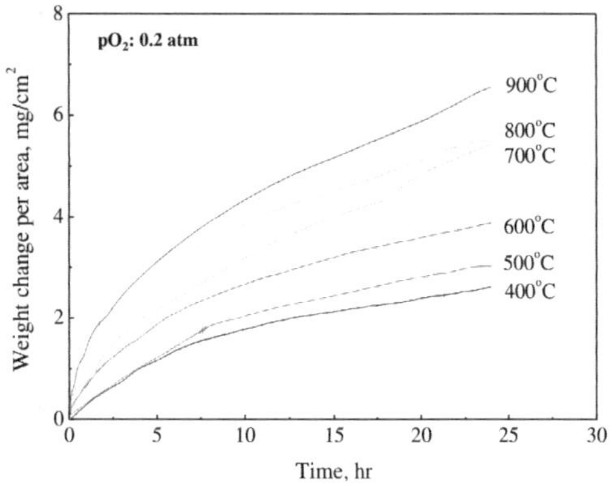

Figure 1. Weight gain with oxidation time on the effect of temperature.

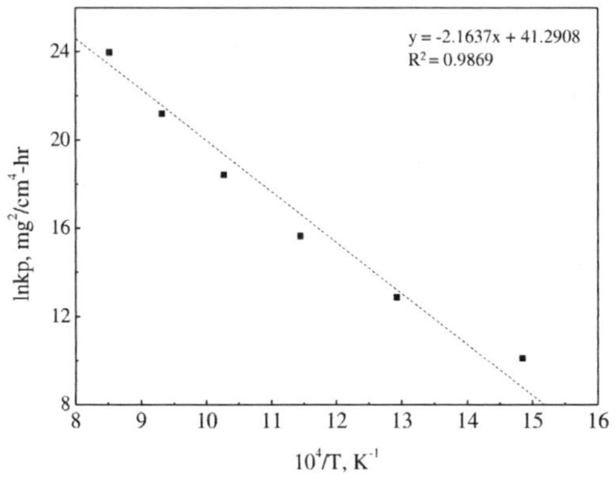

Figure 2. Parabolic rate constants with elevated temperatures.

$$k_p = Ae\frac{-\Delta E^*}{RT}$$

Where ΔE* denotes apparent energy of activation, and A is the frequency factor. The value of ΔE* based on the Arrhenius plot related with caution in oxidation process. 16 Fig. 2 shows the plot of parabolic rate constants on the y-axis and temperature on the x-axis deduced from Arrhenius-type equation. The oxidation rate constants are almost linearly proportional to temperature, which means that the rates were proportionally increased with an increase of temperatures. The activation energy for the oxidation of Fe-77Ni at a 0.2 atm was experimentally calculated to be 17.98 kJ/mol.

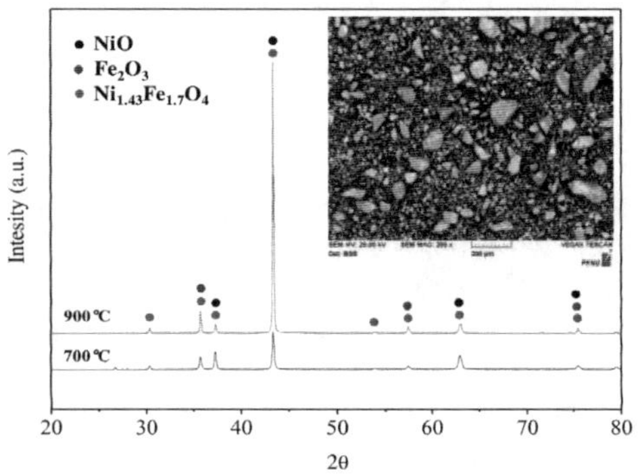

Figure 3. XRD patterns of oxide scales conducted at 700°C and 900°C for 24 hours.

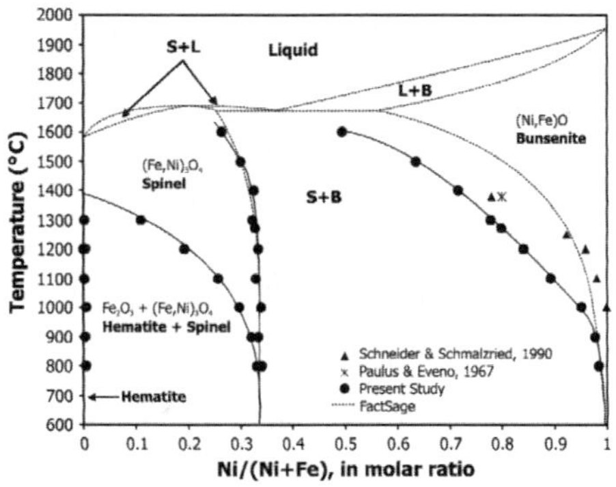

Figure 4. Pseudo-binary phase diagram of 'Fe_2O_3'-NiO in air. The closed-circle data points are from the present study; the dotted curves are calculated using FactSage. Legend: S=spinel, B=bunsenite, L=liquid, and H=hematite [17].

The phase analysis using XRD was conducted and oxide particles were also observed by SEM after oxidation of Fe-77Ni alloy scraps at 700°C and 900°C for 24 hours in given experimental conditions. Fig. 3 shows that three oxide phases, NiO, Fe_2O_3, and $Ni_{1.43}Fe_{1.7}O_4$, were detected, and the peak patterns of the three oxides were not much different regardless of elevated temperatures from 700°C and 900°C. The irregular shape of oxide particles was observed and their size differences are ranging from ±200μm to ±10μm.

According to the previous results of XRD, the following phases might be expected: bensenite-NiO (B), spinel (S), and hematite (H). It can be seen figure 4 that various oxide phases such as hematite, spinel, bunsenite, and complex phases are formed depending on Ni/(Ni+Fe) molar ratio and temperature, and at given experimental conditions spinel $(Fe, Ni)_3O_4$ and bunsenite (Ni, Fe)O are formed in pseudo-binary phase diagram that is a good agreement with data from XRD result. It can be explained that iron ions and electrons might migrate outward through grain boundaries, microcracks, and voids and oxygen anion was also moving inward through mainly spinel and bunsenite structures. It can be explained that especially the spinel structure known as a non-stoichiometric compound of $Ni_{1.43}Fe_{1.7}O_4$ has voids and pores itself, and oxygen anion was therefore relatively easily moving inward.

According to the previous results of XRD, the following phases might be expected: bensenite-NiO (B), spinel (S), and hematite (H). It can be seen figure 4 that various oxide phases such as hematite, spinel, bunsenite, and complex phases are formed depending on Ni/(Ni+Fe) molar ratio and temperature, and at given experimental conditions spinel $(Fe, Ni)_3O_4$ and bunsenite (Ni, Fe)O are formed in pseudo-binary phase diagram that is a good agreement with data from XRD result. It can be explained that iron ions and electrons might migrate outward through grain boundaries, microcracks, and voids and oxygen anion was also moving inward through mainly spinel and bunsenite structures. It can be explained that especially the spinel structure known as a non-stoichiometric compound of $Ni_{1.43}Fe_{1.7}O_4$ has voids and pores itself, and oxygen anion was therefore relatively easily moving inward.

The microstructure of the cross-sectional area of the oxide layer at 900°C for 24 hours and its chemical composition were examined by EDX, and its result is shown in figure 5. It showed the formation of two oxide layers that could be easily verified, and cracks and voids were also easily observed between the outer and inner layers. The chemical constituent of the outside oxide layer was mainly nickel, and nickel oxide (NiO) was, therefore, primarily formed the outer layer (1). It may expected that rate-limiting step was diffusion through either the spinel structure or the NiO layer, both of which were present in this alloy during oxidation at elevated temperatures [16]. The inner oxide layer (2) was found to consist of Ni, Mn, and Fe while three oxide phases, NiO, Fe_2O_3, and $Ni_{1.43}Fe_{1.7}O_4$, coexisted. Also, the Mo content in the inner layer was detected to about 15 wt.%, which may have been influencing the physical properties of the film. In other word, Mo content in oxide layer can be a significant factor in oxidation processes, particularly where thermal cycling is involved [17]. The substrate (3) and (4) only contained Ni and Fe as a metallic alloy, and the Mo and O were not detected.

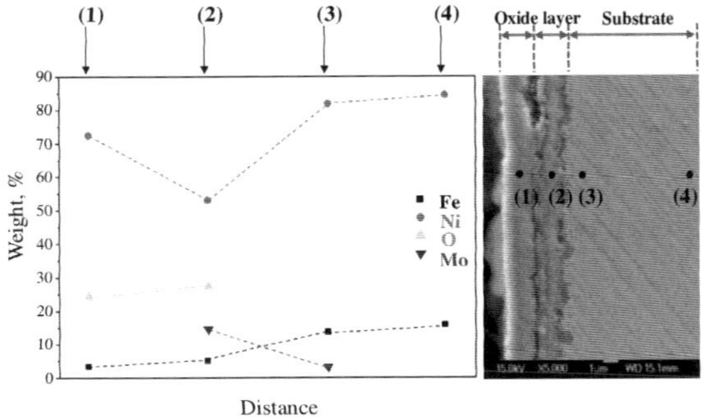

Figure 5. Microstructure of cross-sectional area corresponding to chemical composition of each oxide layer conducted at 900°C for 24 hours.

Based on kinetic data of oxidation of Fe-77Ni alloy scrap, fabrication of oxide powder less than 10 μm was carried out and its overall process is shown in figure 6. The control of particle size is one of the significant factors to enhance powder properties when it comes to powder compaction, density, and sintering. Therefore, final target size of the oxide powder obtained through various processes should be less than 10 μm to apply it to metal injection molding (MIM) process. Scrap foils were oxidized in muffle furnace in air atmosphere at 900°C for 2 hours since oxidation rate was found to be relatively fast at 900°C as well as initial oxidation stage up to 2 hours and cyclic oxidation was sequentially conducted every 2 hours until the foils were fully oxidized. Oxide particles obtained after oxidation process were grinded using the ball-milling machine to control the particle size of the final product. Sieving process was consequently operated to obtain oxide powders less than 10 μm using various levels of sieve for characterizing the particle size distribution of the samples.

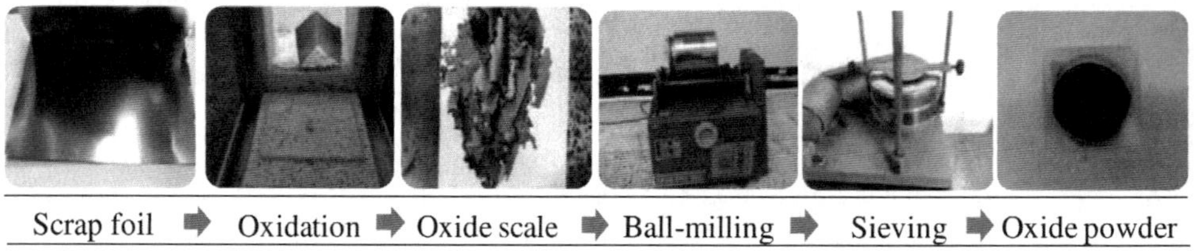

Figure 6. Overall process to manufacture oxide powder from Fe-77Ni alloy scrap.

In the ball-milling process, size control of oxide powder was conducted to obtain target size of the powder less than 10 μm shown in figure 7. It is noted that particle size was found to be reduced with an increase of milling time under a fixed RPM of 84. The particle size was dramatically reduced to about 15 μm from 3 hours to 12 hours and then its size reached to about 10 μm for another 3 hours. It was found that recovery ratio was about 70% for 3 hours at the initial phase and then it started to dramatically increase after 6 hours of ball-milling. The recovery ratio of oxide powder was finally obtained to over 95% for 15 hours, and further milling process might be no longer efficient to raise the recovery ratio.

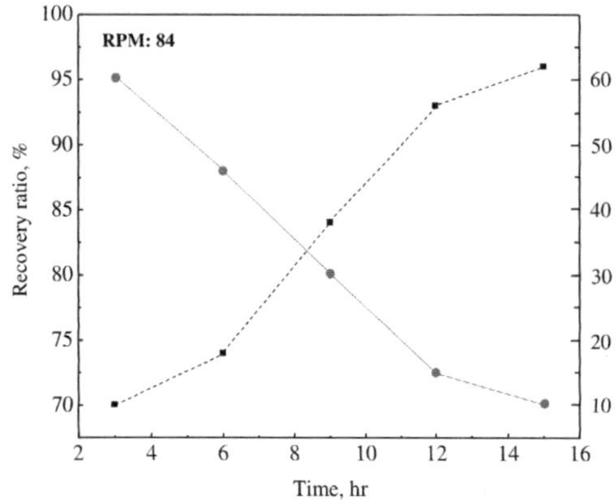

Figure 7. Recovery ratio and size of oxide powder as a function of milling time.

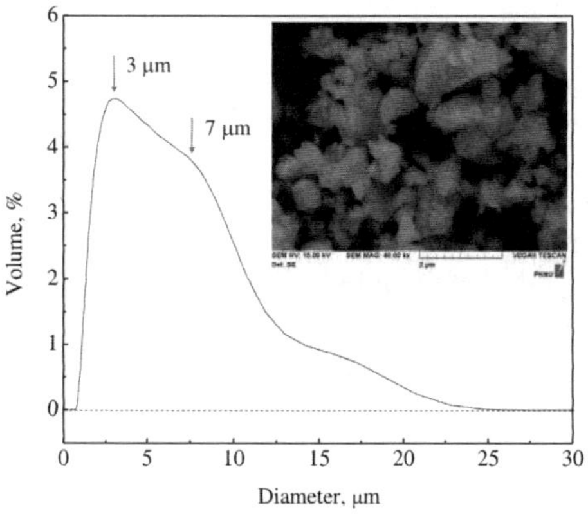

Figure 8. Particle size distribution and shape of oxide powder of Fe-77Ni alloy.

The particle-size distribution (PSD) of a powder can be important in understanding its physical and chemical properties and affect in particular compaction, density, and sintering of powder. The particle-size distribution of oxide powder was analyzed using Matersizer 3000 with a capacity of particle size range of 0.01 μm to 3500 μm and its shape was examined using scanning electron microscopy (SEM) shown in Fig. 8. It shows the average particle distribution, in terms of volume, as a function of particle diameter and particle size might be estimated by the amount of volume of oxide powder. It might be expected that an average particle size is observed to be about 7 μm and the other concentration peak is shown to be approximately 3 μm, which means that the smallest size of oxide powder might be around 3 μm. In company with particle size, the shape of particles can also have a significant influence upon the performance or processing of particulate materials and SEM image of oxide powder illustrates that its shape was observed to irregular and faceted form.

4. Conclusion

The rate of oxidation for Fe-77Ni alloy scrap increased with increasing temperatures in a given oxygen partial pressure. The oxidation behavior of the specimen followed the parabolic rate law regardless of temperature and activation energies were calculated to be 17.98 kJ/mol. The oxidation mechanism may expect that rate-limiting step was diffusion through either the spinel structure or the NiO layer. Particle size distribution of the oxide powder was conducted, and an average particle size was estimated to be around 7 μm with irregular and faceted shape.

Acknowledgment

The work was supported by a grant from the Fundamental R&D Program for Core Technology of Materials funded by the Ministry of Trade, Industry & Energy, Republic of Korea.

References

[1] Choi S H, Yun J Y, Lee H. M, Kong Y. M, Kim B K and Lee K A 2011 High temperature oxidation behavior of Ni based porous metal *J. Korean Powder Metall. Inst.* **18** 122-8

[2] Larring Y, Haugsrud R and Norby T 2003 Corrosion of Cr-5wt%Fe-1wt% Y_2O_3 alloy and conductivity of the oxide scale-effects of water vapor *J. Electrochem. Soc.* **150** 374-9

[3] Douglass D L, Kofstad P, Rahmel A and Wood G C 1996 International workshop on high temperature oxidation-subject area 1: The role of water vapour in the oxidation of metals and alloys *Oxid. Met.* **45** 529-45

[4] Buscail H, Heinze S, Dufour P H and Larpin J P 1997 Water-vapor-effect on the oxidation of Fe-21.5 wt.%Cr-5.6 wt.%Al at 1000°C *Oxid. Met.* **47** 445-64

[5] Al-Badairy H and Tatlock G 2000 The influence of the moisture content of the atmosphere on alumina scale formation and growth during high temperature oxidation of PM2000 *Mater. High. Temp.* **17** 133-7

[6] Nishimura C, Nagai Y, Yanagisawa K and Toshima T 1987 Magnetic properties of NiFe films prepared using ion-beam sputtering *IEEE Trans. Magn.* **23** 2728-36

[7] Boggs W E 1977 The oxidation of iron-aluminum alloys from 450 to 950°C *J. Electrochem. Soc.* **118** 906-13

[8] Onal K, Maris-Sida M C, Meier G H and Pettit F S 2003 Water vapor effects on the cyclic oxidation resistance of alumina forming alloys *Mater. High. Temp.* **20** 81-91

[9] Janakiraman R, Meier G H and Pettit F S 1999 The effect of water vapor on the oxidation of alloys that developed alumina scales for protection *Metall. Mater. Trans. A* **30A** 2905-13

[10] Regina J R, DuPont J N and Marder A R 2004 The effect of water vapor on passive layer stability and corrosion behavior on Fe-Al-Cr Alloys *Oxid. Met.* **61** 69-90

[11] Regina J R, DuPont J N and Marder A R 2005 Gaseous corrosion resistance of Fe-Al based alloys containing Cr additions Part II. Scale morphology *Mater. Sci. Eng. A* **405** 102-10

[12] Wallwlork G R 1976 The oxidation of alloys *Rep. Prog. Phys.* **39** 401-10

[13] Liu C T, Ma J and Sun X F 2010 Mechanism of the oxidation and degradation of the aluminide coating on the nickel-base single-crystals uper alloy DD32M *J. Alloys Compd.* **491** 522-31

[14] Li W Z, Li Y Q, Wang Q M, Sun C and Jiang X 2010 Oxidation of a NiCrAlYSi overlayer with or without a diffusion barrier deposited by one-step arc ion plating *Corro. Sci.* **52** 1753-62

[15] Oku M, Suzuki S, Ohtsu N, Shishido T and Wagatsuma K 2008 Micro-and nano pyramids of manganese-doped indium oxide *Appl. Surf. Sci.* **254** 5141-9

[16] Foley R T and Guare C J 1959 The oxidation of iron-nickel alloys *J. Electrochem. Soc.* **106** 936-40

[17] Rhamdhani M A, Hayes P C and Jak E 2008 Subsolidus phase equilibria of the Fe-Ni-O System *Matell. Mater. Trans. B* **39B** 690-701

Infrared spectroscopy for a quantitative determination of $CH_{3-n}Cl_nCOCl$ in $TiCl_4$

B Yang, Y Wang, H Tan[1], J He and G Song

Guizhou Academy of Testing and Analysis, Guiyang 550002, China

E-mail: 357688237@qq.com

Abstract. Infrared spectroscopy permits a determination of the concentration of $CHCl_2COCl$, $CH_2ClCOCl$ and CCl_3COCl as the organic impurities in $TiCl_4$In this study, an infrared spectrometer with ZnSe window and the cell of PTFE was assembled, and used to determine the concentration of three organic impurities. The detection limits of CCl_3COCl, $CHCl_2COCl$, $CH_2ClCOCl$ were determined as 3.16×10^{-3} μg/g 1.92×10-3 μg/g and 1.55×10-2 μg/g respectively. The new method is applied to quality control of refined $TiCl_4$ from titanium sponge.

1. Introduction

Titanium is a transition metal with various excellent properties, such as low density and high corrosion resistance. Titanium sponge is important for titanium manufacture [1]. An enrichment correlation presented among the impurities in Titanium sponge and that in fine $TiCl_4$. Therefore, the impurities of $TiCl_4$ are the key of titanium quality control [2]. The composition and content of impurities are analyzed quickly and quantitatively to removal of impurities $CH_{3-n}Cl_nCOCl$ (CCl_3COCl, $CHCl_2COCl$, $CH_2ClCOCl$) during the production of Titanium sponge. The quality control standard of $TiCl_4$ is strict in Japan, America and Russia, organic compound was determined by infrared spectral analysis, besides the determination of inorganic compound like As, Sb, Si, Cl, Al, Cu, Fe, Pb, V, Ni and Sn [3-6]. At present, the general determination of impurities in domestic titanium industry just includes $SiCl_4$, $FeCl_3$, $VOCl_3$, and chromaticity analysis [7]. The lack of effective determination method of organic impurities (CCl_3COCl, $CHCl_2COCl$, $CH_2ClCOCl$) reduces the products' hardness. However, CCl_3COCl in fine $TiCl_4$ was determined by infrared semi-quantitative method [8-9] and Fourier-transform infrared spectroscopy (FT-IR) [10].

$SiCl_4$, Fe Cl_3 and $VOCl_3$ are the main impurities in fine $TiCl_4$. It did not have any other interference to the determination of $CH_{3-n}Cl_nCOCl$ due to they are inorganic compounds. FT-IR cannot detect the peak of Fe and Si since the different bond vibration compared to organic compound. Since $TiCl_4$ and $CH_{3-n}Cl_nCOCl$ undergone rapid hydrolysis in the presence of moisture, an infrared liquid cell is damaged, and the results can be influenced. Therefore, a new liquid cell is designed with good tightness and adaptive thickness. Many other IR transparent inert solid window materials for liquid cells are known – some are cheaper and established. A superior waterproofness is the reason for the choice of ZnSe. It is safe, simple, easy removable, washable and reusable. Spiking method or standard addition [11] was done to determine the trace organic impurities due to the content of organic

[1] Address for correspondence: H Tan, Guizhou Academy of Testing and Analysis, Guiyang 550002, China. E-mail: 357688237@qq.com.

impurities are quite low in $TiCl_4$.

FT-IR quantitative analysis was based on absorption strength and it was able to quantify the gas, solid and liquid sample. In the same time, the component of mixture was calculated through the characteristic peak. Therefore, it was able to determine the infrared spectrum of the mixtures by FT-IR without pre-processing. And this measurement method was convenient, visible and quickly with several calculation method. In the study, the component of fine $TiCl_4$ was determined by the Standard addition method.

2. Experimental procedure

2.1. Instrument and reagents
Reagents Titanium tetrachloride was purchased from Zunyi Titanium Company (Guizhou, China), Chloroacetyl chloride, Dichloroacetyl chloride and trichloro-acetic chloride were purchased from Sigma Chemical Co. (St. Louis, MO, USA). Aladdin, dry calcium chloride, absolute ethyl alcohol are purchased from Shanghai Chemical Company.

A Thermo (IR 200) FT-IR was used in the experiment, and it connected with a spectrophotometer operated at a resolution of 4 cm^{-1}. Thirty-two scans were scanning to improve the signal-to-noise ratio.

2.2. IR liquid cell
Make sealed IR liquid cell for avoiding $TiCl_4$ reacts vigorously with moisture. It is made of a Teflon body, a window of ZnSe (25 mm ×2 mm, 7800~440 cm^{-1}) and soft Teflon gaskets. Filling the $TiCl_4$ from these cells into the atmosphere was also found to be a troublesome process for routine analysis. Hence, this IR cell was fabricated. Structure and practicality chart of IR seal cell showed below (figure 1).

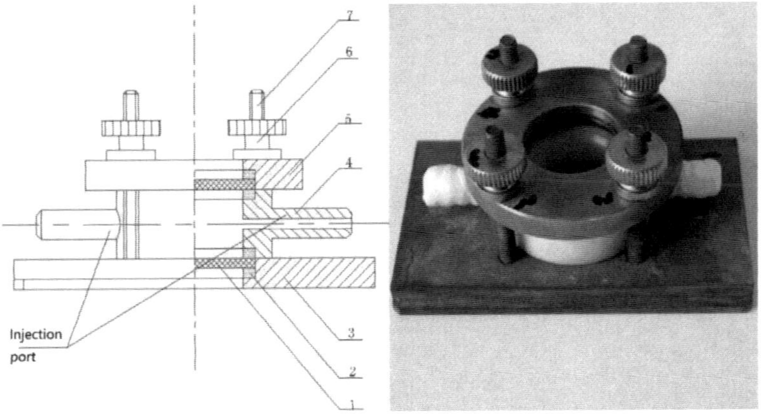

Figure 1. Structure and practicality chart of assembled infrared seal pool.

2.3. Sample preparation
Organic impurities mixture standard stock: Addition of 5 μL CCl_3COCl (98%), 9 μL $CHCl_2COCL$ (98%), 45 μL CHClCOCl (98%) to a 20 mL flask, and dilute it to 20 mL.

Organic impurities mixture standard correction: addition of 150, 180, and 200 μL standard stock to a 20 ml flask respectively, and then diluted to 20 ml. The content of $CH_2ClCOCl$, $CHCl_2COCl$, and CCl_3COCl are $0.857\times10^{-3}\sim2.143\times10^{-2}$, $1.714\times10^{-3}\sim4.286\times10^{-3}$, $0.952\times10^{-4}\sim2.381\times10^{-3}$ mg/g. These solutions were used as calibration.

Determination of organic impurities: The background was scanned while dry nitrogen flow through it at one end and seal the other end with PTFE; then the air tightness was checked by the soap bubble. Injection of 7 mL of organic impurities mixture standard correction and $TiCl_4$ to a vacuum IR liquid cell and the infrared spectrum was determined. The mass fraction of organic impurities in $TiCl_4$ was calculated by standard addition or spiking method.

2.4. Air tightness test

One end of assembled infrared pool sealed with soft silicone rubber, and dry nitrogen was poured into the pool from the other end, then checked tightness with soap bubbles.

2.5. Overlapping peak resolve

The peak of $CH_{3-n}Cl_nCOCl$ was processed by OMNIC software. Set all the parameters and choose the peak position. Then the peaks were separated and it was convenient to quantify the three organic compounds.

3. Results and discussion

3.1. Stability test of IR liquid cell

An organic impurities standard calibration was added to IR liquid cell. And we recorded the absorbance of CCl_3COCl, $CHCl_2COCl$ and $CH_2ClCOCl$ in vacuum space every day for 7 days, and then calculate the content of each impurity. The determination result showed that no obvious difference on organic impurities content with the increase of storage period, and all RSD is less than 10%. It indicated this infrared liquid cell presented a superior inertia and seal.

3.2. Precision test

Table 1 indicated the content of $CH_2ClCOCl$, $CHCl_2COCl$ and CCl_3COCl in $TiCl_4$, and each sample was tested for five times. When RSD was less than 0.04%, it indicated a good precision of IR liquid cell which can meet the study requirement.

Table 1. The results of the same organic impurities mixed standard sample at five measuring times ($\mu g \cdot g^{-1}$).

Measure time	X ($CH_2ClCOCl$)	X ($CHCl_2COCl$)	X (CCl_3COCl)
1	1.7860×10^{-2}	2.1858×10^{-3}	3.629×10^{-3}
2	1.7858×10^{-2}	2.1857×10^{-3}	3.628×10^{-3}
3	1.7859×10^{-2}	2.1859×10^{-3}	3.626×10^{-3}
4	1.7859×10^{-2}	2.1859×10^{-3}	3.626×10^{-3}
5	1.7857×10^{-2}	2.1857×10^{-3}	3.627×10^{-3}
RSD (%)	0.0064	0.0045	0.0359

3.3. Detection limit by standard addition

Calculate detection limit by the lowest content of $CHCl_2COCl$ in $TICl_4$ (absorbance=0.003). The calculated detection limit of CCl_3COCl, $CHCl_2COCl$ and $CH_2ClCOCl$ were 3.159×10^{-3}, 1.917×10^{-3} and 1.554×10^{-2} mg/g, respectively. The measured value of CCl_3COCl, $CHCl_2COCl$ and $CH_2ClCOCl$ were lower than level standard of the compounds in fine $TiCl_4$ in Russia. So the precision can meet the quality control and analysis requirement of impurities in $TiCl_4$ production.

3.4. Sample determination

The content of CCl_3COCl, $CHCl_2COCl$ and $CH_2ClCOCl$ in $TiCl_4$ were tested in IR liquid cell with a standard. Figure 2(a) showed peaks of CCl_3COCl, $CHCl_2COCl$ and $CH_2ClCOCl$ were highly overlapped, which is not able to quantify directly. Carbonyl group (C-O) presented at CCl_3COCl, $CHCl_2COCl$ and $CH_2ClCOCl$ with a sensitive peak at around 1800 cm^{-1}. Inductive effect and other synergistic effect to carbonyl group occurred and it varied from the number of chlorine atoms attached to methyl group. The effects made the difference of sensitive absorption peak, however, overlapping presented when the three organic was determined simultaneously. So the peaks were then separated by OMNIC software [13] for quantifying and the figure was showed in figure 2(b). The infrared spectroscopy characteristic peak of CCl_3COCl, $CHCl_2COCl$ and $CH_2ClCOCl$ are 1801, 1809, and 1820 cm^{-1} respectively. The sample determination result was illustrated in table 2. These formulas are regression equation of the three impurity compound. The X illustrated the content of the impurity and the mean of Y was measure value of the compound by FT-IR. The correlation coefficient is 0.8995 for $CH_2ClCOCl$ standard curve, which is the only one less than 0.996. The average level of $CH_2ClCOCl$, $CHCl_2COCl$ and CCl_3COCl in $TiCl_4$ was 1.7859×10^{-2}, 2.1857×10^{-3} and 3.626×10^{-3} μg/g respectively. Furthermore, the correlation among the chloroacetic acids and their wavelengths of maximum absorbance was also tested. But no obvious correlation (linear or quadratics) can be found through the result.

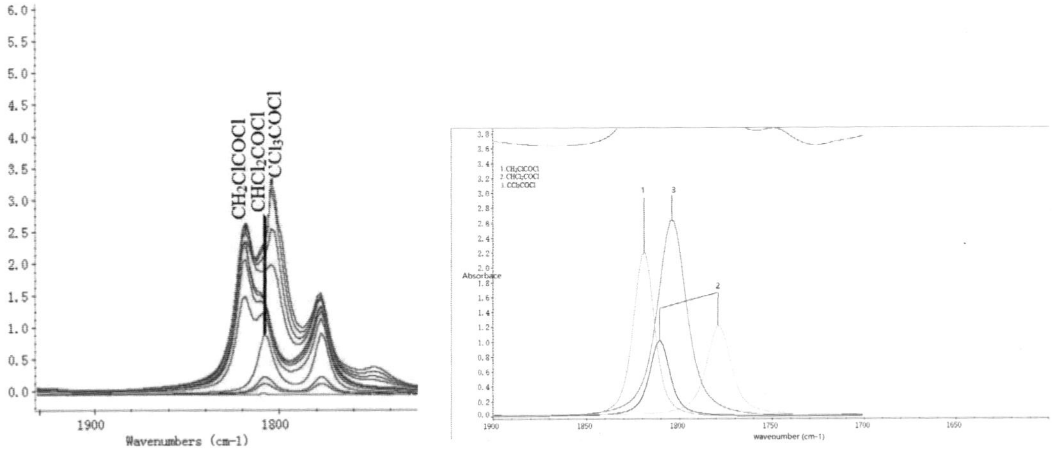

Figure 2. Infrared spectra of the organic impurities in $TiCl_4$ ((a) showed the original results by FT-IR, (b) showed the results processed by OMNIC software).

Table 2. The concentration of the three organic impurities in $TiCl_4$.

Compound	Regression equation	Correlation coefficient(R^2)	Content(μg •g^{-1})
$CH_2ClCOCl$	y=-1.571xx10^{-5}+1.00001785946	0.8995	1.7859x10^{-2}
$CHCl_2COCl$	y=-4.205xx10^{-4}+1.00000223545	0.9999	2.1857x10^{-3}
CCl_3COCl	y=-1.273xx10^{-5}+1.0000036264	0.9962	3.626x10^{-3}

4. Conclusions

A new device with a window film of ZnSe and a body of PTFE can be used to determine the content of impurities in $TiCl_4$. The regression equation of $CH_2ClCOCl$ was y=-1.571x$\times10^{-5}$+1.00001785946, R^2=0.8995; the regression equation of $CHCl_2COCl$ was y=-4.205x$\times10^{-4}$+1.00000223545, R^2=0.9999; and the regression equation of CCl_3COCl was y=-1.273x$\times10^{-5}$+1.0000036264, R^2=0.9963. The new infrared measuring device showed various advantages, including safe, simple, easily removable, washable and reusable. The process of peak by OMNIC software made the research convenient due to the measurement of different compound simultaneously. This new method prevented the influence of added solution on the matrix, which is applicable for multi-component analysis.

Acknowledgments
The research of this paper was financially supported by the Youth Fund in Academy of Guizhou Science. The test was done in Guizhou Academy of testing and analysis with FT-IR. In developing the ideas presented here, I have received helpful input from Tan H and He J L. I have to thank my co-author, Wang Y J.

References

[1] Deng G Z and Wang X D 2007 Promote technological progress is key to maintaining sustainable development of titanium sponge production *Development of Titanium Production* **24** 9-14

[2] Ouyang Q S, Zhao Z W and Zhu Y H 2004 Advances in technology and the new titanium sponge production Rare Metals and Cemented Carbides **32** 47-52

[3] Mo W, Deng G Z and Luo F C 1998 *Titanium Alloys* 1st edn. (Beijing Metallurgical Industry Press， No.39, Zhuyuan North Street, Dongcheng District, Beijing)

[4] Tan H, He J and Song G L 2012 Determination of trace organic impurities in refined $TiCl_4$ by infrared spectroscopy *Journal of Instrumental Analysis* **31** 1325-9

[5] Rolf B and Johnson J 1993 Infrared spectroscopy applied to the determination of impurities in titanium tetrachloride *Technology of Zunyi Titanium* **Sup** 10-5

[6] Liu W J 1994 Qualitative discussion about infrared spectra of impurities in titanium tetrachloride *Vanadium and Titanium* **21** 26-30

[7] China Nonferrous Metals Committe (CNSMQ) 2007 Fine titanium tetrachloride quality standards (CN. Q/WBH001-2007)

[8] Chen W S and Lan W 1996 Determination of fine titanium tetrachloride titanium chloride and oxychloride infrared semi-quantitative method *Guizhou Science* **14** 56-9

[9] Chen W S and Lan W 2001 Quantitative determination for the trichloro-acetyl chloride in titanium tetrachloride by means of standard addition extrapolation *Journal of Guizhou Normal University (Natural Sciences)* **19** 56-8

[10] Song Y P, Tan H, Sun J P and He J L 2008 Infrared quantitative analysis of titanium tetrachloride trichloroacetyl chloride *Chinese Determination Technology* **34** 95-7

[11] Yang L, Huang H and Cheng X L 1999 New calculation method of standard addition method in infrared spectroscopy quantitative analysis *Analytical Laboratory* **18** 79-82

[12] Griffiths P R and Haseth J A D E 2007 *Fourier Transform Infrared Spectroscopy* 2nd edn. (Hoboken: John Wiley & Sons Ltd)

[13] Cai X L 2000 Identification of infrared difference spectroscopy for mixing hypnotics *Spectrosc and Spectral Analysis* **20** 329-32

The experiments for mechanical properties of 20Cr2Ni4 steel and the coefficient definition of constitutive equation

L Pang[1], G C Liu and J P Lu

School of Mechanical Engineering, Beijing Institute of Technology, Beijing 100081, China

E-mail: 18800165868@163.com

Abstract. The 20Cr2Ni4 alloy steel has the properties of high strength, toughness and hardness. It is used in large cross-section carburized parts, such as gears, shafts and components which are required high strength and good toughness. In order to study the static mechanical properties and dynamic mechanical properties of 20Cr2Ni4 steel, the static compression experiment and the Hopkinson Pressure Bar test are conducted. The stress-strain relationship within the scope of 25~400℃ is obtained by experiments, and softening effect of strain rate and strengthening effect of temperature is comprehensively analyzed. The paper has a more comprehensive understanding on mechanical response of 20Cr2Ni4 steel within the scope of 25~400℃. Based on the experiment data the parameters in Johnson-Cook constitutive equation of 20Cr2Ni4 have been gotten. The research results of this paper lay a foundation for the further applications of 20Cr2Ni4 steel.

1. Introduction

The constitutive relation is the inherent nature of the material and the objective law that follows in the deformation process. In the process of machining, metal material has large deformation, which accompanies with strain hardening, reply, recrystallization. The phenomenon is affected by strain, strain rate, temperature and other factors. Therefore, the constitutive relation of metal materials describes the relationship between the factors such as stress, strain and strain rate in the process of deformation, the expression is [1]:

$$\bar{\sigma} = \bar{\sigma}(\bar{\varepsilon}, \dot{\bar{\varepsilon}}, \mathrm{T}) \tag{1}$$

The 20Cr2Ni4 is alloy steel of high strength, good toughness and good hardenability. Table 1 shows its chemical composition, and table 2 shows the basic mechanical property.

Table 1. Chemical composition of 20Cr2Ni4 (mass fraction/%).

C	Si	Mn	Cr	Ni	S	P
0.17~0.23	0.17~0.37	0.3~0.6	1.25~1.65	3.25~3.65	≤0.03	≤0.03

Table 2. Basic mechanical performance of 20Cr2Ni4.

[1] Beijing Institute of Technology, 5 South Zhongguancun Street, Haidian District, Beijing 100081, China. E-mail: 18800165868@163.com.

Tensile strength	Yield strength	Elongation	Shrinkage	Impact strength
≧1175MPa	≧1080MPa	≧10%	≧45%	≧78J

The experiment results under different loading conditions of materials are the most direct and the most effective response to mechanics performance. It can fundamentally show the physical structure of the material properties, and it has incomparable advantages compared with the theoretical research with all kinds of ideal assumption. To study 20Cr2Ni4 mechanics performance, we conduct the tests of static mechanics performance and dynamic mechanical performance respectively. The purpose of the experiment for static mechanical performance is not only to research static mechanics properties of 20Cr2Ni4 at room temperature, but also provide data support for the subsequent experiment for dynamic mechanical performance. While the purpose of the experiment for dynamic mechanical properties is to get the stress-strain relationship, the softening effect of the strain rate and strengthening effect of the temperature, and the parameter in Johnson-Cook constitutive equation of 20Cr2Ni4 steel.

2. The experiment for static mechanical properties

2.1. The scheme

The purpose is to obtain the stress-strain curve of 20Cr2Ni4 steel at 10^{-3} s^{-1} and 10^{-2} s^{-1} respectively at room temperature. Due to the high stability of the static compression experiment, we do two groups under each strain rate, a total of four groups.

2.2. The device and the process of the experiment

We conducted the static compression experiment using the universal testing machine called INSTRON. Figure 1 shows the experimental device. By the requirements for test preparation, the sample is $\phi 5\times 5mm$. Two groups of different strain rate are got by controlling the speed of beam displacement.

Figure 1. INSTRON universal testing machine.

2.3. The results

Figure 2 shows the stress-strain relationship of 20Cr2Ni4 steel at room temperature. Figure 3 shows the contrast figure of the sample before and after the experiment, in which the middle one is the sample at the initial state, the left one is the compressed sample at the strain rate of 10-3 s^{-1}, and the right one is the compressed sample at the strain rate of 10-2 s^{-1}. The size in figure 3 is 2 times of its real size. From the figure, we can see obviously that the deformation of the sample becomes bigger when strain rate increases.

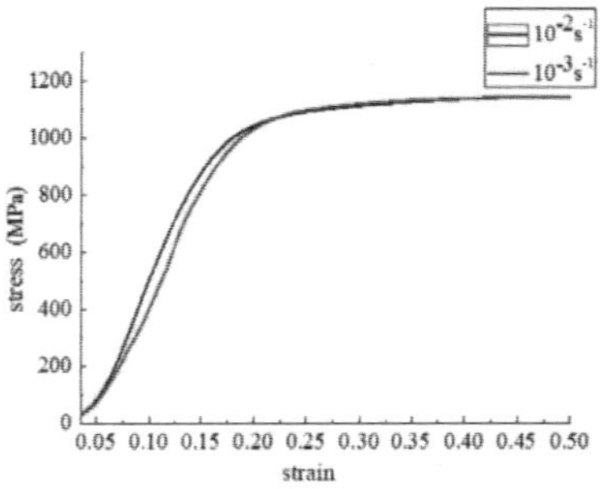

Figure 2. The stress-strain relationship of 20Cr2Ni4 under the static environment at room temperature.

Figure 3. The contrast figure of the sample before and after the experiment (scale: 2:1).

As you can see from the figure 2, the coincidence degree for the stress-strain curves of 20Cr2Ni4 steel at room temperature under different strain rate is high. Under the condition of quasi-static at room temperature, yield strength increases when the strain rate becomes higher. Yield strength of 20Cr2Ni4 steel at the strain rate of 10^{-3} s^{-1} is 1112 MPa, more than 10^{-2} s^{-1} by about 30 MPa. The experiment results show that elastic modulus of the material is not affected by the strain rate under the condition of quasi static, and elastic modulus is about 226 GPa.

3. The experiment for dynamic mechanical properties

3.1. The scheme

In this paper, we study the dynamic mechanical of 20Cr2Ni4 through the Split Hopkinson Pressure Bar (SHPB) experiment. The SHPB technique is considered to be the main experimental method to get the relations between stress and strain within 102 s^{-1}-104 s^{-1} of the material [2-4]. The experiment is made up of dynamic at room temperature and dynamic at high temperature. Because the experiment for dynamic is less stable than the experiment for static mechanical properties, every experiment should take at least three groups of valid data to ensure the reliability of the experimental data.

3.2. The scheme

The SHPB system consists of compressive bar system, heating device, data acquisition and processing system, which is shown in figure 4. The dynamic compression experiment at room temperature achieves loading the samples under different strain rate by controlling the air pressure. The experiment adopts the strain rates of 1000 s^{-1}, 2000 s^{-1} and 3000 s^{-1}. The dynamic compression experiment at high temperature achieves loading the samples under the different temperature by the heating device. The experiment selects four kinds of temperature which are 100℃, 200 ℃, 300 ℃ and 400 ℃, and strain rate is selected as the 2000 s^{-1}, the specific design of the experiment is shown in table 3.

Table 3. The experimental design for dynamic mechanical properties.

Temperature (℃)	Strain Rate (s^{-1})	Experiment Times
25	1000	3
25	2000	3
25	3000	3
100	2000	3

200	2000	3
300	2000	3
400	2000	3

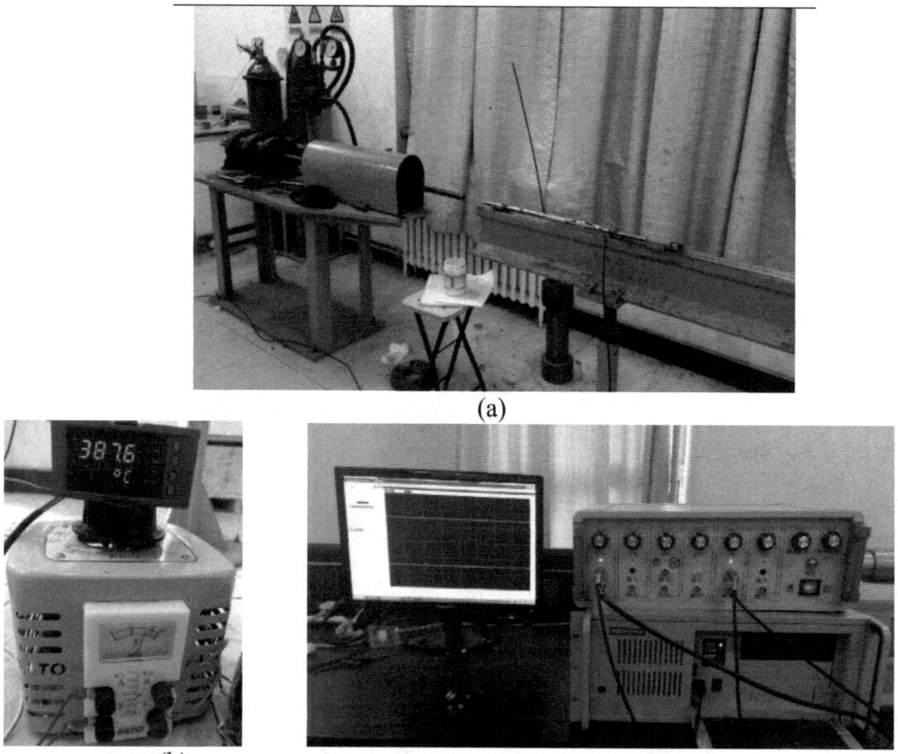

(a)

(b) (c)

Figure 4. Experiment device for SHPB: (a) Pressure bar system; (b) Heating device; (c) Data acquisition and processing system.

3.3. Analysis of the experimental results

Research of mechanical properties of the material under static loading condition has been relatively mature, but the parts do not always work under static conditions, they are likely to be affected by dynamic loading. Especially in cutting process, the whole deformation process of the material will be done in a very short period. Therefore, mechanical properties of the material under dynamic loading are quite different from that under static loading. Strain rate, temperature and many other factors produced in the process of deformation of the materials will have an impact on the mechanical properties of the material, and the most basic influence presents in two aspects: one is that the stress-strain curve moves up with the increase of strain rate, that is strain rate hardening; Second, the stress-strain curve moves down with the increase of temperature, namely thermal softening.

- Dynamic Compression Experiments at Room Temperature

The room temperature is 25℃ during the experiment, the effect of strain rate on dynamic mechanical properties of 20Cr2Ni4 can be found by comparing the data under the condition of different strain rate at the same temperature. Figure 5 is the stress-strain curve of 20Cr2Ni4 under conditions of dynamic loading at 25℃, and figure 6 shows the yield strength of 20Cr2Ni4 under different strain rate at 25℃.

From the experimental results, with the increase of strain rate, the stress-strain curve moves up as a whole. Repeatability of the three sets of data is bad when the strain rate is 1000 s^{-1}, the stress-strain curve fluctuates slightly larger, and the main reason is that Hopkinson Pressure Bar has experimental error when strain rate is low. Since 20Cr2Ni4 has no obvious yield stage under the action of dynamic load, according to experience, the stress when the plastic strain is 0.2% is taken as the Yield Stress. According to the fitting result, the yield strength increases 100MPa when the strain rate is 3000 s^{-1} than when it is 2000 s^{-1}, presenting the strengthening effect of strain rate, but not obvious.

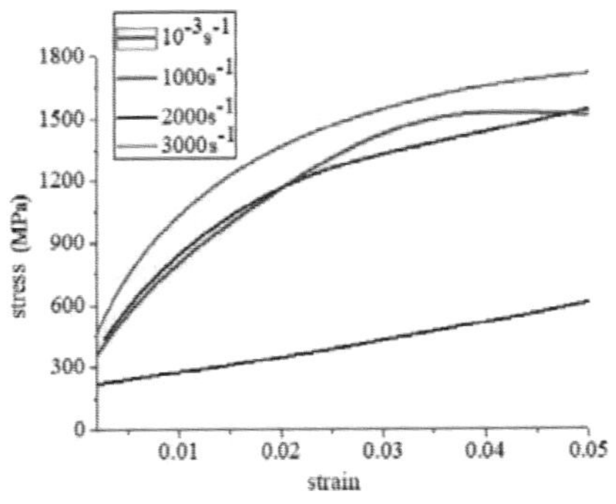

Figure 5. The stress-strain curve of 20Cr2Ni4 under conditions of dynamic loading at room temperature.

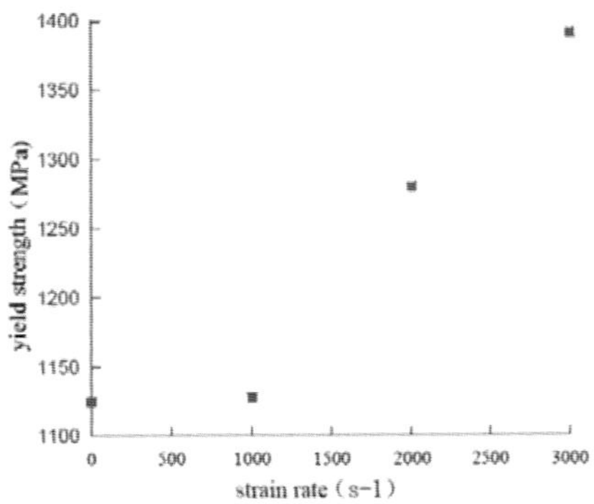

Figure 6. The yield strength of 20Cr2Ni4 under different strain rate at room temperature.

- High-temperature dynamic compression experiment

Take the strain rate of 2000 s^{-1} during the experimental. The effects of temperature on dynamic mechanical properties of 20Cr2Ni4 can be found by comparing the data under the condition of the same strain rate but different temperature. Figures 7 and 8 respectively shows the stress-strain and yield strength curve of 20Cr2Ni4 under the condition of different temperature when the strain rate is 2000 s^{-1}.

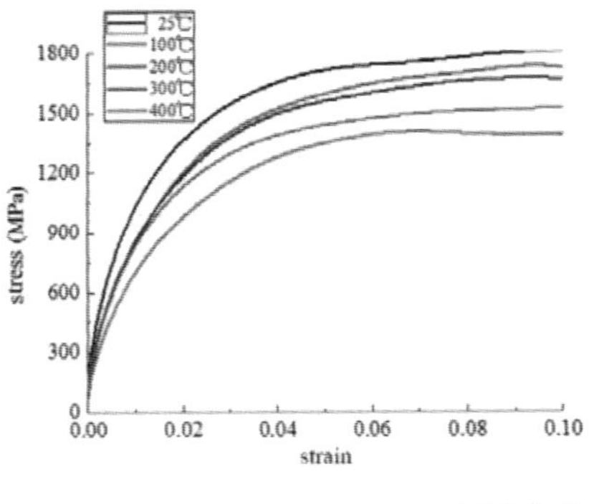

Figure 7. Stress-strain curve of 20Cr2Ni4 under different temperature at 2000 s^{-1}.

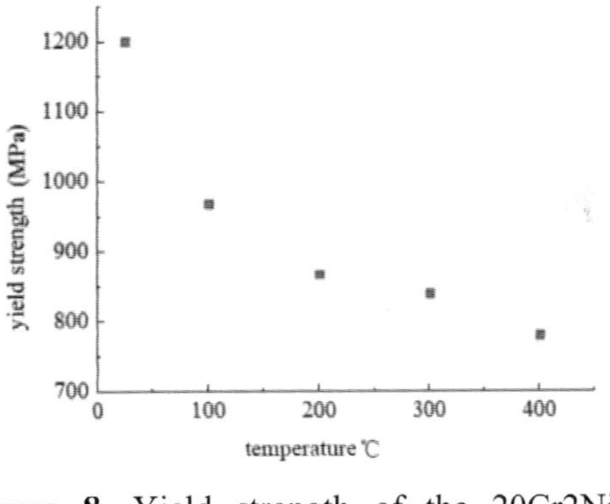

Figure 8. Yield strength of the 20Cr2Ni4 under different temperature at 2000 s^{-1}.

From figure 7, it can be seen that stress-strain curve integrally moves down as temperature rises. When the temperature is 100℃ and 200℃, the difference of stress-strain curve is small, in this way it illustrates that the temperature has a little effect on mechanical properties of 20Cr2Ni4 near the two temperatures. As can be seen from the figure 8, as temperature rises, yield strength comes down. When the temperature is 25℃, the yield strength is 1201 MPa, and when the temperature is 200℃, the yield strength is 867 MPa, reduced by 27.8%, showing obvious softening effect of temperature. When the temperature is 400℃, the yield strength is 780 MPa, reduced by 10% compared with that of 200℃. Thus it can be seen, when the temperature is less than 200℃, the temperature has greater effects on the mechanical properties of the 20Cr2Ni4, the temperature effect on 20Cr2Ni4 decreases gradually when the temperature exceeds 200℃.

4. Construction of constitutive equation

4.1. Johnson-Cook model

It will be able to build a suitable constitutive relation model by getting stress-strain relationship through the above analysis of the test results. In recent years, these constitutive relations are widely used: Johnson-Cook (JC) model, Zerilli Armstrong model, Bodner-Parton model. JC model's form is [5-7]:

$$\sigma = (A + B\varepsilon^n)(1 + C\ln\dot{\varepsilon}^*)(1 - (T^*)^m) \tag{2}$$

In the Eq. (2), σ is stress, ε is the equivalent plastic strain, $\dot{\varepsilon}^*$ is the relative plastic strain rate [8], $\dot{\varepsilon}^* = \dot{\varepsilon}/\dot{\varepsilon}_0$, usually take $\dot{\varepsilon}_0 = 1.0s^{-1}$. T^* is dimensionless temperature, $T^* = (T - T_{room})/(T_{melt} - T_{room})$, and T is the internal temperature of the material, T_{room} is the room temperature, T_{melt} is the melting point, $0 \le T^* \le 1.0$. *A*, *B*, *C*, *n* and *m* are the material constants. $(A + B\varepsilon^n)$, $(1 + C\ln\dot{\varepsilon}^*)$ and $(1 - (T^*)^m)$ respectively describes the material's strain hardening effect, strain rate effect, temperature effect. From the above equation, the JC model can solve the problems of deformation in the process at the condition of high stress, high temperature and high temperature, and can truly reflect constitutive behavior of the flow stress [9]. The paper selects the JC model to express the constitutive relation of 20Cr2Ni4.

4.2. Determination of constitutive equation parameters

JC expression is:

$$\sigma = (A + B\varepsilon^n)(1 + C\ln\dot{\varepsilon}^*)(1 - (T^*)^m)$$

In the equation, A and B are constants related to the materials, n is the influence coefficient of strain hardening, C is the sensitivity coefficient of strain rate, m is softening coefficient of temperature. According to the data of the quasi-static compression experiment, dynamic experiments at normal temperature and high temperature of the 20Cr2Ni4, the coefficients of Johnson-Cook model can be determined, specific steps are as follows [10-13]:

- When the samples are at room temperature ($T = T_{room}$), $\dot{\varepsilon} = \dot{\varepsilon}_0 = 10^{-3}s^{-1}$ the JC expression is simplified as:

$$\sigma = A + B\varepsilon^n \tag{3}$$

According to the stress-strain curve, when the samples are at room temperature and the strain rate is $\dot{\varepsilon} = \dot{\varepsilon}_0 = 10^{-3}s^{-1}$, *A*, *B* and *n* can be determined. First, make sure of *A*. *A* represents the yield strength of material, A is the corresponding value when the plastic strain is zero from the quasi-static experiments. *A* is 1112 MPa obtained by fitting, according to Eq. (3), *B* and n can be calculated. *A*=1112 MPa, *B*=1063 MPa, *n*=0.2.

- when the plastic strain is $\varepsilon = 0$, and temperature is room temperature, namely $T = 25$℃, the relationship between the dynamic stress and the strain rate can be obtained at room temperature based on Eq. (2):

$$\sigma = A(1 + C\ln\dot{\varepsilon}^*) \tag{4}$$

From the fitting result of figure 5, $C = 0.01$.

- when the plastic strain $\varepsilon = 0$, the relationship among dynamical yield stress, strain rate and temperature can be obtained based on Eq. (2):

$$\sigma = A(1 + C\ln\dot{\varepsilon}^*)(1 - (T^*)^m) \tag{5}$$

By data fitting, $m = 0.62$.

After determining the five parameters *A*, *B*, *C*, *n* and *m*, we can get the expression of Johnson - Cook model for the 20Cr2Ni4 as follow:

$$\sigma = (1112 + 1063\varepsilon^{0.2})(1 + 0.01\ln\dot{\varepsilon}^*)(1 - (T^*)^{0.62}) \quad (6)$$

5. Conclusions

According to the experiments for mechanical performance of 20Cr2Ni4 alloy steel, its stress - strain curves at different strain rate and different temperatures are obtained, which confirms the relationship between stress and strain when temperature ranges from 25℃ to 400℃ and the strain rate is within the range of 10^{-3} s^{-1}-3000 s^{-1}. Based on the curves, the strengthening effect of strain rate and softening effect of temperature of 20Cr2Ni4 is analyzed, drawing the conclusions which are shown as follows.

- Under the dynamic condition, Yield Stress of 20Cr2Ni4 steel increases over the strain rate, but it is not obvious. That is to say, the strengthening effect of strain rate of 20Cr2Ni4 is not evident unless the magnitude of strain rate changes.
- The dynamic Yield Stress of 20Cr2Ni4 steel decreases with the increase of temperature when the strain rate is about 2000 s-1. Yield Stress at 400℃ is reduced by 35% than that at 25℃. When the strain rate is constant, the influence of temperature changes on the dynamic yield stress of 20Cr2Ni4 is significant, which means that the softening effect of temperature of 20Cr2Ni4 is notable.
- The parameters of Johnson-Cook constitutive relationship of 20Cr2Ni4 steel are determined by experimental data, which are obtained by experiments under the condition of quasi-static, dynamic at room temperature and dynamic at high temperature.

References

[1] Ozel T and Altan T 2000 *International Journal of Machine Tools & Manufacture* **40** 133-52
[2] Al-Mousawi M M, Reid S R and Deans W F 1997 *Mechanical Engineering Science* **211** 273-92
[3] Seo S, Min O and Yang H 2005 *International Journal of Impact Engineering* **31** 735-54
[4] Hubert W, Meyer J R and David S 2001 *International Journal of Impact Engineering* **26** 509-21
[5] Johnson G R and Cook W H 1983 *Proceedings of Seventh International Symposium on Ballistics* (The Hague, The Netherlands) p 541
[6] Johnson G R and Cook W H 2003 *International Journal of Impact Engineering* **20** 602
[7] Johnson G R and Tim J 1988 *Journal of Applied Physics* **64** 3901-10
[8] Odeshi A G, Al-ameeri S and Bassim M N 2005 *Journal of Materials Processing Technology* **12** 385-91
[9] Lalwani D I, Mehta N K and Jain P K 2009 *Journal of materials processing technology* **209** 5305-13
[10] Dorogoy A and Rittel D 2009 *Exp. Mech.* **46** 881-5
[11] Shrot A and Baeker M 2011 *Proceedings of the 14th International ESA-FORM Conference on Material Forming* (Belfast, UK) pp 29-34
[12] Shrot A and Baeker M 2012 *Comput. Mater. Sci.* **52** 298-304
[13] Bosetti P, Bort C M and Bruschi S 2013 *Journal of Manufacturing Science & Engineering* **135** 5

IOP Conf. Series: Materials Science and Engineering **103** (2015) 012029 doi:10.1088/1757-899X/103/1/012029

Spring-back of flexible roll forming bending process

Y Zhang[1], D H Kim and D W Jung

Jeju National University, 1 Ara 1-dong, Jeju-si, Jeju ,Republic of Korea

E-mail: zhangya777@gmail.com

Abstract. Simulations are now widely used in the field of roll forming because of their convenience. Simulations provide a low cost, secure and fast analysis tool. Flexible roll forming provides the desired shapes with a one time forming process. For roll forming, the velocity of the sheet and friction are important factors to attain an ideal shape. Because it is a complicated process, simulations provide a better understanding of the roll forming process. Simulations were peformed using ABAQUS software linked to elastic–plastic modules which we developed taking into account of interactions between these fields [1]. The application of this method makes it possible to highlight the strain-stress and mechanical behaviour laws and the spring-back. Thus, the flexible roll forming and bending process can bewell described by the simulation software and guide the actual machine.

1. Introduction

Cold roll forming now is widely used in the automotive and construction industries. Cold roll forming is a sheet pass a group of rolls and gets a desired shape. Because of its high productivity and low variable cost, roll forming have a great advantage compare to other forming process. With the roll forming process, we can easily get the shape of the sheet which we want and more suitable for mass production in factory which are very important. Although roll forming is an old technique and have been studied by many researchers but there are still some problems are not well solved such as spring-back. There are many factors can affect the sheet deformation in roll forming process and how to control these factor are very important in the simulation [2].

The roll forming process is operated between roll forming multi-stages continuously. And roll forming a profile is formed in several forming steps from an unreformed strip to a finished profile. The scientific design of passes is worked out by combining the theoretical analysis with the finite element analysis in accordance with the principles of cold roll-forming, thus as the desirable high-quality bending sections achieved through the combination of the theory with practice.

Recently, sheet forming become more and more important in industry production. To find a way for mass production with an easier way is necessary. Compare with the stamping and other process, roll forming is better in mass production and have a cheaper cost [3]. In this study, flexible roll forming bending process is designed to predict the bending process and the sheet mechanical behavior after deformation [4].

Discontinuous manufacturing technologies, such as stamping and stretch-forming, produce a product at one time. The discontinuous flexible forming process, which is based on the usage of reconfigurable die, does not require expensive conventional die and time consuming setup operations,

[1] Address for correspondence: Y Zhang, Jeju National University, 1 Ara 1-dong, Jeju-si, Jeju, Republic of Korea. E-mail: zhangya777@gmail.com.

and thus is suitable and efficient for the small quantity production of doubly curved sheet metals.

2. Objectives and approach

In In this paper we are going to develop a flexible roll forming process for the slope sheet, concave sheet and convex sheet. The mechanical properties are studied and the spring-back is calculated in this paper [5].

We export the CAD model of the rolls and sheets form Solid works and the simulation was calculated using the commercial finite simulation program ABAQUS.

The steps of this investigation were:

- Model a given sheet which has a constant bending width and determine the rolls shape and dimension. the appropriate superscript value just after the surname and just before the address:
- Define the roll forming process and conditions in ABAQUS such as mesh and interactions.
- Determine the rotation center of the roll and establish the boundary conditions for the roll make them moving forward as desired.
- Optimization of the roll forming process with the simulation results and determines better parameters such as velocity and friction.

3. Simulation and analysis

The SGARC 440 material is very common and cheap sheet steel, SGACR 440 with a Young's modulus of 207 (GPa) and Poisson's ratio of 0.28 [6]. The elastic-plastic properties, including the true stress and strain are given in table 1 and the true stress-strain curve derived from these data is display in figure 1.

Table 1. Mechanical characteristics of SGRC 440 steel.

	Young's modulus	Tensile strength	Yield Strength	Poisson 's Ratio
SGARC440	207000 [MPa]	440 [MPa]	300 [MPa]	0.28

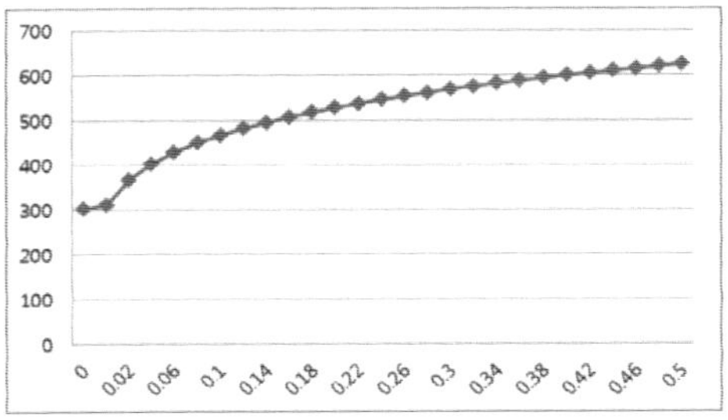

Figure 1. True plastic strain (mm/mm).

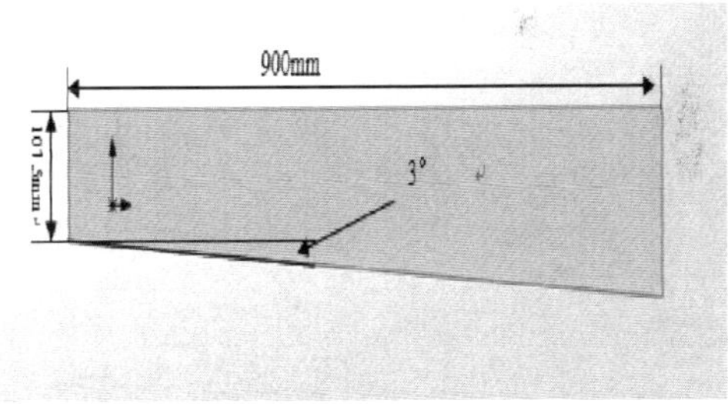

Figure 2. Dimension of sheet with 3degree.

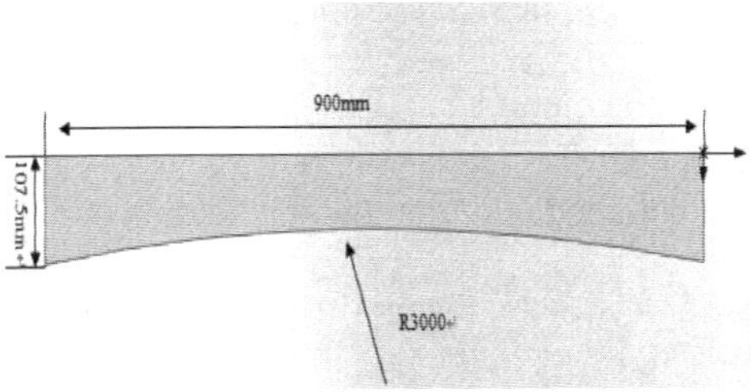

Figure 3. Dimension of concave sheet.

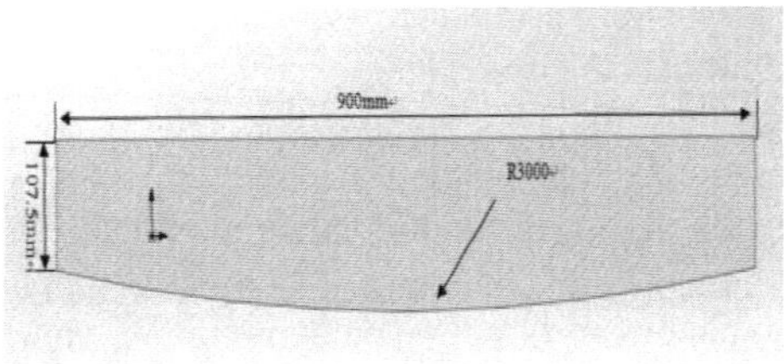

Figure 4. Dimension of convex sheet.

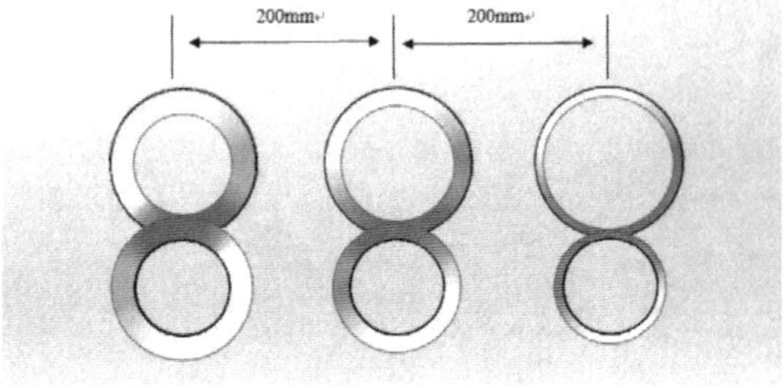

Figure 5. Distance between rolls.

With different design of rolls, sheet could be deformed successively into products with different section.Three experimental setups of flexible roll forming process are modeled.There three kinds of sheets and the dimensions are shown in figures 2, figure 3 and figure 4. For all sheets have a width of 900 mm and the length in longitudinal direction is 107.5 mm. The concave and convex sheets are part of circle which radius is 3000 mm. In this roll forming process, the sheet passes through the three groups of the rolls. The rolls from left to right are 15°, 30° and 45°.The metal sheets with the thickness of 1mm are used as the initial metal strip. The position of the rolls is given in figure 5. There are several of element types in ABAQUS. In this simulation we choose the quadrangular shell element (S4R) volume element with full integration.S4R element is the most stable element type and task less time for the Dynamic-implicit simulation in the roll forming design program [7]. After meshing, there are 3294 elements in the sheet, and the sheet is 900 mm long.

There are many ways in ABAQUS to control the rigid bodies, by referencing the velocity, rotation and displacement or load. We can choose any one to control the movement of the rigid body. For the sheet which has an angle with 3 degree is easier to use the velocity by defining the velocity in X-axis and Y-axis. However, defining the velocity in addition with the sheets which have concave edge or convex edge has some problems. It is better to use the rotation and displacement rotation with period amplitude. Period amplitude is Fourier series in actual.

French mathematician Fourier found that any periodic function can be used to constitute the infinite series of sine and cosine function to represent (sine and cosine function is chosen as the basis function is because they are orthogonal) [8].

In this function, s(x) is a function of the variable x and the function can be shown as below:

$$S_N(X) = \frac{A_0}{2} + \sum_{n=1}^{N} A_n \cdot \sin\left(\frac{2\pi nx}{P} + \Phi_n\right) \tag{1}$$

By which we can definite move path of the roll can be a periodic function. Therefore the roll will move as a circle with no displacement in third direction.

The sheet passes the rolls and bending continually. So for the boundary conditions of the sheet the rotation in the Y direction is fixed. The movement in X and Y direction are fixed. The boundary conditions are performed by defining the centreline of the sheet. Other directions are flexible to get a bending along the edge.

After introducing the boundary conditions, contact were defined between the rigid bodies and deform bodies. The interactions between them were surface to surface contact and definite by the step. It has been assumed that the friction between the roll and sheet is penalty and the friction coefficient is small enough.

The geometry of the strips after rolling deformation are shown in figure 6, figure 7 and figure 8. The experiment results are shown in figure 9, figure 10 and figure 11. The forming result is influenced by many factors such as the material property, the thickness, the velocity of the sheet and the friction coefficient of between the sheet and the roll. According to the stress distribution below we can know that the minimum stress occurs at the middle of the sheet which is unperformed zone. This is corresponding to the deformation theory.

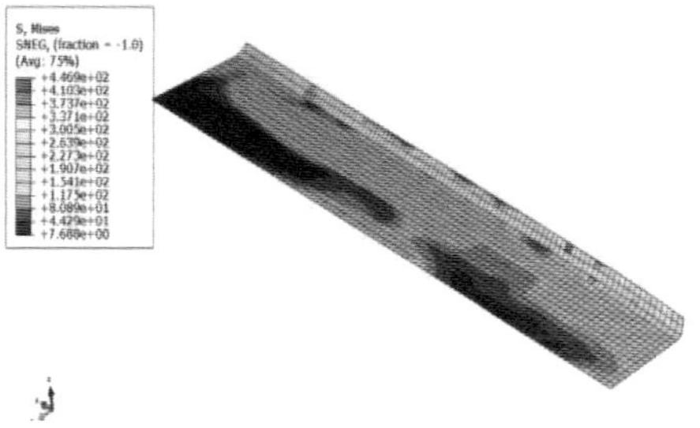

Figure 6. Simulation result of Sheet with 3°

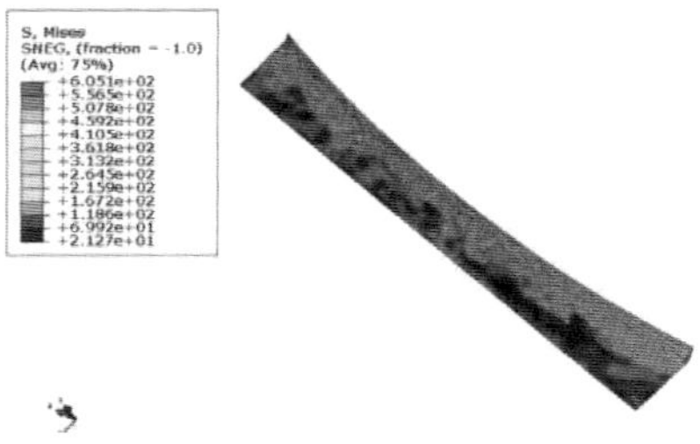

Figure 7. Simulation result of concave sheet°

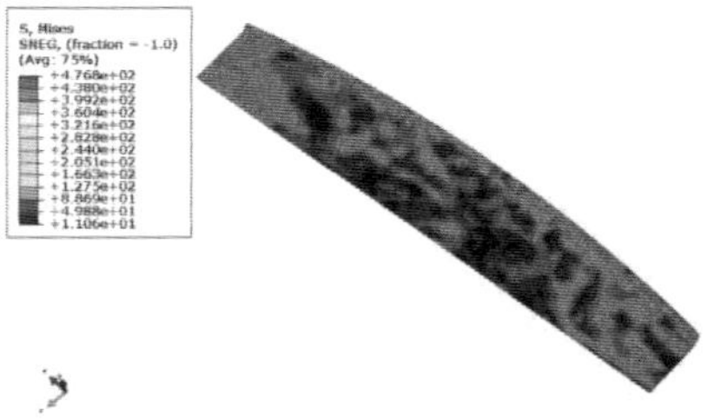

Figure 8. Simulation result of convex sheet

Figure 9. Experiment result of sheet with 3°

Figure 10. Experiment result of convex sheet

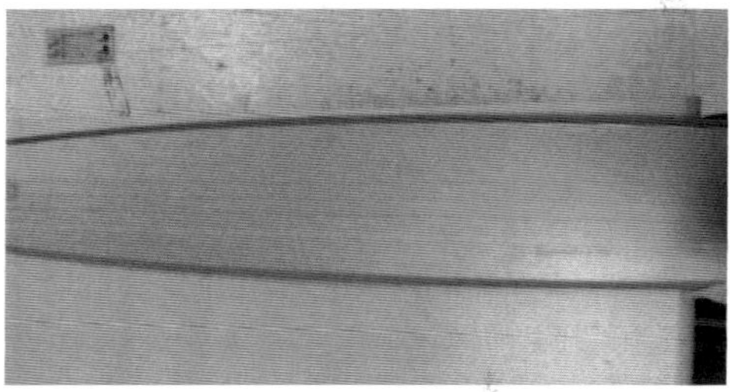

Figure 11. Simulation result of convex sheet

The comparison of the stress of the whole model is shown in figures 9, figure 10 and figure 11. What we can know from the figure is that the simulation results have a good agreement with the experiment results. For every simulation, the comparison of the simulation results and experiment results are very important. The experiment results can verify the simulation results. As shown in the figures above the biggest stress of the sheets is at the edge of the sheet and we can know the stress of the concave sheet is too big at some point .This is because of mesh distortion around the forming process.

Since the proposed process generates the bending dominant deformation, it is necessary to investigate the effect of spring-back. In order to analyse the spring-back, a Static, General step is carried out based on the problem defined. In order to compare the effect of spring-back through the roll forming process, nine points along the X-axis direction are calculate and illustrated in figure 12.

Spring-back is inherent to the material forming process especially in the bending process. These kinds of phenomenon have a great impact on the size of precision and productivity.

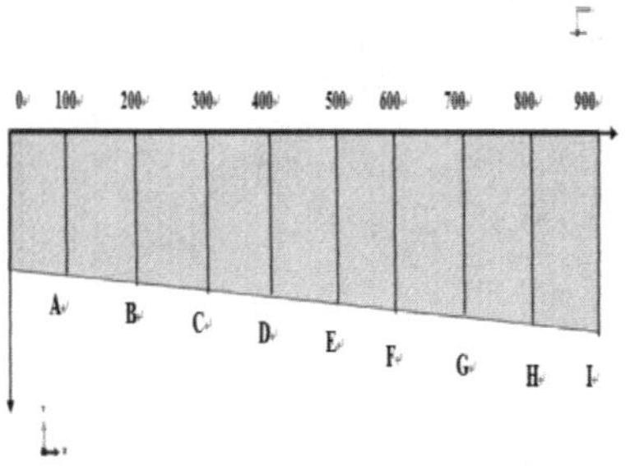

Figure 12. Measurement point for spring-back.

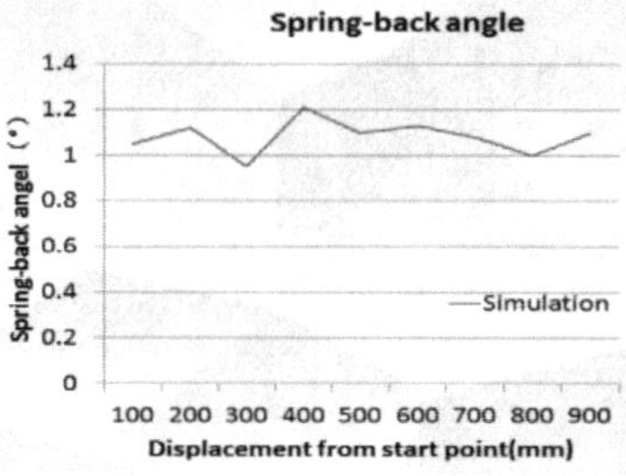

Figure 13. Spring-back angle of sheet with 3°.

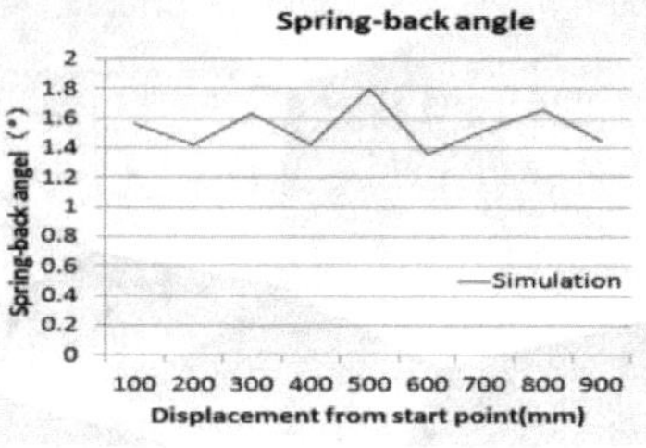

Figure 14. Spring-back angle of concave sheet.

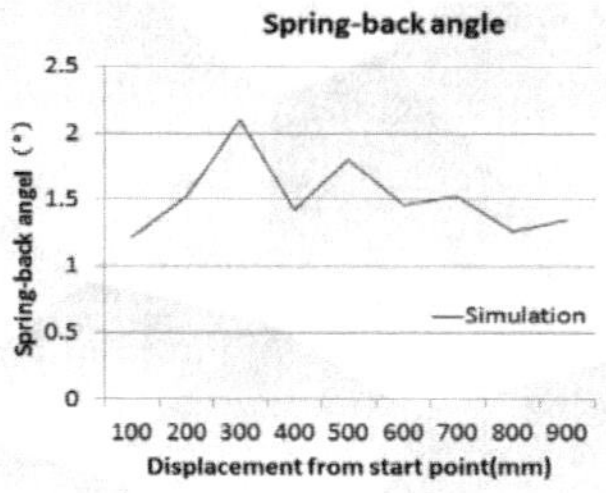

Figure 15. Spring-back angle of convex sheet.

Therefore, to get an accurate estimate of the amount of the spring-back angle in bending process is very important in roll forming design.

From the figure 13, figure 14 and figure 15, we can know that the spring-back angle is not consistency according the sheet the edge. For the sheet with 3°, the maximum spring-back angle occurs at the distance=400 mm. The maximum spring-back angle of the concave sheet occurs at the distance=500 mm. As for the convex sheet, the maximum spring-back angle occurs at the distance=300 mm.

4. Conclusions

In this paper, a new kind design of flexible roll forming bending process with Fourier series. Different kinds of sheet were analysed and we can conclude that the model can effectively predict the trends in the distribution of the strain-stress and energies. Through the simulation program, the important parameters are determined by taking account of roll stand distance and roll forming speed. With the study above, we can know that

- With the finite element simulation in ABAQUS we can get a better understanding of the roll forming process with an easy way and less time.
- The spring-back angles have been learned in this paper, from which we can analysis the deformation and find out how to reduce it.
- The roll forming process is analysis in this paper and the stress and distribution are learned. The simulation above can guide the optimization of the roll forming process in real produce.

From the process analysis above, we can design a reasonable roll forming process. But there are still many works need to do such as the buckling and flange wrinkling have not been studied in this paper. These kinds of defects are very important in roll forming process and should be considered to get better results.

Acknowledgments

This research was supported by the Basic Science Research Program through the National Research Foundation of Korea (NRF) funded by the Ministry of Education (NRF-2014R1A1A4A01009199).

References

[1] Cusack S and Miller A 1979 Determination of the elastic constants of collagen by Brillouin scattering *J. Mol. Biol.* **135** 39-51

[2] Gundiah N, Ratcliffe M B and Pruitt L A 2007 Determination of strain energy function for arterial elastin: Experiments using histology and mechanical tests *Biomech.* **40** 586-94

[3] Kiuchi M 1973 Analysis study on cold roll forming process *Report of the Inst. of Ind Sci.* **23** 1-23

[4] Kiuchi M and Koudobashi T 1984 Automated design system of optimum roll profiles for cold roll forming proc *3rd Int Conf. on Rotary Metal Working Process* (Kyoto) pp 423-7

[5] Farzin M, Salmani M and Shameli E 2002 Determination of buckling limit of strain in cold roll forming by the finite element analysis *Materials Processing Technology* **125-126** 626-32

[6] Rossi B, Afshan S and Gardner L 2013 Strength enhancements in cold-formed structural sections-Part II: Predictive models *Journal of Constructional Steel Research* **83** 189-96

[7] Rhodes J 1996 A semi-analytical approach to buckling analysis for composite structures *Composite Structures* **35** 93-9

[8] Jeong S H and Lee S H 2007 3D modeling of a cold roll-forming process using FEM *Fall Conference KSMTE* (Korean society of Manufacturing Technology Engineers) pp 733-8

Experimental study on improving cement quality with oxygen-enriched combustion technology

Y Q Liu[1,2], A M Zhang[3], S Qing[1], F S Li[1,4], S P Yang[1] and Z F Yang[1]

[1] State Key Laboratory of Complex Nonferrous Metal Resources Clean Utilization, Department of Metallurgical and Energy Engineering Faculty, Kunming University of Science and Technology, Kunming 650093, China
[2] Department of Media and Information Engineering Faculty, Yunnan Open University, Kunming 650000, China
[3] Kunming Sino-platinum Metals Catalyst Co. Ltd. Kunming 650093, China

E-mail: 609263134@qq.com

Abstract. With the intensification of the global energy crisis, the production cost of enterprises is continuously increasing because of the rising fuel prices and high requirements for environmental protection. As result, energy savings and environmental protection are vital considerations for a variety of enterprises. As a practical energy-saving technology, oxygen-enriched combustion has played a major role in energy saving and emissions reduction as its application in industrial furnaces has been popularized in recent years. This experiment was conducted in a cement rotary kiln with a capacity of 4000 t/d in a factory in China. Based on measured data in the oxygen-enriched combustion experiment, we determined the patterns of variation in the main parameters of the cement rotary kiln under oxygen-enriched production conditions. The results provide important theoretical and practical base for the cement building materials industry in energy saving and emissions reduction.

1. Introduction

Resources, energy and environmental problems have recently become bottleneck in the process of economic development. In particular, energy saving and emissions reduction are critical concerns of sustainable development. As one of the foremost energy-consuming and pollutant-emitting industries, cement production must adapt to save resources, mitigate the energy crisis and protect environment [1-4].

Oxygen-enriched combustion technology is one of the most promising approach to energy saving and environmental protection for cement kilns [5, 6]. Research results show that adopting oxygen-enriched combustion technology can not only save fuel, reduce emissions and lessen discharge of harmful gases, but also can improve the yield and quality of cement and extend the kiln's furnace life. Therefore, oxygen-enriched combustion technology has broad prospects for application in the cement rotary kiln [7-10].

2. Experiment

[4] Address for correspondence: F S Li, State Key Laboratory of Complex Nonferrous Metal Resources Clean Utilization, Department of Metallurgical and Energy Engineering Faculty, Kunming University of Science and Technology, Kunming 650093, China. E-mail: 326720228@qq.com.

This experiment was conducted in a cement rotary kiln with a capacity of 4000 t/d located in a factory in China. Enriched oxygen obtained from gasification and decompression of liquid oxygen was combined proportionally in a primary air muffler pre-filter after which is entered the primary air system. According to theoretical calculations and production practices, this test of local oxygen-enriched combustion led the oxygen enrichment rate to rise to 30%.

Figure 1 shows the test process flow diagram.

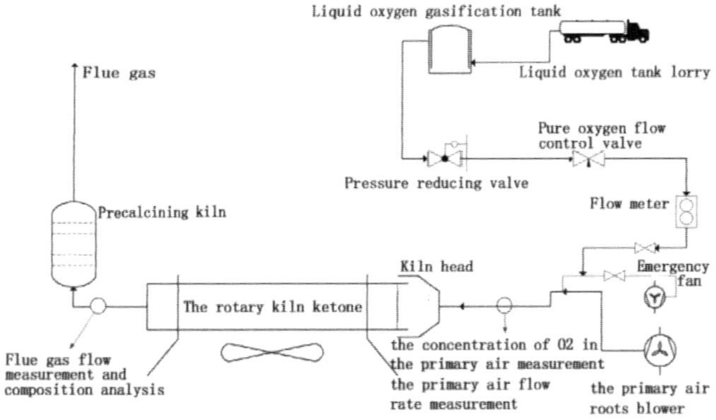

Figure 1. Test process flow diagram.

3. Test results and analysis

3.1. The influence of oxygen-enriched combustion on flame temperature inside the kiln

Cement rotary kilns reach extreme combustion temperature, but is smoke takes away a great deal of heat, leading to energy waste and wall damage. High flame temperature guarantees normal clinker burning, which is connected with fuel type, fuel supply, kiln body heat loss and other factors [11-14]. Table 1 shows the test data.

Table 1. Relationship between oxygen concentration and clinker temperature.

Time	Oxygen concentration in primary air pipe /%	Flame temperature/℃	Temperature of the clinker /℃
D2 14:00	23.0	1,381	1,202
D2 18:00	24.0	1,322	1,312
D2 22:00	25.4	1,431	1,378
D3 02:00	24.7	1,523	1,349
D3 06:00	25.8	1,435	1,335
D3 10:00	25.2	1,420	1,353
D3 14:00	25.5	1,429	1,359

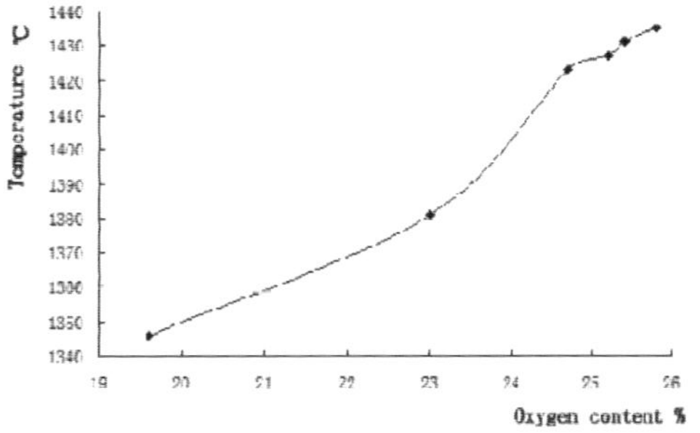

Figure 2. Relationship between oxygen concentration and the temperature of the clinker.

As oxygen content increases in the primary air system, oxygen molecules come into full contact with combustible and achieve complete combustion. Then the flame temperature increases, increasing heat transfer from the flame to the clinker, which in turn improves the clinker's kiln temperature and the heating effect. Figure 2 shows the relationship between oxygen concentration and clinker temperature.

3.2. Impact of oxygen-enriched combustion on energy consumption of clinker sintering

Table 2. The highest energy saving rate calculation table in the steady oxygen-enriched rate stage.

Items	Raw material t/h	coal in the kiln front end t/h	Tailings t/h	Conversion of clinker t/h	Total coal consumption t/h	Receive net calorific kCal/kg	Clinker sintering energy consumption kCal/ kg	Production of clinker per ton of coal t/t
23~25%The optimal value	310.4	9.82	21.37	188.12	31.19	3999.43	663.09	6.03
25~27%The optimal value	297.6	9.69	20.11	180.36	29.8	3999.43	660.79	6.05
Contrast value	304.6	10.2 2	22.32	184.61	32.54	4052.55	714.33	5.67
The improvement rate of index %						23~25%	7.17	6.31
						25~27%	7.49	4.16

The combustion temperature increases with the increase in oxygen content. At the same time, the feed coal quantity of the kiln head and kiln tail decreases. Pulverized coal fully burning in oxygen-enriched conditions improves the combustion temperature and enhances heat transfer in the furnace, which increases productivity. As oxygen content in the combustion air increase, the amount of pulverized coal required decrease, and the combustion-supporting air quantity significantly decrease. The decrease of feed coal and air volume leads to a decrease in smoke gas and carbon monoxide content. In turn, the heat loss of chemical flue gas greatly decreases, which improves the thermal efficiency and reduces energy consumption, as shown in table 2. Experiments show that an obvious coal saving effect results from adopting oxygen-enriched combustion.

3.3. The influence of enriched oxygen on the primary fan's air supply and energy savings

The amount of air needed per kilogram of coal burning under natural conditions exceeds that required by oxygen-enriched burning. Oxygen-enriched combustion also allows for reducing the primary air fan speed and the total air supply. The higher the oxygen content in the oxygen-enriched air, the lower the primary air fan speed and air supply required, which can be seen from table 3.

Table 3. The pure oxygen flow and primary air calculating table.

Primary air rate	Oxygen content in primary air %	Total amount of oxygen-enriched primary air Nm^3/h	Oxygen flow Nm^3/h	Primary air Nm^3/h
24.6%	21	12,200	0	12,200
	23	11,994	325.2	11,650
	25	11,035	566.5	10,500
	27	10,218	776.3	9,500

3.4. The influence of oxygen-enriched combustion on the quality of cement

The most important parameter affecting the quality of cement is the quality of clinker. Controlling clinker quality is the most important step in the process of cement calcinating. Parameters representing the quality of the clinker are free calcium oxide (F-CaO) content in cement and liter weight and strength.

3.4.1. Free calcium oxide (F-CaO). Due to materials' composition and structure, raw materials ratio, fineness and uniformity, clinker calcination temperature and time and the clinker cooling system, a small amount of CaO can not be combined with acidic oxides like i.e.SiO_2, Al_2O_3, Fe_2O_3 to form free oxides (F-CaO) during the process of cement clinker calcination. Hydration of F-CaO in cement occurs slowly, leading to poor stability in the cement; therefore, it is necessary to control the content of F-CaO in cement clinker. F-CaO content directly affects the stability and other properties of the cement. It is difficult to control during the production of cement and exerts great influence on cement quality, which serves as an important index by which to judge the quality of clinker.

The lower the content of F-CaO in the clinker, the better the cement quality. Generally, production requirements of F-CaO fall at 1.5% or less. As the content of F-CaO increases, clinker strength significantly decreases, while its stability and qualified rate drop dramatically. When ventilation and oxygen are deficient in the kiln, pulverized coal combustion in the kiln head remains incomplete, and the continuously reducing atmosphere reduces the temperature in the kiln. If the kiln burning temperature is low, the liquid phase quantity of the clinker is insufficient or the liquid phase reaction is incomplete, high F-CaO content results [15-17].

F-CaO content in the cement clinker is different when calcinated under normal production conditions versus oxygen-enriched combustion. When the oxygen in kiln head is sufficient and the pulverized coal burns fully, F-CaO content displays different degrees of decline in oxygen-enriched combustion compared with normal production conditions at the same time point. As oxygen concentration increases, F-CaO in clinker gradually decreases, indicating that the quality of cement clinker is much better under conditions of oxygen-enriched combustion.

Figure 3 shows the relationship between oxygen concentration in air and F-CaO content in cement clinker. Under conditions of oxygen-enriched combustion, the average content of F-CaO in cement linker is 1.15%, or 0.197% lower than the average F-CaO content under normal conditions, which is 1.347%. Therefore, the cement clinker quality improves with oxygen-enriched combustion. The flame temperature increases as oxygen content in the air increases, which follows oxygen-enriched combustion in the rotary kiln; this improves the temperature of the whole kiln, resulting in greatly increased thermal efficiency and decreased F-CaO content.

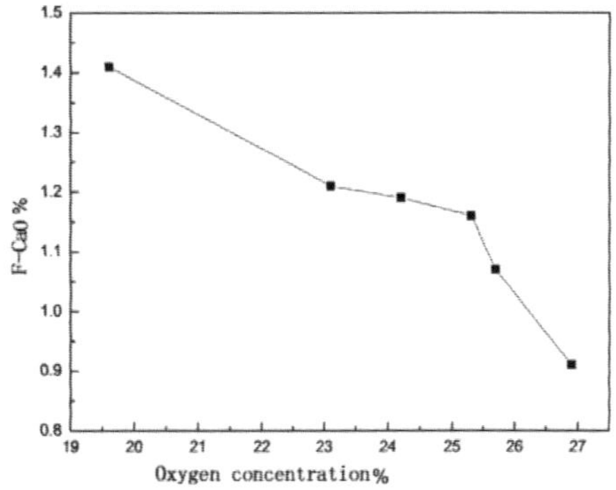

Figure 3. Relationship between oxygen concentration and F-Cao content.

3.4.2. Clinker liter weight. Clinker liter weight is the quality of a liter of clinker expressed in g/L; it is one of the parameters that determine clinker quality and kiln temperature. A material's particle size and uniformity, allows us to conclude whether the firing temperature is normal. Low liter weight indicates less clinker burning; on the contrary, the clinker burnt. At normal furnace temperatures, clinker has high yield, uniform particle size, compact appearance, smooth surface, small spherical approximation, and higher liter weight. If the burning temperature is too high or the material remains in firing too long, the clinker burns too much, leaving the liter weight too high and reducing the clinker quality. As the chemical reaction of material in the kiln remains incomplete, most of the clinker particles are small with a fine powder and low liter weight indicating that the temperature in the kiln is low. Therefore, liter weight should be controlled within a certain range [18, 19]. The liter weight of rotary kiln clinker is generally exceeds 1,240 g /L. Figure 4 shows clinker liter weights in normal production and oxygen-enriched combustion.

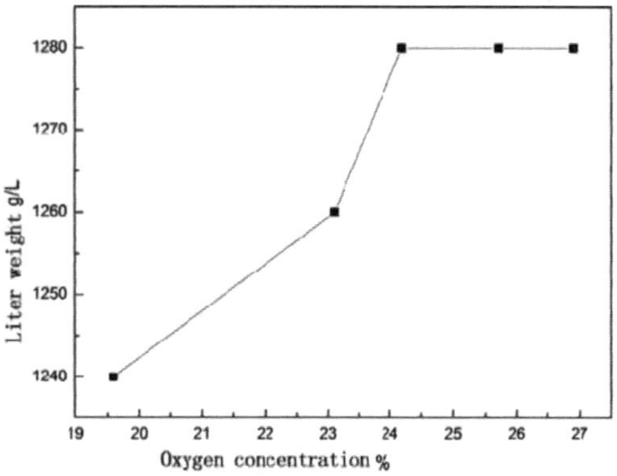

Figure 4. Relationship between liter weight and the oxygen concentration.

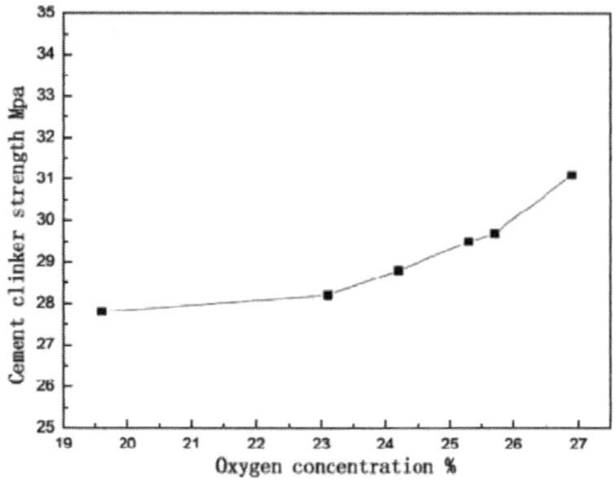

Figure 5. Relationship between oxygen concentration and cement clinker strength.

As the test data show, little difference exists between the average liter weight of 1,249 g/Lin the oxygen-enriched experiment compared with the average liter weight of 1,248 g/L in the normal production period. Both reached the requirement. Figure 4 shows the relationship between liter weight and the oxygen concentration in the oxygen-enriched air.

According to liter weight data measured under the conditions of oxygen enriched combustion, the average clinker liter weight is 1,267 g/L, which is 1.52% higher than the value under normal production conditions 1,248 g/L. Therefore, oxygen-enriched combustion can improve the quality of cement clinker.

3.4.3. Clinker strength index. Cement clinker strength is an important index by which to evaluate cement quality, and it forms the basis for determining cement strength grade. Cement clinker strength refers to the ability of a cement mortar-hardened specimen to withstand external damage, as expressed in Mpa. It is one of the most important physical and mechanical properties of cement.

According to the test results, average cement clinker strength under the conditions of the oxygen-enriched combustion is 29.46, which is 5.97% higher than the value under normal conditions 27.8. Cement clinker strength increases as oxygen concentration increase. Oxygen-enriched combustion in a rotary kiln increases the flame temperature, causing full calcination of the clinker, which improves its strength. Figure 5 shows the relationship between oxygen concentration and cement clinker strength.

3.5. The influence of oxygen-enriched combustion on fuel gas emissions

Oxygen-enriched combustion reduces the required amount of air, and the exhaust heat drops. Usually only one fifth of the total oxygen is in combustion; the rest of the air, rather than providing fuel, absorvs a great deal of heat from burning and emits fuel gas. This results in large energy wastes. The use of oxygen-enriched combustion reduces the amount of air needed for combustion, reaches complete combustion, reduces emissions and meets the requirements of environmental protection [20, 21]. Experimental results show that energy consumption of clinker sintering decreases by about 3.7 kgce of standard coal in the primary air oxygen-enriched test. If the daily output is about 4,000 kg of clinker and the reduction of CO_2 emissions per kilogram of standard coal is 2.493 kg, the daily CO_2 emission reduction will be 36.9 tons. Accounting for 320 working days, CO_2 emissions can decrease by 11,086 tons per year, yielding both environmental and social benefits.

4. Conclusions

Oxygen-enriched combustion technology can improve the fuel combustion conditions of cement production, increase flame temperature, shorten the time needed for combustion and achieve complete combustion, thus increasing the flame radiation heat ability of a material and improving the whole system's thermal efficiency. It can also reduce waste gas and dust and harmful gas emissions, which is beneficial to energy conservation and emissions reduction. Finally, it can improve the production efficiency and quality of cement. Therefore, applying oxygen-enriched combustion technology in cement production can yield economic, social and environmental benefits.

- Without any increase in fuel, oxygen-enriched combustion can improve furnaces flame temperatures by 100-300℃, yielding a remarkable energy-saving effect. According to the results, flame temperature can rise 200-300℃ when oxygen content increases by 4-5%. Higher flame temperatures promote increases in the whole furnace temperature, and the furnace-heated material is easier to heat, increasing the thermal efficiency.
- Oxygen enrichment improves the combustion of pulverized coal, so that the the kiln burns more fully, the combustion rate of coal improves, the kiln temperature improves, F-CaO content in cement clinker decreases, stability and strength increase, better quality clinker results and the cement quality improves.
- Because the oxygen content in combustion air exceeds 20%, combustion occurs more fully, the amount of air needed reduces, less smoke production results and dust pollution and environmental pollution decrease greatly.
- The successful application of oxygen-enriched combustion technology in the rotary kiln cement production industry will bring huge economic and social benefits, and it has broad application prospects.

Acknowledgments

This work was supported by National Natural Science Foundation of China under Contract (No. 51064015), to which the authors are greatly obliged.

References

[1] Wang Z P and Zeng X W 2009 Theoretical calculation of energy-saving eficiency and the key issues in technology of combustion in rich oxygen atmosphere *Glass* **36** 36
[2] Wang H and Wen Z 2006 Research on the application of oxygen-enrichment combustion technology to rotary lime kilns and its economic prediction *Modern Chemical Industry* **26 (S2)** 21-6
[3] Lin F, Ross M and Wang S 1995 Energy efficency of China's cement industry *Energy* **7** 669-81
[4] Li H Y and Wang H 2003 Performance compare analyses of high temperature air combustion and oxyboosted combustion technology *Industrial Heating* **32** 9-12
[5] Anon 1993 Developments and prospects of the Chinese cement industry *World Cement* **24** 1-18
[6] Li J and Xie J L 1997 Theory discussion of the application of oxygen-enriched combustion in cement rotary kiln production *Journal of Wuhan University of Technology* **19** 47-9
[7] Zhu G J, Liang Z Y, Wu M Q, Deng N Y and Zhang S Q 2005 Effects analyses of coal oxygen-rich combustion to energy saving and environment *Industrial Heating* **34** 34-7
[8] Claude P 2000 Cements of yesterday and today, Concrete of tomorrow *Cement and Concrete Research* **30** 1349-59
[9] Chen R 2005 Dissipation rate at extinction and the effects of oxygen-enriched combustion *Combustion and Flame* **142** 62-71
[10] Bejarano P A, Levendis Y A 2007 combustion of coal chars in oxygen-enriched atmospheres *Combustion Science and Technology* **179** 7-9
[11] Murphy J J, Shaddix C R 2006 Combustion kinetics of coal chars in oxygen-enriched

environments *Combustion and Flame* **144** 710-29

[12] Yang S P, Qing S, Yang N, Zhang A M, Liao B and Liu T S 2015 Study on combustion and kinetics of pulverised coal in mixed O_2/CO_2 atmosphere *Materials Research Innovations* **19** 1263-7

[13] Su J L, Pan L and Zu C M 2008 Research status and development of oxygen-enriched combustion technology *Industrial Boiler* **3** 1-4

[14] Zhu G J, Liang Z Y, Wu M Q ,Deng N Y and Zhang S Q 2005 Effects analyses of coal oxygen-rich combustion to energy saving and environment *Industrial Heating* **4** 11-3

[15] Yang N, Qing S, Wang H, Zhang A M, Li W T and Wang J Y 2014 Study on the effect of heating rate on combustion and kinetics of pulverized coal in mixed O/CO atmosphere *Coal Conversion* **37** 50-4

[16] Liu Q C and Chen S R 2004 Overview of the main factors afffecting the enviroment in the oxygen-enriched combustion *Energy Conservation & Environmental Protection* **9** 21-8

[17] Han C Y, Xu M H and Zhou H C 2001 *Pulverized Coal Combustion* (Beijing: Science Press) pp 67-9

[18] Zhu G J, Liang Z Y and Wu M Q 2004 Analysis of the influence of oxygen-enriched combustion of coal on energy saving and environment *Industrial Heating* **34** 11-3

[19] Zheng L, Kang Z J, Zhang L and Zhao Q X 2004 The thoery and application of oxygen-enhanced combustion *Industrial Boiler* **3** 10-4

[20] Zheng X F, Feng Y X and Jia M S 2006 The oxygen-enriched combustion energy-saving features and its impact on the environment *Energy Conservation* **7** 26-8

[21] He Q A and Gao C L 2004 The technical features of the cement rotary kiln using inferior coal in plateau area *Development Guide to Building Materials* **3** 55-9

IOP Conf. Series: Materials Science and Engineering **103** (2015) 012031 doi:10.1088/1757-899X/103/1/012031

Numerical simulation of flexible blank drawer formation

L C Li[1,2,3] and M Z Li[1]

[1] Jilin University, 5988 Renmin Street, Changchun, China
[2] Jilin University of Finance and Economics, 3699 Jingyue Street, Changchun, China

E-mail: liliancheng@jl.gov.cn

Abstract. This paper presents a finite element model for a technology of Fexible Blank Drawer Formation (FBDF) for the forming process.In order to verify the feasibility and versatility of the FBDF technology, we performed numerical simulations for multi-point saddle-shaped parts, and for convex curved surface and hemispherical parts. We analysed the effect of different forming methods on wrinkling and cracking, stress and strain distribution as well as circulating and spring back. Also, we studied the effect on the FBDF result caused by the chucking power, blank drawer force, friction coefficient and material parameters. The results showed that under the same conditions, the parts formed by FBDF technology showed uniform stress and strain distribution with little spring back. The blank drawer force can efficiently restrain the defects such as wrinkling and crack.

1. Introduction

As a common sheet metal forming technology, the press forming often causes wrinkles and cracks in the material sheet. By applying the blank holder forming technology, the instability problem can be limited to some extent. The present blank holder forming technologies include the traditional rigid blank holder technique [1], Variable Blank Holder Force (VBHF) technology [2, 3], piecewise rigid blank holder technique [4], elastic blank holder forming technique [5] and multi-point blank holder forming technique [6, 7]. All of these are in the passive control of sheet flow. For example, while forming such curved parts as automobile panel, because of the different sheet flow at every position, wrinkles and cracks may occur, and it is hard to achieve the formed parts with uniform stress and strain distribution. On the other hand, the present blank holder forming techniques have a common problem, that a mold can be used only by designing a corresponded blank holder mechanism. For sheet metal forming with different size and shape, different blank holder mechanisms need to be designed, leading to increased manufacturing cost and extended product development circle. All these problems need to be resolved to meet the requirement of flexible manufacturing in modern industry [8-10].

In this paper, we introduced a Fexible Blank Drawer Forming (FBDF) technology and its device. By using blank drawer technique instead of flexible blank holder technique, the device of FBDF can be used in conjunction with the whole model or multi-point model. As a result, it can simplify the mold structure, reduce the manufacturing cost and shorten the development period of product. Our simulation results indicated that with the FBDF technology, the defect such as wrinkle and crack can

[3] Address for correspondence: L C Li, Information Management Centre, Jilin University of Finance and Economics, Changchun 130117, China. E-mail: liliancheng@jl.gov.cn.

be efficiently reduced, and the flexible manufacturing of large curved surface can be realized with high efficiency and low cost.

2. Experimental arrangement

The device of Fexible Blank Drawer Forming (FBDF) is mainly composed of several clamping and pulling mechanisms and a support frame. The pulling mechanism is a horizontal hydraulic cylinder whose position ends are connected with the clamping mechanism and cylinder ends are connected with the support frame. The material sheet is clipped surrounding the clamping mechanisms and the blank drawer force is applied in four directions. The force can be adjusted according to the size and shape of forming parts. The hydraulic cylinder is big enough and it is conducive to the material loading and unloading operation. Press the upper mold and then the sheet metal is formed. The clamping mechanism has flexible features, as it is capable of moving and swing. Moreover, the blank drawer applied by the pulling mechanism can effectively suppress the defect of wrinkle and crack.

The FBDF device is designed according to the PASCAL's law. The control system is very simple since a hydraulic value can control several pulling mechanisms and produce uniform blank drawer force. Meanwhile, different blank drawer force and clamping force can be applied according to deformations at different parts of sheet metal. Due to the telescopic feature of pulling mechanism, curved parts in various shapes can be obtained by only replacing the mold cavity.

3. Finite element modeling

Thanks to the rapid development of finite element analysis method and computer technology, the element simulation during forming process has become an effective tool to value the quality of metal sheet forming technique. In order to verity the validity and versatility of blank drawer forming technique, have established the finite element for multi-point mold forming of saddle-shaped parts, and for whole mold forming of convex curved surface as well as hemispherical parts. The numerical simulation has been conducted and analyzed as following.

Considering the complexity of the Fexible Blank Drawer Forming (FBDF) device, and to save calculating time and reduce memory occupancy, simplified the forming device of saddle-shaped parts to ¼ finite element model. The multi-point model is indicated by a series of punches; the stretching mechanism is simplified as a clamp and constraints are applied on the clamp to achieve stretching function.

In this research, the complicated equipment was simplified to use the finite element simulation package of Abaqus. The 08Al steel, usually used as automotive coverage, was adopted and the materials parameters are listed in Table 1. The deformable shell element S4K was selected to discrete. Four-node quadrilateral shell element with rigid S4K was chosen as element cushion. The friction coefficient was of general contact and the penalty function method was applied. The friction coefficient between clamp and sheet was set as 0.8, and other friction coefficient was set as 0.1. The sheet dimension was 360 mm×360 mm×0.5 mm. The multi-point model consisted of 30× 30 punches, with the punch radius of 5 mm. The dimension of elastic pad was 300 mm×300 mm×10 mm.

The convex surface blank drawer forming device is simplified to a finite element model, in which OX and OY are two symmetrical axes of the sheet. The stretching mechanism is simplified as a clamp and constraints are applied on the clamp to achieve stretching function.

The main material is 08Al sheet plate, and the deformable shell element S4K is selected to discrete. Four-model quadrilateral shell element with rigid S4K is used for entity model and clamp. The friction coefficient between clamp and sheet is set to 0.8, and the rest is set to 0.1. The curvature radius of convex surface parts is 500 mm, and the gap between each surface is 8 mm. The dimension of sheet is 360 mm×360 mm×0.5 mm, and the effective forming area is 300 mm×300 mm.

The used material is 1010 steel and 2024 aluminum, with parameters listed in Table 1. The deforming shell element S4K is selected to discrete upper and lower mode, and four-model quadrilateral shell element with rigid S4K is chosen for the clamp. The friction coefficient between clamp and sheet is set to 0.8, and others are set to 0.1. The radius of upper mode is 150 mm, and the dimension of sheet is 400 mm×400 mm×1 mm.

Table 1. Material parameters.

Material	Poisson ratio	Young's modulus,	Yield stress,	Density, Kg
08Al steel	309	608	0.35	140300
1010 steel	232	397	0.35	140100
2024-O al	348	627	0.35	151800

4. Results and discussion

To testify the validity and versatility of Fexible Blank Drawer Forming (FBDF), we have performed numerical simulation for the multi-point model forming process of saddle-shaped parts. Both blank holder forming and blank drawer forming are studied to compare the effects of different forming techniques on the forming results. While using the blank holder forming technique, the blank holder pressure is 3.8 MPa and 8 MPa. While using the blank drawing forming technique, the clamping pressure is 3.8 MPa and 8 MPa, and the blank drawer force applied by each clamp is 1200 N.

While both the blank holder pressure and the clamping pressure are 3.8 MPa, as can be seen, the part formed by blank holder technique has obvious wrinkles in both the flange region and the effective forming zone. Also, because the edge pressing ring keeps balance during the whole forming process, obvious defect occurs in the blank holder force transmission region. Especially in the central force transmission area, the sheet may crack easily due to the large curvature change. On the contrary, the part obtained by blank drawer technique displays no defect of wrinkle or crack, neither in the flange region nor in the effective forming zone. The formed part has smooth surface as well as good quality.

When the blank holder pressure and the clamping pressure are 8 MPa, as can be seen, the part obtained by blank holder forming technique still shows wrinkle defects in the flange region and the effective forming zone. Meanwhile, slight crack defect occurs. As comparison, with the same clamping pressure, the part obtained by using blank drawer technique has smooth transition without wrinkle or crack defects both in the flange area and the effective region.

It is speculated that, with the blank drawer technique, the blank drawer force effectively suppresses wrinkling defect. Meanwhile, in the holding area of the clamp, due to the flexible characteristics of the discrete gripper, the sheet in the flange area can change shape in coordinate with clamp and thus the high quality forming parts can be obtained.

The stress distribution of the forming parts was obtained by using blank holder forming technique and blank drawer forming technique. As can be seen, the stress distribution of the part obtained by using blank holder forming technique is not even. The stress value in forming area varies from 202 MPa to 480 MPa. Especially in the transition area such as blank holder area and forming area, the stress is concentrated seriously. The maximum stress value appears in the effective forming area and it leads to crack defect. By contrast, the stress distribution of part obtained by using blank drawer forming technique is much more uniform. The stress value in forming area changes within the range of 170 MPa to 270 MPa. Also, the transition at the clamping and forming area is smooth and there is no stress concentration. Thus good-quality forming parts can be achieved.

The strain distribution of forming parts obtained by using blank holder forming and blank drawer forming technique. As can be seen, the part obtained by using blank holder technique displays a serious thin sheet near the straight force transmission area, due to the flow less sheet in the flange area. The maximum equivalent strain is 0.589, at which the material sheet is close to rapture. On the other hand, the part obtained by using blank drawer forming technique has the maximum equivalent strain value of only 0.0805, much smaller than that of the part by blank holder technique. With comparison, we can conclude that, using blank drawer forming technique, forming parts with high quality can be achieved, even with the large clamping force, blank holder force or the great curvature of the work piece. By using the blank drawer forming technique, some forming parts which can't be obtained by using blank holder technique will become feasible.

The displacement diagram of forming parts was obtained by using blank holder forming technique, as well as blank drawer forming technique. As seen from the picture, the maximum displacement of the part obtained by using blank holder forming technique is 50 mm in the flange area, and 73 mm in the forming area, with the difference of 23. On the contrary, the maximum displacement of the part obtained by using blank drawer forming technique is 73 mm in the flange area, and 75 mm in the formative area, with little difference. Therefore, a conclusion can be drawn that the part obtained by using blank drawer forming technique has a better sheet flow ability. The reason is that while forming, the sheet can have different amounts of displacement according to the mold curvature changes, thus it makes a smooth transition between the flange area and the formative area.

Rebound phenomenon often occurs in the course of forming. The amount of rebound is defined as displacement changes in certain direction after forming. The rebound of the forming parts obtained by using blank holder forming technique as well as blank drawer forming technique. As can be seen, the maximum spring back of the part obtained by using blank holder forming technique is 1.1 mm in OX direction, and 0.31 mm in OY direction. As comparison, the maximum amount of spring back of the part obtained by using blank drawer forming technique is 0.72 mm in OX direction and 0.2 mm in OY direction. Apparently, in blank drawer forming technique, the pre-clamping of the clamp to sheet can effectively cut down the rebound phenomenon.

After studying the numerical simulation of multi-point model forming process of saddle-shaped parts, can conclude that the blank drawer forming technique is more favourable than the blank holder forming technique. By using the blank drawer forming technique, wrinkle and crack defects can be effectively inhibited. Also, the forming parts can be obtained with uniform stress and strain as well as small amount of rebound.

To further testify the validity and versatility of the flexible blank drawer forming technique, carried out numerical simulation for integral mold forming process of convexity surface parts. By using blank holder forming technique and blank drawer forming technique, we studied and compared the effect of different forming technique on forming results. For blank holder forming, the clamping pressure of clamping ring is 4 MPa and 8 MPa. For blank drawer forming, the clamping pressure of each clamp is 4 MPa and 8 MPa, and the stretching force applied by each clamp is 1200 N.

When both the blank holder pressure and clamping pressure are 4 MPa, the forming parts were obtained by using blank holder forming technique as well as blank drawer forming technique. As can be seen, the part obtained by using blank holder technique has obvious wrinkles in the flange area as well as in the effective forming area. This can be attributed to the fact that the blank holder ring is always flat in the whole forming process. The wrinkle defect occurs in force transmission area, especially in four angles. By contraries, the part obtained by using blank drawer forming technique exhibits a smooth surface, without wrinkle either in the flange area or in the effective forming area.

When both the blank holder pressure and clamping pressure are 8 MPa, the forming parts were obtained by using blank holder forming technique and blank drawer forming technique. As can be seen, higher blank holder pressure can effectively inhibit the wrinkle defect in the forming part obtained by using blank holder forming technique. When the blank holder pressure increases to a certain value, wrinkles disappear, but slight crack defect occurs in the four corners of force transmission zone. As comparison, the integral forming part obtained by using blank drawer forming technique demonstrates good quality without any defects. Therefore, the blank drawer forming technique can be used to match not only with the multi-point mold, but also with the integral mold.

Furthermore, we applied numerical simulation for integral mode forming process of hemispherical parts.

The clamping pressure indicates the average force applied in unit area on clamp. With a reasonable clamping force, the metal sheet can be held tightly in the forming process. In addition, the clamp with reasonable force can have pre-clamping to the sheet, reduce the rebound and inhibit wrinkling. Thus, the forming parts with uniform stress and strain distribution can be achieved. When the blank drawer force applied by each clamp is 1200 N, the clamping pressure of each clamp is 8 MPa and 16 MPa. The state of the clamp by using blank drawer forming technique is shown in Figure 1. As can be seen, when the clamping pressure of each clamp is 8 MPa, the clamp is pulled. While the clamping pressure is 16 MPa, there is no such phenomenon.

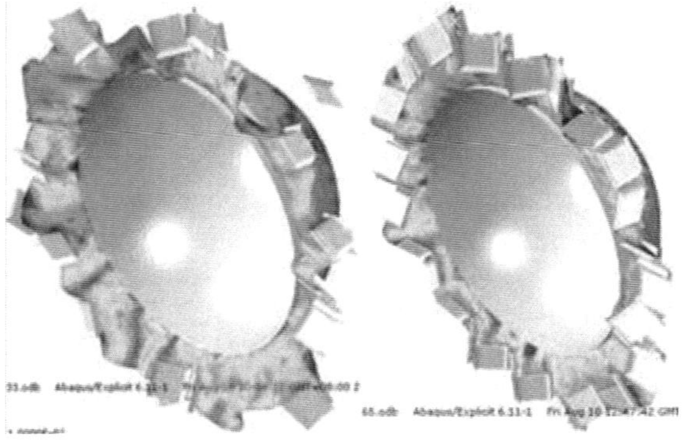

Figure 1. State of clamp after forming.

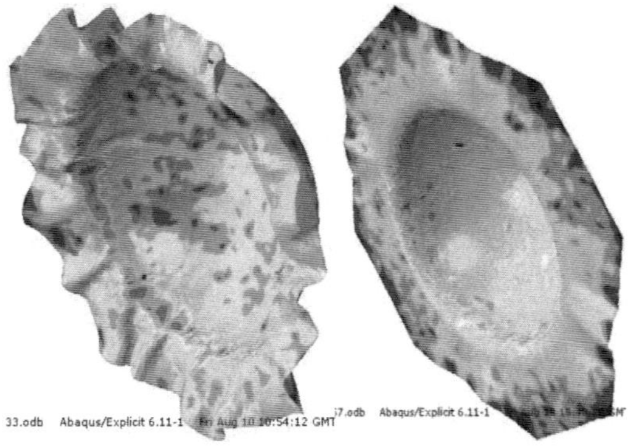

Figure 2. Mises stress distribution.

With the flexible blank drawer forming technique, the blank drawer force is applied to the sheet, which can stretch the sheet, reduce spring back and inhibit wrinkling defect. A reasonable bland drawer force is to make the sheet change. In actual forming process, the clamping force can be adjusted according to the wrinkling situation to reach an optimum value. With the clamping force of each clamp as 2000 N and 3000 N, the Mises stress distribution of forming parts obtained by using blank drawer forming technique is shown in Figure 2. As can be seen, the wrinkling defect appears in the flange area when the blank drawer force applied by each clamp is 2000 N. However, when the force is 3000 N, no wrinkling defect is observed.

The Mises stress distribution of forming parts along X-axis and Y-axis when different friction coefficients are applied. As can be seen, when the friction coefficient between sheet and mold is as small as 0.05, the Mises stress distribution of forming parts is uniform. When the friction coefficient is 0.2, the Mises stress varies more. Usually the sheet surface should be as smooth as possible. In other words, a smaller friction coefficient is better. Since the deformation resistance is small, the better-qualified forming parts can be obtained.

With Fexible Blank Drawer Forming (FBDF) technique, 1010 steel and 2024-O aluminum plates are used to analyze the effect of material parameters on hemispherical parts forming process. When the clamping pressure is 64MPa and the blank drawer force is 3000 N, the forming parts of 1010 steel have no obvious defect. When the clamping pressure is 6 MPa and the blank drawer force is 800 N, the forming parts of 2024-O aluminum also have no obvious defect, as shown in Figure 3. Since the elastic modulus and yield limit of 1010 steel are much higher than those of 2024-O aluminum sheet, the external load capacity needed in 1010 steel sheet forming is much higher than that for 2024-O aluminum. Moreover, greater blank drawer force and clamping pressure are needed to form better-qualified forming parts.

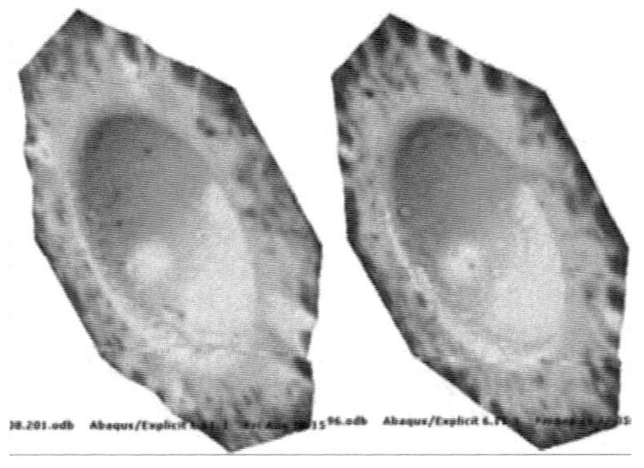

Figure 3. Simulation of hemispherical parts with different materials.

5. Conclusions

In this paper, introduced the Fexible Blank Drawer Forming (FBDF) technique, established the finite element model of saddle-shaped parts, convexity surface parts as well as hemispherical parts, and testified the feasibility and versatility of FBDF technique. The results showed that with the blank drawer forming technique, the blank drawer force can inhibit the defect of wrinkling and crack in the sheet. Meanwhile, the forming parts have uniform stress and strain distribution, well-flow sheet, small rebound and high quality. When the yield limit and elastic modulus of sheet are higher, the clamping pressure and blank drawer pressure needed in forming is higher as well. In addition, smoother sheet surface can result in better quality of forming parts.

References

[1] Sun G, Li M Z, Deng Y S and Li X J 2008 Analysis on flexible and rigid blank-holders technology applying on thin sheet metal in multi-point forming *Journal of Mechanical Engineering* **44(5)** 147-51

[2] Wang D Z, Lou Z L, Zhang Y Q and He D N 2001 Numerical simulation of variable blank holder force in box deep drawing *Journal of Shanghai Jiaotong University* **35(10)** 1543-6

[3] He D N, Wang D Z, Zheng X D, Wang R, Zhang W and Li J 2001 Research about the effect of unit blank holder force on sheet metal forming capability *Metal Forming Technology* **19(1)** 4-7

[4] Siegert K 1993 CNC－hydraulic multipoint blank-holder system for sheet metal forming presses *Ann CIRP* **42(1)** 319-22

[5] Li M Z, Liu Y H and Su S Z 1999 Multi-point forming:a flexible manufacturing method for a 3-D sur- face sheet *Journal of Materials Processing Technology* **87** 277-80

[6] Li M Z, Cai Z Y and Sui Z 2002 Multi-point forming technology for sheet metal *Journal of Materials Processing Technology* **129** 1-3

[7] Kleespieh S S and Crawford R H 1998 Vacuum forming of compound curved surfaces with a variable geometry mold *Journal of Manufacturing Systems* **17** 325-37

[8] Cai Z Y and Li M Z 2002 Multi-point forming of three-dimensional sheet metal and the control of the forming process *International Journal of Pressure Vessels and Piping* **79** 289-96

[9] Walczyk D F and Hardt D E 1998 Design and analysis of reconfigurable discrete dies for sheet metal forming *Journal of Manufacturing Systems* **17** 436-54

[10] Wang S H, Cai Z Y and Li M Z 2010 Numerical investigation of the influence of punch element in multi-point stretch forming process *The International Journal of Advanced Manufacturing Technology* **49** 475-83

IOP Conf. Series: Materials Science and Engineering **103** (2015) 012032 doi:10.1088/1757-899X/103/1/012032

Influence of process parameters to composite interface organization and performance of liquid/solid bimetal

S F Rong[1], Y C Zhu[1,3], Y H Wu[1], P H Yang[1], X L Duan[1] and H T Zhou[2]

[1] Institute of Materials Science and Engineering, Jiamusi University 154007, China
[2] School of Materials Science and Engineering, Harbin Institute of Technology 154001, China

E-mail: jmsdxzyc@163.com

Abstract. The liquid-solid composite technique was used to prepare the high carbon high chromium steel (HCHCS) and low alloy steel (LCS) bimetal composite materials by means of insert casting method. The influence of some process parameters such as liquid-solid ratio, preheat temperature, pouring temperature on the interface microstructure and mechanical properties were studied. Interface microstructure and element distribution were analyzed. The results show that the interface microstructure becomes better, and bonding area becomes thicker with the increase of the volume of liquid to solid ratio, preheating temperature and pouring temperature. When the liquid-solid ratio is 8:1, the preheating temperature is 300℃ and the pouring temperature is 1565℃, a good metallurgical bonding area without any hole can be obtained with the interface combination of diffusion and fusion. The composite interface structure was composed of a core material diffusion layer, a cooling solidification layer, a direction growth layer and some cell particles. The elements of C, Cr and Mn diffuse from the HCHCS side to the alloy steel side. The microhardness increased in the gradient from the LCS side to the HCHCS. The microhardness of the interface is significantly higher than that of LCS.

1. Introduction

Single wear-resisting materials were unable to meet the requirement of sustained load working conditions [1]. Two liquid bimetal casting technique was worthy for being widely researched and used which provided a feasibility way to the crusher hammer between the LCS and the high chromium white cast iron [2, 3]. During the practical application, the hammer handle due to the soft LCS handle was easy to be wear away [4-6], which decreased the service life and limited the application range of relational bimetal crusher hammers [7, 8]. So a prerequisite to realizing the wear-resisting hammer handle based on liquid-solid composite technology between the LCS and the HCHCS can be proposed in view of the soft hammer handle of the LCS [9].

2. Experimental procedures

The experiment of material was composed of HCHCS and LCS. Chemical components were shown in table 1.

The mold of LCS was ladder structure designed into 2:1, 4:1 and 8:1 of outer/inner volume ratio. First LCS after preparation was polished and coated with a borax layer to increase the wettability.

[3] Address for correspondence: Y C Zhu, Institute of Materials Science and Engineering, Jiamusi University, Jiamusi 154007, China. E-mail: jmsdxzyc@163.com.

Table 1. Chemical compositions of HCHCS and LCS (wt%).

Contents	C	Cr	Mn	Mo	Si
HCHCS	1.15-1.18	11.5-12	1.8-2.0	≥0.2	≤0.4
LCS	0.32-0.45	0.6-0.8	0.6-0.7	≤0.2	≤0.4

Then melted metal by induction furnace was raised to the required pouring temperature with the temperature measuring gun. The processed LCS preheated to room temperature, 300℃ and 600℃ was placed into the sand mold. The bimetal was cooled to the room temperature after the poured HCHCS in the sand mould.

The casting was cut into 10 mm×10 mm×55 mm for the impact toughness and the microstructure observation. The austenitizing was at 960℃ for 2 h, and the isothermal quenching based on the mixed sodium nitrate at 320℃ for 1 h. The ZBC-300B automatic metal pendulum tester was used to measure its impact toughness. The Rockwell hardness was tested by the HR-150A durometer. The abrasive resistance was tested by the MLD-10 dynamic load abrasive wear testing machine. The microstructure and composition at the bonding region were examined using the Olympus GX71 optical microscope and the EDAX-FALCON60 energy dispersive spectrometry respectively. The microhardness was tested by the HV-30 Vickers hardness tester.

3. Results and discussion

3.1. Interfacial microstructure

The bonding interface at 1565℃ pouring temperature and 300℃ preheating temperature was obtained under three kinds of liquid-solid volume ratios, and the interfacial microstructures were seen in figure 1. The overall tendency presented the certain variation from mechanical bonding to metallurgical bonding. The long unbonded region at the interface of that was found in figure 1(a), and curved interface bonded line was detected as shown in figure 1(b). The interfacial microstructure with high

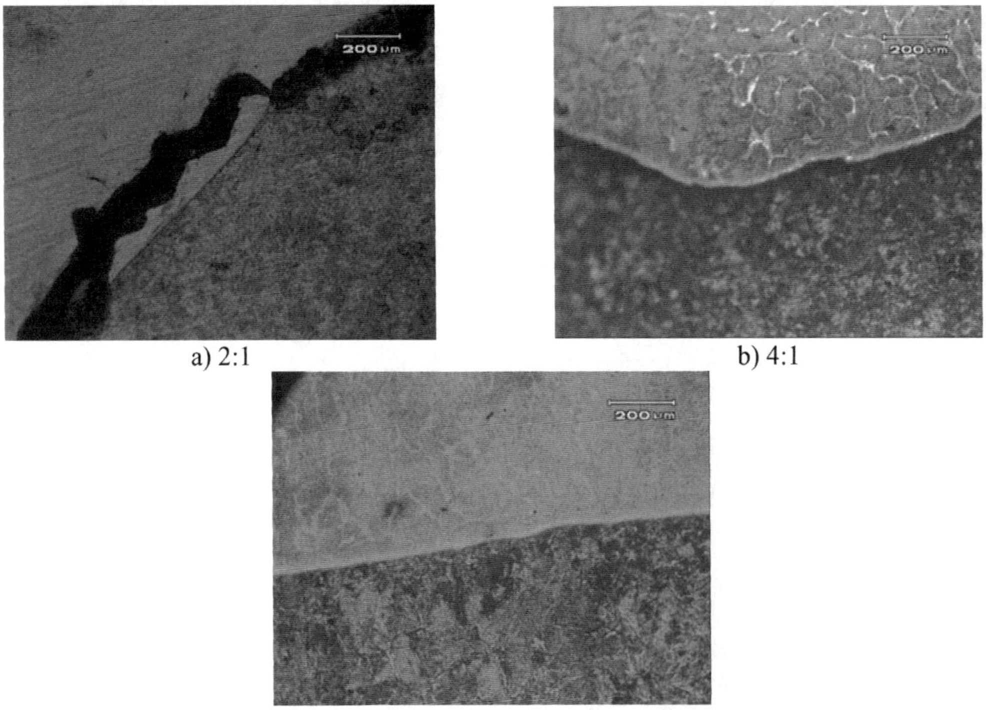

a) 2:1 b) 4:1

c) 8:1

Figure 1. Interfacial microstructures under the different volume ratios.

quality is investigated without any defect such as the unbonded region or the void on the interface in figure 1(c). So the increase of liquid-solid volume ratio was favorable to form the required bonded interface.

The interfacial microstructure at three preheating temperature under the 4:1 liquid-solid volume ratio and at 1565℃ pouring temperature were seen in figure 2.

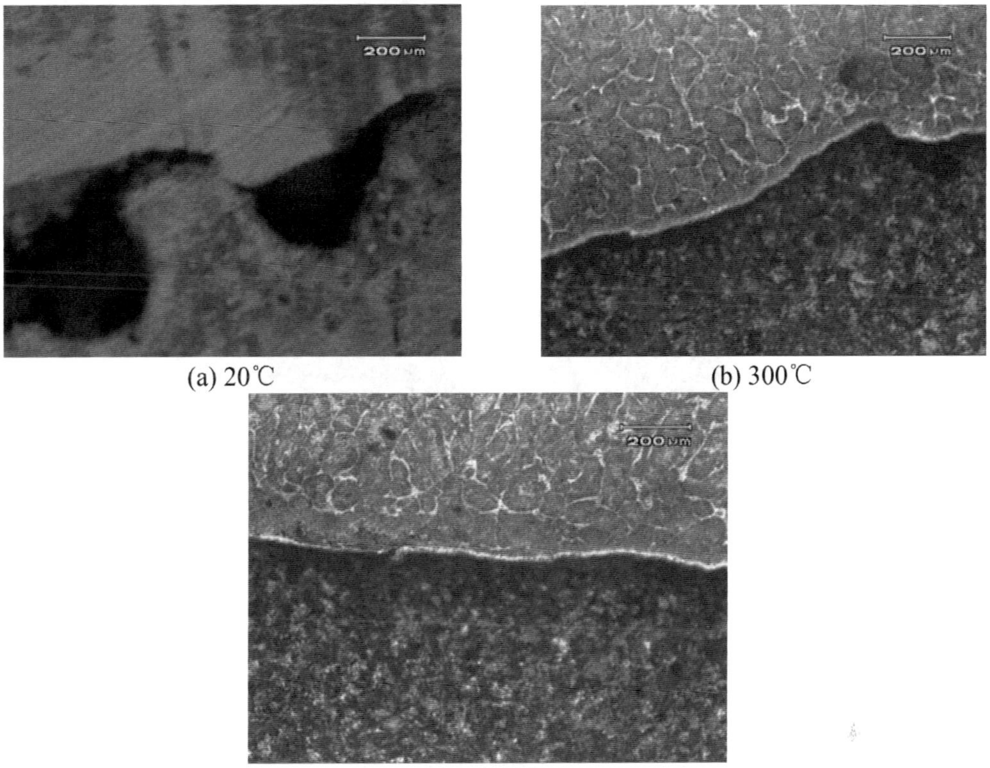

(a) 20℃ (b) 300℃

(c) 600℃

Figure 2. Interfacial microstructures of preheating temperature.

The unbonded region and void can be seen in figure 2(a) because of the unpreheated LCS before pouring HCHCS. The casting was easily cracked because the HCHCS had a strong chilling effect by LCS. The quality of interfacial microstructure was significantly improved because of the higher preheated temperature and was shown in figures 2(b) and 2(c). The difference temperature between the HCHCS and the LCS was decreased by the increase of LCS preheating temperature. So the temperature of LCS was easily to increase higher and form metallurgical bonding.

The interfacial microstructure at three pouring temperature under 4:1 liquid-solid volume ratio and at 300℃ preheating temperature were seen in figure 3.

The unbonded region was seen in figure 3(a) whose interface quality was better than figure 2(a) because LCS was preheated before poured HCHCS, which decreased the temperature difference. The interface quality was increasingly improved when the pouring temperature reached 1620℃. The increase of pouring temperature was favorable to form desirable bonded interface.

3.2. Interface energy spectrum analysis

The spectrum analysis of the bonded interface region at the liquid-solid volume ratio of 8:1, at the preheating temperature of 300℃ and pouring temperature of 1565℃ was shown in figure 4.

There were no pores, inclusions, cracks and other defects in the interface bonding region from the scanning electron microscope. There were mutual diffusion and penetration of bimetal in the

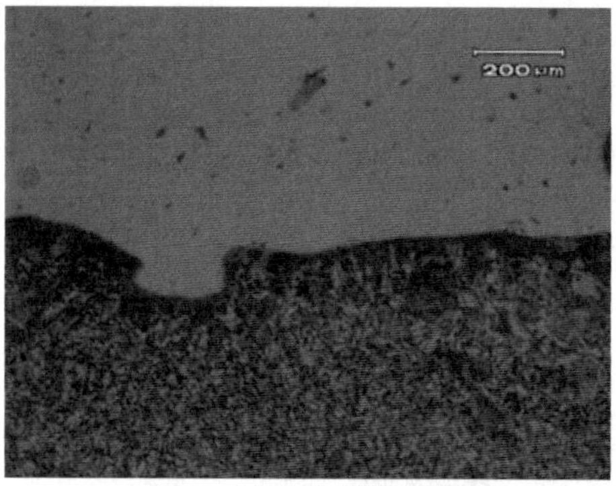

(a) 1490℃

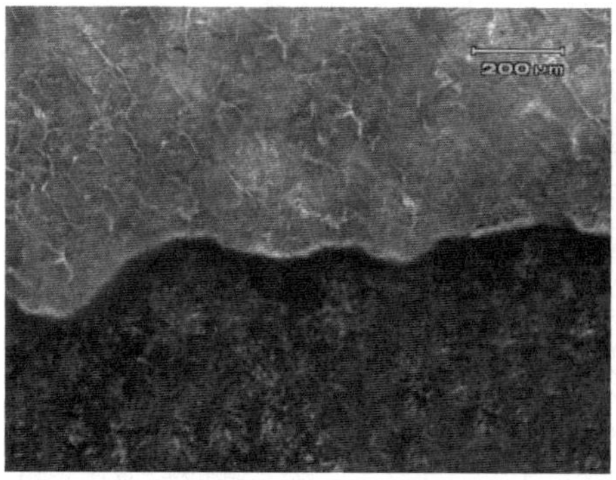

(b) 1565℃

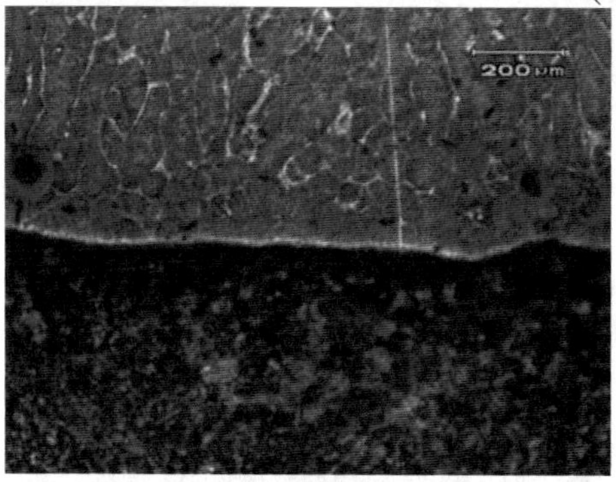

(c) 1620℃

Figure 3. Interfacial microstructures of pouring temperature.

composite process because of the different components between HCHCS and LCS. Diffusion of elements takes place at the interface from the spectrum analysis of bonding interface region. C, Cr, and Mn element diffuse from the HCHCS to the LCS, and Fe diffuse toward the opposite direction, which was important to increase the bonding strength. However, the different condition of diffusion due to the lower temperature of LCS and the characteristic of the different elements. Cr and Mn are decreased from HCHCS to LCS gradually due to the strong diffusibility and the far distance of diffusion. C mainly diffuses from HCHCS to the bonding region because of the poor diffusibility, so the C in bonding region was less. While the element of Si and Mo diffuses from HCHCS to the bonding region due to its strong diffusibility and the content of Si and Mo in the interfacial bonding were closer to that of HCHCS and LCS. The elements mainly diffuse from HCHCS to the interface bonding. The diffusion in the LCS was poor because of the low temperature of LCS. This phenomenon was consistent with the figure 4.

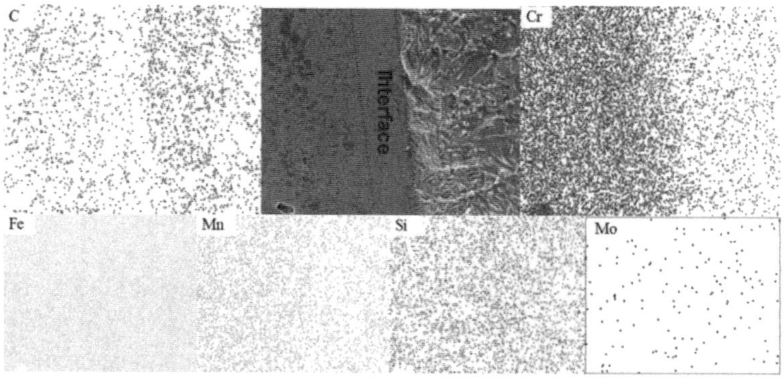

Figure 4. Interface energy spectrum analysis.

4. Mechanical properties

The major mechanical properties of the HCHCS and the LCS after heat treatment were given in table 2. The wear loss of the HCHCS was obviously higher than that of the LCS, so the "eated handle" phenomenon can be solved through the liquid/solid hammer handle between the LCS and the HCHCS substituted for the original LCS hammer handle.

Table 2. Major mechanical properties of HCHCS and LCS.

Materials	Hardness (HRC)	Impact toughness (J/cm^2)	Wear loss (g)
HCHCS	35-38	≥80	0.1087
LCS	60-62	7.9-8.5	0.3325

The microhardness under liquid-solid volume ratio of 8:1, at the preheating temperature of 300℃ and pouring temperature of 1565℃ and 1620℃ was shown in figure 5.

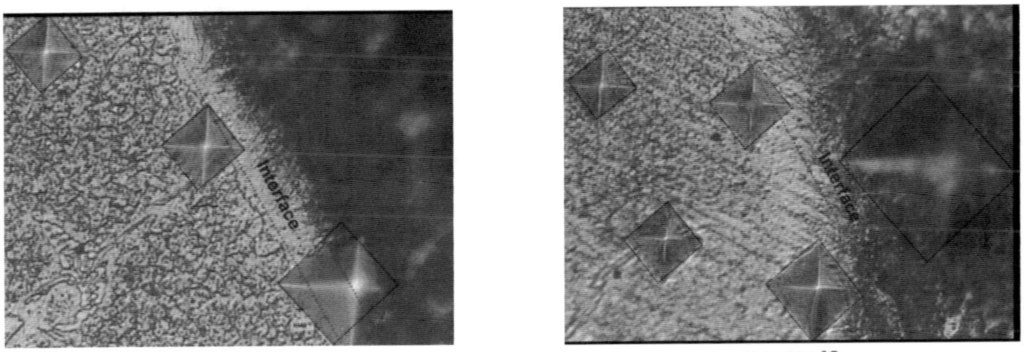

(a) 300℃ (b) 600℃

Figure 5. Microhardness at the preheating temperature.

The microhardness of bimetal near the interface was measured from the HCHCS to the LCS. The left part of the figure was HCHCS and the right was LCS. The concavity at the bonding region was bigger than HCHCS and smaller than LCS, which illustrates the hardness was higher than LCS and lower than HCHCS. The gradient microhardness change proves the good metallurgical bonding at the interface. The thickness of the bonding interface in figure 5(b) was thicker than that in figure 5(a), which can be explained that the higher preheating temperature decreases the difference temperature between the liquid HCHCS and solid LCS. So the LCS will get higher heat from liquid HCHCS, which was obtained better interface bonding.

5. Conclusions

- With the increased volume ratio of liquid to solid, preheating temperature and pouring temperature, the interfacial microstructure was becoming better and the thickness of bonding region was becoming thicker.
- Microhardness was gradually increased from the LCS to the HCHCS, and the microhardness of the bonding interface was in between the LCS to the HCHCS. C, Cr, and Mn diffuse from the HCHCS to the LCS, and Fe diffuses towards the opposite direction.
- The good metallurgical bonding interface without any pore was obtained by liquid-solid casting under the volume ratio of 8:1 through the 300℃preheating temperature and the 1565℃ pouring temperature.

Acknowledgments

This work was financially supported by the National Natural Science Foundation of China under grant Nos. 51371090 and 51203061, and the 2014 University Technology Industrialization Projects Foster Development of Heilongjiang Province No. 1254CGZH08.

References

[1] Danesh Manesh H and Karimi Taheri A 2005 An investigation of deformation behavior and bonding strength of bimetal strip during rolling *Mechanica of Materials* **268** 531-42

[2] Xiong B W, Cai C C and Lu B P 2011 Effect of volume ratio of liquid to solid on the interfacial microstructure and mechanical properties of high chromium cast iron and medium carbon steel bimetal *J. Alloys Compound* **509** 6700-4

[3] Xiong B W, Cai B W and Wan H 2011 Fabrication of high chromium cast iron and medium carbon steel bimetal by liquid–solid casting in electromagnetic induction field *Materials and Design* **32** 2978-82

[4] Xie G L, Sheng H, Han J T and Liu J 2010 Fabrication of high chromium cast iron/low carbon steel composite material by cast and hot rolling process *Materials and Design* **157** 3062-6

[5] Markovich E I, Branovitsky A M and Na Y S 2006 Study on the possibility of continuous-casting of bimetallic components in condition of direct connection of metals in a liquid state *Materials and Design* **48** 1016-26

[6] Xiao X F, Ye S P and Yin W X 2012 High Cr white cast iron/carbon steel bimetal liner by lost foam casting with liquid-liquid composite process *China Foundry* **2** 136-42

[7] Sun J B, Song X Y and Wang T M 2012 The microstructure and property of Al–Si alloy and Al–Mn alloy bimetal prepared by continuous casting *Mater. Lett.* **67** 21-3

[8] Zhu Y C, Wei Z J and Rong S F 2013 Carbon diffusion mechanism of two-liquid bimetal bond interface on dissimilar carbon steel *Mater. Res. Innov.* **18** 350-3

[9] Rong S F 2011 Influence of casting technique on the mechanical properties of bimetal composites *J. Iron Steel Res. Int.* **18** 546-9

IOP Conf. Series: Materials Science and Engineering **103** (2015) 012033 doi:10.1088/1757-899X/103/1/012033

Atomistic modeling determination of placeholder binding energy of Ti, C, and N atoms on α-Fe (100) surfaces

X J Wei, Y P Liu[1] and S P Han

College of Mechanical Engineering, Taiyuan University of Technology, Taiyuan, Shanxi, 030024, P.R. China

E-mail: lyplr@163.com

Abstract. A Fe(100) surface containing Ti, C, and N was constructed and optimized to study the placeholder binding energy of the Ti, C, and N surface atoms; this was achieved by searching the transition state with the LST (linear synchronous transit) method of the CASTEP (Cambridge Serial Total Energy Package) module. Also, the authors analyzed electron structures to determine how Ti, C, and N atoms strengthen the Fe(100) surface. The results show that when Ti, C, or N atoms take placeholder alone, or simultaneously at the Fe(100) surface, the structure stability is at its best. When including Ti, C, and N as solid solutions on the Fe(100) surface, orbital electrons of Fe3d, Ti3d, C2p, and N2p hybridize near the Fermi level; the number of electronic bonding peaks increase and bonding capacity enhances. Also, a large amount of covalent bonds formed. Covalent bonds and metallic bond coexisted.

1. Introduction

Fe is the most widely used metallic material today. As such, improving the strength and abrasion resistance of Fe-based alloy has been a popular research topic. Researchers have investigated surface modification because it can affect an alloy's abilities without affecting its strength, hardness, wear-resistance etc. Liu Y P, et al. [1-7] formed Ti and N double cementation with plasma on mild steel surfaces; the infiltration layer had an average hardness reaching 2500 HV, its corrosion resistance was 14 times better than stainless steel, its friction coefficient was 0.15, and its abrasion resistance also improved. However, the placeholder mechanism of the alloy atoms after element infiltration on the Fe surface is still unknown due to experimental conditions and a lack of supporting theories for calculation. The first-principles method is currently an effective way to solve this problem. Zhu, et al. [8] adopted the VASP (Vienna Ab-initio Simulation Package) to search the transient state, and calculated the diffusion activation energy of C in a nickel substrate. Sorescu [9] also used the same tool to calculate the diffusion activation energy and the ideal placeholder path of H in an Fe substrate, as well as on an Fe(100) surface. He believed that the activation energy of the H atoms when they diffuse from the first layer to the third layer on the Fe surface is the same as when the H atoms diffuse inside the substrate. Liu L F, et al. [10] used the LST/QST method to search the transient state among CASTEP modules, calculated the diffusion activation energy of C, N, and O in α-Fe, and determined their placeholder pathways; however, they did not consider the spin polarization of Fe. Therefore, this report will systematically study the placeholder energy of Ti, C, and N atoms in Fe substrates and on

[1] Address for correspondence:Y P Liu, College of Mechanical Engineering, Taiyuan University of Technology, Taiyuan, Shanxi, 030024, P.R. China. E-mail: lyplr@163.com.

Fe-(100) surfaces by analyzing the electron distribution; the paper also provides a theoretical explanation for why Ti, C, and N surface alloying improves Fe-based alloy.

2. Computing method

A CASTEP (Cambridge Serial Total Energy Package) module [11, 12] in Materials Studio 4.4 was used to perform the calculations. The module was based on the plane pseudo potential method under density functional theory. It replaced ionic potential with pseudo potential, and electron wave functions were expanded with plane wave basis sets. The reaction tendency between electrons was adjusted by generalized gradient approximation (GGA). All atom pseudo potential adopted ultrasoft pseudo potential, and electron wave function truncation energy was set at 300 eV. The K segmentation point of the model in the first brillouin zone was set as fine precision. The interchange correlation function was adjusted by GGA-PBE (Perdew-Burke-Ernzerhof). Before calculating property the supercell all is geometry optimization by BFGS (Broyden, Fletcher, Gold-farb, Shanno) way. Considering spin polarization and obtaining the stable structure in a local area, the admissible error was set as 1×10^{-6} eV. Moreover, at the end of the self-consistent operation process, the genetal energy converged to 1×10^{-5} eV/atom. The force at every atom was lower than 0.03 eV/nm, the tolerance excursion was less than 1×10^{-4} nm, and the stress excursion was less than 0.05 GPa.

3. Placeholder of Ti, C, N on α-Fe surface

3.1. Calculating model

Using transition metal α-Fe, whose group space is cubic system $I-m\bar{3}m$, as the subject and choosing 2.88664Å as an experimental value of the lattice parameter. In regards to the site occupancy of Ti, C and N inside the Fe-base, document [13] states that Ti atoms had slightly larger diameters than Fe atoms. Therefore, solid solution displacement occurs. The diameters of the C and N atoms are shorter; therefore, an interstitial solid solution forms, which occupies the position interstitially. When C and N undergo solid solution in octahedral interstices, there are two adjacent Fe atoms, and there will be four Fe atoms if solid solution occurs inside a tetrahedron. Since the strain energy required to move 4 atoms is greater than that required to move 2 atoms, C and N atoms take precedence in existing inside octahedral interstices.

The α-Fe(100) surface model was constructed for determining the migration of C and N atoms on the α-Fe surface, as well as Ti's effects as a placeholder. To select the appropriate surface structure layer number, the super cell structure of the Fe atom surface was constructed with 5, 6, and 7 layers. When calculating, the two outermost layers were set to be free layers; all layers below were rigid. The 10Å thick vacuum layer was established perpendicular to the lattice surface. Figure 1 shows the atomic layer used to connect the vacuum layers simulated as the actual surface.

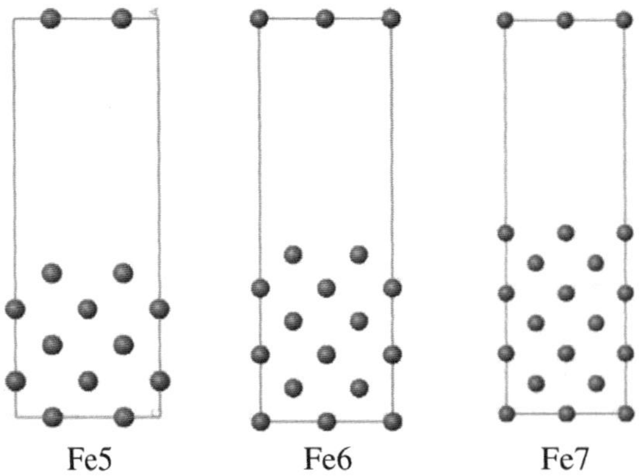

Figure 1. Fe(100) surface structure model of Fe atoms with 5, 6, and 7 atomic layers.

Geometry optimizations were performed for the Fe(100) surface structures of the Fe atoms in the 5, 6, and 7 layer models. Table 1 lists the relaxation results of atoms present in the two outermost layers; in these, $\Delta mn=(dmn-d)/d\times100\%$ (m=1,2, n=m+1), and d is the layer interval value in the matrix unit cell. Comparing the relaxation magnitude of the three models in table 1, the sixth Fe(100) surface layer has the lowest relaxation and the calculation matched with calculations obtained by others [14, 15]; thus, the sixth Fe(100) layer was chosen to be the surface structure model to calculate the placeholder of Ti, C, and N elements on the Fe(100) surface.

Table 1. Fe(100) surface structure relaxation data of Fe atoms with 5, 6, and 7 atomic layers.

Relaxation data	Fe5	Fe6	Fe7	Others [14]
Δ_{12}(%)	-3.51	-2.67	-3.75	-2.62
Δ_{23}(%)	2.14	1.67	0.95	2.33

Figure 2 models single Ti, C, and N placeholder atoms on the Fe(100) surface; this data can be used to calculate the total energy and binding energy of the alloying surfaces with Ti, C, and N solid solution in each Fe(100) surface layer. Figure 3 shows the models of Ti atoms and C(N) atoms when they occupy positions in each Fe(100) surface layer.

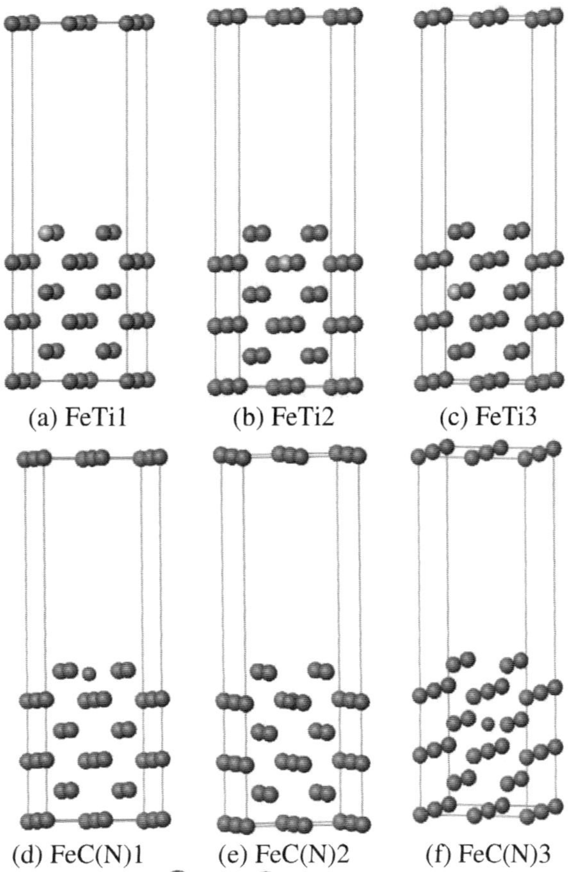

(a) FeTi1 (b) FeTi2 (c) FeTi3

(d) FeC(N)1 (e) FeC(N)2 (f) FeC(N)3

—Fe —Ti —C

Figure 2. Unit cell model of Ti, C, and N placeholder atoms on each Fe(100) surface layer.

3.2. The binding energy of Fe(100) surfaces with Ti, C, and N

The lattice structure strength and stability strongly relates to the binding energy. The lattice binding energy is the energy released by the formation of free atoms into the lattice structure; in other words, it

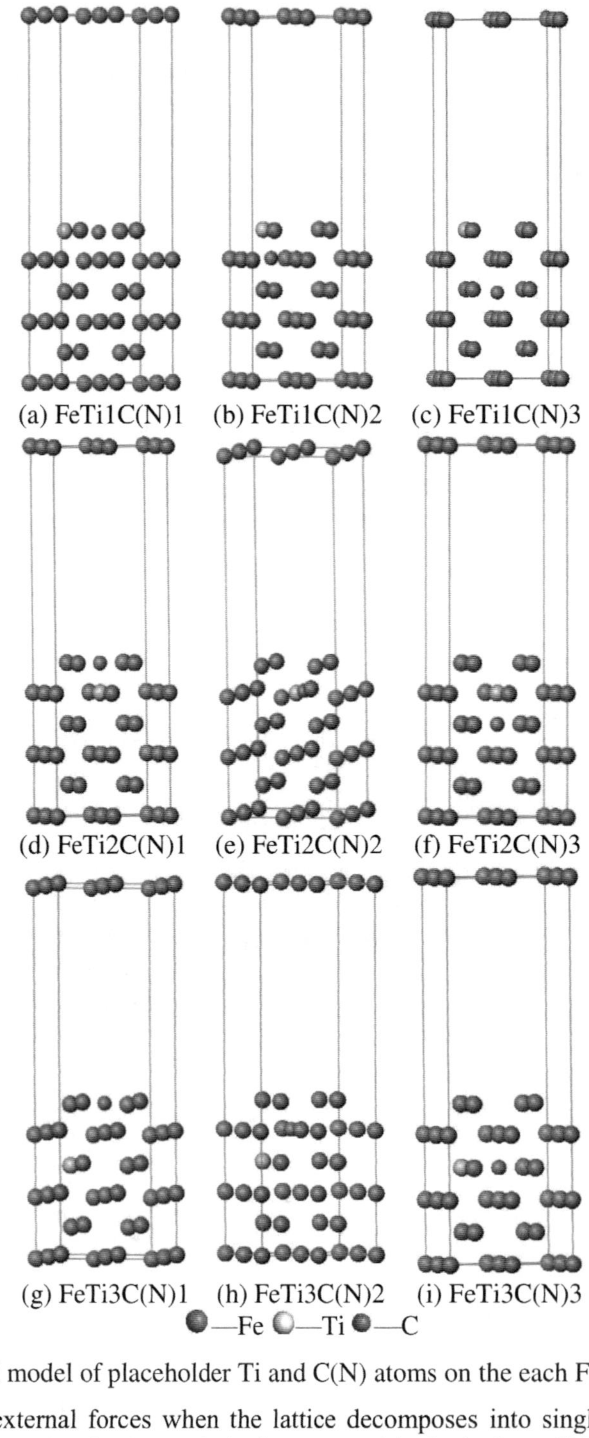

Figure 3. Unit cell model of placeholder Ti and C(N) atoms on the each Fe(100) surface layer.

is the work done by external forces when the lattice decomposes into single atoms. The lower the binding energy (the larger the absolute value), the more stable the lattice will be.

The formula for calculating the binding energy is [15-19]:

$$\Delta E^{ABC} = \frac{1}{N_A + N_B + N_C}\left(E_{tot}\text{-}N_A E^A\text{-}N_B E^B\text{-}N_C E^C\right)$$

In the formula, ΔE^{ABC} is the binding energy, *Etot* is the total energy of the lattice, and E^A, E^B, and E^C are the energies of single atoms A, B, and C respectively. *NA, NB,* and *NC* are the numbers of A, B, and C atoms in the unit cell, respectively. Figures 4-6 show the binding energy of each alloying element shown in the previous figures derived from the above equation.

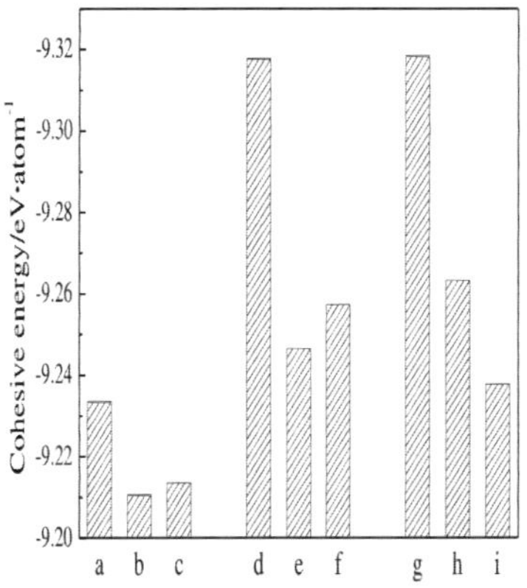

Figure 4. Bonding energy of a Ti, C, and N single atom on an Fe(100) surface.

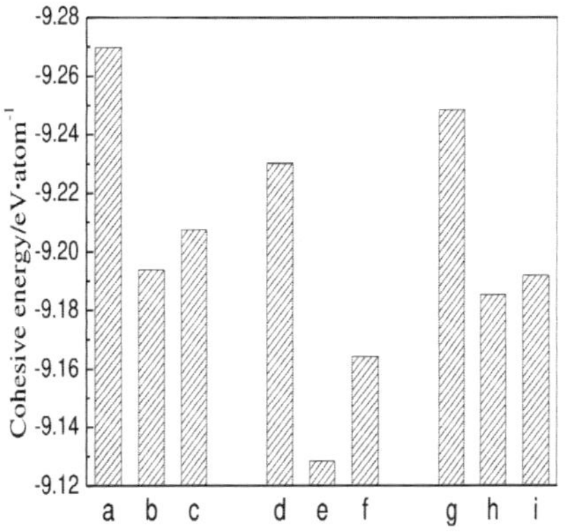

Figure 5. Bonding energy of Ti and C atom placeholders simultaneously on Fe(100) surface.

Figures 4(a)-4(c) represent the binding energy of Ti atoms in the first, second, and third layer of the Fe(100) surface, respectively. d, e, and f represent the binding energy of C atoms in the first, second, and third layer of the Fe(100) surface, respectively. g, h and i represent the binding energy of N atoms in the first, second, and third layer of the Fe(100) surface, respectively.

From figure 4, the binding energies are negative when Fe(100) surface existing single Ti, C, N atoms respectively as solid solution, this means the solid solution could exist stably. When Ti displaces the Fe atoms in the first, third, and second surface layers, the binding energy increases (the absolute value decreases) respectively in order, at the same time the stability decreases gradually. When C is present in the octahedral interstices of the surface first, third, and second layers, the binding energies increase (the absolute value decreases) and stability decreases respectively in order. When N is present in octahedral interstices of the first, second, and third surface layers, the binding energies increase (the absolute value decreases) and stability decreases respectively in order. However, since N migrates to the first layer after being optimized in the second layer, N atoms are unstable when the second octahedral layer on the Fe surface is a solid solution. In conclusion, Ti, C, and N atoms are the most stable when they are in a solid solution respectively in the first layer of Fe(100) surface.

Figures 5(a)-5(c), represent the binding energy of C atoms in the first, second, and third layers when Ti is present only in the first layer. d, e, and f represent the binding energy of C atoms in the first, second, and third layers when Ti is present only in the second layer. g, h, and i represent the binding energy of C atoms in the first, second, and third layers when Ti is present only in the third layer.

From figure 5, when the Fe(100) surface is alloyed by Ti and C, the binding energy is negative, indicating that the both status could exist stably. As can be seen from figures 5(a)-(c), when Ti displaces first layer Fe atoms, the binding energy is the smallest (the absolute value is the biggest) when C is present in the first layer octahedral interstices; this produces the most stable state. The same applies for d, e, and f in figure 5, and for g, h, and i in figure 5 when Ti displaces the second and third layer Fe atoms. Comparing the lowest binding energy of the above mentioned statues shown in figures 5(a), 5(d) and 5(g), the following conclusion can be reached: The binding energy is smallest (the absolute value is biggest) and shows the most stability when Ti displaces the first layer of Fe atoms and C is present in the first layer octahedral interstices.

Figures 6(a)-6(c) respectively represent the binding energy of N atoms in the first, second, and third layers when Ti is present only in the first layer. d, e, and f respectively represent the binding energy of N atoms in the first, second, and third layers when Ti is present only in the second layer. g, h, and I respectively represent the binding energy of N atoms in the first, second, and third layers when Ti is present only in the third layer.

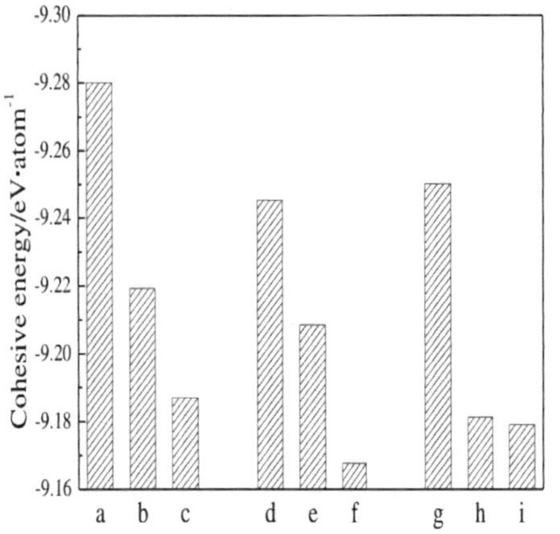

Figure 6. Bonding energy of Ti and N atom placeholder simultaneously on an Fe(100) surface.

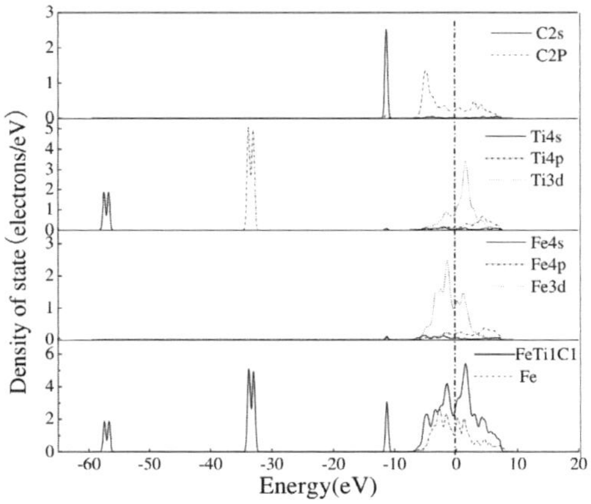

Figure 7. Alloy state densities.

From figure 6, when Fe(100) surface taking place Ti and N alloying, the binding energy is negative, indicating that the both status could stably exist. As can be seen from a, b, and c in figure 6, when Ti displaces the first layer Fe atoms, the binding energy is the smallest (the absolute value is the biggest) when N is present in the first layer octahedral interstices; this produces the most stable state. The same applies when Ti displaces the second (from figures 6(d)-6(f)) and third (from figure 6(g)-6(i)) layer Fe atoms. Comparing the lowest binding energy of the above mentioned statues shown in figures 6(a), 6(d) and 6(g), the following conclusion can be reached: The binding energy is smallest (the absolute value is biggest) and shows the most stability when Ti displaces the first layer of Fe atoms and N is present in the first layer octahedral interstices.

3.3. The density of α-Fe(100) surface states including Ti, C, and N atoms

To a more particular knowledge of the interactions of Ti, C, and N on an Fe(100) surface, the density of states of FeTi1C1and undoped Fe(100) was respectively calculated and shown in figure 7. This shows that the bonding electrons of FeTi1C1 were mainly distributed in the four sections, such as -58.4~ -55.8 eV, -34.8~ -32.0 eV, -12.3~ -10.3 eV, and -8.7~ 8.2 eV. The reactions between Ti, C, N, and Fe atoms were mainly shown as the inter-hybridization between Ti3d, C2p, N2p, and Fe3d orbits near the Fermi level. Also, at a lower Fermi level, the bonding peaks of FeTi1C1 are more two than the peaks without Fe doped. FeTi1C1 has a relatively large pseudogap and its wave is at the Fermi level; Femi levels are closer to the wave crest in Fe state density. This shows that when Ti and C atoms are a solid solution in Fe(100) at the same time, a large amount of covalent and metallic bonds form; without Fe doping, the majority are metallic bonds and only a small amount of covalent bonds are present [20, 21].

In conclusion, the placeholder of Ti, C, and N elements on an Fe(100) surface allow the surface to be hybridized at the electron orbit in the Fermi level, causing the electron bonding peak to increase; this enhances the material's strength and creates stronger pseudogaps near the Fermi level. This shows that Ti, C, and N form solid solutions of the Fe substrate where metallic and covalent bonds could coexist.

4. Conclusion

When single Ti, C, or N atoms occupy positions near the Fe(100) surface, the displacement of Ti atoms with the first layer Fe atoms will result in the best stability. When C or N atoms are present in the octahedral interstice in the first layer on the surface, the stability is at its best. When the Ti and C atoms are accommodated at the same time to the Fe(100) surface, it exhibits the best stability when Ti and C atoms occupy the first layer together. When Ti and N atoms are accommodated to an Fe(100) surface at the same time, it exhibits the best stability when they both occupy the first layer.

After Ti, C, and N form solid solutions on the surface of Fe(100), the orbit electrons of Fe3d, Ti3d, C2p, and N2p will hybridize at Fermi energy, causing the electron bonding peak to increase; this strengthens the binding energy and creates pseudogaps near the Fermi level. That is to say, surface alloying of Ti, C, and N on the Fe substrate will produce covalent bonds, and the metallic bonds will be able to coexist with the covalent bonds.

References

[1] Liu Y P, Xu J Y, Kui X Y, Wang J Z, Gao Y and Xu Z 2005 Study on Microstructure and performance of TiN Multi-permeating Coatings by glow discharge plasma *Transactions of Materials and Heat Treatment* **26** 109-12

[2] Kui X Y,Wang J Z, Liu Y P, Xu J Y, Gao Y and Xu Z 2006 Microstructure and corrosion resistance of TiN coating synthesized by plasma discharge on carbon steel *Corrosion Science and Protection Technology* **18** 12-5

[3] Liu Y P, Xing F and Guo H X 2007 Study on corrosion and wear resistance of ceramic multi-permeating coatings by plasma discharge technique *China Surface Engineering* **20** 19-26

[4] Gao Y, Wang C L, Liu Y P and Xu J Y 2011 Corrosion resistance of TiN layer prepared by double glow plasma surface alloying process *Transactions of Materials and Heat Treatment* **32** 143-7

[5] Liu Y P, Xue J X and Han P D 2010 Electrochemical corrosion behaviors using double glow forming TiN multi-permeation layer *Materials Science Forum* **654** 1968-71

[6] Liu Y P, Xu J Y, Gao Y and Xu Z 2005 A study on TiN/Ti multi-permeations alloying coatings by glow discharge plasma *Materials Review* **19** 112-4

[7] Liu Y P, Xu J Y, Kui X Y and Xu Z 2005 Ceramic alloying with Ti-N deposition and diffusion of multi-layers by plasma surface alloying technique *Trans. Nonferrous Met. Soc. China* **15** 415-9

[8] Zhu Y A, Dai Y C, Chen D and Yuan W K 2007 First-principles study of carbon diffusion in bulk nickel during the growth of fishbone-type carbon nanofibers *Carbon* **45** 21-7

[9] Sorescu D C 2005 First principles calculations of the adsorption and diffusion of hydrogen on Fe(100) surface and in the bulk *Catalysis Today* **105** 44-65

[10] Liu L F, Zhou S Q, Huang Y T and Zhou A R 2008 Computation of diffusion activation energies of C, N and O in α-Fe *Materials Review* **22** 120-2

[11] Segall M D, Lindan P J D, Prober M J, Pickard C J, Hasnip P J, Clark S J and Payne M C 2002 First-principles simulation: Ideas, illustrations and the CASTEP code *J. Phys.: Condens. Matter* **14** 2717-44

[12] Chamati H, Papanicolaou N I and Mishin Y 2006 Embedded-atom potential for Fe and its application to self-diffusion on Fe(100) *Surface Science* **600** 1793-804

[13] Wang X T 1987 *Metal Material Science* (Beijing: Machine industry Press) p 8

[14] Tao M L and Tan X C 2009 Study on first-principles of carbon absorption on Fe(100) surface *Journal of Atomic and Molecular Physics* **26** 349-56

[15] Sorescu D C, Thompson D L, Hurley M M and Chabalowsk C F 2002 First-principles calculations of the adsorption, diffusion, and dissociation of a CO molecule on the Fe(100) surface *Phys. Rev. B* **66** 035416

[16] Zhou D W, Peng P, Hu Y J and Liu J S 2006 Study on structural stability of Mg-Ce intermetallic compounds based on the pseudopotential plane-wave method *Rare Metal Materials and Engineering* **35** 871-5

IOP Conf. Series: Materials Science and Engineering **103** (2015) 012033 doi:10.1088/1757-899X/103/1/012033

[17] Zhang C L, Li J M, Han P D, Chi M, Yan L Y, Liu X G and Xu B S 2008 Study on alloying effects of Ni3Al with Cr based on the pseudopotential plane-wave method *Rare Metal Materials and Engineering* **37** 1705-9

[18] Huang Z W, Zhao Y H, Hou H, Wang Z, Mu Y Q, Niu X F and Han P D 2011 Point defects structure and alloying effects of V atoms into Ni3Al alloy: A first-principles study *Rare Metal Materials and Engineering* **12** 2136-41

[19] Zhao Y H, Huang Z W, Li A H, Mu Y Q, Yang M H, Hou H, Han P D and Zhang S Y 2011 First principles study on substitution behavior and alloying effects of Nb in Ni_3Al *Acta Phys. Sin.* **60** 047103-49

[20] Chen Y, Shang J X and Zhang Y 2007 Effects of alloying element Ti on α-Nb5Si3 and Nb3Al from first principles *Phys: Condens Matter* **19** 016215

[21] Hu M Q, Yang R, Xu D S, Hao Y L and Li D 2003 Geometric and electronic structure of Ti_2AlX(X=V, Nb, or Ta) *Phys. Rev. B* **68** 054102

IOP Conf. Series: Materials Science and Engineering **103** (2015) 012034 doi:10.1088/1757-899X/103/1/012034

Artificial neural networks application for modeling of friction stir welding effects on mechanical properties of 7075-T6 aluminum alloy

E Maleki[1]

Mechanical Engineering Department, Sharif University of Technology-International Campus, Kish Island, Iran

E-mail: maleki_erfan@kish.sharif.edu/ maleky.erfan@gmail.com

Abstract. Friction stir welding (FSW) is a relatively new solid-state joining technique that is widely adopted in manufacturing and industry fields to join different metallic alloys that are hard to weld by conventional fusion welding. Friction stir welding is a very complex process comprising several highly coupled physical phenomena. The complex geometry of some kinds of joints makes it difficult to develop an overall governing equations system for theoretical behavior analyse of the friction stir welded joints. Weld quality is predominantly affected by welding effective parameters, and the experiments are often time consuming and costly. On the other hand, employing artificial intelligence (AI) systems such as artificial neural networks (ANNs) as an efficient approach to solve the science and engineering problems is considerable. In present study modeling of FSW effective parameters by ANNs is investigated. To train the networks, experimental test results on thirty AA-7075-T6 specimens are considered, and the networks are developed based on back propagation (BP) algorithm. ANNs testing are carried out using different experimental data that they are not used during networks training. In this paper, rotational speed of tool, welding speed, axial force, shoulder diameter, pin diameter and tool hardness are regarded as inputs of the ANNs. Yield strength, tensile strength, notch-tensile strength and hardness of welding zone are gathered as outputs of neural networks. According to the obtained results, predicted values for the hardness of welding zone, yield strength, tensile strength and notch-tensile strength have the least mean relative error (MRE), respectively. Comparison of the predicted and the experimental results confirms that the networks are adjusted carefully, and the ANN can be used for modeling of FSW effective parameters.

1. Introduction

The AA7075-T6 aluminum alloy is one of the strongest nonferrous metals, used in various industries nowadays. High strength-to-weight ratio and its property of natural aging have led to the ever-growing use of this aluminum alloy in aerospace and high-tech industries.

7XXX series aluminum alloys are typically composed of zinc (Zn), magnesium (Mg), and copper (Cu) elements combined with aluminum (Al) as the predominant metal. The superior strength of this aluminum alloy is due to the presence of the above-mentioned alloying elements [1]. The existence of these elements in the composition of this aluminum alloy has led to the formation of AL_2CuMg and Mg_2Zn phases in the final structure. The reason for overall strength of this structure is existence of

[1] Address for correspondence: E Maleki, Mechanical Engineering Department, Sharif University of Technology-International Campus, Kish Island, Iran. Email: maleki_erfan@kish.sharif.edu/ maleky.erfan@gmail.com.

such phases [2]. The simultaneous appearance of copper and zinc increases strength, but also reduces the weld ability of this structure.

There are many articles that investigate AA7075 behavior in different conditions. Much of these articles are the studies about the investigation of mechanical properties, metallurgical properties, how to the crack growth, fatigue behavior, fatigue life and its increasing methods [3-7]. Majzoobi, et al. [8-10] surveyed the fretting fatigue behavior of Al7075-T6 considering the effects of coating and temperature in their work.

Due to having the aforementioned elements (Zn, Cu, Mg and Al) in their structure, 7XXX series aluminum alloys are highly sensitive to heat cracking in the freezing process and also to liquation cracking in the heat effected area [1, 2]. Moreover, the probability of zinc oxidation or evaporation during fusion welding of these alloy types, can be the main reason of serious problems, such as porosity or partial melting [11]. Low boiling temperatures of some elements in the structure, namely zinc and magnesium, often cause numerous problems in laser beam welding [12, 13].

Friction stir welding (FSW) is one of the non-fusion techniques of welding which has eliminated many of the defects associated with fusion welding techniques in aluminum alloy joints. In the process of friction stir welding, a tool creates frictional heat by moving along the joint line of two plates and rotating, simultaneously. It eventually creates a joint from the plasticized material. The process of welding by the FSW technique has various effective parameters including welding speed, rotational speed and axial force of the tool, tool hardness, and sizes of pin and shoulder.

Many experimental works have been done to study the usage of FSW on AA7075-T6 and investigate its parameters and characteristics. These works have survived the effect of welding parameters on microstructure, thermal, mechanical and metallurgical properties of AA7075-T6 welded joints in various conditions [14-19]. Azimzadegan, et al. [20] have investigated the effects of high rotational speeds. Rajakumar, et al. [21] and DebRoy, et al. [22] have studied the effect of tool in friction stir welded joints. Bahrami, et al. [23] have explored this process on AA7075-T6 in nano-scale.

In the case of modeling and simulation of FSW process, it can be noted to applications of numerical methods and the heuristic algorithms, which are the most-used methods in simulating and optimizing this process. The finite element method (FEM) as the most common numerical method and the simulated annealing (SA) algorithms and artificial neural networks (ANN) as heuristic algorithms have been widely employed in the simulation of FSW process. Fratini, et al. [24-26] havw provided the FEM models of FSW in their studies. He, et al. [27] have reviewed the numerical analysis of friction stir welding in their work.

Some works have employed heuristic algorithms for simulation of FSW; Babajanzade Roshan, et al. [28] have used the adaptive neural fuzzy inference system (ANFIS) models and SA algorithm to optimize friction stir welding process of AA7075 and achieve desirable mechanical properties. As mentioned, the ANNs have also been used in predicting and obtaining optimum parameters in the process of friction stir welding. Specifically, in the context of employing the artificial neural networks in predicting and optimizing the FSW process, in addition to how the network trained, the main difference is in the inputs and outputs of neural network that they have been regarded to training networks. Okuyuca, et al. [29] have evaluated the ability of neural networks in predicting the parameters of friction stir welding for aluminum plates. In their study, they took the parameters of welding speed and rotational speed of the tool as network inputs, and the parameters of yield strength, ultimate length variation of specimens, and ultimate strength of the weld as network outputs. Eventually, the results of their work showed an acceptable correspondence between the results of the neural network and those of experimental tests. In a similar study, Shojaeefard, et al. [30] have predicted the mechanical properties of non-cognate friction stir welding on AA5083 and AA7075 aluminum alloys. In their study, they used the speeds of welding and tool rotation as the inputs of the neural network and the final mechanical and metallurgical properties of the weld as the outputs of the neural network. In the end, the results of their work, likewise, showed proper abilities of neural networks for predicting the parameters of FSW process. Similarly, Yousif, et al. [31] have regarded the welding speed and rotational speed as inputs and the tensile strength, yield strength and elongation as outputs in simulation of the FSW parameters on aluminum alloy plates. Lakshminarayanan, et al. [32] gathered the welding speed, rotational speed and axial force as inputs and tensile strength as an output to survey the FSW on AA7039. Ghetiya, et al. [33] have considered tool shoulder diameter, welding speed, rotational speed, and axial force as inputs whereas the output of their modeling is the tensile strength of AA8014 welded joint. For better comparison, the used inputs and outputs of aforementioned studies about ANN application (references [29-33]) have been shown in table 1.

Table 1. The used inputs and outputs of previous works [29-33].

Reference No.	Inputs	Outputs
29	Welding speed Rotational speed	Yield strength Length variation
30	Welding speed Rotational speed	Tensile shear force hardness
31	Welding speed Rotational speed	Tensile strength Yield strength Elongation
32	Welding speed Rotational speed Axial force	Tensile strength
33	Welding speed Rotational speed Axial force tool shoulder diameter	Tensile strength

As the table 1 illustrates, the authors have regarded 2-4 inputs and 1-3 outputs in their works, while the influential parameters of friction stir welding are much more; among which tool hardness and tool pin diameter can be mentioned as inputs. Similarly, the notch-tensile strength can be considered as output. Although using fewer parameters makes the network training process faster and gives more accurate answers, it simultaneously makes the neural network responses more unrealistic in examining parameters of friction stir welding. Accordingly, this study tries not to ignore any of the influential parameters of the process, while lowering the error of network answers to its minimum by applying correct and proper values for neural networks structures. In this paper, six parameters including: welding speed, rotational speed of tool, axial force, shoulder diameter, pin diameter and tool hardness regarded as inputs of the ANN. Furthermore, four parameters including: yield strength, tensile strength, notch-tensile strength and hardness of welding zone gathered as outputs of ANN.

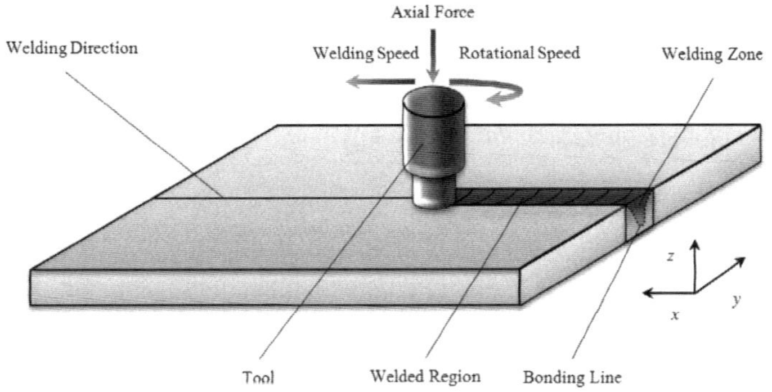

Figure 1. A Schematic of FSW.

2. Friction stir welding

Friction stir welding was invented by England's The Welding Institute (TWI) in 1991 [34]. Figure 1 shows the tools sample and a piece being welded by FSW.

In the process of friction stir welding, a tool creates frictional heat by moving along the joint line of two plates and rotating simultaneously and eventually creates a joint from the plasticized material [35].

As figure 1 demonstrates, the process of welding by the FSW technique has various effective parameters. These parameters are including the welding speed, rotational speed and axial force of the tool, tool hardness and sizes of pin and shoulder. Defects in the joints created by this technique are due to behavior of the material when it flows under welding. The material flow is in turn under the influence of the mechanical properties of the material and the tool, and the effective parameters of friction stir welding. Using friction stir welding eliminates the difficulties of fusion welding and also reduces residual stress and distortion created in the welding process [36, 37]. Table 2 presents the effective parameters in the process of friction stir welding and their mechanisms of influence.

Table 2. Parameters of friction stir welding and their influence mechanisms.

Welding Parameter	Influence Mechanism
Welding Speed	As the welding speed increases, higher work hardening is done on the joint area.
Rotational Speed	An increase in the rotational speed, increases the created heat and the intensity of stir.
Axial Force	Based on the equation $f=\mu_k F_n$,, as the axial force increases, friction increases as well, and consequently, the created heat is increased.
Tool Hardness	As the tool hardness increases, the friction coefficient between the tool and the material is increased. According to the equation $f=\mu_k F_n$, this leads to an increase in the created friction and heat.
Tool Geometry	An increase in the size of pin and shoulder, increases the contact area between tool and material. As a result, friction and heat are increased while the tool is rotating.

Investigation of welding zone and its surrounding microstructure areas demonstrate that the FSW process in welding zone affects its near areas. Figure 2 shows the typical friction stir weld in its cross-section, consisting of four main zones [38]:

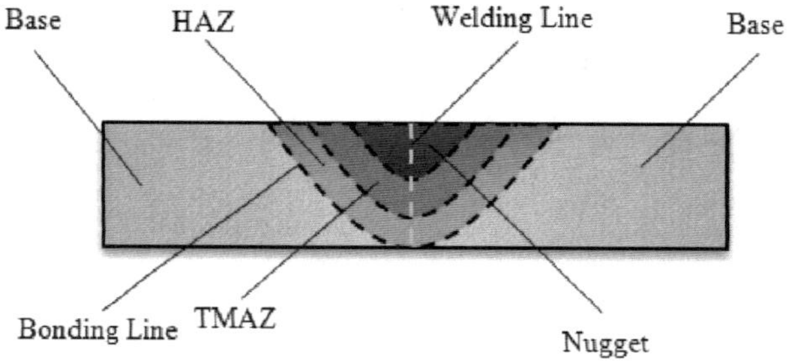

Figure 2. Typical friction stir weld in its cross-section.

- Weld Nugget (WN): stirring zone by pin rotation, fully recrystallized area.
- Heat Affect Zone (HAZ): material zone which is close to welding area, have a thermal effect but no plastic deformation.
- Thermo-mechanically Affected Zone (TMAZ): transition zone between HAZ and WN that have plastic deformation without recrystallization and thermal effect (It is usually difficult to distinguish the precise boundary between TMAZ and nugget).
- Base material (BM): the material zone that is remote from the welding area. It nearly keeps the original microstructure and mechanical properties.

Studying the influence mechanisms of the parameters, one can conclude that an increase in any of the parameters enhances the flow of material as a result of the heat increase. However, since other factors, such as the reduction of cooling rate due to the temperature rise at the beginning of the process, cannot be ignored, joint properties do not necessarily improve. It can be seen that at first increasing each of the parameters improves mechanical and metallurgical properties of the joint, and after reaching the optimum value, any further increase causes a decline in the properties.

It is observed that grain size reaches its lowest point under 5kN force, where mechanical properties are determined based on Hall-Petch equation, which is a relation between yield stress and grain size. Hall-Petch equation has been determined as follows [39]:

$$\sigma_y = \sigma_0 + k_y/d^{1/2} \tag{1}$$

Where σ_y is the yield stress, σ_o is material constant for the starting stress of dislocation movement, k_y is strengthening coefficient and d is average grain diameter.

3. Artificial neural networks

Artificial neural networks (ANNs) have found applications in many optimization and prediction problems in the last decade. ANNs are computational models inspired by an animal's central nervous system, in particular the human's brain, which is capable of machine learning as well as pattern recognition [40]. ANN obtains a nonlinear relation among the influential factors of a process and their corresponding outputs that, in this study, are mechanical and metallurgical properties of a joint. Therefore, a neural network enables the researcher to determine outputs for other arbitrary inputs with high accuracy. Employing ANN eliminates any need to carry out experimental tests that are costly and time-consuming. These networks are essentially simple mathematical models defining a function $\boldsymbol{f}: \boldsymbol{x} \rightarrow \boldsymbol{y}$ or a distribution over $\boldsymbol{x}$ or both $\boldsymbol{x}$ and $\boldsymbol{y}$.

Figure 3 represents a neural network. In this network, each input consists of r parameters and each output comprises s parameters. Furthermore p, w, b, f and a represent the inputs, weight matrixes, bias vectors, transfer function in neurons, and outputs, respectively.

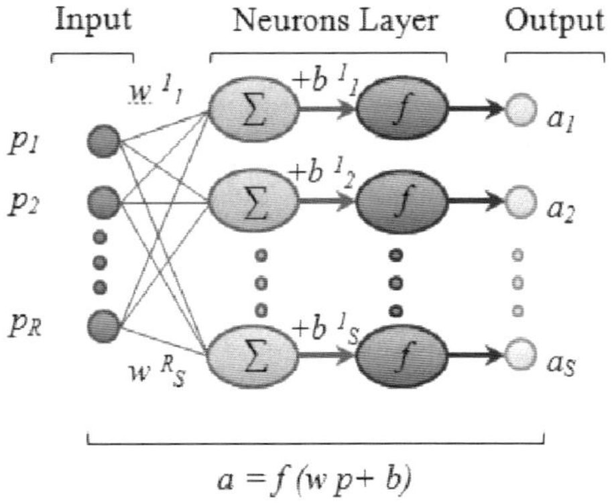

Figure 3. One layer network with R inputs and S neurons.

Based on their functions, neural networks need to be trained using a set of inputs and their corresponding outputs. Assuming one layer of neuron, training is accomplished through multiplying each input with a variant called the weight and summing it with another variant called bias before entering the neuron. The initial value of both variants can be predetermined. The obtained values from each input are summed up to form the input of the transfer function, and its output is the neuron's output that needs to compare with the experimental output value. The tangent sigmoid (Tansig) $\phi(x)$, logarithmic sigmoid (Logsig) $\psi(x)$ and linear $\chi(x)$ transfer functions are described as follows [41]:

$$\phi(x) = \frac{2}{1+e^{-2x}} - 1 \tag{2}$$

$$\psi(x) = \frac{1}{1+e^{-x}} \tag{3}$$

$$\chi(x) = linear(x) \tag{4}$$

Through the comparison, an error value is obtained. If it exceeds the permissible error value, the obtained answer is returned to the network for corrections in the weights and biases values until the desired answer is obtained. The network described above consists of one layer of neurons; however, in practice more neurons are needed to interact in parallel constructions to implement a neural network. Neural networks compose of several layers containing the neurons.

3.1. Training of ANN

The ANNs are trained with a training set of input and known output data. There is no exact available formula to decide what architecture of ANN and which training algorithm will solve a given problem and the best solution is obtained by trial and error. In the present study, the feed forward error back propagation (BP) algorithm is used for ANN training, which is the most common and efficient algorithm for training. The BP algorithm defines a systematic way to update the synaptic weights of multi-layer feed forward supervised networks. The networks composed of an input layer that receives the input values, an output layer, which calculates the neural network output, and one or more intermediary layers, so-called hidden layers. The basis of BP supervised learning process, is the gradient descent method that usually minimizes the sum of squared errors between the target value and the output of the neural network [42].

3.2. Implementation of ANN

In the implementation of ANN, there are many parameters that changing them causes alteration of the operation method, speed and accuracy of the network [42]. The number of ANN layers, the number of each layer neurons and rate of network training are some of these parameters [43]. Number of ANN layers and neurons of each layer are important parameters in deployment of ANN, since they are so effective in its operation. The significant point in using ANN is the number of layers and neurons, which are processing units of network. Increasing these parameters does not necessarily lead to speed and accuracy improved. The more complex the structure of the network is, the more time it takes. Although training time may be reduced by increasing the rate of training, if it is lower than a specific limit, speed of the network to find desirable answer will be reduced or if it is higher than a specific limit the network will become unstable. Furthermore, a considerable point in the training of the network is scattering rate of input and output parameters. If scattering of parameters is high, the network will be in trouble. In such cases by employing reasonable changes in the data, scattering is decreased and problem is resolved. In this research all values of each parameter are divided to maximum value of that parameter and normalized [44]. This action decreases scattering rate and distribution of all values between 0 and 1.

In this paper, welding speed, rotational speed of tool, axial force, shoulder diameter, pin diameter and tool hardness regarded as inputs of ANN. Yield strength, tensile strength, notch-tensile strength and hardness of welding zone considered as outputs of neural networks. Figure 4 represents an example of the conceptual structure for this ANN: 4 layers with full interconnection. Six input parameters are logged into input layer to determine four outputs.

3.3. Performance evaluation of ANN

The performance of the ANN models in predicting the FSW effective parameters are statistically evaluated using prediction score metrics. A large number of statistical criteria are available to compare the adequacy of any given model. The performance evaluation statistics used for ANN training in the present study are Pearson coefficient of correlation (PCC) and mean relative error (MRE). These parameters have been determined using the following equations:

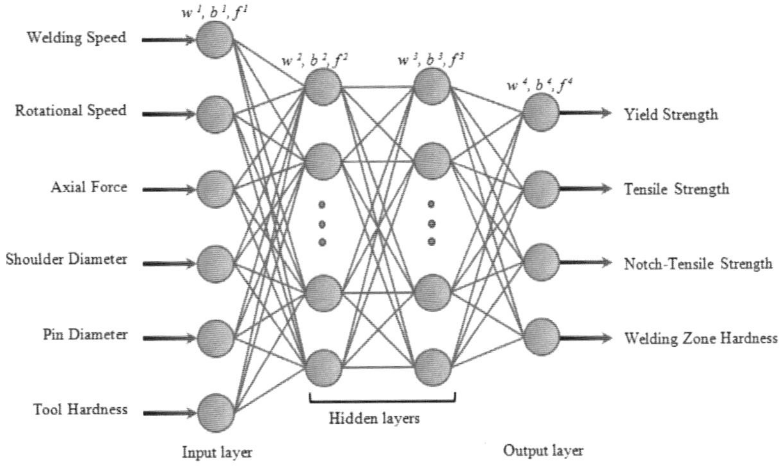

Figure 4. An example of the conceptual structure for a network.

$$PCC = \frac{\sum_{i=1}^{n}\left(f_{EXP,i} - F_{EXP}\right)\left(f_{ANN,i} - F_{ANN}\right)}{\sqrt{\sum_{i=1}^{n}\left(\left(f_{EXP,i} - F_{EXP}\right)^2\left(f_{ANN,i} - F_{ANN}\right)^2\right)}} \tag{5}$$

$$MRE = \frac{1}{n}\sum_{i=1}^{n}\frac{\left|f_{ANN,i} - f_{EXP,i}\right|}{f_{EXP,i}} \times 100 \tag{6}$$

Where:

$$F_{EXP} = \frac{1}{n}\sum_{i=1}^{n} f_{EXP,i}\,,\; F_{ANN} = \frac{1}{n}\sum_{i=1}^{n} f_{ANN,i}\,,\; f_{EXP} = Experimental\,,\; f_{ANN} = \mathrm{Pr}\,edcited$$

3.4. Generating model function

After the neural network is trained successfully with four layers as it is mentioned above, the values of the four parameters of the network (*p*, *b*, *w*, and *f*) can be obtained. The function which correlates the inputs to the corresponding outputs can be calculated applying the aforementioned parameters. Finally, the model function can be determined as below:

$$\begin{aligned} a^1 &= f^1(w^1 p + b^1) \\ a^2 &= f^2(w^2 p^1 + b^2) \\ a^3 &= f^3(w^3 p^2 + b^3) \\ a^4 &= f^4(w^4 p^3 + b^4) \end{aligned} \tag{7}$$

$$G\,(g(1),\, g(2),\, g(3),\, g(4)) = a^4 = f^4(w^4 f^3\,(w^3 f^2(w^2 f^1(w^1 p + b^1) + b^2) + b^3) + b^4) \tag{8}$$

Where a^1, a^2 and a^3 are outputs of the first, second and third layer, respectively. a^4 is the fourth layer output which is equal to the function *G(g(1), g(2), g(3), g(4))*. The function *G* gets the values of six input parameters: welding speed, tool rotational speed, axial force, shoulder diameter, pin diameter and tool hardness. Functions of *g(1)*, *g(2)*, *g(3) and g(4)* represent yield strength, tensile strength, notch-tensile strength and hardness of welding zone as output parameters, respectively. The methodology used for neural network application in this work is as follows:

- Start
- Normalize the data (inputs & outputs)
- Feed the data to artificial neural network
- Find network optimum parameters

- Execute network training
- Obtain Pearson correlation coefficient
- If PCC ≥ 0.99 go to 8, if not go back to 4 with revising the parameters of network
- Continue processing until obtaining desired convergence between experimental and predicted values
- Obtain weights & biases values
- Create the model function
- Conduct analysis based on model function
- Verify the results using experimental values
- Calculate the error for each answer
- End

4. Experimental tests

The experimental data obtained from Rajakumar, et al. [45] work. In their study, rolled plates of 7075-T6 aluminum alloy with the thickness of 5 mm were used. Prior to the welding process, the plates were cut to the size of 150×300 mm. The welding was done perpendicular to the rolling direction. Welding was carried out in a single pass and with non-consumptive tools of several hardnesses. The hardness of tools was determined by quenching them in different materials (water, oil, air, and saltwater) in the final step of heat treatment. The chemical composition of the base metal and its mechanical properties are presented in tables 3 and 4. The joint dimensions and welding and rolling directions are shown in figure 5. Figure 6 shows the tool of FSW process.

Table 3. Chemical composition of 7075-T6 aluminum alloy (*wt %*) [45].

Material	Mg	Mn	Zn	Fe	Cu	Si	Cu	Al
AA 7075-T6	2.1	0.12	5.1	0.35	1.2	0.58	1.2	Balance

Table 4. Mechanical properties of the AA7075-T_6 [45].

Material	Yield Strength (MPa)	Ultimate Strength (MPa)	Elongation (%)	Vickers Hardness (Hv 0.05)
AA 7075-T6	410	485	12	160

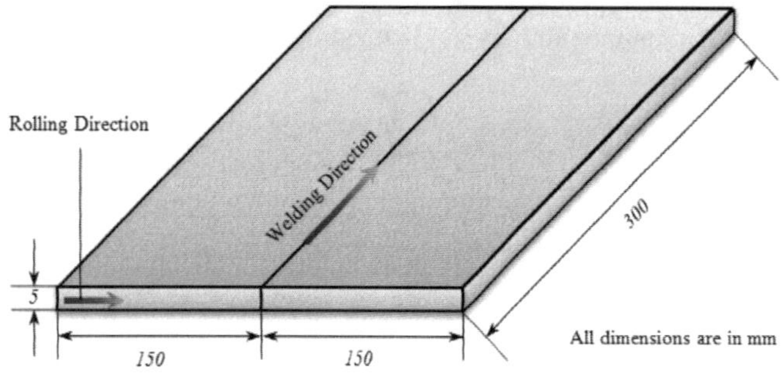

Figure 5. Square butt joint configuration.

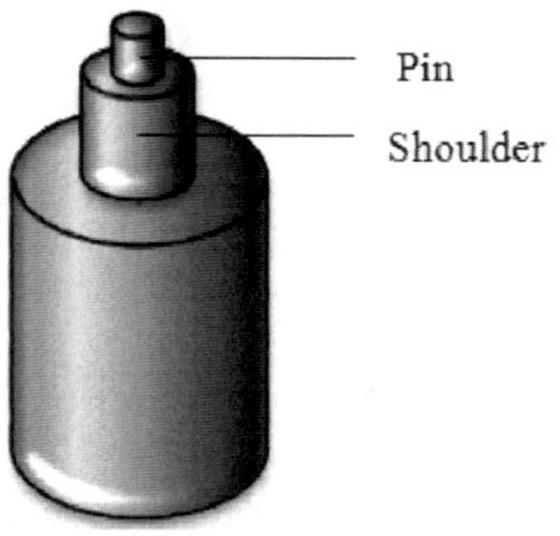

Figure 6. FSW tool.

After the welding process, the joints were cut and then machined to the desired size. In this study, American Society for Testing of Materials (ASTM E8M-04) guidelines was followed for preparing the test specimens. Two different tensile specimens were prepared to evaluate the transverse tensile properties. The smooth (un-notched) tensile specimens were prepared to evaluate yield strength and tensile strength in the cross-sectional area. Notched specimens were prepared to evaluate notch tensile strength of the joints. Figure 7 illustrates samples of these specimens with their sizes; figure 7(a)

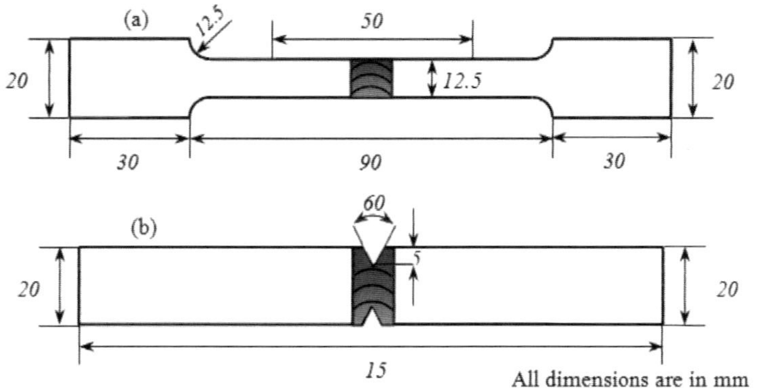

Figure 7. A machined specimens based on the ASTM standard for a) simple-tensile test, b) notch-tensile test.

Table 5. Results of experimental tests on thirty different samples [45].

Experiment No.	Rotational speed *(RPM)*	Welding speed *(mm/min)*	Axial force *(kN)*	Shoulder diameter *mm*	Pin diameter *mm*	Tool hardness *(HRc)*	Yield strength *(MPa)*	Tensile strength *(MPa)*	Notch-tensile strength *(MPa)*	Welding zone hardness *(VHN)*
1	1400	60	8	15	5	45	314	372	397	203
2	1800	60	8	15	5	45	310	334	321	185
3	1400	40	8	15	5	45	279	363	339	194
4	1400	60	8	15	5	45	310	372	393	198
5	1400	80	8	15	5	45	308	371	392	197
6	1400	60	7	15	5	45	282	321	344	180
7	1400	60	8	15	5	45	314	372	393	199
8	1400	60	8	12	5	45	280	310	333	193
9	1400	60	8	15	5	45	310	372	393	198
10	1400	60	8	18	5	45	256	271	296	197
11	1400	60	8	15	4	45	292	331	361	194
12	1400	60	8	15	5	45	310	372	393	198
13	1400	60	8	15	6	45	300	340	360	197
14	1400	60	8	15	5	40	261	283	302	186
15	1400	60	8	15	5	45	313	372	393	198
16	900	60	8	15	5	45	245	270	310	175
17	1200	60	8	15	5	45	290	310	360	191
18	1400	20	8	15	5	45	255	280	320	180
19	1400	100	8	15	5	45	245	261	283	179
20	1400	60	6	15	5	45	263	298	320	173
21	1400	60	10	15	5	45	285	286	356	171
22	1400	60	8	9	5	45	242	266	284	178
23	1400	60	8	21	5	45	296	350	312	187
24	1400	60	8	15	3	45	264	281	302	181
25	1400	60	8	15	7	45	284	321	346	178
26	1400	60	8	15	5	33	271	299	324	178
27	1400	60	8	15	5	56	282	312	336	178
28	1400	60	9	15	5	45	301	361	383	190
29	1400	60	8	15	5	50	310	368	391	192
30	1600	60	8	15	5	45	314	367	397	202

shows the smooth tensile specimen and figure 7(b) illustrates the notched specimen. Tensile tests were carried out with a 100 kN force and speed of 0.5 mm/min. Vickers micro-hardness tester was used for measuring the hardness of the weld metal with a 0.05 kg load. Table 5 presents the results obtained from the experimental tests on 30 specimens.

After the experimental tests, it was observed that increasing the tool rotational speed from 900 RPM to 1400 RPM improves the mechanical properties of the joint. Value of this parameter faces relative decline over 1400 RPM. Moreover, increasing the welding speed from 20 to 60 millimeters per minute increases the strength of specimens, and any further rise of this parameter over 60 millimeters per minute reduces specimen strength. This situation observed for all of the examined welding parameters. Optimum values for parameters of rotational speed, welding speed, axial force, tool hardness, shoulder diameter, and pin diameter were obtained as 1400 (*RPM*), 60 (*mm/min*), 8 (*kN*), 45 (*HRc*), 15 (*mm*) and 5 (*mm*), respectively.

5. Results and discussion

In this work there are two main networks: called "training network" and "testing network". The first one is used to generate the model function. The second one used the model function to give the results of ANN. In the training network, 15 experimental test results (data of experiments 16-30) are used from the total of 30, as data sets to networks training, while in the testing network 15 different experimental data (data of experiments 1-15) which are not used during training, are used as networks testing. So the whole experimental results did not include in the training sets for modeling of FSW process on Al7075. All the input and output data for training and testing networks are normalized between 0 and 1. Table 6 shows the sample data used for networks training. For investigative purposes, 50% training data sets against 50% test data sets is obtained.

Table 6. Normalized samples data used for network training.

Sample No.	Rotational speed	Welding speed	Axial force	Shoulder diameter	Pin diameter	Tool hardness	Yield strength	Tensile strength	Notch-tensile strength	Welding zone hardness
1	0.50	0.60	0.80	0.71	0.71	0.80	0.7803	0.7258	0.7809	0.8621
2	0.66	0.60	0.80	0.71	0.71	0.80	0.9236	0.8333	0.9068	0.9409
3	0.77	0.20	0.80	0.71	0.71	0.80	0.8121	0.7527	0.8060	0.8867
4	0.77	1.00	0.80	0.71	0.71	0.80	0.7803	0.7016	0.7128	0.8818
5	0.77	0.60	0.60	0.71	0.71	0.80	0.8376	0.8011	0.8060	0.8522
6	0.77	0.60	1.00	0.71	0.71	0.80	0.9076	0.7688	0.8967	0.8424
7	0.77	0.60	0.80	0.42	0.71	0.80	0.7707	0.7151	0.7154	0.8768
8	0.77	0.60	0.80	1.00	0.71	0.80	0.9427	0.9409	0.7859	0.9212
9	0.77	0.60	0.80	0.71	0.42	0.80	0.8408	0.7554	0.7607	0.8916
10	0.77	0.60	0.80	0.71	1.00	0.80	0.9045	0.8629	0.8715	0.8768
11	0.77	0.60	0.80	0.71	0.71	0.58	0.8631	0.8038	0.8161	0.8768
12	0.77	0.60	0.80	0.71	0.71	1.00	0.8981	0.8387	0.8463	0.8768
13	0.77	0.60	0.90	0.71	0.71	0.80	0.9586	0.9704	0.9647	0.9360
14	0.77	0.60	0.80	0.71	0.71	0.89	0.9873	0.9892	0.9849	0.9458
15	0.88	0.60	0.80	0.71	0.71	0.80	1.0000	0.9866	1.0000	0.9951

To increase the accuracy of the ANNs results and the network performance, separate different networks are trained for output parameters. The input parameters in all of the networks training are the same. All of the considered input parameters, including: welding speed, the rotational speed of the tool, axial force, shoulder diameter, pin diameter and the tool hardness are used in each modeling for different output parameters.

As it is mentioned, several networks with different architectures have been trained to find the optimum structure (OS) of ANN. OS is used to predict the considered parameters with the least mean relative error possible. Results of the networks have been investigated. Table 7 demonstrates the relevant information of network structure and some related results for training networks for each considered output parameters.

Table 7. Related information of three different training networks for each considered output parameters.

Output parameter	Modeling No.	Rate of Training	Layers Structure	Hidden Transfer Function	Output Transfer Function	PCC	MRE (%)
Yield strength	**1**	**0.175**	**6×6×10×4**	**Logsig**	**Logsig**	**0.99899**	**0.9315**
	2	0.170	6×6×4×4	Tansig	Linear	0.99811	1.1261
	3	0.137	6×6×8×4	Logsig	Tansig	0.99835	0.9867
Tensile strength	**1**	**0.156**	**6×8×10×4**	**Logsig**	**Logsig**	**0.99976**	**0.9746**
	2	0.148	6×8×8×4	Logsig	Logsig	0.99913	1.0012
	3	0.124	6×6×8×4	Logsig	Tansig	0.99922	1.1553
Notch-tensile strength	**1**	**0.160**	**6×8×12×4**	**Logsig**	**Logsig**	**0.99837**	**1.2127**
	2	0.165	6×6×8×4	Tansig	Linear	0.98891	1.6696
	3	0.150	6×6×10×4	Tansig	Logsig	0.99018	1.5134
Welding Zone Hardness	**1**	**0.112**	**6×6×12×4**	**Logsig**	**Logsig**	**0.99984**	**0.7397**
	2	0.098	6×8×10×4	Logsig	Tansig	0.99979	0.7766
	3	0.120	6×6×8×4	Logsig	Linear	0.98764	0.9891

To find the ANNs optimum response, after implementation of multiple networks, some of them are noted in table 7, and investigation of them, the structures shown in table 8 are selected and used for simulation and obtaining the results.

Table 8. Information of selected networks.

Output parameter	Rate of Training	Layers Structure	Hidden Transfer Function	Output Transfer Function	Training PCC	Training MRE (%)
Yield strength	0.175	6x6×10×4	Logsig	Logsig	0.99899	0.9315
Tensile strength	0.156	6x8×10×4	Logsig	Logsig	0.99976	0.9746
Notch-tensile strength	0.160	6x8×12×4	Logsig	Logsig	0.99837	1.2127
Welding Zone Hardness	0.112	6x6×12×4	Logsig	Logsig	0.99964	0.7398

Results of table 8 show that the values of PCC and MRE are both acceptable for each output parameters, so it is concluded that networks are trained finely.

The selected networks employed to networks testing. Figure 8 demonstrates the predicted values of each output parameters by means of the obtained model function in comparison to the experimental values.

Percentage of relative error values for each fifteen testing samples, at the whole considered output parameters are shown in figure 9. Table 9 shows the achieved values of PCC and the mean relative error for testing samples by use of the selected training networks and their related model function.

As it can be seen the obtained values of PCC from testing in comparison to training, decreased; and the achieved mean relative error values from testing in comparison to training, increased. The PCC values are more than 99.9% for the hardness of welding zone and more than 99.4%, 99.5% and 99.8% for notch-tensile strength, yield strength, and tensile strength, respectively. In addition, the obtained

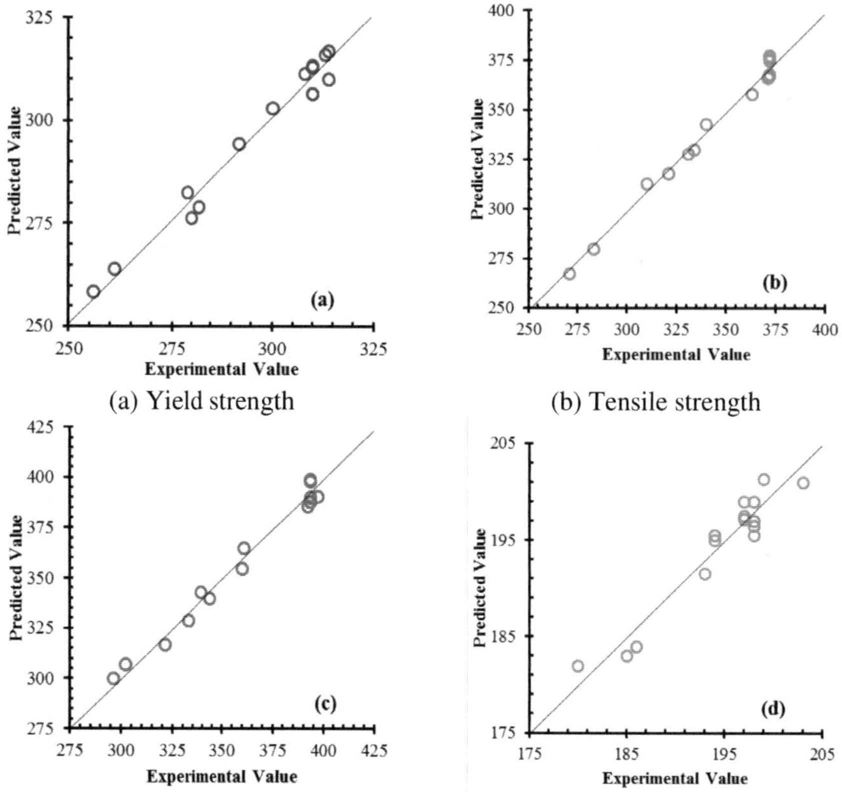

(a) Yield strength (b) Tensile strength

(c) Notch-tensile strength (d) welding zone hardness

Figure 8. Predicted values in comparison to experimental values for each fifteen testing samples for different considered output parameters.

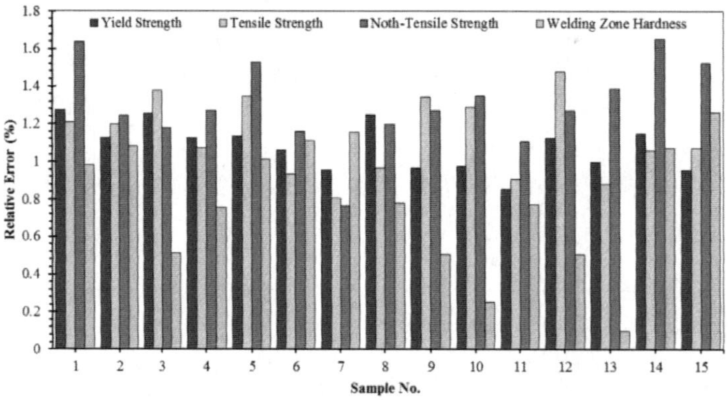

Figure 9. Obtained relative error values for each fifteen experiments of testing networks.

Table 9. Achieved values of PCC and MRE for testing networks.

Output parameter	Testing PCC	Testing MRE (%)
Yield strength	0.99543	1.0820
Tensile strength	0.99875	1.1303
Notch-tensile strength	0.99403	1.3046
Welding Zone Hardness	0.99918	0.7917

mean relative errors are in small range (0.7917-1.3046), and the predicted values for the hardness of welding zone, yield strength, tensile strength and notch-tensile strength have the least mean relative errors, respectively. According to the obtained results, it is observed that, predicted values form the response of ANN and the experimental values are in admirable agreement. So it can be concluded that the ANNs are tuned finely to predict the FSW effective parameters, and the ANNs can be used in prediction and optimization of this process parameters.

6. Conclusion

In the present study, the artificial neural networks were employed to predict the friction stir welding parameters on 7075-T6 Aluminum Alloy. The rotational speed of the tool, welding speed, axial force, shoulder diameter, pin diameter and tool hardness were regarded as inputs of the ANN. Four effective parameters of FSW process including: yield strength, tensile strength, notch-tensile strength and hardness of welding zone modeled. The obtained values of MRE are 1.0820, 1.1303, 1.3046 and 0.7917 for each parameter, respectively. Also, the values of PCC for all of the parameters are more than 99%. According to the achieved results, it can be concluded when the artificial neural networks are tuned finely, the modeling results are in admissible agreement with the experimental results. Therefore, using ANNs instead of experiments, decreases costs and the need for special testing facilities; and the ANNs can be employed to optimize and predict the effective parameters for FSW process.

References

[1] El-Shennawy M, Omar A and Masoud M 2005 Effect of Cu and Mg contents on similar and dissimilar welding of 7XXX series aluminum alloys *AEJ-Alexandria Eng. J.* **44** 715-29

[2] Balasubramanian V, Ravishankar V and Reddy G M 2008 Effect of post weld aging treatment on fatigue behavior of pulsed current welded AA7075 aluminum alloy joints *J. Mater. Eng. Perform.* **17** 224-33

[3] Jha A K and Sreekumar K 2008 Metallurgical studies on cracked Al–5.5 Zn–2.5 Mg–1.5 Cu aluminum alloy injector disc of turbine rotor *J. Fail. Anal. Prev.* **8** 327-32

[4] Nayan N, Murty S N, Mittal M and Sinha P 2009 Optimization of homogenizing mode for aluminum alloy AA7075 using calorimetric and microstructural studies *Met. Sci. Heat Treat.* **51** 330-7

[5] Ghosh R, Venugopal A, Sankaravelayudham P, Panda R, Sharma S, George K M and Raja V 2015 Effect of thermomechanical treatment on the environmentally induced cracking behavior of AA7075 alloy *J. Mater. Eng. Perform.* **24** 1-11

[6] Wang L, Daniewicz S, Horstemeyer M, Sintay S and Rollett A 2009 Three-dimensional finite element analysis using crystal plasticity for a parameter study of microstructurally small fatigue crack growth in a AA7075 aluminum alloy *Int. J. Fatigue* **31** 651-8

[7] Paturi U M R, Narala S K R and Pundir R S 2014 Constitutive flow stress formulation, model validation and FE cutting simulation for AA7075-T6 aluminum alloy *Mater. Sci. Eng. A* **605** 176-85

[8] Majzoobi G, Azadikhah K and Nemati J 2009 The effects of deep rolling and shot peening on fretting fatigue resistance of Aluminum-7075-T6 *Mater. Sci. Eng. A* **516** 235-47

[9] Majzoobi G, Nemati J, Rooz A N and Farrahi G 2006 Modification of fretting fatigue behavior of AL7075-T6 alloy by application of titanium coating and shot peening *AMPT2006 Conference* (Submitted)

[10] Majzoobi G, Sori M, Farrahi G and Hojjati R Talemi 2010 The effect of temperature on fretting fatigue behavior of Al7075-T6 *6th International Symposium on Fretting Fatigue* p 31

[11] Gupta R K, Ramkumar P and Ghosh B R 2006 Investigation of internal cracks in aluminum alloy AA0704 forging *Eng. Fail. Anal.* **13** 1-8

[12] Kou S 1987 *Welding Metallurgy* (Cambridge University Press, Cambridge)

[13] Cross C E, Olson D L and Liu S 2003 *Aluminium Welding, Handbook of Aluminum* (Dekker,

New York) vol 1

[14] Goloborodko A, Ito T, Yun X, Motohashi Y and Itoh G 2004 Friction stir welding of a commercial 7075-T6 aluminum alloy: Grain refinement, thermal stability and tensile properties *Mater. Trans.* **45** 2503-8

[15] Rafi H K, Ram G J, Phanikumar G and Rao K P 2010 Microstructure and tensile properties of friction welded aluminum alloy AA7075-T6 *Mater. Des.* **31** 2375-80

[16] Lotfi A H and Nourouzi S 2014 Effect of welding parameters on microstructure, thermal, and mechanical properties of friction-stir welded joints of AA7075-T6 aluminum alloy *Metall. Mater. Trans. A* **45** 2792-807

[17] Venugopal T, Rao K S and Rao K P 2004 Studies on friction stir welded AA 7075 aluminum alloy *Trans. Indian Inst. Met* **57** 659-63

[18] Givi M and Asadi P 2014 *Advances in Friction Stir Welding and Processing* (Elsevier, Amesterdam)

[19] Mishra R S, De P S and Kumar N 2014 *FSW of Aluminum Alloys in Friction Stir Welding and Processing* (Springer, Switzerland) pp 109-48

[20] Azimzadegan T and Serajzadeh S 2010 An investigation into microstructures and mechanical properties of AA7075-T6 during friction stir welding at relatively high rotational speeds *J. Mater. Eng. Perform.* **19** 1256-63

[21] Rajakumar S, Muralidharan C and Balasubramanian V 2011 Influence of friction stir welding process and tool parameters on strength properties of AA7075-T 6 aluminum alloy joints *Mater. Des.* **32** 535-49

[22] DebRoy T, De A, Bhadeshia H, Manvatkar V and Arora A 2012 Tool durability maps for friction stir welding of an aluminium alloy *Proceedings of the Royal Society A: Mathematical, Physical and Engineering Science* **468** 3552-70

[23] Bahrami M, Nikoo M F and Givi M K B 2015 Microstructural and mechanical behaviors of nano-SiC-reinforced AA7075-O FSW joints prepared through two passes *Mater. Sci. Eng. A* **626** 220-8

[24] Buffa G, Fratini L and Shivpuri R 2008 Finite element studies on friction stir welding processes of tailored blanks *Comput. Struct.* **86** 181-9

[25] Buffa G, Hua J, Shivpuri R and Fratini L 2006 A continuum-based fem model for friction stir welding—model development *Mater. Sci. Eng. A* **419** 389-96

[26] Fratini L, Buffa G and Shivpuri R 2010 Mechanical and metallurgical effects of in process cooling during friction stir welding of AA7075-T6 butt joints *Acta Mater.* **58** 2056-67

[27] He X, Gu F and Ball A 2014 A review of numerical analysis of friction stir welding *Prog. Mater Sci.* **65** 1-66

[28] Roshan S B, Jooibari M B, Teimouri R, Asgharzadeh-Ahmadi G, Falahati-Naghibi M and Sohrabpoor H 2013 Optimization of friction stir welding process of AA7075 aluminum alloy to achieve desirable mechanical properties using ANFIS models and simulated annealing algorithm *Int. J. Adv. Manuf. Technol.* **69** 1803-18

[29] Okuyucu H, Kurt A and Arcaklioglu E 2007 Artificial neural network application to the friction stir welding of aluminum plates *Mater. Des.* **28** 78-84

[30] Shojaeefard M H, Abdi R, Akbari M, Besharati M K and Farahani F 2013 Modeling and Pareto optimization of mechanical properties of friction stir welded AA0704/AA4703 butt joints using neural network and particle swarm optimization algorithm *J Mater Design* **44** 190-8

[31] Yousif Y, Daws K and Kazem B 2008 Prediction of friction stir Welding characteristic using neural network *Jordan J. Mech. Ind. Eng.* **2** 151-5

[32] Lakshminarayanan A and Balasubramanian V 2009 Comparison of RSM with ANN in predicting tensile strength of friction stir welded AA7039 aluminium alloy joints *Transactions of Nonferrous Metals Society of China* **19** 9-18

[33] Ghetiya N and Patel K 2014 Prediction of tensile strength in friction stir welded aluminium alloy using artificial neural network *Procedia Technol.* **14** 274-81

[34] Thomas W 1991 Friction stir butt welding (International Patent Application No. PCT/GB92/0220)

[35] Nandan R, DebRoy T and Bhadeshia H 2008 Recent advances in friction-stir welding–process, weldment structure and properties *Prog. Mater Sci.* **53** 980-1023

[36] Bussu G and Irving P 2003 The role of residual stress and heat affected zone properties on fatigue crack propagation in friction stir welded 2024-T351 aluminium joints *Int. J. Fatigue* **25** 77-88

[37] Jata K, Sankaran K and Ruschau J 2000 Friction-stir welding effects on microstructure and fatigue of aluminum alloy 7050-T7451 *Metall. Mater. Trans. A* **31** 2181-92

[38] Threadgill P, Leonard A, Shercliff H and Withersd P 2009 Friction stir welding of aluminium alloys *Int. Mater. Rev.* **54** 49-93

[39] Hall E O 1951 *Proc. Phys. Soc. Lond. B* **64** 747-53

[40] Chen S C, Lin S W, Tseng T Y and Lin H C 2006 Optimization of back-propagation network using simulated annealing approach *IEEE International Conference on Systems, Man and Cybernetics* (IEEE, Taiwan) pp 2819-24

[41] Vogl T P, Mangis J K, Rigler A K, Zink W T and Alkon D L 1988 Accelerating the convergence of the back propagation method *Biol. Cybern.* **59** 257-63

[42] Niyati M and Moghadam A M E 2009 Estimation of products final price using bayesian analysis generalized poisson model and artificial neural networks *J. Ind. Eng.* **2** 55-60

[43] Maleki E and Sherafatnia K 2016 Investigation of single and dual step shot peening effects on mechanical and metallurgical properties of 18CrNiMo7-6 steel using artificial neural network *Int. J. Mater. Mech. and Manufac.* **4** 100-5

[44] Mendelsohn L 1994 MATLAB neural network toolbox *Technical Analysis of Stocks & Commodities* pp 12

[45] Rajakumar S, Muralidharan C and Balasubramanian V 2011 Influence of friction stir welding process and tool parameters on strength properties of AA7075-T6 aluminum alloy joints *J Mater Design* **32** 535-49

IOP Conf. Series: Materials Science and Engineering **103** (2015) 012035 doi:10.1088/1757-899X/103/1/012035

Study on springback of high tension steel following various temperature and local pre-heating in the roll forming process

D H Kim, D Won Jung[1], D H Yoon and Y Zhang

[1] Department of Mechanical Engineering, Jeju National University, 1 Ara 1-dong, Jeju-si, Jeju, 690-756, Republic of Korea

E-mail: jdwcheju@jejunu.ac.kr

Abstract. In this study, we studied spring-back phenomenon following pre-heating and various temperatures on forming sheet in roll forming process using finite element method (FEM) simulations. Material of forming sheet was high tension steel (POSTEN 60), and we analyzed spring-back phenomenon using material properties from room temperature to 500℃. For spring-back phenomenon analysis, FEM simulation results were thermal stress and strain according to temperature distribution on forming sheet. Assumed pre-heating methods were induction heating and laser heating, also this study assumed that pre-heating is same temperature conditions. Finally, the spring-back phenomenon was analyzed by induction pre-heating and laser pre-heating method.

1. Introduction

Roll forming process consists of several pairs of roll die and it is the progressive bending forming of metal sheet for desired cross section. Also, the process is a continuous process for mass production and mold change and setup process time takes less time. This process is well used for its effectiveness in the mass production of long sheet metal products with desired cross section. Another advantage of this process is a material cost savings due to the high material recovery [1]. For this reason, the application range of the roll forming process is wide, it has recently been applied to the automotive industry.

The automotive industry is tendered to replace the high tensile steel for lightweight vehicles. Because high tensile steel is due to have a greater rigidity and lower weight than other materials in the same thickness. However, high tensile steel has a low elongation and a high yield strength, also press working process method is difficult to forming. But forming of high tensile steel in the roll forming process has a high precision of product and abrasion resistance of roll die due to incremental forming. Roll forming process occurs a lot of spring-back phenomenon than other materials because of the low forming force of high strength steel. In order to solve this problem, many engineers did over-bending to predict empirically spring back phenomenon. Until now, spring back prediction focus on design process such as over-bending, but this method has limitations when using high tensile steel. So, to solve this problem, this paper was applied to the pre-heating process in the roll forming process. In the roll forming process, there are many cases using heat such as a tube welding process, but research to pre-heat the sheet with improved formability has not been investigated to date. Many roll forming

[1] Address for correspondence: D Won Jung, Department of Mechanical Engineering, Jeju National University, 1 Ara 1-dong, Jeju-si, Jeju, 690-756, Republic of Korea. E-mail: jdwcheju@jejunu.ac.kr.

process studies have been carried out relative to the other sheet material, roll forming process study using high strength steel is not much [2-6].

In research methods suggested above, similar study is V or U shape bending using stamping process. Sheet bending studies in stamping was done much. Xie H Y, et al. [8] studied through electrical pulse-assisted (EPA) V-bending at 373 K. It was applied to the electrical (pulse) energy to the sheets, and a study of the spring back phenomenon of the V-bending in the stamping process. Fang Z, et al. [9] were studied the spring back effect that occurs after the micro V-bending a classic processing method to manufacture microparts. FEM simulation has been conducted to investigate the size effect in terms of Voronoi tessellation and spring back. Roll forming process has many process parameters such as the forming speed, the material of the forming sheet, the rotational speed of the roll die, friction between the roll and the sheet, so it should be considered a number of process parameters. Many studies have been made on the roll forming process as above. However, roll forming process applied the pre-heating process has not yet been investigated, so this process should be performed by the basic research.

This study was carried out V-bending in order to analyze the phenomenon in 3-pass roll forming process [7], this process is assumed to apply induction and laser pre-heating. The study deduced from FEM simulation results about V-banding sheet in accordance with various preheating temperature after forming. Through the FEM simulation results, a spring-back phenomenon of the pre-heating was analyzed and pre-heating process was checked to ensure it can be applied to the roll forming process.

2. Design of roll forming process

In the roll forming process, V, U, hat shapes of cross-section were basic shapes to analysis the spring back phenomenon. This study was processed V-shape through the roll forming process, configurations of roll forming devices for forming of V-bending are shown in figure 1.

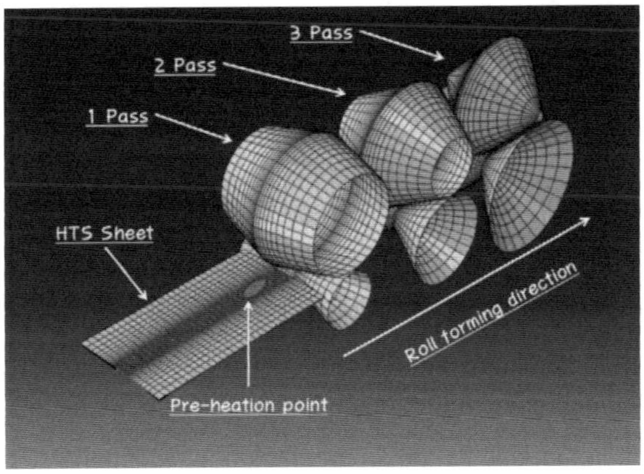

Figure 1. Set-up of the roll forming process in this study.

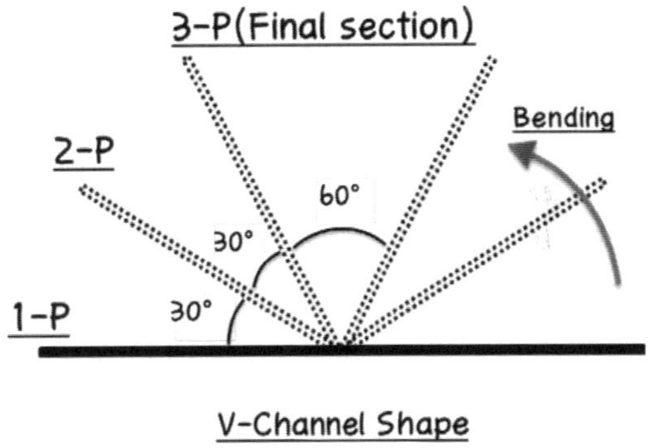

Figure 2. Bending angle of the 3-Pass roll forming.

In order to process the basic V-banding, roll forming machine is designed in 3 passes. Bending angles of 3 passes were 30°, 60° and 90° respectively (figure 2). Horizontal distance to the previous roll stand is 105 mm and total forming distance is 550 mm. Forming speed of forming sheet is 200 mm/s. The diameters of the upper roll die and the lower roll die were 50 mm and 10 mm respectively, the R value of the roll dies is applied to 0.5 mm to prevent excessive deformation. The width and length size of the forming sheet are each 80 mm, 150 mm, and thickness of the forming sheet is 1 mm.

Material for forming sheet using roll forming process is high tension steel and the product name of sheet material is POSTEN 60. The POSTEN60 was studied in the previous studies for mechanical properties and the study analyzed high-temperature tensile properties and the residual stress characteristic [10-12]. The paper was applied to POSTEN60 properties with reference to previous studies, and table 1 shows the properties of POSTEN60.

Table 1. Mechanical properties of POSTEN 60 at room temperature.

Young`s modulus/GPa	215
Yield strength/MPa	532
Ultimate tensile strength/MPa	625
Ductility/%	20
Poisson`s ratio	0.3

3. Finite element model

In this study, roll forming process thought FEM simulation performed using Abaqus version 13.1 simulation software. FEM simulation of Roll forming process carried out a thermal-structure interaction analyses, spring-back analysis performed static analysis using analysis result of forming sheet.

Pre-heating of the overall of the sheet needs a lot of thermal energy, and it causes many problems such as warping and bowing. Therefore, we consider a local pre-heating method for a small energy consumption and precise forming. This study analyzed the spring back effect on the local pre-heating in the forming sheet. It assumed that the friction between the roll die and the forming sheet, the boundary conditions were given in the sheet to move the forming direction. Pre-heating method is considered the induction type and the laser type on the upper surface of forming sheet. Pre-heating shapes of induction type and laser type were referred to previous studies. Pre-heat shape and size is shown in figure 3 [13, 14].

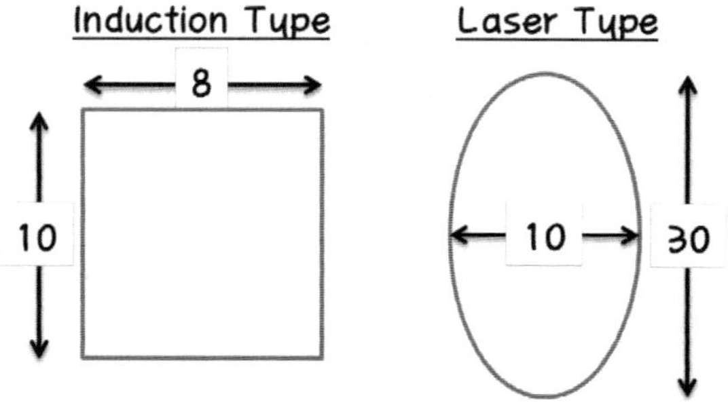

Figure 3. Shape and size of pre-heating sources [Unit: mm].

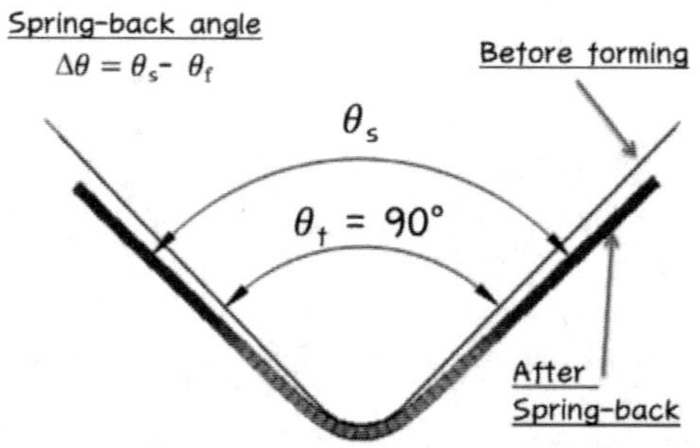

Figure 4. Definition of the spring-back angle.

Pre-heating temperatures were applied at room temperature, 100℃, 200℃, 300℃, 400℃ and 500℃. Pre-heating position during the roll forming process is a red area in figure 1, and this area is pre-heated before the 1-Pass. The Pre-heating area is moving heat source that moved along the center line of the forming sheet. To analyze the accurate temperature distribution in the forming sheet after forming, Film coefficient of forming sheet is 0.28 $W/m^2{\cdot}K$, and ambient temperature is 25℃. Heat loss of the sheet due to heat radiation is considered. The heat transfer coefficient of forming sheet is 0.47 W/mm℃.

Forming sheet was solid model and the mesh was concentrated in the center line of forming sheet for accurate bending behavior. The roll dies were a shell model for fast analysis. The spring back calculation method of the molded sheet is shown in figure 4 and the final spring back value was calculated by the static analysis.

4. Results of the FEM simulation

FEM simulation analysis is not possible continuous process such as the actual roll forming process. So the length of the sheet is effective for the optimization analysis to a minimum length of sheet, for the interaction of the two roll pass, length of the forming sheet was designed to 150 mm. For an efficient analysis of the forming sheet of the roll forming process, analysis results of the formed sheet were measured by dividing the three locations as shown in figure 5.

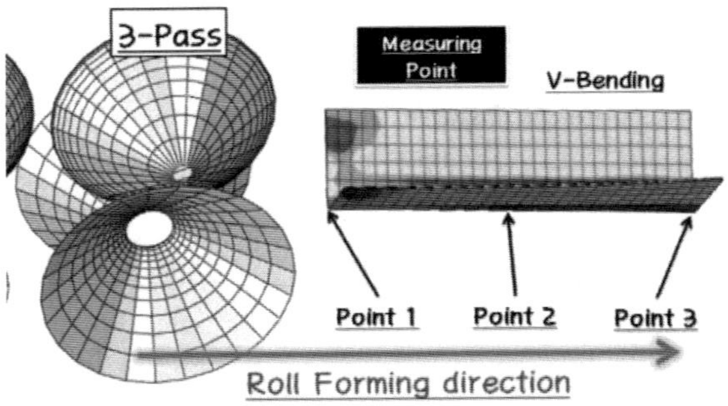

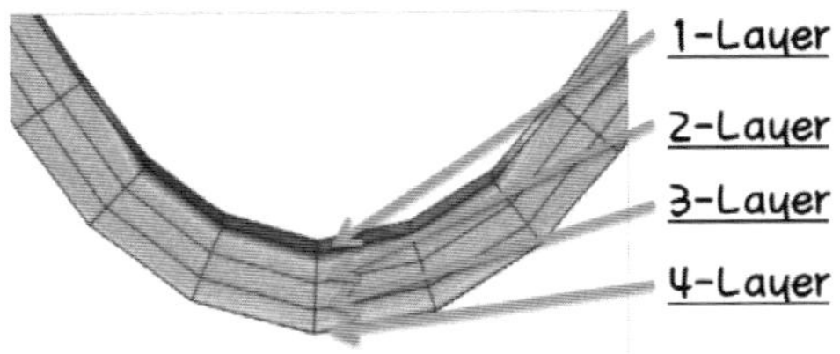

Figure 5. Measuring point on the forming sheet. **Figure 6.** Layer definition of forming sheet.

Because the stress distribution of sheet is different from Point-1 to Point-3, it was selected as the measurement point position such as figure 5 [15].

Forming sheet occurred tension and compression at the same time in bending process. So, to determine the characteristics of the tension and compression, forming sheet was analyzed by dividing the layer in the thickness direction from the center line, as shown in figure 6. And the forming sheet was divided into three layers in the thickness direction.

4.1. Temperature distribution of thickness direction

Figure 7 shows the temperature curves of each layer according to the roll pass. The temperature of the forming sheet exhibits a rapid rise in the preheating, and then it was confirmed that the temperature is gradually reduced after preheating. After the final forming, temperature of pre-heating was reduced to about 300℃. Maximum Temperature dwelling Time is different from the induction type and laser type because of two types is due to differences in the pre-heating area.

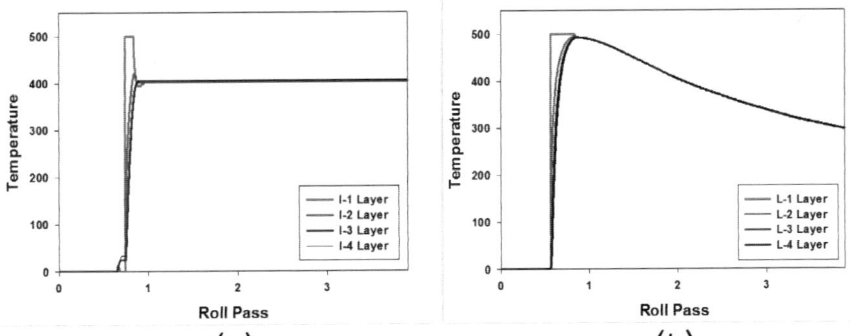

Figure 7. Temperaure distribution of forming sheet(500℃); (a) Induction type, (b) Laser type.

Directly the preheated 1-layer has reached the maximum temperature, and then 1 layer was soon converged with the other layers. Because the thickness of the formed sheet is 1 mm, and the sheet has a high heat transfer coefficient.

Figure 8 shows the temperature curves in the width direction of the forming sheet. Lee, et al. [16] was proposed concept for the cumulative distance, and this study applied the cumulative distance from the width direction of the sheet. The cumulative distance means away length from the centerline of the sheet. A positive value is the right side of the forming plate and a negative value is the left side of the forming plate. Figure 8 was measured the temperature of the sheet at Point-2 after forming. Heat affected zone is from the centerline (cumulative distance: 0) of the forming plate to 10 mm. The bending area and the heat affected zone are almost the same range and the variables influence the spring back value.

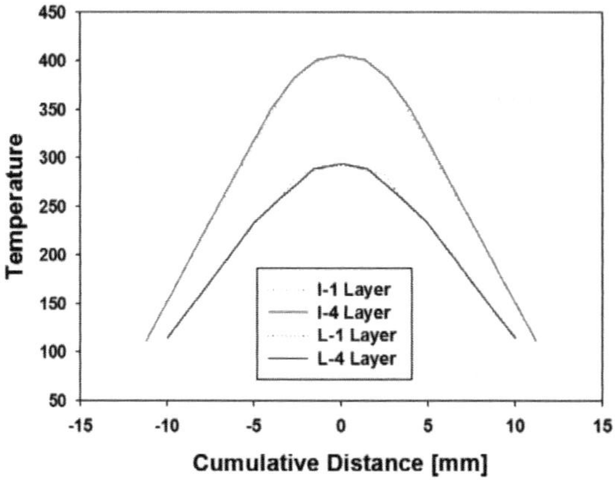

Figure 8. Cumulative distance of forming sheet according to pre-heating (1, 4-layer, Point-2).

4.2. Stress according to measurement point in induction pre-heating type

Figure 9 shows the von-mises stress of 1, 4-layer in each measurement point. The von-mises stress is showing a tendency to increase gradually from Point-1 to Point-3. 1-layer of compression behavior exhibits higher von-mises stress than 4-layer tensile of behavior. The tendency of the graph curves of 1,4-layer is similar, Point-3 shows the von-mises stress difference in the other layer. After final forming, von-mises stress value was derived about 200~300 MPa results of stress at each measurement point.

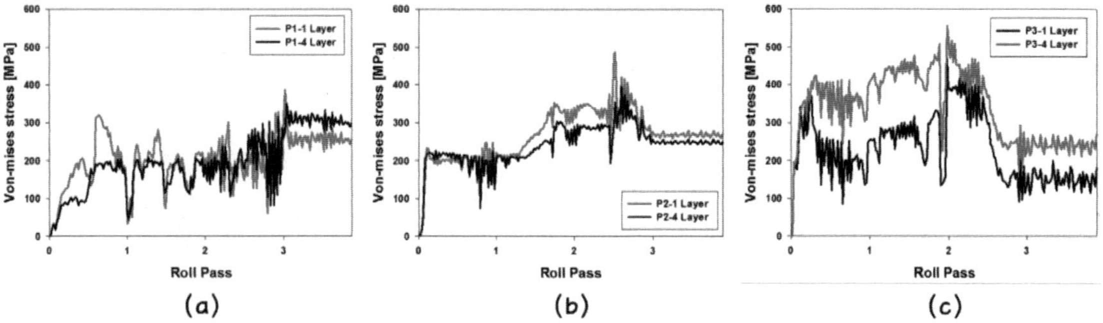

Figure 9. Von mises stress according to 1, 4-layer; (a) Point-1, (b) Point-2, (c) Point-3.

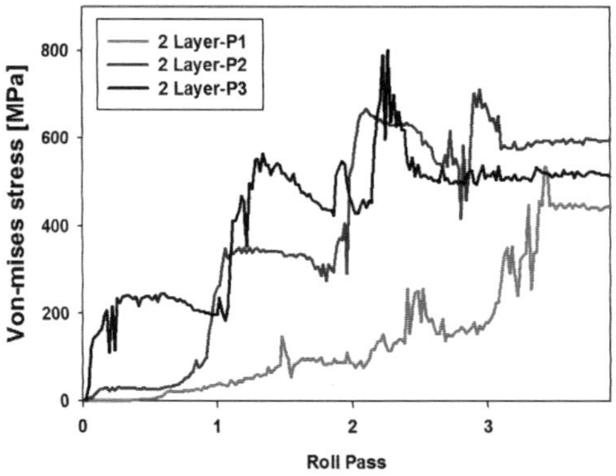

Figure 10. Von mises stress according to measurement point.

Figure 10 is a graph for comparing the von-mise stress of the measuring point in the 2-layer. 3 points were derived the highest stress value, Point-1 was derived the lowest stress value. Also, Compared with the figure 9, 2-layer exhibit a higher stress value than 1,4-layer and the final stress value according to measure point in 2-layer was obtained between 400~600 MPa. This result is approximately two times higher than the stress value of 1,4-layer. The reason for this phenomenon is predicted that the local maximum stress value was obtained due to the slip of the tension and compression in the width direction of the interior of the high tensile steel.

In order to observe the thermal stress by temperature, figure 11(a) and 11(b) shows the results of the thermal stress in 1,4-layer of the Point-2. The above results show a different trend in figure 9(b), and figure 11 is not greater thermal stress value of the difference due to the initial thermal stress in the roll forming process in the early and late. Figure 9(b) is not a stress difference in 1,4-layer, but figure 11 shows a stress difference in 1,4-layer. This result is determined to be higher thermal stress by pre-heating before final forming. 1-layer directly affected in a pre-heating, the final thermal stress was 200~250 MPa and 4-layer is subjected to the indirect heat effect of 1layer, the final thermal stress value was obtained in 380~400 MPa.

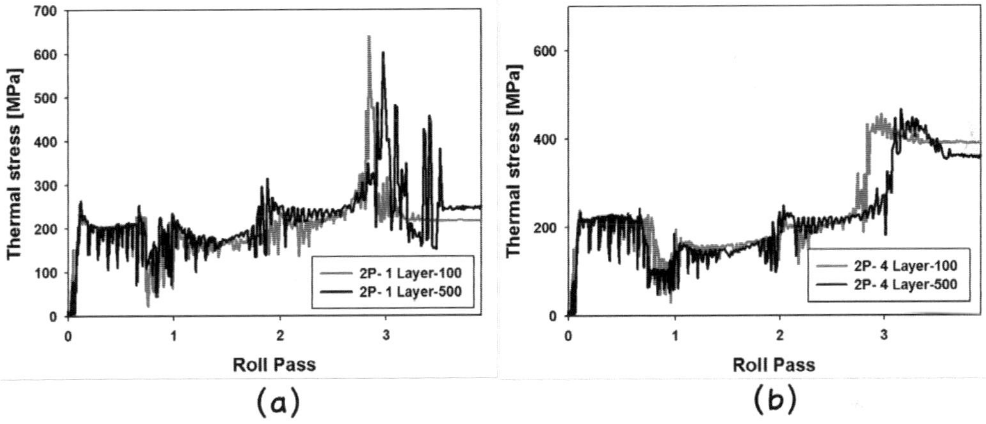

Figure 11. Thermal stress according to sheet layer at Point-2; (a) 1-layer, (b) 4-layer.

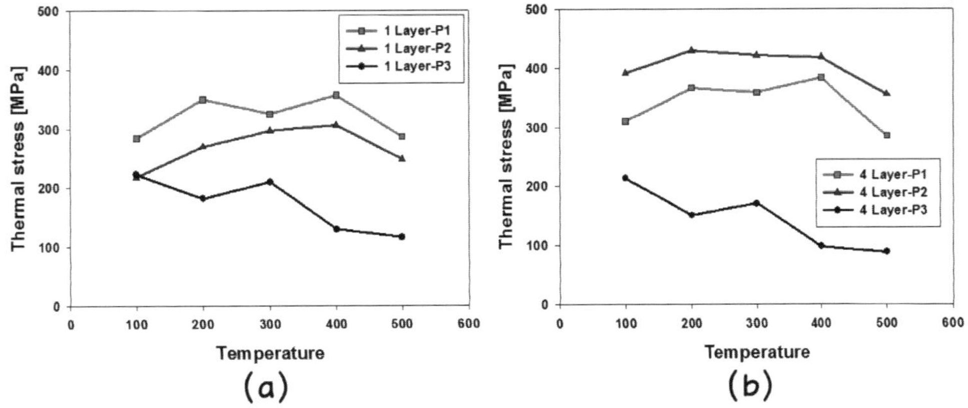

Figure 12. Thermal stress according to measurement point; (a) 1-layer, (b) 4-layer.

Figure 12 shows the thermal stress of 1,4-layer of each measurement point in accordance with various temperatures in the final formed high tensile steel sheet. Point 3 shows a lower thermal stress value than other points. The tendency of the graph in figure 12 is not constant. But if tendency of the graph in figure 12 significantly divided into three parts, it can be classified such as 100℃, 200~400℃, 500℃. The simulation result is larger the influence of the physical properties of a high tension steel in this FEM simulation. This FEM simulation analysis was applied to mechanical properties of POSTEN60 from the previous study [10-12]. Applied POSTEN60 gradually decreases tensile strength from room temperature to 200℃ and then tensile strength increases due to the blue shortness at higher than 200℃. POSTEN60 has the maximum value in 300℃ and then decreased gradually to 600℃. In general steel, blue shortness increases tensile strength and hardness in the vicinity of 250℃ and it increased brittle due to reduced elongation. Blue shortness is shown at about 250℃ in the tensile test and temperature of blue shortness is different depending on the load rate. In a previous study, POSTEN60 is similar to the yield stress of the tensile strength tends to decrease and yield strength and tensile strength of the POSTEN60 study was found to increase rapidly even after 600℃. Also, the value of the elastic modulus is gradually decreased to 400℃ and it has been studied to be rapidly reduced even after 400℃. In this study, for accurate analysis, it was applied to the above properties. Due to this mechanical property of POSTEN60, curves of figure 12 were derived.

4.3. Strain according to measure point in induction pre-heating

Figure 13 shows the equivalent strain of 1, 4-layer for each measurement point. Equivalent strain value is the highest in Point-3. The equivalent strain values of 1,4-layer is almost identical in Point-1 and 2. But the equivalent strain is showing the value of 1,4-layer a big difference in Point-3. The equivalent strain value of 4-layer is 0.233.

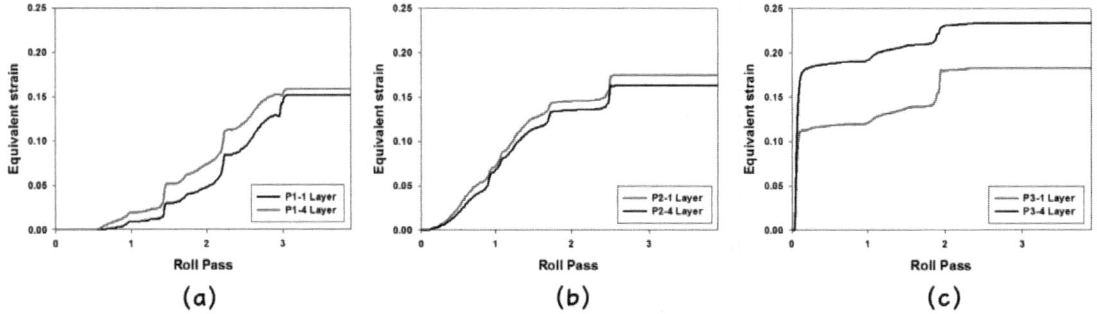

Figure 13. Equipvalent strain according to 1, 4-layer; (a) Point-1, (b) Point-2, (c) Point-3.

Figure 14 shows the equivalent stress value of the three measurement points on the 2-layer. Each measured point has shown a rapid increase deformation each time passing the roll pass and 2-layer was found to be obtained at a high strain value than 1,4-layer. The interior of the forming sheet was confirmed that is a greater effect of bending force than 1,4-layer for the surface of the sheet.

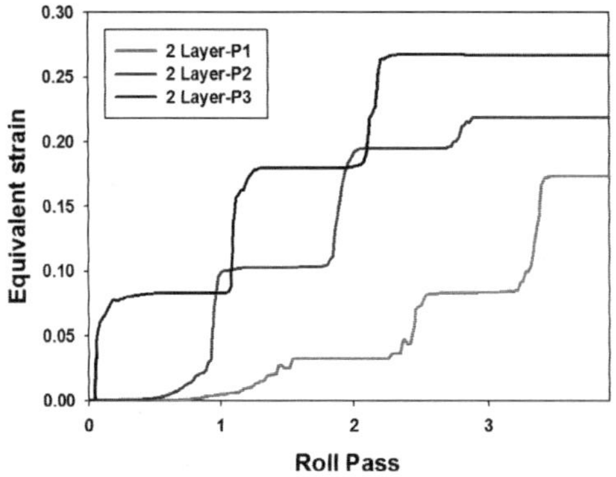

Figure 14. Equipvalent strain of 2-layer according to measurement point.

Figure 15 shows the thermal strain value according to the temperature of pre-heating and thermal strain was compared with respect to 100℃ and 500℃ at Point-2. In final forming, Heat strain was 500℃ higher than 100℃ and heat strain at 500℃ is 0.5 higher than the equivalent strain at room temperature. Heat strain is invisible whenever clear trend over each roll pass, as shown in figure 14. And figure 15 of the curve could be confirmed when only significantly increased over the 3-Pass.

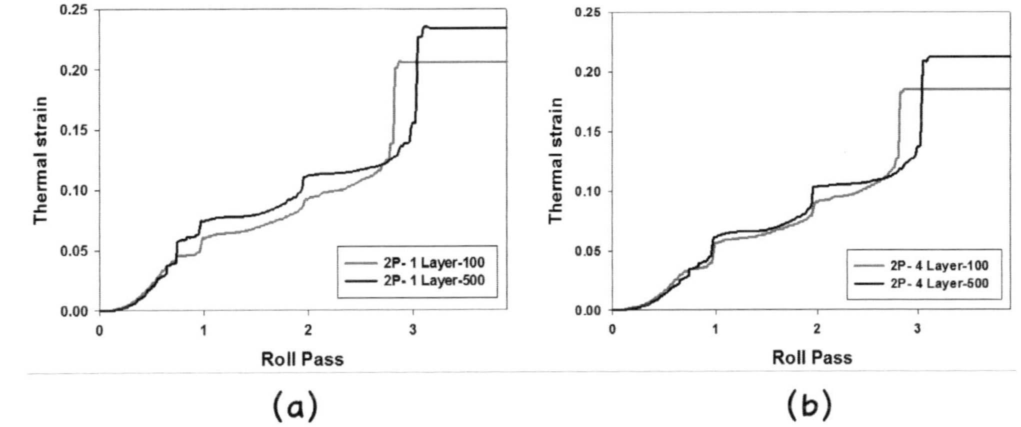

Figure 15. Thermal strain at 2-Point; (a) 1-layer, (b) 4-layer.

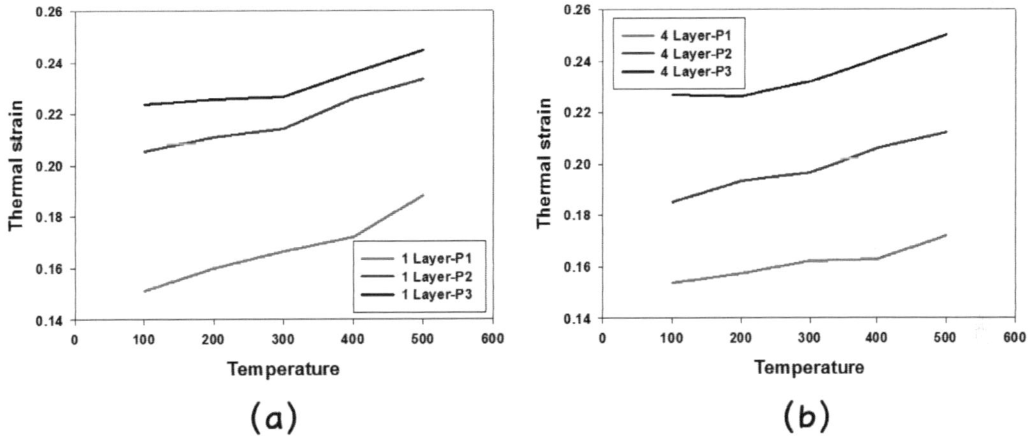

Figure 16. Thermal strain according to measurement point; (a) 1-layer, (b) 4-layer.

Figure 16 is a thermal stress graph of the respective temperature. Thermal strain values were obtained point-3 higher than point-1. Although stress is lower as the temperature increases such as figure 12, the thermal stress was increased. With reference to the simulation results in high temperature, the stress is reduced when the temperature increases but the strain is increasing.

4.4. Compare with induction type and laser type

Figure 17 shows the thermal stress according to pre-heating type. The thermal stress in figure 17(a) is significantly increased when the forming sheet passes over a roll die. The area of the induction type and laser type graph is identified as similar. In addition, the result was similarly derived.

Figure 18 shows the thermal stress according to roll pass. The strain is increased when the forming sheet passes over a roll die. The results of the two types of pre-heating show a difference in 4-layer. As mentioned above, pre-heating has less influence about the stress, but the thermal strain of effects is large. In this study, the main difference of Induction of type and laser type is area of the pre-heating shape. Because of the laser type is approximately 11.8 times larger than the induction type, the simulation results shown in figure 18 was derived.

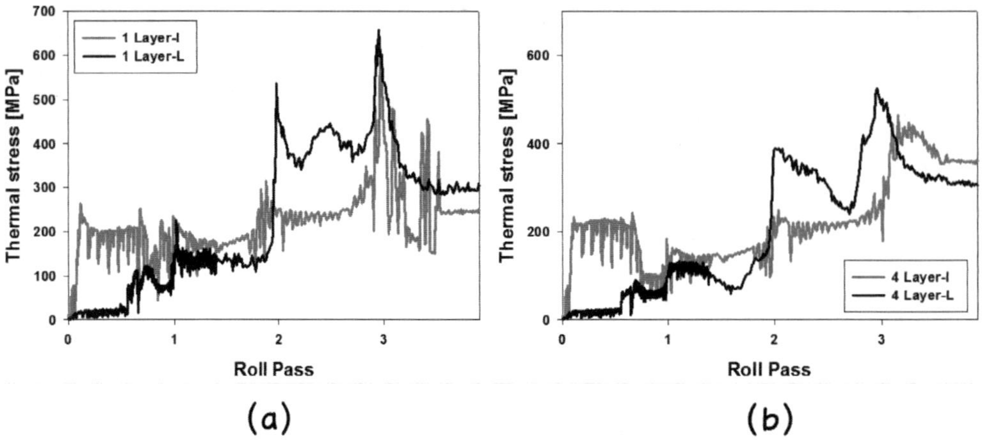

Figure 17. Thermal stress according to type of pre-heating; (a) 1-layer, (b) 4-layer.

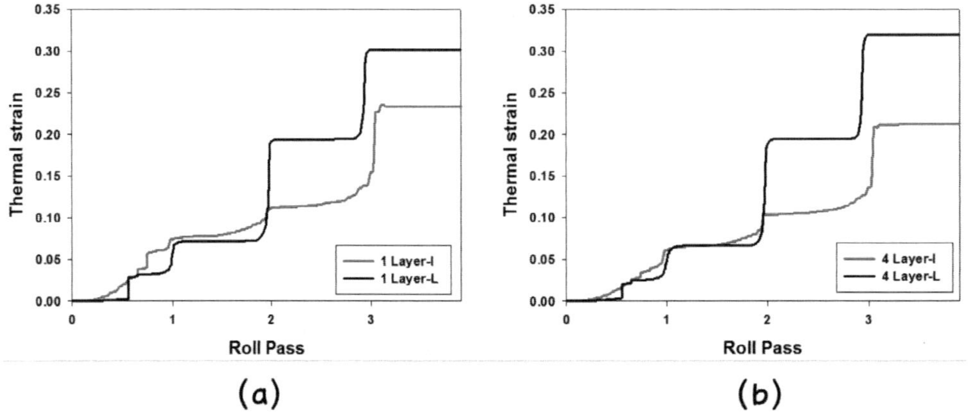

Figure 18. Thermal stress according to type of pre-heating; (a) 1-layer, (b) 4-layer.

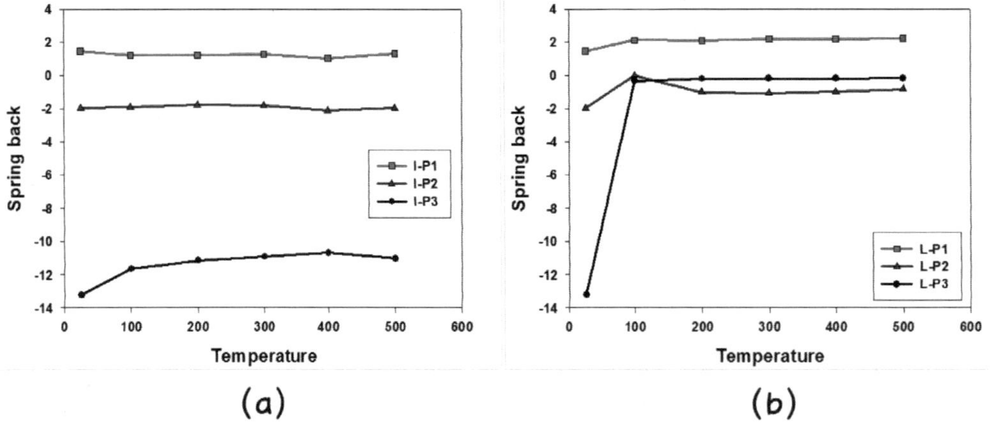

Figure 19. Spring back value according to measurement point; (a) induction pre-heating type, (b) Laser pre-heating type.

4.5. Result of spring back

Figure 19 shows the spring-back of each measurement point according to the pre-heating temperature about the two kinds of heat sources. A positive value means increase spring back and a negative value means over-bending in figure 19. In previous research, V- bending has no concept of a negative value, but this study has a negative value because the roll forming process is a progressive process. The measurement result shows a very similar result to the spring back value of the room temperature, the induction pre-heating and laser pre-heating at the Point-1 and 2. In the laser heating, the spring back of Point-3 is much improved.

Spring-back effect according to the pre-heating temperature is not a significant difference. The spring back phenomenon is determined that the effect on a large area of pre-heating shape than the temperature of pre-heating temperature. Also, the spring back value is predicted to differ according to the correlation between the heat affected zone and bending area in forming sheet.

5. Conclusion

In this study, the spring back was predicted by the preheated plate in the roll forming process. FEM simulation analysis was derived von-mises stress, equivalent strain in room temperature, and thermal stress, thermal strain in high temperature.

The conclusions of this study are as follows:

- Forming sheet has a high temperature convergence in the thickness direction and temperature reduction depends on the pre-heating method in after forming.
- Stress behavior of 1,4-layer is similar and von-mises stress tends to increase with the point in 2-layer at room temperature. In the high temperature, the thermal stress difference in temperature is not large by the blue shortness effect of high tensile steel.
- In the room temperature, the equivalent strain is increased gradually when the bending force is applied and equivalent strain exhibits a rapid increase in processing by the die. Thermal strain is increased higher the preheating temperature and thermal strain of 1,4-layer is not a big difference in high temperatures.
- Laser pre-heating area is about 11.8 times larger than the magnetic induction pre-heating area. Thermal stress by the pre-heating of the two methods is similar, but the laser type is a larger thermal strain value than the induced type.
- In results of springback analysis, forming a sheet tends to increase the springback value as forming process, and in the laser heating, the spring back of Point-3 is much improved.
- Pre-heating has less influence about the stress, but the thermal strain of effects is large in the roll forming process.

Also, the spring back value is determined to be able to control the spring back value in accordance with the pre-heating shape and area. And through the FEM simulation, defects in the roll forming products are expected to be predicted in the pre-heating roll forming process.

Acknowledgments

This research was supported by Basic Science Research Program through the National Research Foundation of Korea (NRF) funded by the Ministry of Education (NRF-2014R1A1A4A01009199).

References

[1] Wick C 1992 *Tool and Manufacturing Engineers Handbook* (Society of Manufacturing Engineers) vol 6 4th edn. chapter 8 10-14

[2] Bhattacharyya D, Smith P D, Yee C H and Collins I F 1984 The prediction of deformation length in cold roll forming *J. Mech. Working Technol.* **9** 181-91

[3] Troive L and Ingvarsson L 2008 Roll forming and the benefits of ultrahigh strength steel *Ironmaking Steelmaking* **35** 251-3

[4] Yang M B and Pan F S 2009 Analysis about forming mechanism of equiaxed crystal zone for 1Cr18Ni9Ti stainless steel twin-roll thin strip *Journal of Materials Processing Technology* **209** 2203-11

[5] Jung D W, Kim D H and Kim B C 2012 A study on development of automotive panel of bumper reinforcement with high strength steel using roll forming process *Journal of the*

IOP Conf. Series: Materials Science and Engineering **103** (2015) 012035 doi:10.1088/1757-899X/103/1/012035

Korean Society for Precision Engineering **29** 840-4
[6] Kim D H and Jung D W 2013 Roll forming analysis for high strength steel bumper process *Journal of the Korean Society for Precision Engineering* **30** 797-801
[7] Xie H Y, Wang Q, Liu K, Peng F, Dong X H and Wang J F 2015 Investigation of influence of direct-current pulses on springbackduring V-bending of AZ31B magnesium alloy sheet *Journal of Materials Processing Technology* **219** 321-7
[8] Fang Z, Lu H N, Wei D B, Jiang Z Y, Zhao X M, Zhang X M and Wu D 2014 Numerical study on springback with size effect in micro V-bending *Procedia Engineering* **81** 1011-6
[9] Duggal N, Ahmetoglu M A, Kinzel G L and Altan T 1996 Computer aided simulation of cold roll forming-a computer program for simple section profiles *Journal of Materials Processing Technology* **59** 41-8
[10] Chang K H, Lee C H and Shin Y E 2004 A study on the high temperature tensile property and the characteristics of residual stress in welds of high strength steels *Journal of the Korea Welding and Joining Society* **22** 50-8
[11] Chang K H, Jung G C, Stiemer S and Loewen N 2013 Dynamic hysteretic characteristics of high-strength steels (POSTEN60, POSTEN80) and application of a dynamic hysteresis model to FE analysis *Journal of Materials in Civil Engineering* **25** 1549-57
[12] Chang K H and Lee C H 2006 Characteristics of high temperature tensile properties and residual stresses in weldments of high strength steels *Materials Transactions* **47** 348-54
[13] Lee K R, Yang Y S, Hwang C Y and Yoo Y T 2012 Characteristics of surface hardening of Nd:YAG laser according to temperature changes of SM45C *Journal of the Korean Society of Manufacturing Technology Engineers* **21** 988-97
[14] Hong S K, Lee S H, Heo Y M and Kang J J 2007 Induction heating apparatus for rapid heating of flat-type metallic mold in hot embossing *Transactions of Materials Processing* **16** 282-7
[15] Kim D H, Yoon D H, Yu H and Jung D W 2014 Study on spring-back effect with various temperature of magnesium alloy in roll forming process *Advanced Materials Research* **1016** 75-9
[16] Lee T S and Kim G W 2009 Finite element analysis for precision roll forming process of stainless slide rail *Journal of the Korean Society for Precision Engineering* **26** 96-103

Accurate prediction model of bead geometry in crimping butt of the laser brazing using generalized regression neural network

Y M Rong, Y Chang, Y Huang[1], G J Zhang and X Y Shao

State Key Lab of Digital Manufacturing Equipment and Technology, School of Mechanical Science and Engineering, Huazhong University of Science and Technology, Wuhan 430074, China

E-mail: yuhuang7208@163.com

Abstract. There are few researches that concentrate on the prediction of the bead geometry for laser brazing with crimping butt. This paper addressed the accurate prediction of the bead profile by developing a generalized regression neural network (GRNN) algorithm. Firstly GRNN model was developed and trained to decrease the prediction error that may be influenced by the sample size. Then the prediction accuracy was demonstrated by comparing with other articles and back propagation artificial neural network (BPNN) algorithm. Eventually the reliability and stability of GRNN model were discussed from the points of average relative error (ARE), mean square error (MSE) and root mean square error (RMSE), while the maximum ARE and MSE were 6.94% and 0.0303 that were clearly less than those (14.28% and 0.0832) predicted by BPNN. Obviously, it was proved that the prediction accuracy was improved at least 2 times, and the stability was also increased much more.

1. Introduction

In recent years, the research on the bead geometry (BG) has gradually become a hot issue in the welding field because the mechanical properties and distortion can be reflected by BG [1, 2]. From the point of prediction and optimization of the bead geometry (POBG), the research methods of BG mainly include the intelligent algorithm and the mathematic model, which both are detailed as follows.

Especially, artificial neural network (ANN) is one of the most widely intelligent algorithms that have been applied in POBG problems. Back propagation artificial neural network (BPNN) was used to forecast penetration depth, weld-seam width of thermoplastics and the four BG sizes of the crimping butt [3-5]. Shojaeefard, et al. built ANN model to simulate the correlation between the friction stir welding parameters and mechanical properties [6]. Singh, et al. developed a model using ANN and genetic algorithm (GA) to study BG and hardness profile [7]. Omajene, et al. applied ANN to optimize BG welding process parameters of the underwater wet welding and the influence of the water environment [8]. In general, it was the composite application of BPNN and GA that was used more widely in POBG field [5, 9-11]. In addition, some other intelligent algorithms, such as simulated annealing algorithm, adaptive neuro-fuzzy inference system, and hybrid artificial bee colony with

[1] Address for correspondence: Y Huang, State Key Lab of Digital Manufacturing Equipment and Technology, School of Mechanical Science and Engineering, Huazhong University of Science and Technology, Wuhan 430074, China. E-mail: yuhuang7208@163.com.

sequential kriging algorithm, were also used in POBG [12-14]. On the other hand, mathematical model was usually developed to research POBG based on different welding methods. Response surface methodology (RSM) was used to build math model to optimize the important process parameters while BG was selected as one of several output goals [15-17]. Lin et al established the grey relational analysis model to demonstrate the influences of process parameters on BG in gas metal arc welding [18]. Magudeeswaran et al used analysis of variance to optimize welding parameters of tungsten inert gas welding [19, 20].

However, there are few researches about POBG problem of laser brazing, especially scarcely any using GRNN to crimping butt. As shown in figure 1, a typical laser brazing system consists of fiber laser, welding robot, wire feeder machine, laser welding header, workbench and so forth, while the BG of crimping butt is composed of the efficient length of connection on the left side (ELL), top width of bead (WT), the efficient length of connection on the right side (ELR) and bottom width of bead (WB) [5]. In this paper, generalized regression neural network (GRNN) algorithm is used to predict BG of the crimping butt of the laser brazing, and the prediction accuracy is analyzed by comparing with other articles and BPNN model.

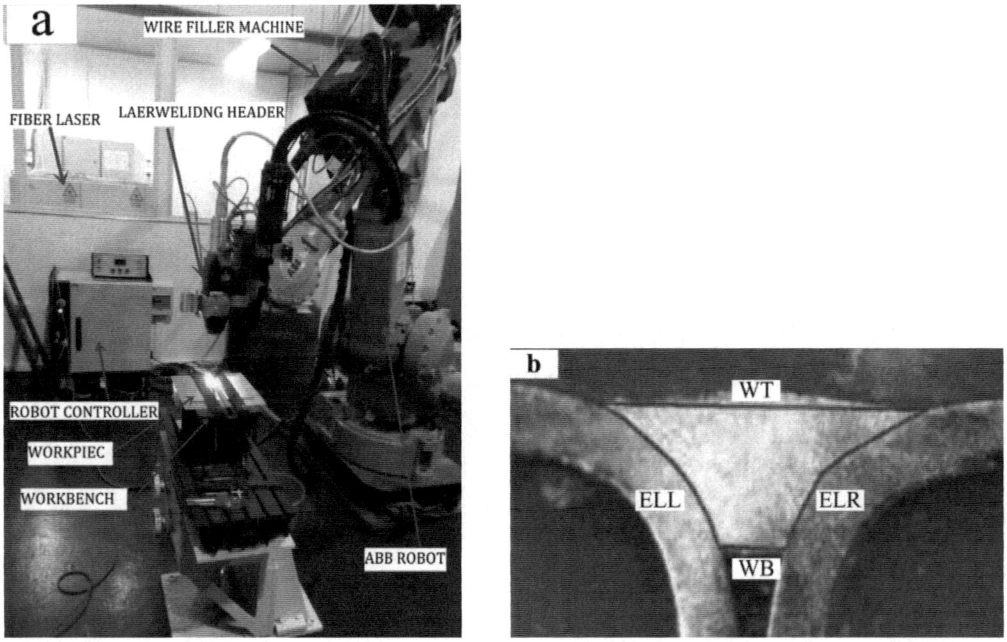

Figure 1. A typical laser brazing system: (a) laser brazing setup and (b) a typical crimping butt.

2. Prediction methodology

2.1. Generalized regression neural network

GRNN is a kind of radial basis function (RBF) networks with a highly parallel structure, which was developed in 1991 by Specht [21], and has been widely used to achieve prediction and estimation results in many fields such as plant disease, power load, the maintenance cost of construction equipment, intermittent flow and so on [22-25]. The basic function of GRNN is nonlinear regression surface dominated by independent variable $\mathbf{X}$, given the dependent variable y . The procedure of GRNN can be given by equation (1) [24]

$$\hat{\mathbf{Y}}(\mathbf{X}) = E(y \mid \mathbf{X}) = \frac{\int_{-\infty}^{\infty} y f(\mathbf{X}, y) dy}{\int_{-\infty}^{\infty} f(\mathbf{X}, y) dy} \tag{1}$$

where $\hat{\mathbf{Y}}(\mathbf{X})$ is the expected value of the output y , $\mathbf{X}=[x_1,x_2,\ldots,x_n]^T$ is a n dimensional input vector, $f(\mathbf{X},y)$ is the joint probability density function of $\mathbf{X}$ and y .

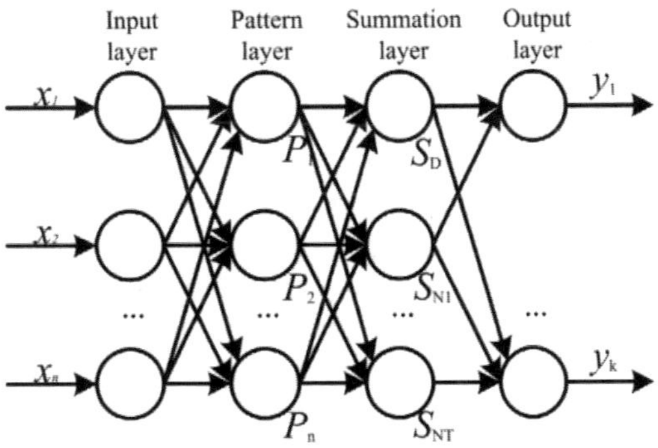

Figure 2. GRNN block diagram [21].

As shown in figure 2, the GRNN is composed of four layers (input layer, pattern layer, summation layer and output layer). The input layer is distribution layer, which provides the information of input vector $\mathbf{X}$ to the second layer. The pattern layer possesses the nonlinear transformation from input space to pattern space after the data fed back to the pattern layer from the input layer using the input neurons. The pattern neurons can memorize the map between the input neurons and the proper response of pattern layer, and the number of neurons equals to the number of input variables. The transfer function of pattern neurons is Gaussian function of p_i as shown equation (2) [22]. The pattern neuron outputs are passed on to the third layer.

$$p_i=\exp\left[-\frac{(\mathbf{X}-\mathbf{X}_i)^T(\mathbf{X}-\mathbf{X}_i)}{2\sigma^2}\right] \quad i=1,2,\ldots,n \tag{2}$$

where $\mathbf{X}_i$ is a specific learning sample of the neuron i, σ indicates the smoothing parameter.

Then, two kinds of calculation methods are used in summation layer. Just as shown in equation (3) [21], one is simple summation S_D that is applied to compute the arithmetic sum of the second layer outputs, and the interconnection weight is 1. The other is weighted summation S_{Nj} that is calculated as seen in equation (4) [21], while the interconnection weight between the i th neuron of the pattern layer and the j th neuron of the summation layer is the j th unit of the i th ouputY_i.

$$S_D=\sum_{i=1}^{n}P_i \tag{3}$$

$$S_{Nj}=\sum_{i=1}^{n}y_{ij}P_i \quad j=1,2,\ldots,k \tag{4}$$

Finally, the summations of neurons are fed into the output layer, and the j th output shown as equation (5) [21] corresponds with the j th unit of the estimation result $\hat{\mathbf{Y}}(\mathbf{X})$. It is noted that the number of neurons in the output layer is equal to the k -dimension of output vector $\mathbf{Y}$.

$$y_j=\frac{S_{Nj}}{S_D} \quad j=1,2,\ldots,k \tag{5}$$

Through the above simple description of the GRNN basic principle, it is easy to find that there is only one tunable parameter (the smoothing value, σ) of GRNN model. The parameter σ is very important for GRNN model in process of the prediction, and determines the generalization capability of the GRNN. Generally, the prediction results of this network are closer to the sample data with a smaller σ. Meanwhile, the approximation process from the GRNN prediction network to the sample data presents much smoother when the σ value turns bigger.

2.2. The prediction of laser brazing BG using GRNN

GRNN has the capacity to solve the complex nonlinear problems due to its advantages of good fault tolerant performance, high robustness and good ability to deal with the unstable data. Through article surveys, it can be found that some abnormal errors were occasional and even the individual error unexpectedly reached to 247.75% during BG prediction procedure using ANN [5, 26, 27]. In this article, the GRNN model was developed and trained considering cross-validation method to decrease the prediction error that may be influenced by the sample size [28], and the σ parameter was achieved through cycle structure to search the best value. Considering the characteristics of the crimping butt of the laser brazing, the prediction procedure is planned in detail as below:

- Preparation work. The task at this stage is to identify the inputs, outputs, train set and test set. The inputs include welding speed (WS, 0.8 m/min-1.6 m/min), wire feed rate (WF, 2.6 m/min-3.4 m/min) and gap (GAP, 0 mm-0.8 mm). The outputs consist of ELL, ELR, WT and WB. About two-thirds of experiment data are selected as train set and the rest are as test set.

Table 1. Divisions of experiment data to train set and test set.

NO.	Welding parameters			Bead geometry			
	WS m/min	WF m/min	GAP mm	ELL mm	ELR mm	WT mm	WB mm
Train set							
1	0.8	2.6	0.0	3.43	2.48	4.12	0.61
2	1.0	3.4	0.0	2.61	1.86	3.82	0.83
3	1.4	3.0	0.0	2.1	1.33	3.5	1.09
4	1.6	2.8	0.0	2.23	1.01	3.44	1.25
5	1.0	2.6	0.2	2.38	2.29	3.84	3.11
6	1.2	3.4	0.2	2.46	1.68	3.64	1.11
7	1.6	3.0	0.2	1.92	1.443	3.31	1.12
8	1.0	2.8	0.4	2.56	1.93	3.71	0.96
9	1.4	3.4	0.4	2.07	1.81	3.37	1.27
10	1.6	3.2	0.4	2.16	1.12	3.26	1.24
11	1.0	3.0	0.6	2.88	1.93	4.0	0.93
12	1.2	2.8	0.6	2.59	1.72	3.63	1.28
13	1.6	3.4	0.6	2.39	2.07	2.41	0.86
14	0.8	3.4	0.8	3.1	2.26	3.8	1.25
15	1.2	3.0	0.8	3.02	2.94	3.65	1.01
16	1.4	2.8	0.8	2.37	1.76	3.62	1.3
Test set							
17	1.2	3.2	0.0	2.36	1.33	3.42	1.03
18	0.8	2.8	0.2	2.55	2.07	3.61	0.95
19	1.4	3.2	0.2	1.92	1.6	3.41	1.12
20	0.8	3.0	0.4	2.37	1.79	4.07	1.26
21	1.2	2.6	0.4	2.73	2.53	3.7	0.85
22	0.8	3.2	0.6	3.32	1.98	4.2	1.03
23	1.4	2.6	0.6	1.91	0.92	3.24	1.41
24	1.0	3.2	0.8	2.62	1.86	3.88	1.24
25	1.6	2.6	0.8	2.4	1.24	3.5	1.35

- Training the GRNN network. Firstly, the best smooth parameter σ can be determined using cross-validation method with a cycle structure, and then GRNN network is developed and trained by invoking the best smooth parameter and train set.
- Prediction through GRNN testing. Each output of the test set would be fed to GRNN network that has been trained, and finally the prediction results and errors can be obtained by comparing them with the experiment data.

3. Results and discussion

3.1. The division of data set

The welding experiment of galvanized thin double phase steel 590 (DP590) with thickness about 0.8mm was designed as a 3-factor, 5-level problem using Taguchi L_{25} method. The welding parameters were WS, WF and GAP, and the output goals were acquired as ELL, ELR, WT and WB [5]. The train set and test set were randomly decided as table 1.

3.2. The prediction results by GRNN

The program of GRNN model was run in the MATLAB2013a. Just as shown in table 1, 16 groups of experiment data were selected to create and train the network, and the rest 9 sets of data were used to conduct the validity of the GRNN model. The error (Err) of each prediction result was calculated by equation (6), while $y_{\exp}(i)$ is the i th actual result, and $y_{pre}(i)$ is the i th prediction result. As shown in table 2, the prediction results of GRNN and errors were listed as specific as possible.

$$Err = \frac{y_{\exp}(i) - y_{pre}(i)}{y_{\exp}(i)} \times 100\% \qquad i = 1, 2, \ldots, 9 \tag{6}$$

Table 2. Forecast results and errors of BG by GRNN model.

NO	ELL			ELR			WT			WB		
	Result	Err (%)	σ	Result	Err (%)	σ	Result	Err (%)	σ	Result	Err (%)	σ
17	2.344	0.683	1.0	1.425	-7.157	0.7	3.433	-0.383	0.9	1.083	-5.119	0.8
18	2.498	2.049	1.2	2.108	-1.851	1.1	3.719	-3.011	1.1	0.872	8.192	0.3
19	2.071	-7.875	0.7	1.603	-0.192	0.8	3.392	0.533	1.1	1.134	1.489	0.5
20	2.386	-0.685	0.1	1.734	3.110	0.7	3.967	2.544	0.9	1.211	3.901	2.0
21	2.628	3.750	2.0	2.125	16.017	1.0	3.672	0.750	1.3	0.790	7.058	2.0
22	3.034	8.623	0.1	1.951	1.474	1.4	3.979	5.267	0.8	0.960	6.796	0.1
23	2.253	-17.963	0.1	0.982	-6.685	0.8	3.421	-5.587	0.8	1.447	-2.605	2.0
24	2.667	-1.779	0.8	1.895	-1.871	0.9	3.903	-0.580	1.0	1.185	4.428	2.0
25	2.493	-3.866	1.4	1.539	-24.078	0.5	3.530	-0.859	0.9	1.230	3.718	0.5

3.3. Discussion about the results

Figure 3 showed the comparison analysis among experiment values, BPNN predictions and GRNN predictions. The results of the ELL of the crimping butt were shown in figure 3(a), and the maximum prediction errors of the BPNN and GRNN were 38.34% and 17.96%. Figure 3(b) showed that the maximum prediction errors of the ELR through BPNN and GRNN were 60.35% and 24.08%. As shown in figures 3(c) and 3(d), the maximum prediction errors of the WT using BPNN and GRNN were 9.10% and 5.27%, and these of the WB were 22.16% and 8.19%. Thus the maximum errors of ELL, ELR, WT and WB reduced half or even more. In addition, comparing the prediction results of bead profile with other articles [5, 26, 27], as shown in table 3, the maximum error (24.08%) was obviously less than that of other articles (60.35%, 247.75% and 29.125%). The prediction error of BG occasionally exceeds 20% (1 out of 36 outputs, and only about 2.78%), still the prediction precision of this article was clearly better than that of other papers (11.11%, 11.67% and 25%) and hence GRNN was able to accurately predict bead profile of the crimping butt.

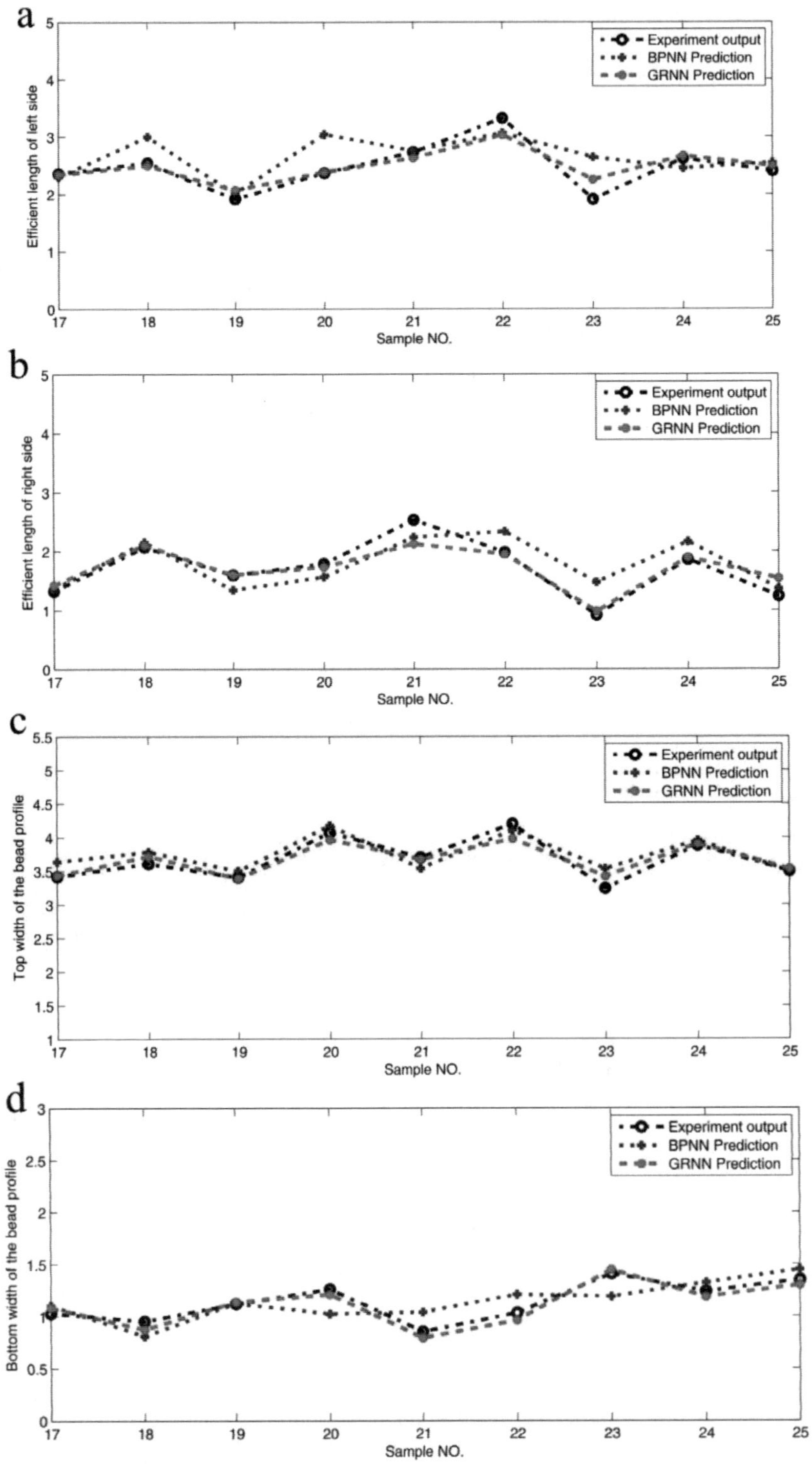

Figure 3. Comparison of results using BPNN and GRNN: (a) ELL, (b) ELR, (c) WT and (d) WB.

Table 3. Comparison of the prediction accuracy with other articles using ANN.

Comparison	Method	Maximum Error (%)	Number of exceeded 20%	
			Number	Percentage (%)
Present	GRNN	**24.08**	1/36	**2.78**
Ref. 5	BPNN	60.35	4/36	11.11
Ref. 28	BPNN	247.75	7/60	11.67
Ref. 29	BPNN	29.125	3/12	25.00

The reliability and stability of the prediction results can be weighed by average relative error (ARE, equation (7)), mean square error (MSE, equation (8)) and root mean square error (RMSE, equation (9)). Through comparing prediction results from the point of ARE and MSE in table 4, ARE values of crimping butt using GRNN model were 5.25%, 6.94%, 2.61% and 4.81%, while these results calculated by BPNN were 10.45%, 14.28%, 3.64% and 12.53%. The MSE of each BG dimensions of the crimping butt through GRNN (0.0275, 0.0303, 0.0119 and 0.0030) were all better than these predicted by BPNN. Meanwhile, as shown in table 4, evaluation indicator of RMSE for GRNN model is clearly better than that of BPNN model. The prediction accuracy and stability of bead geometry of the crimping butt using GRNN model were apparently better than that by BPNN model, especially the WB group. In short, the GRNN model we had developed was credible and stable, and would be used in the further welding process optimization and actual manufacturing.

$$ARE = \frac{1}{n}\sum_{i=1}^{n}\frac{y_{\exp}(i) - y_{pre}(i)}{y_{\exp}(i)} \times 100\% \tag{7}$$

$$MSE = \frac{1}{n}\sum_{i=1}^{n}\left[y_{\exp}(i) - y_{pre}(i)\right]^2 \tag{8}$$

$$RMSE = \sqrt{\frac{1}{n}\sum_{i=1}^{n}\left[y_{\exp}(i) - y_{pre}(i)\right]^2} \tag{9}$$

Table 4. Calculation of ARE, MSE and RMSE of the BPNN and GRNN models.

Bead shape	BPNN model			GRNN model		
	ARE (%)	MSE	RMSE	ARE (%)	MSE	RMSE
ELL	10.45	0.1467	0.3830	5.25	0.0275	0.1658
ELR	14.28	0.0832	0.2884	6.94	0.0303	0.1741
WT	3.64	0.0249	0.1550	2.16	0.0119	0.1091
WB	12.53	0.0240	0.1549	4.81	0.0030	0.0548

4. Conclusions

In this paper, considering welding parameters (WS, WF and GAP) and the output goals (ELL, ELR, WT and WB), GRNN model was created and trained considering cross-validation method to decrease the prediction error that may be influenced by the sample size. The prediction accuracy was demonstrated by comparison with other articles and BPNN model. Meanwhile, the reliability and stability were discussed from the point of ARE, MSE and RMSE. Eventually, three conclusions can be drawn: (1) The GRNN model can be used to accurately predict the bead shape of crimping butt, while the average errors of ELL, ELR, WT and WB (5.25%, 6.94% 2.16% and 4.81%) were all apparently less than these using BPNN model (10.45%, 14.28%, 3.64% and 12.53%); (2) The reliability and accuracy of prediction results by GRNN model were obviously better than that of other articles and BPNN model from the point of ARE, MSE, RMSE, and maximum error and so forth; (3) The GRNN model is reliable and stable in predicting bead shape of the crimping butt, and would be used to direct the further welding process optimization and actual manufacturing.

Acknowledgments
This research was supported by the National Basic Research Program of China (973 Program, NO. 2014CB046703).

References

[1] Juang S C and Tarng Y S 2002 Process parameter selection for optimizing the weld pool geometry in the tungsten inert gas welding of stainless steel *J. Mater. Process. Technol.* **122** 33-7

[2] Aloraier A, Almazrouee A, Shehata T and John W H Price 2012 Role of welding parameters using the flux cored arc welding process of low alloy steels on bead geometry and mechanical properties *J. Mater. Eng. Perform.* **21** 540-7

[3] Ghosal S and Chaki S 2010 Estimation and optimization of depth of penetration in hybrid CO_2 laser-MIG welding using ANN-optimization hybrid model *Int. J. Adv. Manuf. Technol.* **47** 1149-57

[4] Acherjeea B, Mondal S, Tudu B and Misra D 2011 Application of artificial neural network for predicting weld quality in laser transmission welding of thermoplastics *Appl. Soft. Comput.* **11** 2548-55

[5] Rong Y M, Zhang Z, Zhang G J, Yue C, Gu Y F, Huang Y, Wang C M and Shao X Y 2015 Parameters optimization of laser brazing in crimping butt using Taguchi and BPNN-GA *Opt. Lasers Eng.* **67** 94-104

[6] Shojaeefard M H, Behnagh R A, Akbari M, Givi M K B and Farhani F 2013 Modelling and Pareto optimization of mechanical properties of friction stir welded AA7075/AA5083 butt joints using neural network and particle swarm algorithm *Mater. Des.* **44** 190-8

[7] Singh A, Cooper D E, Blundell N J, Pratihar D K and Gibbons G J 2014 Modelling of weld-bead geometry and hardness profile in laser welding of plain carbon steel using neural networks and genetic algorithms *Int. J. Comput. Integr. Manuf.* **27** 656-74

[8] Omajene J E, Martikainen J, Wu H and Kah P 2014 Optimization of underwater wet welding process parameters using neural network *Int. J. Mech. Mater.* **9** 26

[9] Sathiya P, Panneerselvam K and Soundararajan R 2012 Optimal design for laser beam butt welding process parameter using artificial neural networks and genetic algorithm for super austenitic stainless steel *Opt. Laser Technol.* **44** 1905-14

[10] Kumar A, Jadoun R S and Bist A S 2014 Optimization of MIG welding parameters using artificial neural network (ANN) and genetic algorithm (GA) *Int. J. Eng. Sci. Res. Tech.* **3** 614-20

[11] Sathiya P, Panneerselvam K and Abdul Jaleel M Y 2012 Optimization of laser welding process parameters for super austenitic stainless steel using artificial neural networks and genetic algorithm *Mater. Des.* **36** 490-8

[12] Kolahan F and Heidari M 2010 Modeling and optimization of MAG welding for gas pipelines using regression analysis and simulated annealing algorithm *J. Sci. Ind. Res.* **69** 177-83

[13] Subashin L and Vasudevan M 2012 Adaptive neuro-fuzzy inference system (ANFIS)-based models for predicting the weld bead width and depth of penetration from the infrared thermal image of the weld pool *Metall. Mater. Trans. B-Proc. Metall.* **43B** 145-54

[14] Fang J G, Gao Y K, Sun G Y, Xu C M, Zhang Y T and Li Q **2014** Optimization of spot-welded joints combined artificial bee colony algorithm with sequential kriging optimization *Adv. Mech. Eng.* **11** 1297-319

[15] Kiaee N and Aghaie-Khafri M 2014 Optimization of gas tungsten arc welding process by response surface methodology *Mater. Des.* **54** 25-31

[16] Korra N N, Vasudevan M and Balasubramanian K R 2015 Multi-objective optimization of activated tungsten intert gas welding of duplex stainless steel using response surface methodology *Int. J. Adv. Manuf. Technol.* **77** 67-81

[17] Lafdan M H K, Zahrani E G and Galloway A M 2014 Optimization study of weld geometry in

the tandem submerged arc welding process *Proc. Inst. Mech. Eng. Part B-J. Eng. Manuf.* doi: 10.1177/0954405414535578

[18] Lin H L and Yan J C 2014 Optimization of weld bead geometry in the activated GMA welding process via a grey-based Taguchi method *J. Mech. Sci. Technol.* **28** 3249-54

[19] Magudeeswaran G, Nair Sreehari R, Sundar L and Harikannan N 2014 Optimization of process parameters of the activated tungsten inert gas welding for aspect ratio of UNS S32205 duplex stainless steel welds *Def.Tech.* **10** 251-60

[20] Kumar V 2014 Optimization of weld bead width in tungsten inert gas welding of austenitic stainless steel alloy *Ame. J. Mech. Eng.* **2** 50-53

[21] Specht D F 1991 A general regression neural network *IEEE Trans. Neural Netw.* **2** 568-76

[22] Chtioui Y, Panigrahi S and Francl L 1999 A generalized regression neural network and its application for leaf wetness prediction to forecast plant disease *Chemometrics Intell. Lab. Syst.* **48** 47-58

[23] Li H Z, Guo S, Li C J, Sun J Q 2013 A hybrid annual power load forecasting model based on generalized regression neural network with fruit fly optimization algorithm *Knowledge-Based Syst.* **37** 378-87

[24] Yi H, Fan H and Chiang Y 2014 Predicting the maintenance cost of construction equipment: Comparison between general regression neural network and Box–Jenkins time series models *Autom. Constr.* **38** 30-8

[25] Cigizoglu H K 2005 Application of generalized regression neural networks to intermittent flow forecasting and estimation *J. Hydrol. Eng.* **10** 336-41

[26] Nagesh D S and Datta G L 2010 Genetic algorithm for optimization of welding variables for height to width ratio and application of ANN for prediction of bead geometry for TIG welding process *Appl. Soft. Comput.* **10** 897-907

[27] Kanti K M and Rao P S 2008 Prediction of bead geometry in pulsed GMA welding using back propagation neural network *J. Mater. Process. Technol.* **200** 300-5

Effects of process parameters on microstructural evolution and properties of AZ61 alloy during hot extrusion

Y L Lu[1,2,3], X C Li[1], W T Xu[2], D Wu[2] and M Yang[2]

[1] Key Construction Laboratory of Green Forming and Equipment from Jiangsu Province, Changzhou, 213001, China
[2] Jiangsu University of Technology, Changzhou, 213000, China

E-mail: luyalin@163.com

Abstract. Extrusion testing of AZ61 alloys were conducted at deformation temperatures of 300℃ to 410℃ and extrusion ratios of 3, 10 and 16 respectively to optimize the process parameters. The experimental results show that deformation process parameters significantly affect microstructures and properties. Optical microscope observation shows that grains are refined greatly during hot extrusion and the mechanical properties are improved with increasing the extrusion ratio. For AZ61 alloy, the optimal extrusion temperature is 370℃. When deformation temperature increases, more slip systems participate in the plastic deformation, which leads to the decrease of the cleavage surface. After severe plastic deformation, a remarkable improvement of ductility of AZ61 alloy has been found. The ductile fracture mechanism is gradually dominant instead of brittle fracture with increasing the extrusion ratio.

1. Introduction

Magnesium alloys are widely used due to a series of advantages, such as the biological compatibility, the specific strength and stiffness, excellent process performance and corrosion resistance [1-3]. Due to the close-packed hexagonal structure and fewer slip systems, the magnesium alloy always exhibits low strength and limited ductility when produced by the traditional casting process and semi-solid process [4, 5]. Therefore, some plastic deformation process, such as extrusion, forging, and rolling has been applied for obtaining the high-quality products of magnesium alloy [6-16].

In this work, the extrusion process of AZ61 alloy was implemented under different deformation parameters. Based on the analysis of microstructures and mechanical properties, the optimal parameters for extrusion has been chosen, which provides the guide for its application in practice.

2. Experimental procedures

The as-cast AZ61 alloy with chemical composition (wt.%) of Mg-5.93Al-0.66Zn-0.3Mn-0.14Si was used for extrusion test. The cylinder samples with a diameter of 40 mm and height of 60 mm were cut first and then homogenized at the temperature of 400℃ for 12 hours followed by furnace cooling.

After homogenization, the samples were extruded according to the given scheme. A schematic drawing of extrusion process was shown in figure 1. The extrusion process was carried out at

[3] Address for correspondence: Y L Lu, Key Construction Laboratory of Green Forming and Equipment from Jiangsu Province, Changzhou, 213001, China. E-mail: luyalin@163.com.

temperatures of 300℃, 350℃, 370℃, 390℃, and 410℃ and extrusion ratio of 3, 10, and 16 respectively. Before the extrusion process, die was lubricated with MoS_2 and preheated. The die temperature was below 20℃ compared with extrusion temperature. After the extrusion tests, the samples were mechanically polished and eroded in a solution including distilled water of 192 ml, oxalic acid of 4ml and nitrate of 4 ml. Hardness was tested by HVS21000 type digital microhardness tester.

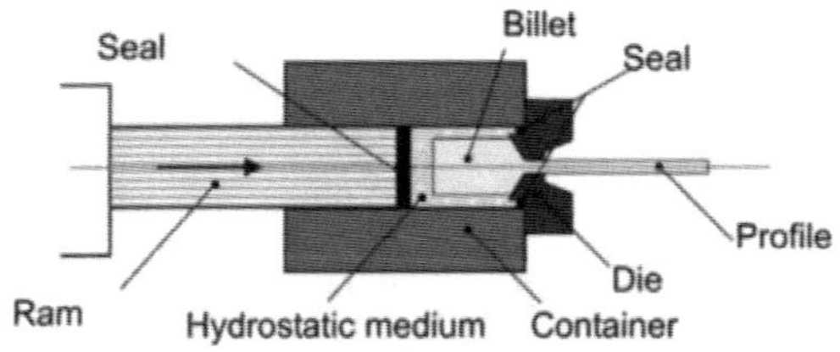

Figure 1. Schematic drawing of extrusion process.

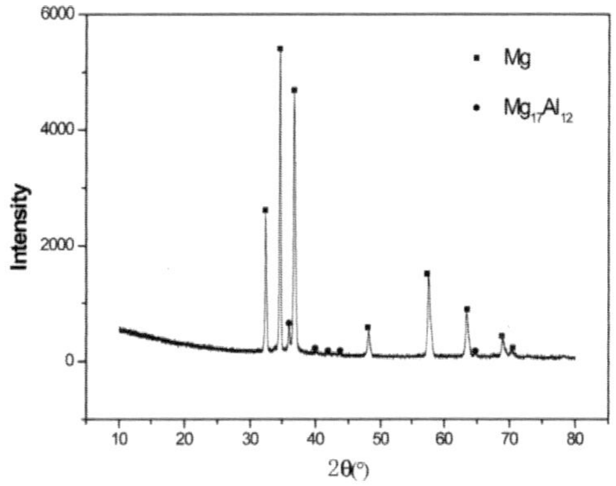

Figure 2. XRD pattern of AZ61 alloy.

3. Experimental results and discussion

3.1. Homogenization process for AZ61 alloy

Figure 2 shows the XRD pattern of AZ61 alloy, which is mainly composed of the primary α-Mg and secondary eutectic $Mg_{17}Al_{12}$.

Figures 3(a) and 3(b) shows the microstructure of the as-cast AZ61 alloy and the microstructure after homogenization at temperature of 400℃ for 12 hours respectively. In figure 3, the white structure represents α-Mg phase while the black is the eutectic $Mg_{17}Al_{12}$ phase. From figure 3(a), grains have a characteristic of coarse and dendritic structure with the average size of about 200 μm. The grain boundary is not clear. After homogenization, the morphology and the size of grain vary apparently as shown in figure 3(b).

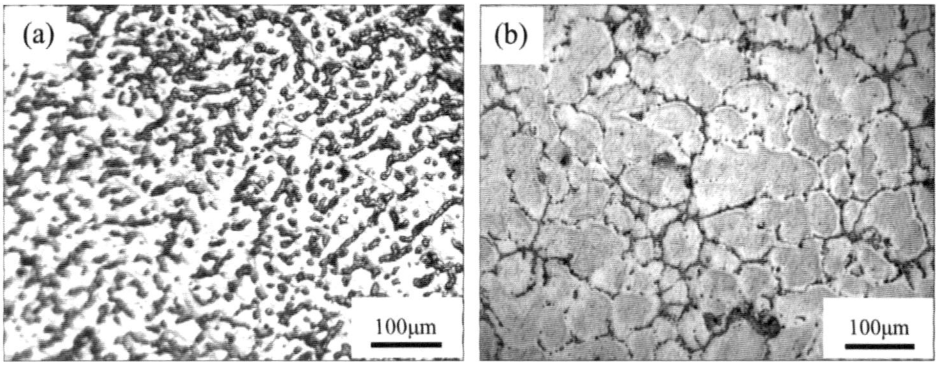

Figure 3. Effect of homogenization on microstructures of AZ61 alloy (a) before homogenization; (b) after homogenization.

3.2. Effect of deformation temperatures on microstructures and mechanical properties

Optical photomicrographs for AZ61 alloy at different deformation temperatures are shown in figure 4. It can be observed in figure 4(a) that the phenomenon of dynamic recrystallization appears when deformation temperature is lower than 300℃. Meanwhile, the microstructures consist of partially recrystallized grains and coarse primary grains. From figure 4(b), the coarse primary grains at deformation temperature of 350℃ are gradually replaced by dynamic recrystallization grains. The distribution of grains tends to be more homogeneous when the deformation temperature increases. In figure 4(c), the microstructures of samples extruded at deformation temperature of 370℃ are fully recrystallized. As shown in figures 4(c) and 4(d), the recrystallized grains coarsen gradually with increasing of deformation temperature.

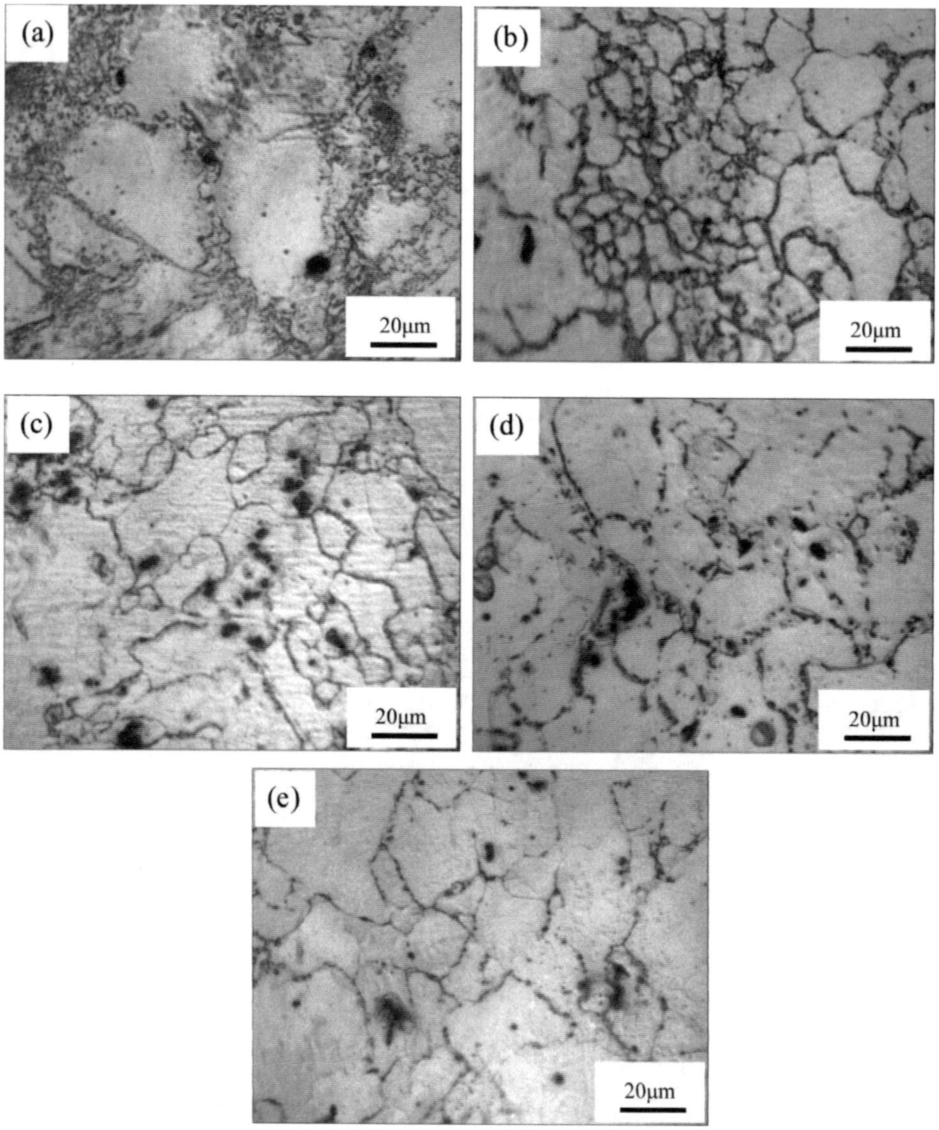

Figure 4. Microstructures of AZ61 alloy at different extrusion temperature(a) 300℃; (b) 350℃; (c) 370℃; (d) 390℃; (e) 410℃.

Figure 5(a) shows the strength of samples extruded at different extrusion temperatures. It can be seen from the curve that tensile strength increases with increasing of deformation temperature when extrusion temperature is lower. Tensile strength reaches optimal value at extrusion temperature of 370℃. When the extrusion temperature is above 370℃, the strength of AZ61 alloy decreases.

Effects of extrusion temperatures on the elongation for AZ61 alloy are illustrated in figure 5(b). It can be found that elongation is low at temperatures lower than 300℃ and increases gradually with increasing of the deformation temperature. The elongation of AZ61 alloy reached 12% at extrusion temperature of 350℃ and then decreases quickly when the deformation temperature increases above 370℃.

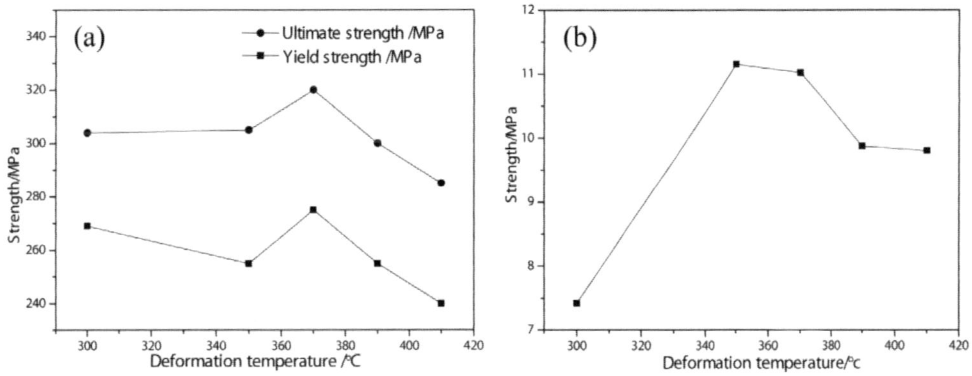

Figure 5. Curves of the mechanical properties vs. deformation temperatures for AZ61 alloy (a) strength; (b) elongation.

Compared figure 4 with figure 5, it can be concluded that the ideal mechanical properties are attributed to refined grains. The sample keeps finer and equiaxed grains at deformation temperature of 370℃ and exhibits optimal strength and plasticity. In other words, for AZ61 alloy, the optimal extrusion temperature is 370℃.

3.3. Effect of extrusion ratio on microstructure and mechanical properties of AZ61 alloy

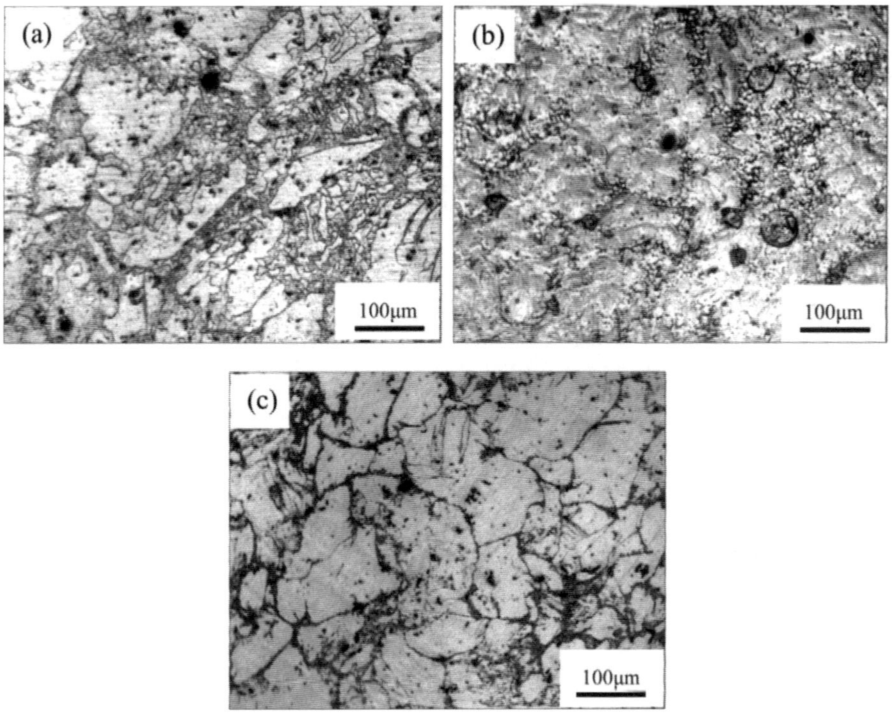

Figure 6. Microstructures of AZ61 alloy at different extrusion ratio (a) extrusion ratio of 3; (b) extrusion ratio of 10; (c) extrusion ratio of 16.

Figure 6 shows the optical photomicrographs of AZ61 alloy at different extrusion ratio. From figure 6, it may be noted that dynamic recrystallization occurs and the grains become finer and equiaxed when the extrusion ratio increases. In other words, the microstructure of AZ61 alloy can be effectively refined by means of hot extrusion process.

Figure 7 illustrates the changing of the strength and elongation of AZ61 alloy with the extrusion ratio. It is clear that the strength and elongation increase gradually with increasing of the extrusion ratio when extrusion ratio is less than 10. When the extrusion ratio gets to 16, ultimate strength and elongation reach the ideal value of 345 MPa and 14% respectively.

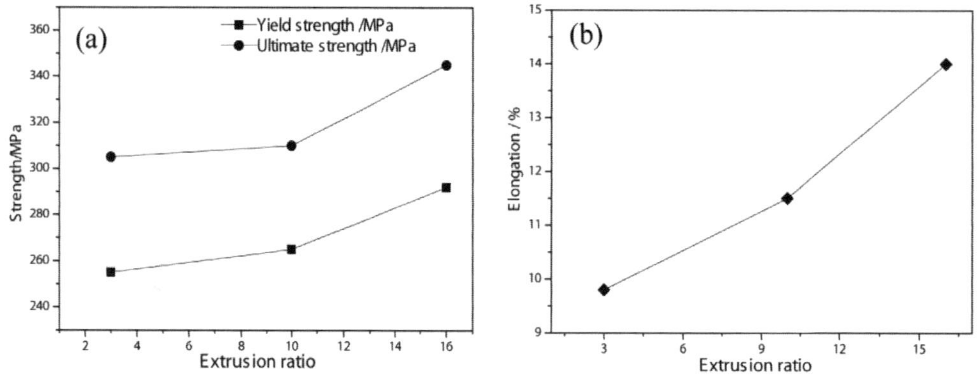

Figure 7. Curves of the mechanical properties vs. extrusion ratios for AZ61 alloy (a) strength; (b) elongation.

Effect of extrusion ratio on the strength and elongation of AZ61 alloy is mainly related to extrusion parameter and characteristics of microstructures. When extrusion ratio is higher, grain refinement is enhanced, which leads to improvement of the mechanical property.

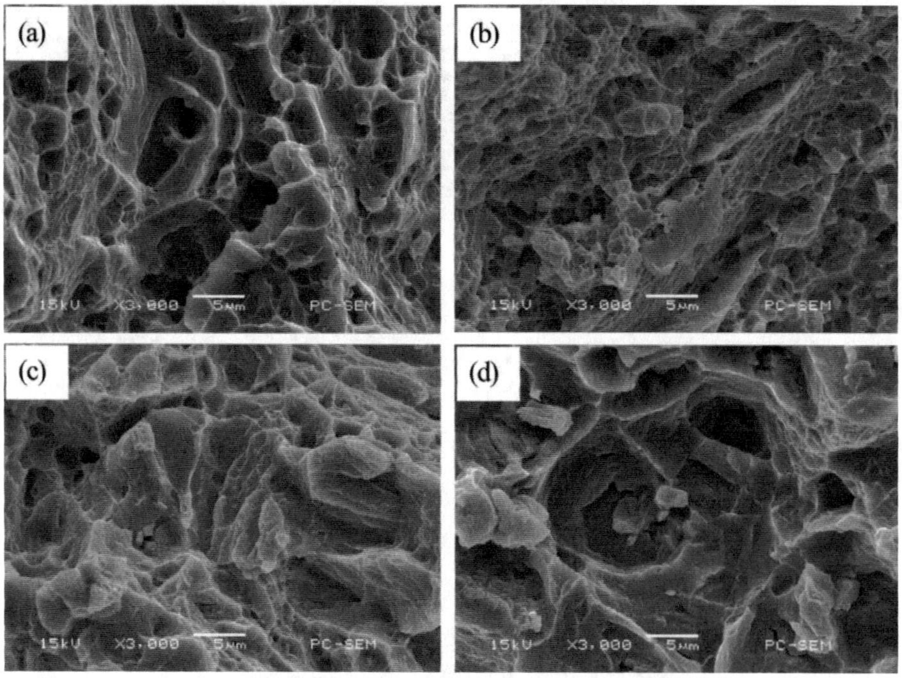

Figure 8. SEM morphology of AZ61 alloy at different extrusion parameters (a) 300℃ at extrusion ratio of 3; (b) 370℃ at extrusion ratio of 3; (c) 410℃ at extrusion ratio of 10; (d) 410℃ at extrusion ratio of 16.

3.4. SEM observation for fracture morphology

The SEM fracture morphology of AZ61 alloy after extrusion under different process parameters is shown in figure 8. At low temperature in figure 8(a), the cleavage surface and dimples, which are taken as mixed-rupture characteristics of brittle fracture and ductile features, can be seen on fracture surfaces. With increasing of deformation temperature, more slip systems participate in the plastic deformation, which leads to decreasing of the cleavage surface. Then, many fine dimples can be observed as shown in figure 8(b).

The number and depth of dimples increase apparently with increasing of extrusion ratio as shown in figures 8(c) and 8(d). The tear ridge of dimple is very small and uniform, which is typical characteristics of ductile features. Therefore, the remarkable improvement of ductility of AZ61 alloy can be obtained after severe plastic deformation. The ductile fracture mechanism is gradually dominant instead of brittle fracture with increasing of extrusion ratio.

4. Conclusions

The microstructure and mechanical properties for AZ61 alloy are affected by process parameters of extrusion temperature and extrusion ratio. Due to dynamic recrystallization, grain refinement has been enhanced, which is helpful to the improvement of mechanical property of AZ61magnesium alloy. With increasing of the extrusion ratio, grains have been refined apparently and ultimate strength and elongation increase greatly as well. Then, the ductile fracture mechanism is gradually dominant instead of brittle fracture with increasing of extrusion ratio. So, it is necessary to extrude at high extrusion ratio to obtain ideal mechanical property of AZ61 alloy. It can be concluded that the optimal extrusion temperature for AZ61 alloy is 370℃.

Acknowledgments

This research was supported by an International Science and Technology Cooperation Program of China (2012DFG70640) and a Pre-research Project (KYY13037) from Jiangsu University of Technology.

References

[1] Chen Z H 2005 Deformed magnesium alloys (Beijing: Chemical Industry Press) pp 10-30

[2] Pan F S and Han E H 2007 Processing technology of high-performance and wrought magnesium alloy (Beijing: Science Press) pp 32-46

[3] Zhang D F, Geng Q M, Yang X S, Yu D L and Pan F S 2014 Effect of temperature in the new compound extrusion on microstructure and properties of AZ61 *Journal of Materials Engineering* **7** 1-4

[4] Hsiang S H and Kuo J L 2005 Application of ANN to the hot extrusion of magnesium alloy sheets *International Journal of Advanced Manufacturing Technology* **25** 292-300

[5] Liu B C and Xiong S M 2011 High pressure die casting process of magnesium alloys and its modeling and simulation for automobile industry *Journal of Automotive Safety and Energy* **2** 1-11

[6] Zhang Z, Wang M P, Li S M, Jiang N and Hu H L 2010 Evolution of microstructure and texture of AZ31 magnesium alloy during hot rolling process *Transaction Nonferrous Metals Society of China* **20** 1447-52

[7] Xv G L, Chen G Q, Zhou W L, Fu X S, Ren X and Sun Z G 2011 Effect of equal channel angular extrusion on the microstructure and mechanical properties of AZ31 magnesium alloy *Journal of Materials Engineering* **2** 69-72

[8] Ogawa N, Shiomi M and Osakada K 2002 Forming limit of magnesium alloy at elevated temperatures for precision forging *International Journal of Machine Tools & Manufacture* **42** 607-14

[9] Xu Y, Hu L X, Sun Y, Jia J B, Jiang J F and Ma Q G 2015 Microstructure and mechanical properties of AZ61 magnesium alloy prepared by repetitive upsetting-extrusion *Transaction*

Nonferrous Metals Society of China **25** 381-8

[10] Wang L F, Huang G S, Li H C and Zhang H 2013 Influence of strain rate on microstructure and formability of magnesium alloy sheets *Transaction Nonferrous Metals Society of China* **23** 916-22

[11] Ding X Z, Liu T M, Chen J, Zhang Y and Lu L W 2013 Effect of twin boundary on static recrystallization of AZ31 magnesium alloy *Transaction Nonferrous Metals Society of China* **23** 1-8

[12] Samman T A and Gottstein G *2008* Dynamic recrystallization during high temperature deformation of magnesium *Materials Science and Engineering* **490** 411-20

[13] Zhang H andHu Z H 2014 Study and analysis on hot extrusion process of high-plastic magnesium alloy *Hot Working Technology* **43** 132-4

[14] Yang M, Yang Y S and Yang D D 2011 Effect of deformation temperature on microstructures and properties of AZ80 magnesium alloy *Hot Working Technology* **40** 24-6

[15] Jiang Y B, Guan L, Tang G Y and Zhang Z H 2011 Influence of electropulsing treatment on microstructure and mechanical properties of cold-rolled Mg-9Al-Zn alloy strip *Materials Science and Engineering A* **528** 5627-33

[16] Liang H C, Feng L J and Cui Z L 2012 Effects of extrusion ratio on microstructure and properties of hot extruded AZ80 magnesium alloy *Special Casting and Nonferrous Alloys* **32** 20-2

FEM analysis of dynamic behavior of asphalt pavement structure weakened by grassroots with account of hydraulic and vehicle load coupling effects

J Huang[1,2], X Pan[1], S B Dai[1,2] and Y Cai[1,3]

[1]School of Civil Engineering and Architecture, Wuhan University of Technology, Wuhan 430070, China
[2]Design & Research Institute, Wuhan University of Technology, Wuhan 430070, China

E-mail: clovya@whut.edu.cn

Abstract. This paper establishes the finite element model of saturated asphalt pavement with ANSYS. The thickness and elastic modulus of pavement surface layers and grassroots are taken as influencing factors to analyze the influence of these factors' change on the pavement mechanical behavior. The results indicate that the increase of surface thickness leads to the shear stress's decrease that may relieve the internal shear stress concentrated phenomenon effectively. The increase of elastic modulus of surface makes the pore pressure decrease, and result in the shear stress's increase. The grassroots thickness has no obvious effect on the pore-water pressure and shear stress; the grassroots elastic modulus' increase may increase the internal shear stress and decrease the pore pressure.

1. Introduction

As a kind of porous medium, the asphalt mixture is shape-complex, *randomly filled* and has porosity. Therefore, the pavement structure of asphalt mixture is of a certain permeability [1, 2]. Existing results indicate that the asphalt pavement in saturated water state is the most vulnerable to damage [3, 4]. When the vehicle passes through the water-covered pavement, water, which failed to be discharged in time, will produce enormous pore-water pressure in a short period. This kind of pore-water pressure will cause washing and squeeze to pavement structure that may cause the asphalt mixture in the pavement structure to lose the adhesive strength and peel off easily, causing the pavement damage [5]. Also, the presence of the shear stress also can induce pavement rutting damage [6]. Therefore, It's necessary to understand the dynamic behavior of asphalt pavement structure under the action of hydraulic and vehicle coupling effect. Also, in-depth study of the changing rule of the pavement mechanical response to pavement material parameters changes is essential. That can provide theoretical basis for damage prevention and control of pavement structure.

The research of moisture-loading coupling effect of asphalt pavement started in the 1960s.

Hunter has carried on the related research [7] about the hydraulic seepage law of asphalt pavement and arrived at a conclusion that when the diameter of the permeation apparatus is small or the grassroots is thick the deviation of the permeability of the pavement will be great.

[3] Address for correspondence: Y Cai, School of Civil Engineering and Architecture, Wuhan University of Technology, Wuhan 430070, China. E-mail: clovya@whut.edu.cn.

Kringos adopted a new modeling approach to study the phenomenon that the pore-water pressure of the asphalt pavement may result in the loss of the cohesive force of the aggregate. Through the study, Kringos made a great achievement about the interrelation between the cohesive force of the aggregate and the pore-water pressure

Krishnan studied the interrelation between permeability coefficient and porosity of the asphalt pavement. He held that the movement of the mixed particles in the asphalt leads to water damage of the pavement and established the particle motion equation [8].

Although, many splendid achievements about this field have been achieved, the current researches about seepage field and stress field coupling effect are incomplete. For instance, multi physical field coupling and more precise simulation of the pavement are still remained to be explored.

This paper, by using ANSYS finite element software to simulate saturated asphalt pavement which is under the vehicular action, analyzes the influence exerted by thickness' change of the surface and grassroots and elastic modulus's variation on pavement dynamic behavior.

2. The establishment of finite element model

2.1. Analysis model and boundary conditions

Because the pavement depth and length of the driving direction is infinite, the pavement in this case cannot be simulated by software. To guarantee the accuracy of the calculation, avoid a tedious calculation, the paper adopts 2 dimension plane model of 3 x3 m as the analysis model. As shown in figure 1, in structural layer thickness direction, the model from top to bottom are upper layer, mid-surface, lower layer, grassroots, sub-base and subgrade, and the layer thickness are respectively 0.04 m, 0.06 m, 0.08 m, 0.36 m, 0.20 m and 2.26 m.

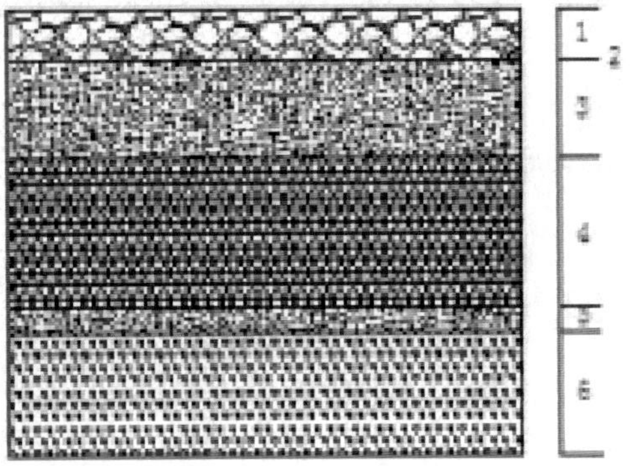

Figure 1. The Schematic diagram of the model.

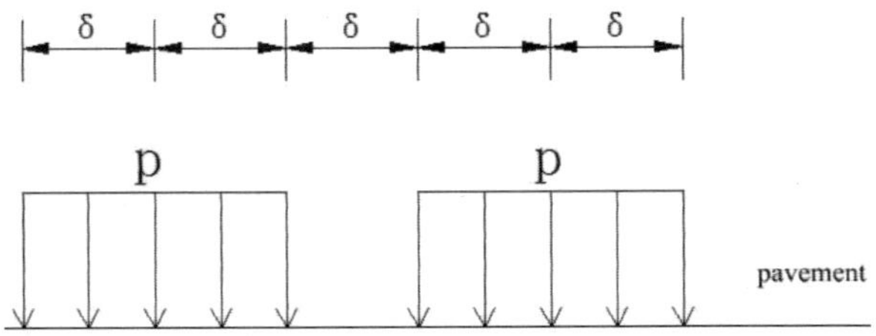

Figure 2. The model of wheel load.

Boundary conditions of the model are as follows: The boundary of the model in the horizontal direction has been established, but the boundary can be mobilized in the vertical direction and is waterproof. The upper boundary of the model is established in the horizontal direction but free in the vertical direction. Except the load region, the rest region is permeable. The lower boundary of the model is waterproof, and established in the horizontal direction and the vertical direction. Based on the schematic diagram of the model, this paper established the finite element model with ANSYS.

2.2. Basic parameters of pavement structure material

The density, Poisson's ratio, elastic modulus, saturated permeability coefficient and other parameters of each layer of the pavement structure model are shown in table 1.

Table 1. Parameters of each layer of the structure.

Name of each layer	Density/(g/cm^3)	Poisson ratio	Modulus of elasticity/MPa	Saturation permeability coefficient (m/s)
upper layer	2.64	0.30	1400	4.8×10^{-5}
mid-surface	2.55	0.35	1200	1.75×10^{-5}
lower layer	2.52	0.35	900	1.25×10^{-5}
grassroots	2.2	0.20	1600	1.23×10^{-7}
sub-base	2.2	0.30	350	2.65×10^{-4}
subgrade	1.65	0.40	40	3.32×10^{-9}

2.3. The dynamic load of the driving vehicle

The vehicle's dynamic load model is an approximate simulation of the process of vehicle running, Due to the circular uniformly distributed load is accordance with the actual situation of ordinary vehicle wheel's incentive. Therefore, according to the current rules and regulations, the vehicle load is simplified as double circular uniformly distributed vertical load [9] which is presented in figure 2.

When the vehicle drive down the road, the load's acting time on the pavement is very short., sinusoidal-load is adopted to simulate the driving load, the maximum circular uniformly distributed vertical load is P_{max}=0.7MPa, the time of the single acting driving is T. During the analysis time, the uniformly distributed load value function is as equation (1) [10]. The driving speed is 90 km/h that is fixed and immovable, the load time is 0.05 s, and the analysis time is T=0.25 s.

$$p=\begin{cases} P_{max}\sin\left(\frac{\pi}{T}t\right) & 0\le t\le 0.05\text{s} \\ 0 & 0.05\text{s}\le t\le 0.25\text{s} \end{cases} \tag{1}$$

3. The influence of the surface layer material

3.1. The influence of surface layer thickness

In this section, surface's thickness is taken as influencing factor. The paper considers five groups of layer thickness to analyze the pavement surface's mechanical response to different layer thickness's influence. Specific combinations situations are shown in table 2, except the layer thickness values, other parameters are presented in table 1.

Table 2. Surface layer thickness

	combination 1	Combination 2	combination 3	combination 4	combination 5
upper layer thickness/m	0.03	0.04	0.05	0.06	0.07
mid-surface thickness/m	0.05	0.06	0.07	0.08	0.09
lower layer thickness/m	0.07	0.08	0.09	0.10	0.11
Total thickness of surface layer/m	0.15	0.18	0.21	0.24	0.27

As shown in figures 3 and 4, the figures are respectively the changing rule of the maximum pore water pressure and shear stress in the surface layer structure under the condition of different thickness of surface layer. From figure 3, we can see the maximum pore-water pressure value of the surface layer's bottom increases monotonically with the increase of the surface layer's thickness. The maximum pore-water pressure of the bottom layer increased from 72.875 to 103.1 KPa when the surface layer's thickness increased from 0.12 to 0.3 m, rising by about 41% or so.

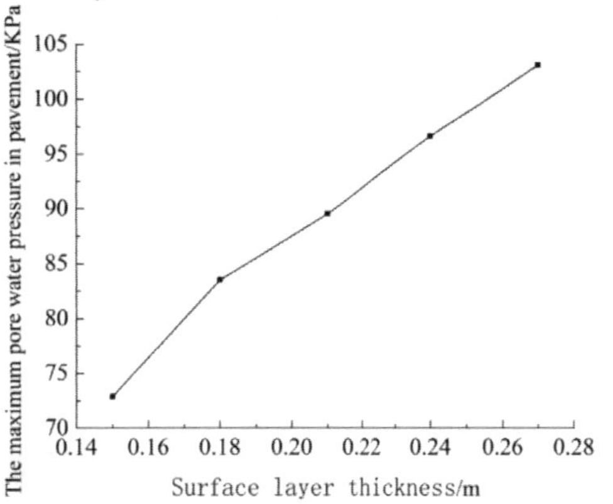

Figure 3. The influence of surface layer thickness on the maximum pore-water pressure in pavement.

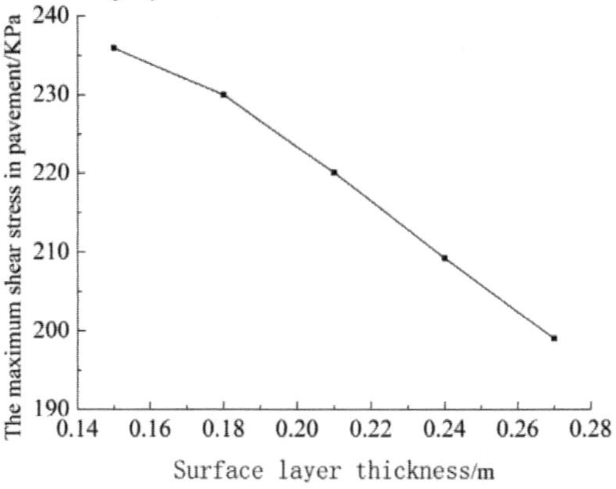

Figure 4. The influence of surface layer thickness on the maximum shear stress in pavement.

From figure 4, it can be seen that the thicknesses of the surface layer has a great effect on the shear stress, and the maximum shear stress decreases with the increase of the surface layer's thickness. When surface layer's thicknesses are 0.15 m and 0.27 m, the maximum shear stress of pavement structure is 235.92 KPa and 199 KPa. Stress concentration phenomenon in the pavement is effectively relieved. This suggests that the surface layer can hinder the force's wave-form transfer, and layer's thickness is the key factor of hindering effect.

3.2. The influence of surface layer modulus

In this section, five groups of surface layer modulus are selected as contrast factor to analyze the influence of modulus change on the mechanical behavior of Asphalt Pavement. Its surface layer modulus are in turn 800 MPa, 1000 MPa, 1200 MPa, 1400 MPa and 1600 MPa. Except the surface layer modulus values, other parameters are presented in table 1.

As shown in figures 5 and 6, the figures are respectively changing rule of the maximum pore-water pressure and shear stress in the surface layer structure under the condition of different surface layer modulus. As shown in figure 5, the maximum pore-water pressure of the bottom layer decreased from 92.046 to 73.45 KPa when the surface layer modulus values increased from 800 to 1600 MPa. When modulus is small, the effect of modulus change on pore-water pressure change is more obvious. The maximum shear stress in surface layer increases with the modulus, but the increase is slight, the shear stress only increases from 223.16 to 241.28 KPa.

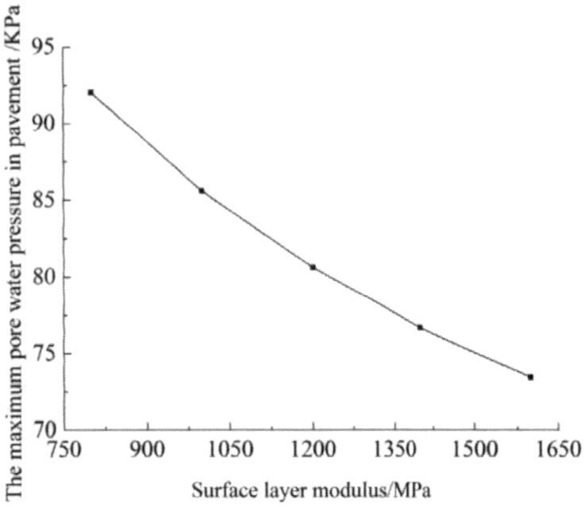

Figure 5. The influence of surface layer

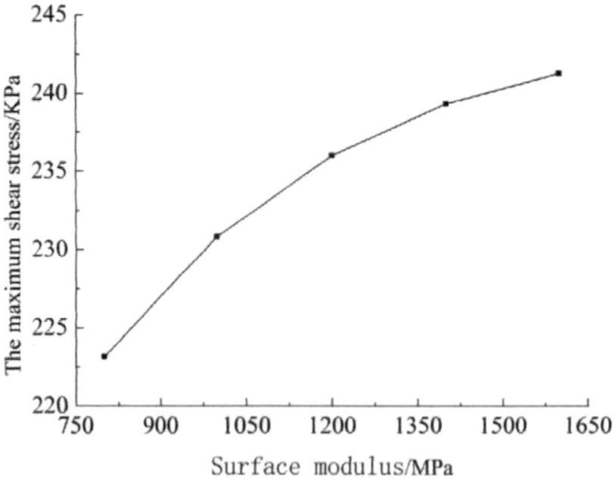

Figure 6. The influence of surface layer modulus

modulus on the maximum pore-water pressure in pavement. on the maximum shear stress in pavement.

4. The influence of grassroots materials

4.1. The influence of grassroots thickness

In this section, grassroots thickness is taken as influencing factor to analyze the influence of different grassroots thicknesses on pavement surface structure's mechanical response. The grassroots thicknesses respectively are 0.32 m, 0.36 m, 0.40 m, 0.44 m and 0.48 m. Except grassroots thickness values, other parameters are presented in table 1.

As shown in figures 7 and 8, the figures are respectively changing rule of the maximum pore-water pressure and shear stress in the surface layer structure under the condition of different thickness of grassroots. Figure 7 shows that when the grassroots thickness increased, the change range of the maximum pore-water pressure is basically between 80 and 85 KPa, the change is almost negligible. This is mainly because the permeability coefficient of grassroots is much smaller than surface layer's permeability coefficient. Thus, water is still concentrated in the surface layer structure and will not penetrate more moisture to the grassroots. Therefore, the change of grassroots thickness will not affect the pore-water pressure in the surface layer structure significantly. As shown in figure 8, when grassroots thickness increases from 0.32 to 0.48 m, the maximum shear stress of surface layer decreases from 239.94 to 200.6 KPa, and the variation of shear stress is consistent with the change of grassroots thickness.

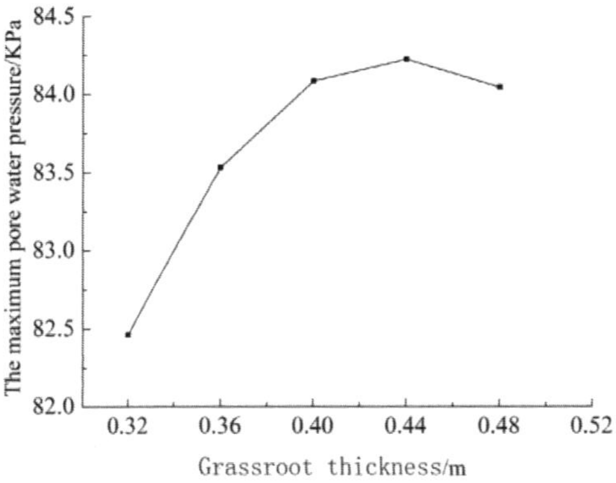

Figure 7. The influence of grassroots thickness on the maximum pore-water pressure in pavement.

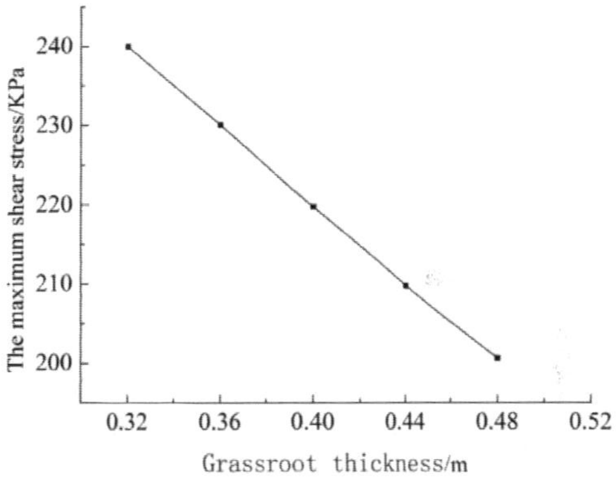

Figure 8. The influence of grassroots thickness on the maximum shear stress in pavement.

4.2. The influence of grassroots modulus

In this section, the grassroots modulus is taken as the influencing factor to analyze the influence of modulus change on the mechanical behavior of Asphalt Pavement. The grassroots modulus are respectively 1000 MPa, 1200 MPa, 1400 MPa, 1600 MPa and 1800 MPa. Except the surface layer modulus values, other parameters are presented in table 1.

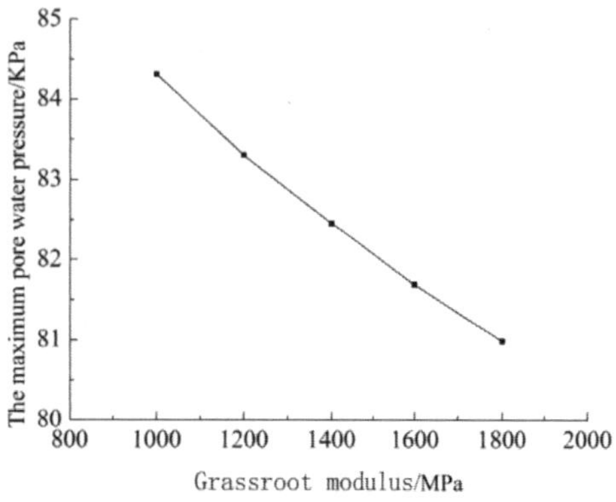

Figure 9. The influence of grassroots modulus on the maximum pore-water pressure in pavement.

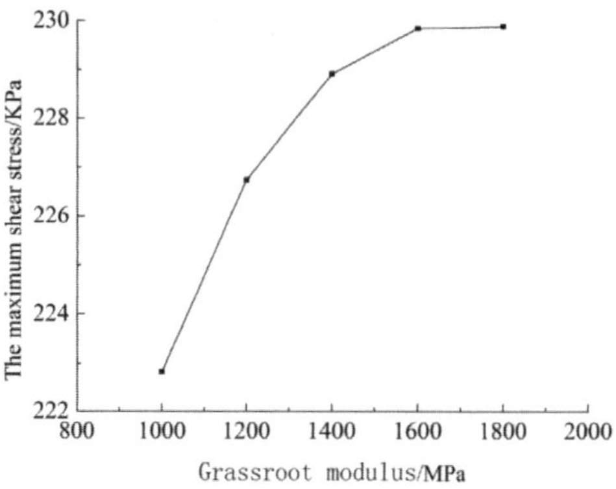

Figure 10. The influence of grassroots modulus on the maximum shear stress in pavement.

As shown in figures 9 and 10, the figures are respectively changing rule of the maximum pore-water pressure and shear stress in the surface layer structure under the condition of the different grassroot modulus. As shown in figure 9, the maximum pore-water pressure of the bottom layer decreases from 84.31 to 80.98 KPa when the grassroots modulus increases from 1000 to 1800 MPa, the variation of pore-water pressure is small. Also, the influence of grassroots modulus change on the maximum shear stress of the surface layer bottom is not obvious. When grassroots modulus increases, the shear stress increases from 222.82 to 229.88 KPa, and, along with the grassroots modulus's increase, the shear stress gradually tends to saturation.

5. Conclusions

By using ANSYS finite element software to simulate saturated asphalt pavement that is under vehicular action, the paper analyzed the influence of thickness' change of the surface layer and grassroots and elastic modulus's variation on pavement dynamic behavior. The following conclusions can be drawn:

- When the surface layer thickness increases, the pore-water pressure in the surface layer correspondingly increases which is disadvantageous for pavement resistance to water-induced damage. The surface layer can hinder the force's wave-form transfer, and, layer's thickness is the key factor of the hindering effect.
- The increase of modulus of the surface layer reduces the pore-water pressure. Therefore, the control of surface layer materials' elastic modulus is important for the pavement to resist the erosion of the pore-water.
- The change of grassroots thickness mainly affects the shear stress of pavement structure, but, the change of pore-water pressure is not obvious. The maximum shear stress in the structure of the surface layer decreases with the increase of the grassroots thickness.
- The change of grassroots modulus has little effect on the shear stress of pavement structure and the pore-water pressure.

Acknowledgments

This material is based upon work supported by the open subject foundation (grant no. DQJJ201301) of the Key Laboratory of Road Bridges and Structural Engineering of Hubei Province (Wuhan University of Technology).

References

[1] Wei P K 2012 Damage of concrete asphalt pavement under hydrodynamic pressure (Chongqing Jiao tong University, Chongqing, China: PhD Thesis)

[2] Dong Z J, Tan Y Q and Cao L P 2007 Research on pore pressure within asphalt pavement under the coupled moisture-loading action *Journal of Harbin Institute of Technology* **39** 1614-7
[3] Xu B 2003 *Theory and Practice of Drainage Asphalt Pavement* (Shanghai: Tongji University Press)
[4] Liu R J 2013 Dynamic response analysis of asphalt pavement under moving load (Shijiazhuang Tiedao University, Shijiazhuang, China: PhD Thesis)
[5] Sun L J 2003 *Asphalt Pavement Structure Behavior Theory* (Shanghai: Tongji University Press)
[6] Chen F F, Huang X M and Shan J S 2008 Mechanical analysis of different asphalt pavement under over loading *Shanghai Highways* **2** 11-5
[7] Hunter A E and Airey G D 2005 Numerical modeling of asphalt mixture site permeability *The 84th Annual Meeting of the Transptrtation Research Board* (Washington)
[8] Krishnan J M and Rao C L 2001 Permeablility and bleeding of asphalt concrete using mixture theory *International Journal of Engineering Science* **39** 611-27
[9] Sha A M 2011 *Roadbed Engineering* (Beijing: Higher Education Press)
[10] Yu L 2013 Study on porous asphalt pavement structures for urban road (Hunan University, Changsha, China: PhD Thesis)

IOP Conf. Series: Materials Science and Engineering **103** (2015) 012039 doi:10.1088/1757-899X/103/1/012039

The effect of the model posture on the forming quality in the CNC incremental forming

H Zhu[1], W Zhang, J L Bai, C Yu and Y F Xing

College of Mechanical and Electrical Engineering, Shenyang Aerospace University, Shenyang, Liaoning 110136, China

E-mail: zhuhu20@126.com

Abstract. Sheet rupture caused by a sheet metal thickness non-uniformity persists in CNC (Computer Numerical Control) incremental forming. Because the forming half cone angle is determined by the orientation of the model to be formed, so is the sheet metal's uniformity. The finite element analysis models for the two kinds of the postures of the model were established, and the digital simulation was conducted by using the ANSYS/LA-DYNA software. The effect of the model's posture on the sheet thickness distribution and the sheet thickness thinning rate were studied by comparing the simulation results of two kinds of the finite elements analyzes.

1. Introduction

The sheet metal CNC incremental forming is a flexible dieless forming technology adopting the idea of layered manufacturing [1-3]. The fundamental principle of sheet metal CNC incremental forming consists of the model to be formed being put in a certain direction (pressing direction) and the sheet metal being pressed point by point along the forming tool path [4, 5]. However, the problem of the thickness non-uniformity of the sheet metal part and easy rupture in the part of the big curvature is still relevant for the CNC incremental forming. In the forming process, the final thickness t_1 is dependent on the initial blank thickness t_0 and the forming half cone angle a, as given by $t_1 = t_0 \sin a$ [6-8], moreover, the half cone angle a is dependent on the posture of the model to be formed. Therefore, the posture of the model has a significant impact on the thickness thinning and distribution of the sheet metal. The optimization of the model posture, however, is also of a great importance [9, 10].

In this paper, the effect of the model posture on the forming quality has been discussed through the CNC incremental forming digital simulation process of the different posture models using the digital simulation technology based on the finite element analysis (FEA).

2. The establishment of the finite element analysis model

In order to compare and analyze the thickness distribution and thinning rate of the sheet, the ANSYS/LS-DYNA software is used for digital simulation. Figure 1(a) shows the original model posture. Figure 1(b) shows the model with a posture optimized according to the method of Zhu, Zhang and Ju [9], and with a surface reconstructed according to the method of Zhu, Zhang and Ju [11].

The two types of finite element analysis models of the CNC incremental forming process are

[1] Address for correspondence: H Zhu, College of Mechanical and Electrical Engineering, Shenyang Aerospace University, Shenyang, Liaoning 110136, China. E-mail: zhuhu20@126.com.

established for the two types of models, i. e. the original posture of the model and the optimized posture of the model as shown in Figure 2. The analysis model is composed of the following elements: the pressing tool heads, the sheet metals, and supports. The sheet metals are two square blank sheet metals the length and width of which are 100 mm×80 mm and 120 mm×100 mm respectively. Ball head part of the pressing tool contacting sheet metal is reserved and simplified as a sphere with a radius of 3 mm. The necessary CAD model of the support is obtained by offsetting the outside surface of the sheet metal model 1 mm inwards. Figure 3(a) and 3(b) show the support models generated on the basis of the original model (the green model) and the posture optimized model (the green model) respectively.

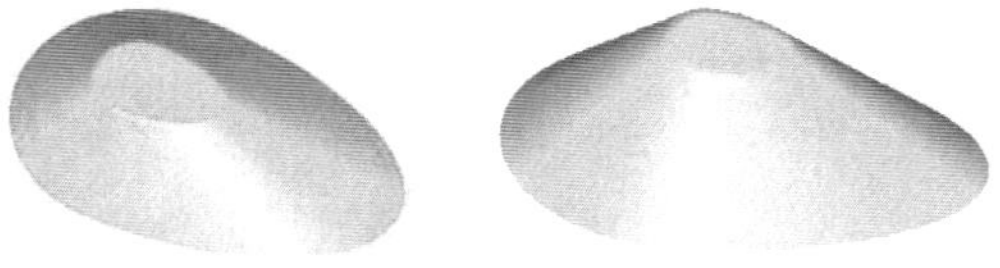

(a) (b)
Figure 1. Case study model: (a) Original model; (b) Posture optimized model.

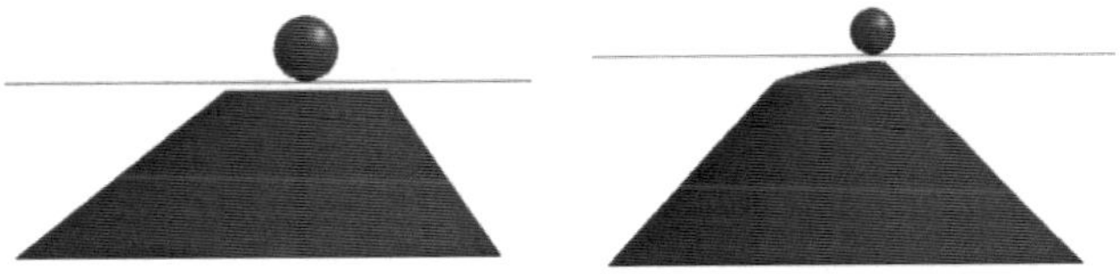

(a) For the original model (b) For the optimized model
Figure 2. The finite element analysis model.

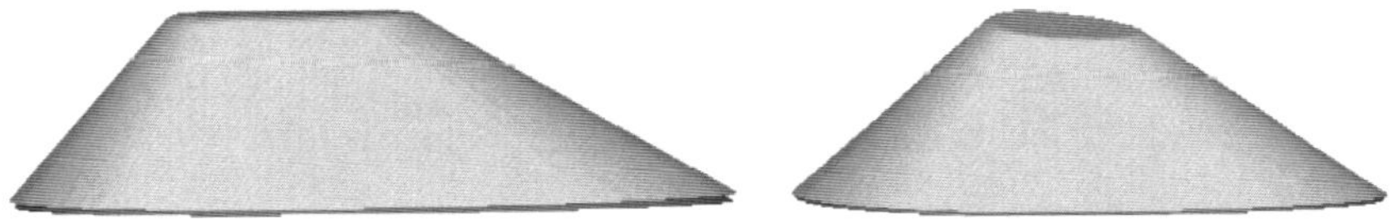

(a) For the original model (b) For the optimized model
Figure 3. Support model.

2.1. The finite element analysis condition setting
The SHELL 163 element was used in order to define the sheet metal, and the SOLID 164 element was adopted in order to define the support and the pressing tool [12]. The sheet was defined as a transversely anisotropic hardened plastic model, namely Al1060 aluminum sheet. The rigid body material model was used for the support and the pressing tool head, namely bearing steel GCr15. The sheet was meshed by the mapped meshing with a size of 1 mm, the pressing tool head was meshed by the free meshing with a size of 1 mm, and the support was meshed by the free meshing with a size of 4 mm.

2.2. Description of pressing movement
In this paper, the vertex offset method of Qu and Stucker [13] was used. In order to generate an equidistant surface for the forming tool path, the outer surface of the sheet part model was offset 3 mm towards the outside equaling the distance of the pressing tool radius. Then, in order to obtain the contour line path, the equidistant surface was cut with the plane vertical to Z axis with a spacing distance of 1 mm, and it was dispersed into a serious of path points which were put into the ANSYS system to load. The generated forming path is shown in Figure 4.

3. Results of the finite element analysis

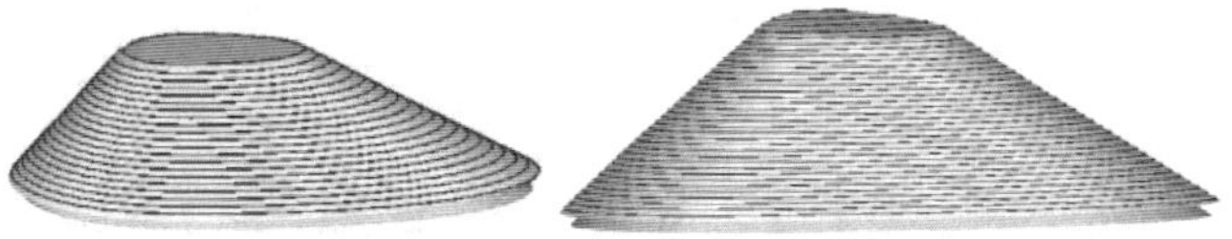

(a) Original model (b) Posture optimized model

Figure 4. Generation of the forming path.

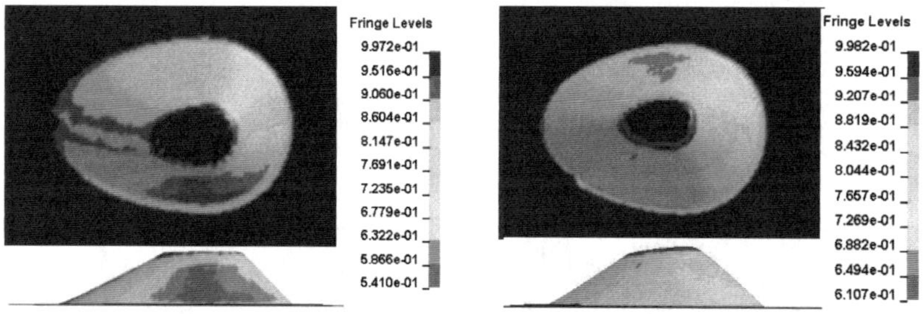

(a) the original model (b) the optimized model

Figure 5. Cloud pictures of the sheet thickness distribution.

3.1. Comparative analysis of the sheet thickness distribution

Figure 5(a) and 5(b) show the cloud pictures for the sheet thickness distribution of the finite element analysis. Figure 5(a) shows the digital simulation result in the sheet CNC incremental forming process based on the original model, the minimum final sheet thickness is 0.5140 mm and concentrates on the area of the higher pressing angle. However, the digital simulation result in the sheet CNC incremental forming process based on the optimized model shows the uniform sheet thickness distribution, the minimum final sheet thickness is 0.6107 mm. It is evident that the minimum sheet thickness has been increased by 0.0697 mm due to the model posture optimization (Figure 5(b)).

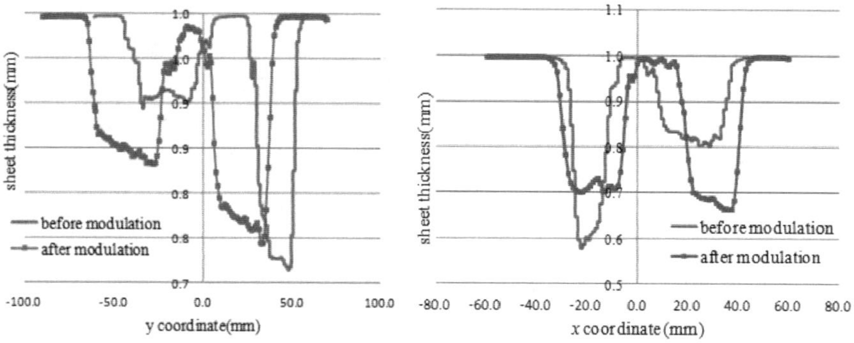

(a) Direction of Y axis (b) Direction of X axis

Figure 6. Sheet thickness.

As shown in Figure 6(a), on the outline of the middle surface parallel to the Y axis: for the original posture of the model, the minimum sheet thickness is 0.718 mm at the section of x>0; for the optimized posture of the model, the minimum sheet thickness is 0.745 mm at the same section of x>0, the minimum sheet thickness has increased by 0.027 mm, therefore, the thickness of the sheet is more uniform. As shown in Figure 6(b), on the outline of the middle surface that is parallel to the X axis: for the model's original posture, the minimum sheet thickness is 0.579 mm at the section of y<0; for the model's optimized posture, the minimum sheet thickness is 0.665 mm at the section of y>0, the minimum sheet thickness has increased by 0.086 mm, therefore, it can be concluded that the sheet thickness of the posture optimized model is more uniform than the posture of the non-optimized model. It is evident from the positive and negative sides of the X and Y axes that the thickness difference on both sides of the model has decreased after the posture of the model was optimized.

Before the model's posture optimization, the model thickness difference is 0.183 mm along both sides of the X axis and 0.226 mm along both sides of the Y axis. After the model's posture was optimized, the model thickness difference is 0.088 mm along both sides of the X axis and 0.041 mm along the both sides of the Y axis.

As the forming angle's difference is large along the direction, the thickness uniformity along the Y axis is observable. But the effect of model-posture-optimization is not significant along the X axis because the forming angle's difference along the X axis is not big enough.

3.2. Comparative analysis of the sheet thickness thinning rate

The sheet thickness thinning rate's cloud pictures are shown inFigure 7. The CNC incremental forming's digital simulation results, based on the original model, show that the thickness thinning is concentrated around the area with the bigger forming angle, which has a maximum sheet thickness thinning rate of 45.81% (Figure 7(a)). However, the digital simulation result of the CNC incremental forming, based on the posture optimized model, portray a more uniform sheet, with a maximum sheet thickness thinning rate of 37.67%. The sheet thickness thinning rate decreased by 8.2% due to the model's posture optimization (Figure 7(b)).

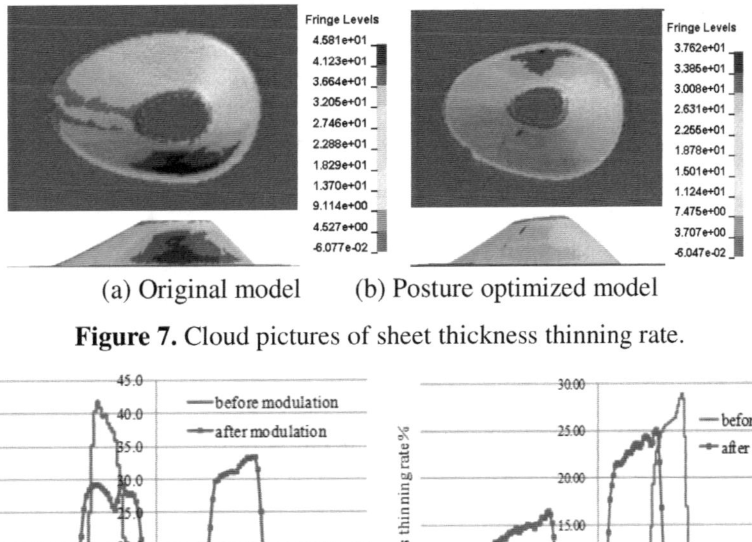

(a) Original model (b) Posture optimized model

Figure 7. Cloud pictures of sheet thickness thinning rate.

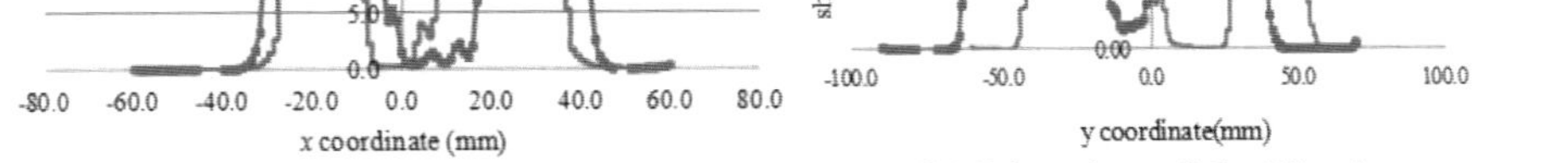

(a) Direction of the Y axis (b) Direction of the Y axis

Figure 8. Sheet thickness thinning rate.

As shown inFigure 8(a), on middle surface's outline that is parallel to the Y axis, before the posture optimizing, the maximum sheet thickness thinning rate is 28.91% at the section of x>0, while after the posture optimization, the maximum sheet thickness thinning rate is 24.86% at the section of x>0. The of the maximum sheet thickness thinning rate's difference is 4.05%.Figure 8(b) portrays, on the outline of the middle surface that is parallel to the X axis, that before the posture was optimized, the maximum sheet thickness thinning rate lays at 41.74% for<0, while after the posture optimizing, the maximum sheet thickness thinning rate scores 28.98% at the section of y<0 and 33.23% for y>0. Therefore, the maximum sheet thickness thinning rate is 33.23% and the difference between them is 8.54%. Before the model's posture optimization, the model thinning rate's difference on both sides lays at 19.805% on the outline of the middle surface which is parallel to X axis. After posture optimizing, the thinning rate's difference on both sides of the model is 8.598%. Before model posture optimizing, the maximum difference for the thinning rate on both sides of the model is 22.671% on the outline of the middle surface which is parallel to the Y axis. After posture optimizing, the model thinning rate's maximum difference on both sides is 4.368%. Thus, the sheet thickness thinning rate distribution is more uniform after the model posture optimizing on the both sides of the X and Y axes.

4. Conclusions

The finite element analysis results of the CNC incremental forming show that the model posture has a direct effect on the sheet thickness distribution and the thickness thinning rate. The effect of the model posture is especially important when the curative difference of the surfaces is large. The sheet thickness difference and distribution non-uniformity can be decreased by optimizing the model posture reasonably, which can also help avoiding the rupture problem caused by the large differences of the sheet thickness and distribution non-uniformity.

Acknowledgments

The authors gratefully acknowledge the financial support from the Natural Science Foundation of Liaoning Province (No. 2014024011).

References

[1] Tisza M 2012 General overview of sheet incremental forming *J. of Achiev. in Mater. and Manuf. Eng.* **55** 113-9
[2] Mo J H and Han F 2008 State of the arts and latest research on incremental sheet NC forming technology *China Mecha. Eng.* **19** 491-497
[3] Micari F, Ambrogio G and Filice L 2007 Shape and dimensional accuracy in single point incremental forming: State of the art and future trends *J. of Mater. Process. Technol.* **191** 390-5
[4] Azaouzi M and Lebaal N 2012 Tool path optimization for single point incremental sheet forming using response surface method *Simu. Modell. Prac. and Theory* **24** 49-58
[5] Bahloul R, Arfa H and Bel Hadj Salah H 2014 A study on optimal design of process parameters in single point incremental forming of sheet metal by combining Box–Behnken design of experiments, response surface methods and genetic algorithms *Int J Adv Manuf Technol* **74** 163-85
[6] Tekkaya A E, Shankar R, Sebastiani G, Homberg W and Kleiner M 2007 Surface reconstruction for incremental forming *Production Eng.* **1** 71-8
[7] Young D and Jeswiet J 2004 Wall thickness variations in single point incremental forming *J. of Eng. Manuf.* **218** 1453-9
[8] Hirt G, Ames J, Bambach M and Kopp R 2004 Forming strategies and process modelling for CNC incremental sheet forming *Ann CIRP* **53** 203-6
[9] Zhu H, Zhang W and Ju J 2015 Determination of the model posture in the CNC incremental forming based on STL model *Forg. & Stamp. Technol.* **43** 74-7
[10] Meng L, Gao L, Shi X F, Zhang Q L and Ren Y X 2013 Lofting face of NC incremental forming *Mecha. Eng. & Autom.* **1** 134-6
[11] Zhu H, Zhang W and Ju J 2015 Research on the model surface reconstruction for the optimization of the forming direction in CNC incremental forming *Hot Work. Technol.* **44** 1-5
[12] Lin W W 2014 Study on CAD/CAM system for NC incremental reverse forming (Shenyang, Shenyang Aerospace University: PhD Thesis)
[13] Qu X Z and Stucker B 2003 A 3D surface offset method for STL-format models *Rapid Prototyping J.* **9** 133-41

Simulation and design of ECT differential bobbin probes for the inspection of cracks in bolts

S W Ra[1], K H Im[2,7], S G Lee[3], H J Kim[3], S J Song[3], S K Kim[4], Y T Cho[5], Y D Woo[2] and J A Jung[6]

[1]Research Center, Seoul Metal Co. Ltd., Incheon, 440-310, Korea
[2]Department of Mechanical and Automotive Engineering, Woosuk University, Jeonbuk, 565-701, Korea
[3]Department of Mechanical Engineering, Sungkyunkwan University, Kyung-do 440-746, Korea
[4]Division of Mech. System Engineering, Chonbuk Natl. University, Jeonbuk, 561-756, Korea
[5]Department of Manufacturing Design Engineering, Jeonju University, Jeonbuk, 560-759, Korea
[6]Department of Mechanical and Automotive Engineering, Songwon University, Gwangju 501-210, Korea

E-mail: khim@woosuk.ac.kr

Abstract. All Various defects could be generated in bolts for a use of oil filters for the manufacturing process and then may affect to the safety and quality in bolts. Also, fine defects may be imbedded in oil filter system during multiple forging manufacturing processes. So it is very important that such defects be investigated and screened during the multiple manufacturing processes. Therefore, in order effectively to evaluate the fine defects, the design parameters for bobbin-types were selected under a finite element method (FEM) simulations and Eddy current testing (ECT). Especially the FEM simulations were performed to make characterization in the crack detection of the bolts and the parameters such as number of turns of the coil, the coil size and applied frequency were calculated based on the simulation results.

1. Introduction

Oil filters for a use for vehicular parts are being used under the high temperature and cooling of the engine and defects could be generated under repetition in the operation environment as well as shape changes such as very high internal loss for the bolts [1]. Also, it is impossible to check the defect caused inside the bolt as shown in figure 1 visually. These defects could make engine efficiency dropped when operating the engine, and may lead to accelerated wear and damage to the engine parts by many abrasive particles contained in the lubricating oil, if not prevented in advance upon finding in its initial stage. Such things may affect the life and efficiency of auto engine leading to serious economic problems. Therefore, as an applicable method for detecting such fine surface defects of a few hundred μm inside the bolt, ECT techniques are known as the best among the non-destructive

[7] Address for correspondence: K H Im, Department of Mechanical and Automotive Engineering, Woosuk University, Jeonbuk, 565-701, Korea. E-mail: khim@woosuk.ac.kr.

evaluation methods [2-9].

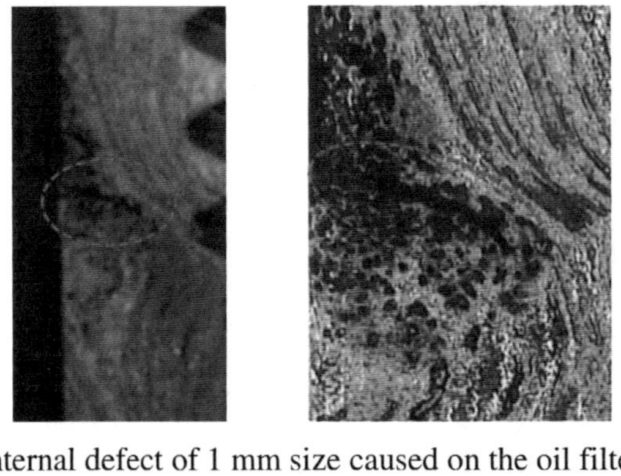

Figure 1. Internal defect of 1 mm size caused on the oil filter bolts

Oil filter bolts have high possibility of causing wears in the lubrication phase. Types of wear are of fusion, wear and burning respectively. By high temperature, cooling and high speed operation of engine, scant oil flow, foreign material from outside and particles in the oil may cause wear [2]. Defects developed on the surface were classified into circumferential crack, axial crack and angular crack respectively, and the work in this study has focused on the development of differential bobbin eddy current sensor that is applicable to detect the circumferential crack.

In this study, standard specimens with a rod type were prepared and differential bobbin eddy current sensor was designed, which can detect fully circumferential cracks on the surface of the specimen. Using the designed sensor, experiment was conducted for the standard specimen, and the results of the experiments were compared against each other. It was found that the differential bobbin eddy current sensor thus developed was appropriate for detecting cracks on the bolt.

2. Related theory

2.1. Eddy current testing

When AC current flowing coil is brought near to the conductor of test specimen, the primary magnetic field generated by the current flowing in the coil induces secondary magnetic field on the conductor. An eddy current coil developed from AC current can be obtained roughly by AC Circuit that includes resistance and inductance. Impedance, Z, the ratio between the voltage and current, is obtained by use of Ohm's Law. Impedance, Z, is shown as follows.

$$Z = \frac{V}{I} \tag{1}$$

When an AC current flow through coil inductance, L, at frequency f, impedance of coil is identical to the induced reactance of the circuit, X_L. Likewise, impedance of AC current flowing in the coil of resistance R and inductance L at operating frequency f is shown as the following formula.

$$Z = R + jX_L = R + \omega jL = R + j2\pi fL \tag{2}$$

Magnitude of impedance in the coil and phase angle can be expressed as follows.

$$Z = \sqrt{R^2 + {X_L}^2} \tag{3}$$

$$\theta = Tan^{-1}\left(\frac{X_L}{R}\right) \tag{4}$$

2.2. Standard penetration depth

When eddy current probe is placed on the test specimen, the eddy current induced into the test specimen is not distributed evenly in the test specimen. Density of the eddy current is highest in the surface of the test specimen, decreasing by the depth below the surface of the specimen. This is called skin effect. Equation of standard penetration depth is shown as below;

$$\delta = \frac{1}{\sqrt{\pi f \mu_0 \mu_r \sigma}} \tag{5}$$

By Eq. (5), standard penetration depth for frequencies has been calculated as shown in table 1 below.

Table 1. Standard depth of penetration.

Frequency (kHz)	δ(mm)	2δ(mm)
1	0.6426	1.2851
5	0.2874	0.5747
10	0.2032	0.4604
15	0.1659	0.3318
20	0.1437	0.2974
30	0.1173	0.2346
40	0.1016	0.2032
50	0.0909	0.1817
100	0.0643	0.1285

2.3. Wheatstone bridge

Two general systems are used; i.e. electrical bridge circuit and filter circuit. These two systems make impedance, Z, the value related with basic signal, electrically equilibrium. As shown in figure 2, Wheatstone Bridge was employed in the experiment. Formula of equilibrium is as follows.

$$\frac{Z_1}{Z_2} = \frac{Z_3}{Z_4} \tag{6}$$

Once the condition of Eq. (6) is reached, the system becomes equilibrium and the volt meter reads zero (0). Therefore, phase value and amplitude are measured using Wheatstone Bridge [5].

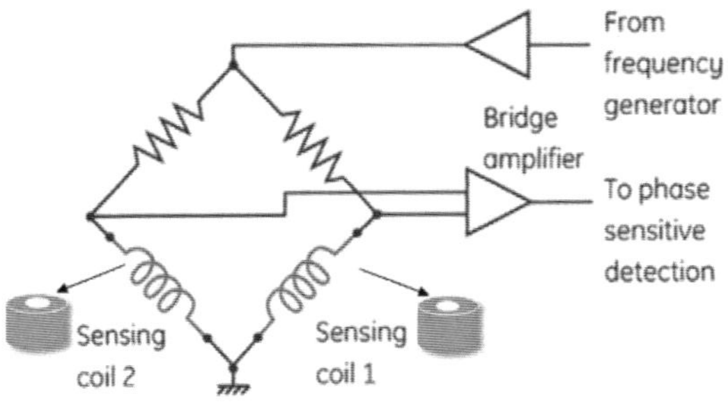

Figure 2. Wheatstone bridge.

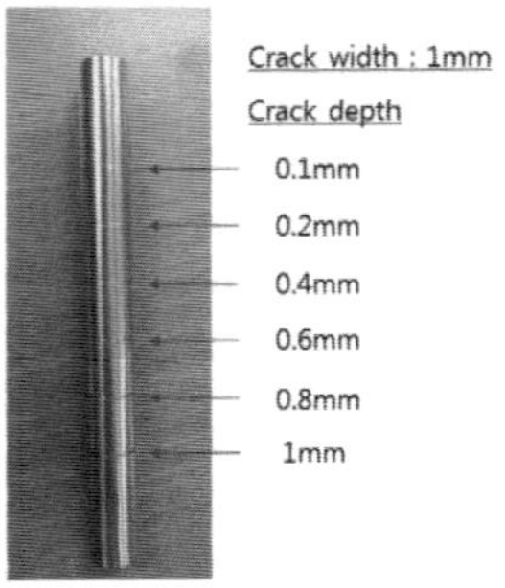

Figure 3. Standard Test specimen.

3. Developing eddy current sensor for bolts

3.1. Eddy current design and manufacture

In this study, the ultimate objective can be realized by use of interior eddy current sensor when detecting the surface defect developed in the oil filter bolts. Interior eddy current sensor is structurally identical to the exterior eddy current sensor for detecting exterior surface defect in the simulation test specimen as shown in figure 3.

Design parameters required for designing eddy current sensors can be thought of coil wire diameter, coil gap, coil width, height, coil turns, lift-off, frequency and others as shown in figure 4(a).

In this work, coil gap, coil width and frequency were selected as design parameters. For coil wire, coil of 0.1 mm in diameter was used, with 100 coil turns and 0.5 mm lift-off. With consideration of other parameters, a most optimum eddy current sensor was developed. Lift-off is one of important design parameters in designing eddy current sensor. Smaller lift-off makes the magnetic field induced in the test specimen stronger. So, lift-off was set to 0.5 mm. Using the bobbin eddy current sensor thus manufactured, defect signal of fully circumferential cracks were detected by moving the test specimen in axial direction. Experiment was conducted for change of test frequency and for change of coil width and gap.

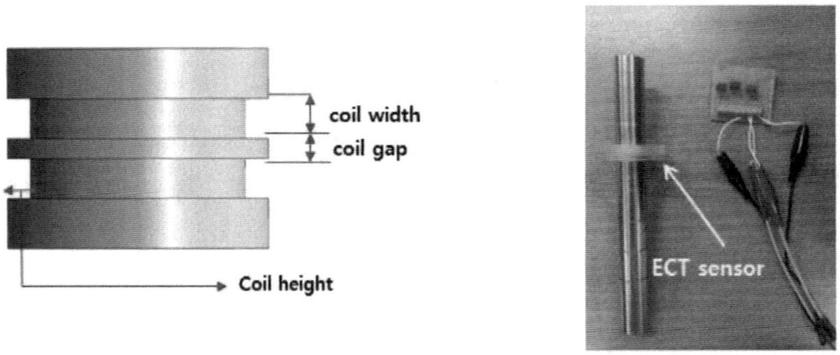

(a) Coil parameter (b) Manufactured Probe

Figure 4. Differential bobbin probe.

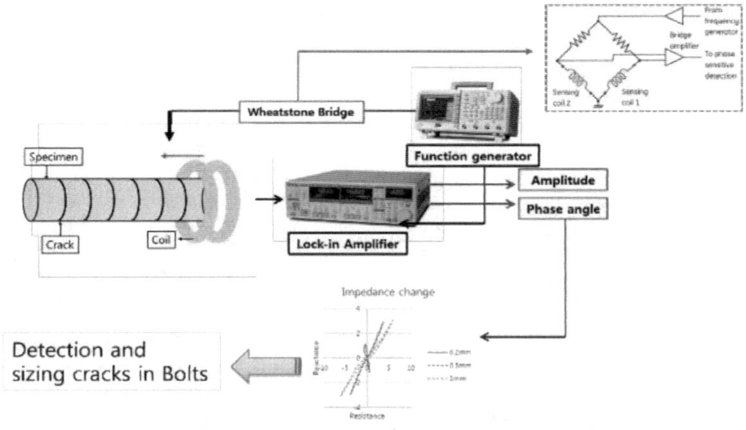

(a) Eddy current testing system configuration for bobbin probe

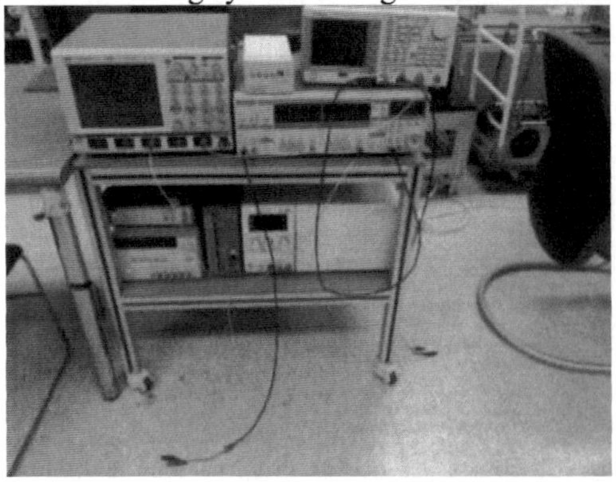

(b) Experiment setup

Figure 5. Eddy current testing system.

3.2. Making standard test specimen

Simulating oil bolt of 25 mm inner diameter as shown in figure 2, rod shape standard test specimen of 25 mm outer diameter and 231 mm length was made for inspecting exterior surface defects. Among the defects that may be generated outside, circumferential cracks of 0.5 mm width and 0.1, 0.2, 0.4, 0.6, 0.8 and 1 mm depths respectively that may be possible to inspect by the bobbin were artificially notched by EDM on the test specimen. Standard Test Specimen was of AISI 1045 Steel.

3.3. Eddy current test systems

In order to evaluate the signal property of the differential bobbin eddy current sensor applicable to detect defects in the test specimen, eddy current system was developed by application of Wheatstone Bridge as shown in figure 5(a).

In order to make Bobbin Eddy Current Sensor possible to evaluate defect detecting property of test specimen having fully circumferential cracks, frequency generating system has employed Function Generator of Tektronix having frequency band of 100 MHz. Through these Lock-in amplifier and Oscilloscope, change of amplitude value, i.e. amplitude change of impedance and change of phase value could be obtained. Experiment instruments used for sensor design and performance evaluation were as shown in figure 5(b).

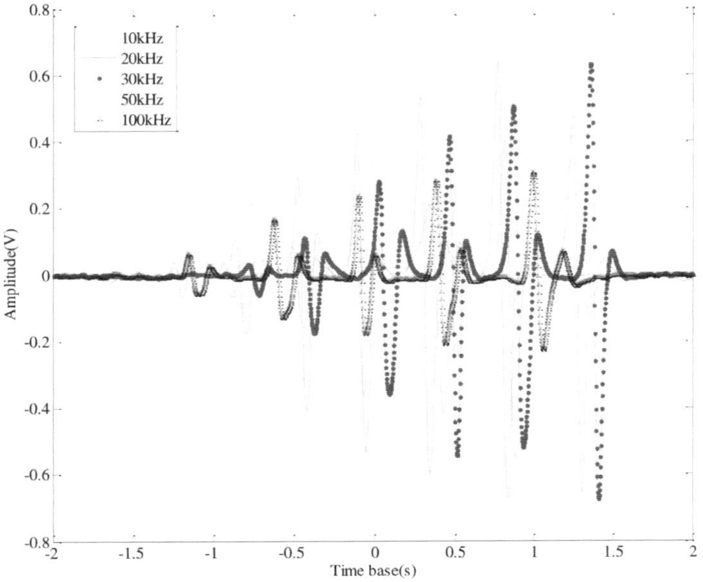

(a) Comparison of amplitude value for test frequency

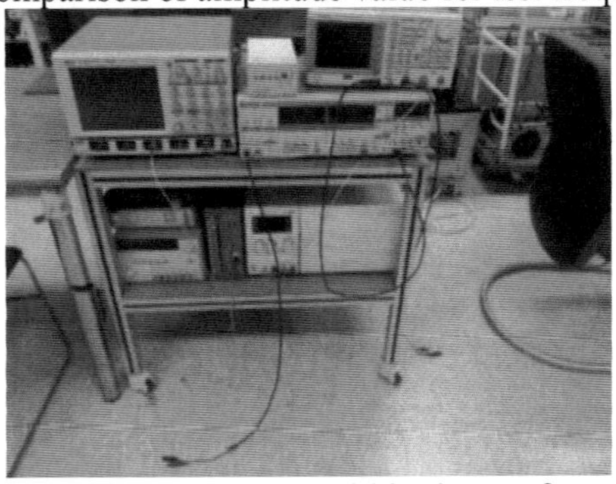

(b) Comparison of phase valves within the test frequency range

Figure 6. Comparison of amplitude and phase value ((a), (b)) for frequencies of 10, 20, 30, 50 and 100 kHz.

3.4. Evaluation of ECT property by change of frequency

For frequencies used in this work, amplitude and phase value were obtained as test results by lock-in amplifier and oscilloscope as shown in figure 6. The experiment was carried out with frequencies of 10, 20, 30, 50 and 100 kHz, and values of amplitude and phase were shown in figures 6(a). It could be observed that deeper the depth of defect in the specimen made bigger the change of amplitude and phase value. Frequencies used in the experiment were overlapped as shown in figures 6(a) and 6(b). In figure 6(a), it was found that signals from 20 kHz and 30 kHz were of similar strength each other and stronger than from other frequencies. This indicates that these frequencies are capable to detect defects of 0.2 mm depth since the values of 2 δ for frequency of 20 and 30 kHz are 0.29 mm and 0.23 mm respectively in table 1.

Therefore, in this study, frequencies of 20 kHz and 30 kHz could be selected as finally applicable for depth of defect minimum 0.2 mm [10]. In this study, frequencies of 10, 20, 30, 50 and 100 kHz were used in the experiment, which is the frequency range decided by Eq. (5) and table 1. With changing of frequencies, surface defect signals from the test specimen were obtained. As shown in figure 7, differences of amplitude and phase value by change of defect depth could be observed. From the experiment, it could be found that frequency of 50 and 100 kHz had lower sensitivity of detecting defect signals than the other 3 frequencies of 10, 20 and 30 kHz.

In figure 7, 20 kHz and 30 kHz showed similar level of change, and phase value became smaller by the order of 10, 20, 30, 50 and 100 kHz. Even though 10 kHz made the biggest change of phase value, it was excluded in further experiment since this is not adequate for detecting defect of 0.2 mm depth, the final objective of this study. So, it was found that frequency of 20 kHz could be a best case for detecting cracks due to higher sensitivity.

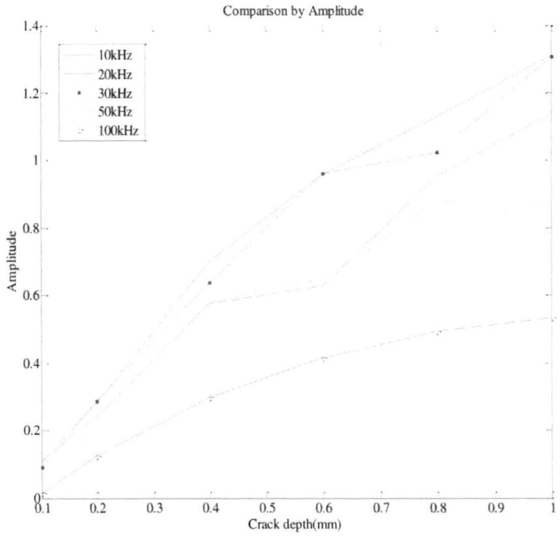

Figure 7. Comparison of amplitude value difference of bobbin eddy current sensor by change of test frequency.

3.5. Evaluation of ECT Characteristics by Change of Coil Gap

Experiment was conducted with change of Coil Gaps for test frequency of 20 kHz and for coils of 1 mm and 2 mm width respectively. With change of Coil gaps in 0.5 mm levels by 3 mm, 3.5 mm, 4 mm and 4.5 mm respectively, change of signal characteristics for 4 coil gaps could be obtained as shown in figure 8, and made analysis.

When compared at test frequency of 20 kHz as shown in figure 9, ΔPhase value of coil gap for coil width of 1 mm and 2 mm respectively, it was found that the ΔPhase value was the biggest at coil gap of 3.5 mm, and the value became bigger when coil width was 1 mm [10].

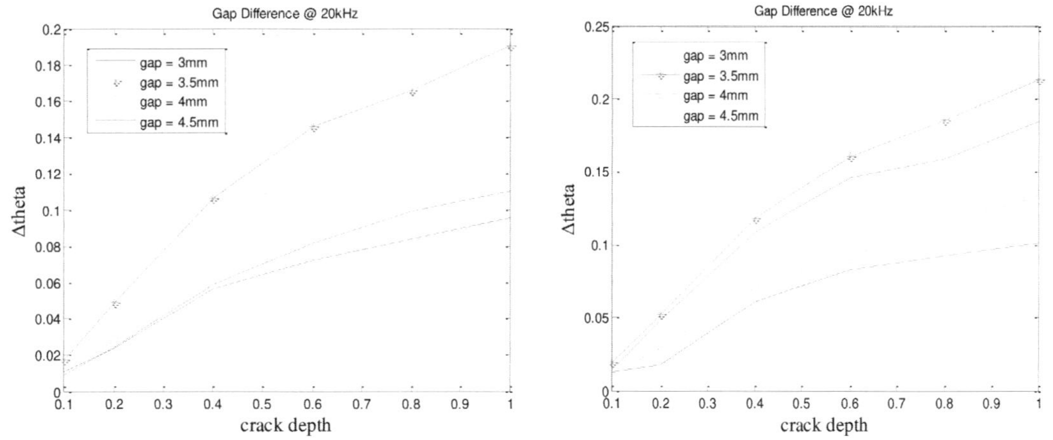

(a) coil width 1 mm @ 20 kH (b) coil width 2 mm @ 20 kHz

Figure 8. Comparison of ΔPhase value by change of coil gap (Frequency 20 kHz).

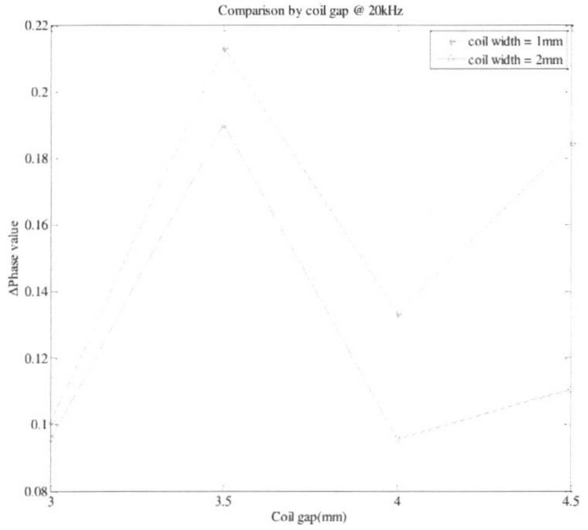

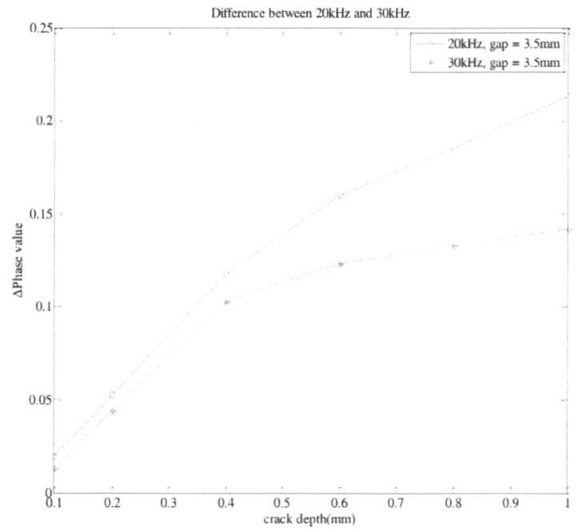

Figure 9. 20 kHz, Comparison of ΔPhase values for coil width of 1mm and 2 mm and coil gaps when depth of defect is 1 mm.

Figure 10. Comparison of ΔPhase value when coil width is 1 mm and 3.5 mm gap for frequency of 20 and 30 kHz.

3.6. Evaluation of ECT Characteristics by changes of frequency for coil gap

Since good performance was shown with 1mm coil width and 3.5 mm coil gap in Charter 3.5, values analysed there were made base for obtaining the graph of figure 10. In order to find which frequency shows bigger phase value change among frequencies of 20 kHz and 30 kHz, graph of figure 10 was obtained. From figure 10, bigger change of phase value was observed for frequency of 20 kHz than that of 30 kHz [10].

4. ECT Sensor Design Parameters based on FEM Simulation

ECT sensor design parameters were set up in order to simulate the signals of the eddy current based on FEM-based eddy current simulation as shown in figure 11 and simulation was carried out for the defect signal of eddy current probe. Exciter and receiver were utilized as shown in figure 11(a) and the variation of signals in the eddy current was simulated to arrow direction in bolts. Figure 11(b) shows a case of mesh generation in ECT simulation. Figure 12 shows the simulation results of the distribution and signals of the eddy currents for the bolt internal defects with 0.15 mm, 0.2 mm, 0.5 mm and 1 mm in defect depth under the difference coil gaps (3.5 mm, 4.0 mm and 4.5 mm). It was found that higher impedance variations of Lissajous plane (relation between resistance and reactance) were generated in case of the gap of 3.5 mm as shown in figure 12(a). This result in the gap of 3.5 mm is the most reasonable signal and it could be applied to design parameter in differential probe. Therefore, the design parameters for bobbin-types were optimized under Eddy current FEM simulations.

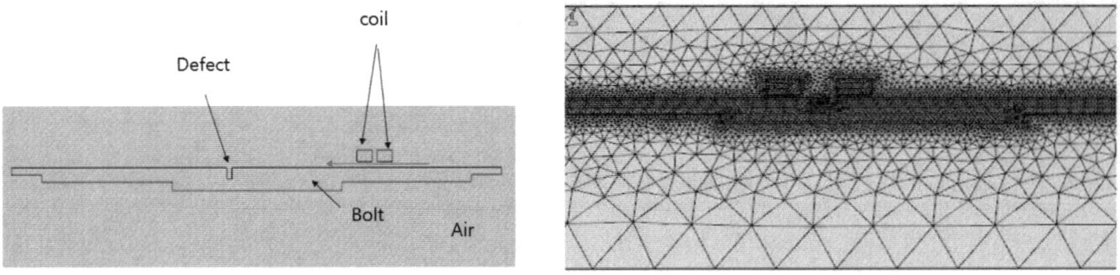

(a) Simulation setup (b) Mesh generation

Figure 11. Simulation setup of bobbin ECT probe.

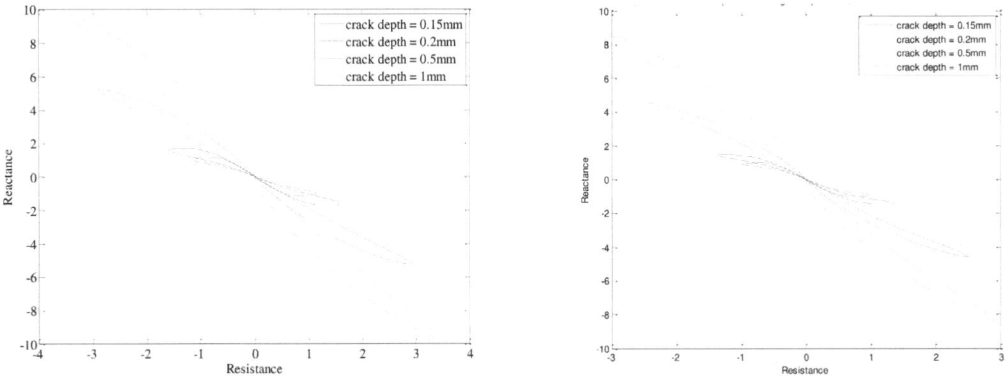

(a) Gap of 3.5 mm (b) Gap of 4.0 mm

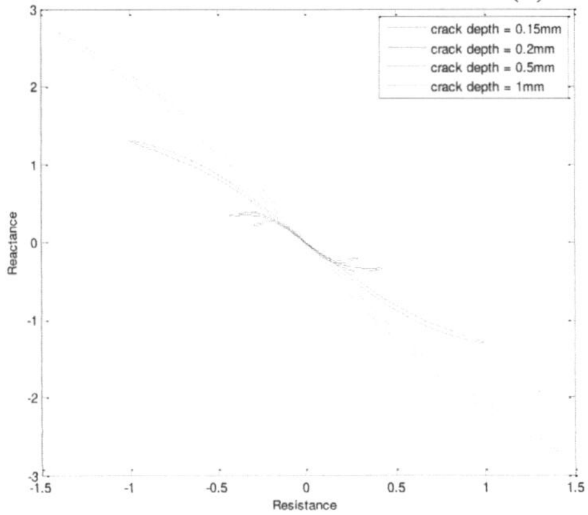

(c) Gap of 4.5 mm

Figure 12. Simulation results of Bobbin ECT probe under frequency of 20 kHz and coil width of 1 mm.

5. Performance evaluation of bobbin eddy current sensor

From the results of above experiment, optimum coil width, coil gap and operating frequency were selected. Based on these selected parameters, differential bobbin probe was designed and manufactured finally as shown in figure 13. Performance of the eddy current sensor designed under the same experimental condition done as above using the designed probe was verified. Result of performance evaluation for the bobbin eddy current sensor is shown in figure 14. Figure 14(a) shows resistance of differential bobbin eddy current sensor at frequency of 20 kHz, and (b) is reactance. From (a) and (b) and Eqs. (2) and (3), graph plotted with resistance in X-axis and reactance with Y-axis gives impedance plane of figure 14(c). As can be seen in figure 14(c), magnitude of impedance becomes bigger by bigger size of defect. Therefore, it was found that the differential bobbin eddy current sensor designed through the process of selecting design parameters as above is suitable for detecting the target surface defect of 0.2 mm depth at frequency of 20 kHz [10].

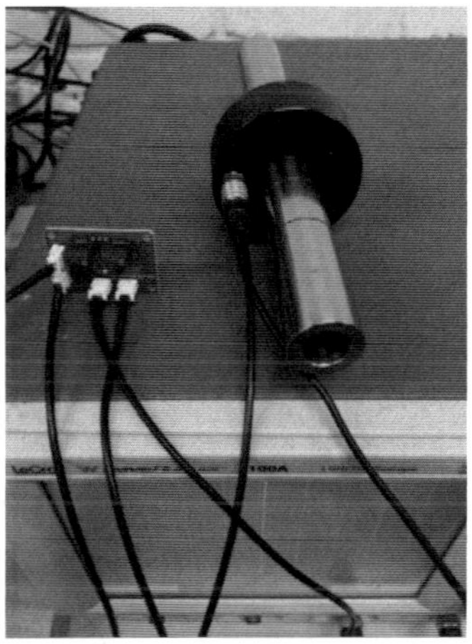

Figure 13. Finally designed differential bobbin probe.

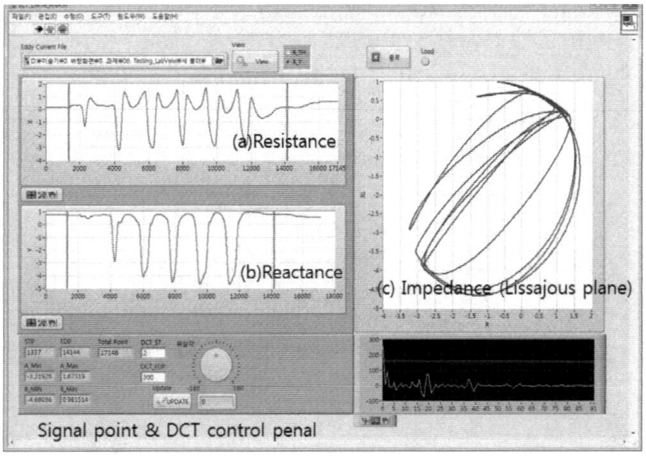

Figure 14. Signal characteristics of bobbin eddy current sensor at frequency of 20 kHz.

6. Conclusions

In this study, a differential bobbin eddy current sensor for detecting surface defects on the test specimen was developed through experiment, and the base for manufacturing interior bobbin eddy current sensor was worked out. Introduced conclusions are as follows;

- It was found that the phase value and amplitude becomes gradually bigger with depth of the test specimen deeper at the respective test frequencies as based on the experiment for standard test specimen of 25 mm outside diameter AISI 1045 Steel and differential bobbin probe having 0.1 mm diameter and 1 mm coil width in 100 turns of winding conducted with 3.5 mm coil gap at test frequency of 10, 20, 30, 50 and 100 kHz respectively.
- In an effort to make the frequency band to be applied on the differential bobbin eddy current sensor narrower, pertinent amplitude and phase value were obtained by use of lock-in amplifier. Frequency of 20 kHz was found most adequate followed by 30 kHz next, Frequency of 100 kHz was found making defect signal saturated for defect larger than 0.2 mm.
- Among coil width of 1 mm and 2 mm, change of phase value was bigger in case of 1 mm, and 3.5 mm coil gap was the best for change of phase value among the 4 coil gaps. With the defect becoming bigger, 20 kHz frequency was found showing better sensitivity for signal characteristics in both cases of coil width. Therefore, it was decided to select the coil of 1 mm width and 3.5 mm gap as the coil parameter for final design. The coil thus fabricated is judged applicable to experiment at frequency of 20 kHz.
- Parameter finally selected in this study was 1 mm coil width, 3.5 mm coil gap and 20 kHz operating frequency. Under this condition, it was possible to design differential bobbin eddy current sensor suitable for detecting 0.2 mm defect successfully, the final objective defect. FEM simulations show the impedance becomes bigger in the impedance plane with depth of defect going deeper as ECT experimental results.
- It is possible to design interior eddy current probe for inspecting oil filter bolt based on the results and ECT simulation of this work.

Acknowledgments
This work (Grants No. S2175645) was supported by Global Innovations R&D between Industry, Academy, and Research Institute funded Korea Small and Medium Business Administration in 2015.

References

[1] Hellier C J *2001 Handbook of Nondestructive Evaluation* (New-York : McGRAW-HILL)
[2] Blitz J 1996 *Ultrasonic Methods of Non-destructive Testin* (London : Chapman & Hall)
[3] Mooer P O 1986 *Nondestructive Testing Handbook* 3rd ed. (New-York : ASNT)
[4] Shull P J 2002 *Nondestructive Evaluation: Theory, Techniques and Applications* (New-York: CRC Press)
[5] Ahn Y S, Gil D S and Park S G 2011 *Korean Power Plant Journal* **15** pp 37-41
[6] Heo T H, Cho Y, Lee J H, Oh J S and Lee K H 2009 *Journal of the Korean Society for Nondestructive Testing* **29** pp 450-457
[7] Nam M W, Lee H J, Cho C H and You H J 2010 *Journal of the Korean Society for Nondestructive Testing* **30** pp386-395
[8] Lee H J, Choi C H, Nam M W, Kim Y S, Yoon B S, Yoo H J, Jo Y B, Lee J S, Moon G Y, Kim L C, Chun G Y and Lee T H 2013 *Final Research Report* (KHNP CRI) pp 250-321
[9] Lee H J, Choi S N, Cho C H, Yoo H J and Moon G Y 2014 *Journal of the Korean Society for Nondestructive Testing* **34** pp 254-259
[10] Im K H, Lee S G, Kim H J, Song S J, Woo Y D, Na S W and Lee H H 2015 *Journal of the Korean Society of Manufacturing Technology Engineers* **24** pp 294-300

IOP Conf. Series: Materials Science and Engineering **103** (2015) 012041 doi:10.1088/1757-899X/103/1/012041

Load and wear experiments on the impact hammer of a vertical shaft impact crusher

J H Yang[1], H Y Fang and M Luo

Huaqiao University, Xiamen, Fujian, 361021, China

E-mail: yjhong@hqu.edu.cn

Abstract. Impact hammers are important components of impact crushers, and are often short-lived due to the high-impact nature of their use. Wear-resistant alloys are welded to the surface of impact hammers to prolong their service life. In this paper, a simulation model of the rotor and impact hammers in impact crushers was designed to utilize the Discrete Element Method (DEM). The wear-resistant alloy on each impact hammer was divided into twenty-two action regions. The load distribution on each alloy block is affected by the structural and manufacturing parameters of the impact crusher. The wear distribution of the impact hammer was measured by shape morphology according to relative impact crushers. The results demonstrated that the real measurements of wear distribution on the impact hammer were similar to simulated load distribution measurements on the same surface. The study of load distribution of impact hammers by DEM established a theoretical foundation on which to base the optimal design of impact crushers.

1. Introduction

With the rapid development of infrastructure and real estate in China, the demand for concrete is increasing quickly, of which sand is a necessary component, and a limited natural resource. It is inevitable that artificial sand will eventually replace natural sand to a large extent in the future of concrete manufacturing. Rock breakage is the key technology in artificial sand manufacturing systems, widely achieved by use of vertical shaft impact crushers (VSI) [1, 2], which consist of a rotor, crushing chamber, and center distributor plate. High silicon dioxide content in the raw materials used to create artificial sand leads to high hardness of the material, and the rocks must be accelerated to a high speed in order to be broken into artificial sand materials. High hardness and high speed are the two most important aspects that affect the contact parts of VSI, and which result in the fast and significant wear and tear of VSI parts. The highly priced wear-resistant parts must be changed frequently because of excessive wear and tear caused by strong impact [3, 4], which decreases the efficiency of sand production and machine reliability, and increases production costs. It is necessary to study the internal relationships between the wear of VSI and related parameters to improve the performance and efficiency of VSI [5].

A vertical shaft impact crusher has two operating modes: rock-to-steel and rock-to-rock. Rocks are accelerated by a quickly rotating rotor. They are then broken into pieces in the crushing chamber [6]. In order to improve the crushing effect, many manufacturers equip the rotor with impact

[1] Address for correspondence: J H Yang, College of Mechanical Eng. & Auto. Huaqiao University, Jimei Ave. 668, Xiamen, Fujian, 361021, P.R. China. Email: yjhong@hqu.edu.cn.

hammers at the edges of rotor outlets, which are usually covered with carbon tungsten alloy blocks to extend their service life [7]. This study primarily investigates the impact load of a unit area on carbon tungsten alloy blocks by the Discrete Element Method (DEM). The internal relationship among rotor structure, manufacturing parameters and impact load is studied by simulation. Optimized manufacturing parameters and VSI structure are obtained by simulation so as to extend the service life of the impact hammer. Comparison between simulation results and practical measurements are in agreement, demonstrating that DEM is an efficient method to decrease the production costs and development cycle of artificial sand manufacturing.

2. Crusher simulations

2.1. Establishment of 3D rotor model

The VSI rotor primarily consists of a trail plate, impact hammer and center distributor plate. Impact hammers withstand strong impact loads, which leads to severe wear and tear. Therefore, this paper analyzes the wear and load distribution of the impact hammer. The VIS rotor used in this study has five channels, and the 3D solid model of the rotor is shown in figure 1(a). There are 22 colored sections made of carbon tungsten alloy at the edges of the channels, on the surface of the impact hammer. Each colored section is a carbon tungsten alloy block; each carbon tungsten alloy block represents an action unit. The 3D solid modeling IGS data (i.e. STEP data) was imported into the Discrete Element Modeling (DEM) software. Figure 1(b) displays the rotor model as it appears in the DEM analysis software. The shape of the particle used in the DEM model was spherical, which can reduce simulation time. The rotor in figure 1(a) was simplified in the DEM simulation model shown in figure 1(b). A particle factory was established in the DEM simulation model.

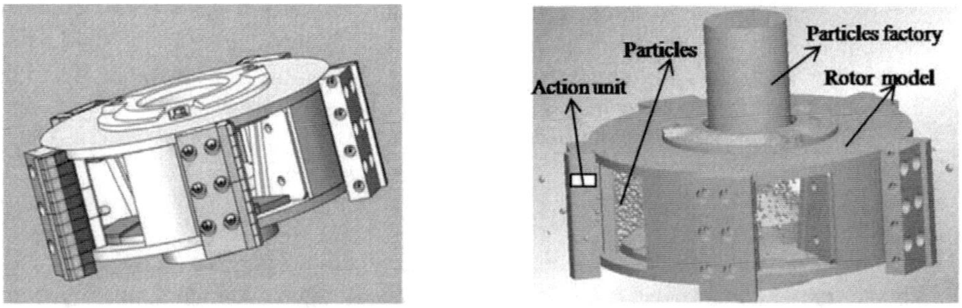

(a) Three-dimensional entity model (b) Discrete element simulation model

Figure 1. Rotor model of impact crusher.

2.2. Simulation parameters for DEM simulations

Before simulation, parameters must be set in the DEM software. Some model parameters were obtained from design manuals and previous literature [8-11]. The rotor material was set as steel, Poisson's ratio = 0.28, shear modulus = 7.5e+7 Pa, and rotor density = 7800 kg/m^3. The crushed material was set to granite. The material properties of the crushed particles were set as follows: Poisson's ratio = 0.25, shear modulus = 1.4e+ 7 Pa and density = 2800 kg/m^3. The particle-particle coefficient of restitution was set to 0.45, and particle-geometry coefficient of restitution was set to 0.5. The particle-particle sliding friction coefficient was set to 0.545, and particle-geometry sliding friction coefficient was set to 0.5. The particle-particle rolling coefficient and the particle-geometry rolling friction coefficients were set to 0.3. According to the particle size distribution of the practical rocks, the distribution of particle size in the simulation was set to normal distribution. A cylinder factory coaxial to the rotor was created to simulate the feed hopper, and the initial velocity of the particles was set to zero.

3. Simulation experiments of load on impact hammer

An impact hammer consists of 22 carbon tungsten alloy blocks, which are welded to the steel surface by specialty technology. A simulation test was created to determine the maximum impact load area among the 22 carbon tungsten alloy blocks. This was accomplished by analyzing the influences of various practical manufacturing and structural parameters on the impact load of each block.

3.1. Determination of the active zone

In order to determine the greatest wear area, the 22 carbon tungsten alloy blocks were labeled by number, as shown in figure 2. The rotating speed of the rotor was set to 1550 r/min, the particle feeding rate was set to 7.5 kg/s and the particle size range was set to 0-13 mm with normal distribution. After the simulation test, the number of collisions between particles and carbon tungsten alloy blocks were counted. Comparing the collision numbers of the 22 blocks, the block with the greatest number of collisions was chosen as the load analysis object to investigate by a series of simulation tests with varying parameters. Table 1 displays the numbers of collisions on each of the 22 blocks; blocks 5, 6, and 7 experienced more collisions than others, which is consistent with the actual processing results.

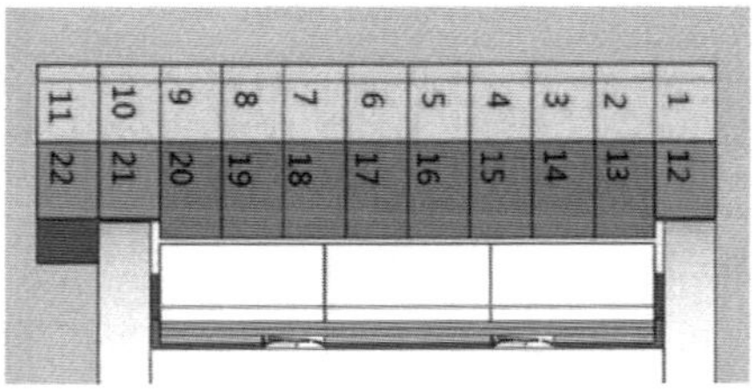

Figure 2. Action regions on impact hammer.

Table 1. Simulated number of particle collisions on action regions of impact hammer.

Region Number	1	2	3	4	5	6	7	8	9	10	11
Collisions number	25	284	711	933	1022	1013	1017	984	363	53	8
Region Number	12	13	14	15	16	17	18	19	20	21	22
Collisions number	5	37	65	67	51	56	78	86	64	15	1

3.2. Relationship between manufacturing parameters and load on the impact hammer

A series of simulation tests was conducted on the 22 blocks. Rotor speed, particle size distribution, and feeding rate were manipulated to investigate the relationship between manufacturing parameters and the total force on each carbon tungsten alloy block. The simulation results are shown in figures 3-5. As was illustrated by Table 1, the impact loads on blocks 1-11 were stronger than on blocks 12-22. Therefore, analytical results primarily investigated the influences of different parameters on blocks 1-11. First, simulation tests were conducted with three different rotating speeds: 1440, 1550, and 1700 r/min. The feeding rate was set to 7.5 kg/s, and particle size ranged from 0-13 mm with a normal distribution. After the simulation, the total forces exerted on the first 11 blocks were separately imported to DEM software. Figure 3 illustrates the relationship between rotating speed and total force on each carbon tungsten alloy block. As shown in figure 3, total force increases with increasing rotor speed. Additionally, the total force on blocks 5, 6, 7, and 8 are far larger than exerted on other blocks, indicating that more significant wear and tear would occur on these blocks.

Next, simulation tests were conducted with three different feeding rates: 6.5, 7.5, and 8.5 kg/s. The rotor speed was set to 1550 r/min, and particle size was set to the same value as in the previous three tests. Figure 4 shows the relationships between different feeding rates and the total force exerted on each of the 11 blocks. Figure 4 demonstrates that total force exhibits no obvious increase with increasing feeding rate, and different feeding rates result in different maximum impacts on load blocks. Finally, simulation tests were conducted with three different particle sizes: 5.5, 6.5 and 7.5

mm. As shown in figure 5, different particle sizes significantly influence the force exerted on block 5, 6, and 7, particularly on block 7.

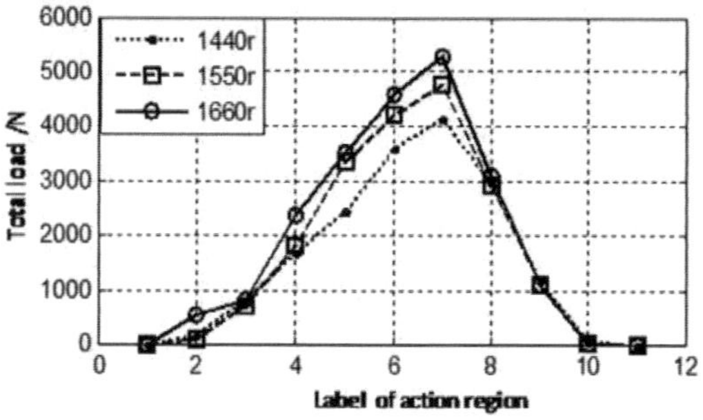

Figure 3. Relationship between rotation speed and total load.

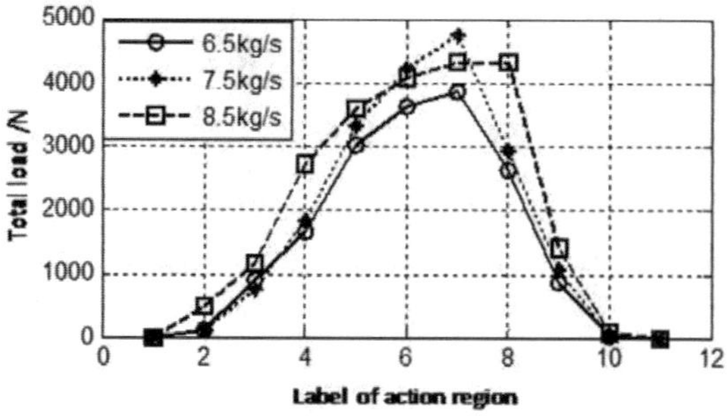

Figure 4. Relationship between production speed and total load.

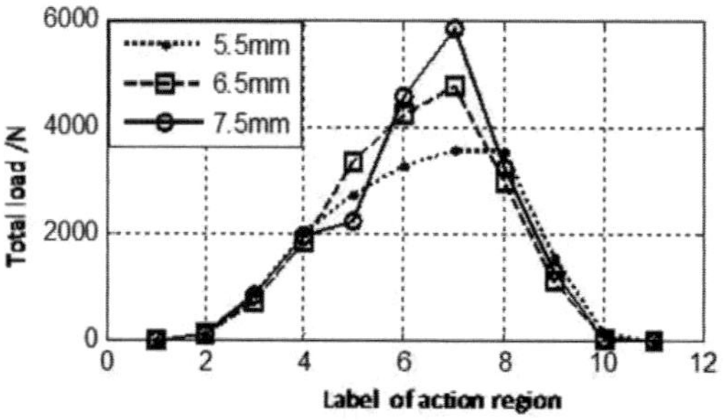

Figure 5. Relationship between particle size and total load.

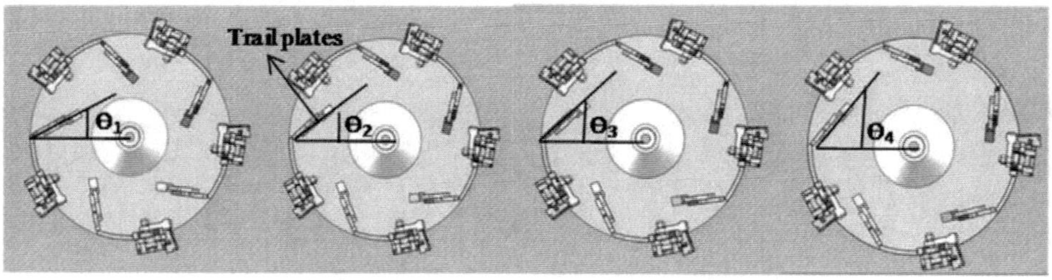

(a) $\theta_1=0°$, $\theta_2=30°$, $\theta_3=40°$, and $\theta_4=50°$

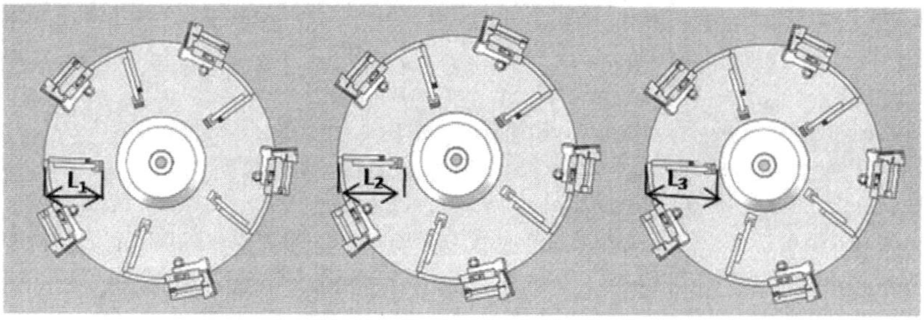

(b) $L_1=112$ mm, $L_2=122$ mm and $L_3=132$ mm

Figure 6. Changed structural parameter of rotor.

3.3. Relationship between structural parameters and load on the impact hammer
In a vertical shaft impact crusher, the main factors which influence the velocity of particles from the channel outlets are the length and setting angles of trail plates. Rotor models of different setting angles and trail plate lengths in 3D software (Solid Works) are shown in figure 6. The trail plate length of

VSIs was originally 122 mm, with an install angle of 0°. Simulation tests were conducted with different trail plate lengths L_1, L_2 and L_3, which were 112 mm, 122 mm and 132 mm, respectively, as shown in figure 6(b). The feeding rate was set to 7.5kg/s, rotor speed was 1550 r/min, and particle size ranged from 0-13 mm with normal distribution. Figure 7 shows the relationship between different trail plate lengths and the total force exerted on each of the 11 blocks. As shown in figure 7, the impact load on the carbon tungsten alloy blocks decreased with increasing trail plate length. When the trail plate length increases, particles take longer to travel, which increases the friction action time between the trail plate and the material particles? The kinetic energy of the particles declines, and the velocity of the particles that hit the impact hammer drops accordingly. Shorter trail plates can cause a stronger impact load on block 7.

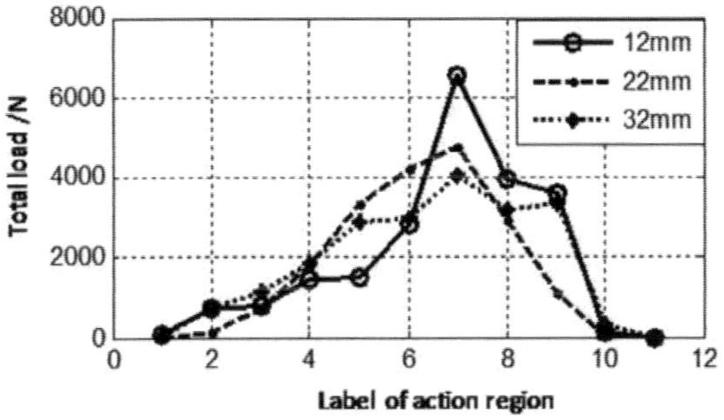

Figure 7. Relationship between length of retaining plate and total load.

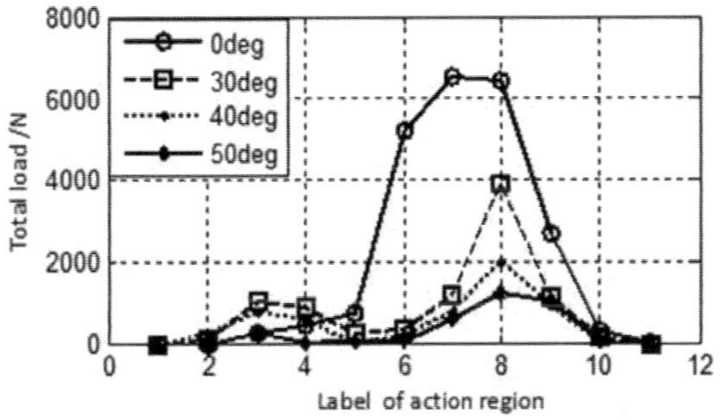

Figure 8. Relationship between angle of retaining plate and total load.

Simulation tests were then conducted with increasing install angles. The simulated install angles of the trail plate were Θ_1=0°, Θ_2=30°, Θ_3=40°, and Θ_4=50°. Figure 8 shows the relationship between the install angle of the trail plate and the total force exerted on each of the 11 blocks. When the installation angle was 0°, the motion path of the particles was perpendicular to the trail plate, causing the particles to hit the impact hammer without much friction. With the increasing install angle of the trail plate, the friction on particles coming from the trail plate would increase, and the particles' kinetic energy loss would increase accordingly, and the impact load from particles on the carbon tungsten alloy blocks would decrease. Designing different installation angels according to different rocks sizes can not only ensure an optimal crushing effect, but also decrease the impact wear and tear on the trail plate.

4. Analyses and experiment comparison of throwing material hammer

In order to verify the actual wear distribution characteristics of the impact hammer, real experiments were conducted using the same type of VSI. In the field experiment, the rotation speed of VSI was 1440 r/min, rotor diameter was 1660 mm, the rotor weight was 2000 kg, and the crushing chamber diameter was 3070 mm. After 10,000 hours of crushing, the impact hammer was removed from the VSI. Figure 9(a) shows the scanned picture of the carbon tungsten alloy blocks on the impact hammer. In the simulation process, rotor size identical to that used in the actual VSI in the field experiment. The feeding rate was 7.5 kg/s, rotor speed was 1440 r/min, particle size ranged from 0-13 mm with normal distribution, the install angle of the trail plate was 0°, and trail plate length was 122 mm. Figure 9(b) shows the result of the simulation experiments: the distribution and number of collisions between the particles and the impact hammer. The axes represent the length of the hammer, width of the hammer and the number of particle collisions, respectively. The gradient projection in figure 9(b) demonstrates the impact load distribution of the hammer. Figure 9 demonstrates that the actual and simulation experiments shows similar results: blocks 5, 6, 7, 8 and 9 demonstrate greater impact loads than other blocks, and the wear and tear on blocks 1-11 are generally greater than on blocks 12-22. Results verify that the impact load distribution and actual wear distribution of the hammer were consistent. The DEM and load distribution can be used to optimize structural and manufacturing parameters of VSI and prolong the service life of impact hammers. Figure 9 shows a comparison of the actual wear and load distribution on the impact hammer by simulation.

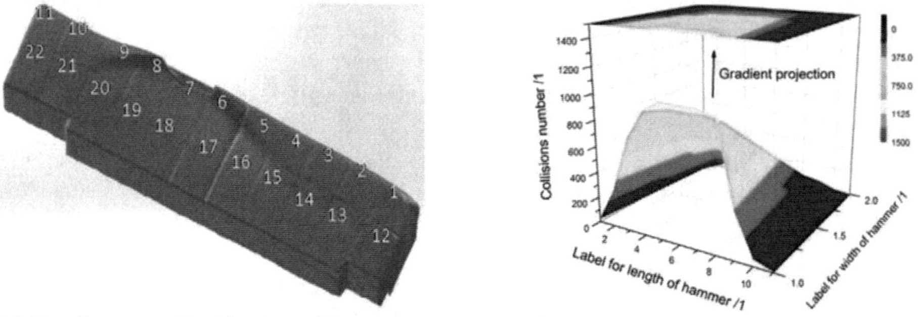

(a) Really wear distribution of hammer (b) Simulation load distribution of hammer

Figure 9. Really wear distribution compare with simulation load distribution of hammer.

5. Conclusions

In this paper, a geometrical model of a vertical shaft impact crusher is created, and 22 carbon tungsten alloy blocks on the impact hammer were labeled. First, DEM was used to study the impact load on the carbon tungsten alloy blocks on the surface of the impact hammer. Secondly, the effects of rotor speed, feeding rate and particle size on load distribution were studied by simulation. Next, the influence of the length and installation angle of the trail blade on the load distribution of the impact hammer was investigated by simulation. Finally, comparing wear and tear on the impact hammer by actual experiments and simulation, it is found that actual wear and tear distribution and the simulation load distribution produce similar results. The simulation of the influences of various parameters on load distribution is an efficient way to optimize the crushing effects of VSI. Meanwhile, optimal parameters can improve the service life of the impact hammer by DEM simulation, which provides a strong tool to optimize the design of vertical shaft impact crushers.

Acknowledgments

This work was financially supported by National and international scientific and technological cooperation in special projects (2015DFA710402), major project foundation of science and technology in Fujian Province (2012H6014 and 2014H6017).

References

[1] Xu L J 2012 Study on influence factors of impact crushing speed of coal rejects *Coal Science and Technology* **40(4)** 82-5
[2] Dong A M and Jiang G S 2014 New method for DEM simulation of particle crushing evolution in soil *Science Technology and Engineering* **14(25)** 277-81
[3] Jiang S P, Yu H L and Rui X T 2014 Dynamic analysis on impact fragmentation of granular systems *Explosion and Shock Waves* **34(2)** 247-51
[4] Zhao F, Zhang J M and Wang W 2011 Research on the load distribution of throwing ends in vertical shaft impact crusher *Advanced Materials Research* **156-157** 1463-6
[5] Shi X S and Cheng Z L 2010 Fractal behavior in crushing of rock fill material *Chinese Journal of Rock Mechanics and Engineering* **29(2)** 38553-7
[6] Zhao Y C, Liu X J and Wang W 2010 Study on the interaction of surface topography based on discrete element method *Journal of Hefei University of Technology (Natural Science)* **33(4)** 487-490
[7] Dmitrieva A I, Popov V L and Psakhie S G 2006 Simulation of surface topography with the method of movable cellular automate *Tribology International* **39** 444-9
[8] Tan Y Q, Song J H and Zhang H 2014 Numerical study of wear process on pipe wall in

concrete pumping using DEM *China Mechanical Engineering* **25(15)** 2090-7

[9] Deng T, Patel M, Hutchings I, et al. 2005 Effect of bend orientation on life and puncture point location due to solid particle erosion of high concentrated flow in pneumatic conveyors *Wear* **258** 426-33

[10] Zhu H P, Zhou Z Y, Yang R Y, et al. 2007 Discrete particle simulation of particular systems: Theoretical developments *Chemical Engineering Science* **62** 3378-96

[11] Andrew H 2009 Simulation of an aggregate dryer using coupled CFD and DEM methods *International Journal of Computational Fluid Dynamics* **23(2)** 199-207

Enhanced mercury removal from fix-bed reactor by lamella manganese oxide sorbents

H W Cheng and C T Yu[1]

Chemistry Division, Institute of Nuclear Energy Research, Longtan 325, Taiwan

E-mail: ctyu@iner.gov.tw

Abstract. Mercury (Hg) is an extremely hazardous metal and attracted more concern because of its high toxicity and bioaccumulation. Several manganese-oxide-containing sorbents prepared by co-precipitation method could exhibit the mercury removal activities toward Hg^0. The mercury removal test at the temperature of 300°C has the highest removal efficiency. Under this temperature, the maximum absorption equivalent of Mg–Al–Mn and Mn–Al were up to 90.9 and 247 μg/g, then gradually decreased at 400°C. The mercury removal efficiency declined in the following sequence: Mn–Al > Mg–Al–Mn > Mg–Al–Mn/AC ≒ $Mn/AC_{(p)}$> $Mn/AC_{(g)}$, due to the manganese-oxide content formed on the sorbents.

1. Introduction

Mercury (Hg) pollutions in the environment effect have received increasing concern worldwide in the last decade. Human health hazards from such pollutants are associated with their potential bioaccumulation in the food chain [1]. The emissions of Hg were estimated to be 1930 tons per year and mainly came from human activities and natural sources, and the emissions of Hg 45.6% directly attributed to thermal power plants [2]. Mercury is present in coal in relatively low ppb levels and as elemental mercury (Hg^0), mercuric chloride ($HgCl_2$) and particle-bound mercury ($Hg_{(p)}$) from coal combustion. Most of the mercury exists in the oxidized form of $HgCl_2$ which is by far the most easily removed, but it's much more complex to remove of Hg^0 from coal power plant [3]. Activated carbon has gained much attention as the most promising adsorbents for mercury removal. However, the traditional thermodynamic absorption efficiency of mercury sorbents as activated carbon is significantly poor at high temp condition [4]. Activated carbon is a well-known adsorbent for the removal of elemental mercury at ambient temperature by physically adsorption [5]. However, due to the weak binding of mercury over room temperatures, they are still limited to temperatures below 120°C [6]. The mercury amalgamation with noble metals such as silver and gold is notable. Some studies showed that porous silver or gold containing samples were effectively removing mercury but expendable gold or silver sorbents were not often been used for some cost reasons. There is a wide variety of adsorbents for mercury removal of which metal oxides such as MnO_2, V_2O_5, Cr_2O_3, MoO_3, TiO_2, CuO and Fe_2O_3 can increase mercury removal performances [7-10]. The surfaces of metal oxides can react with the mercury through the surface complex reaction [11]. The sorbents contained such metal oxide for mercury removal under medium-high temperature conditions actually exhibit with larger adsorption capacity and lower cost.

[1] Address for correspondence: C T Yu, Chemistry Division, Institute of Nuclear Energy Research, Longtan 325, Taiwan. E-mail: ctyu@iner.gov.tw.

Layered double hydroxides (LDHs) are a type of layered solids and contain divalent ions (Mg^{2+}, Ca^{2+}, Zn^{2+}, Mn^{2+}), trivalent ions (Al^{3+}, Fe^{3+}), and anions (CO_3^{2-}, SO_4^{2-}, OH^-) inserted between adjacent layers [12]. The ability of LDHs to adsorb inorganic or organic anions was quite efficiency and make these materials very technological importance for applications in several areas.

2. Experimental procedure

2.1. Sorbent preparation

Both LDH material of Mg–Al–Mn and Mn–Al were synthesized by co-precipitation method using $Mg(NO_3)_2{\cdot}6H_2O$, $Al(NO_3)_3{\cdot}9H_2O$ and $Mn(NO_3)_3{\cdot}4H_2O$ with molar ratio of Mn^{2+}/Al^{3+}at 3:1 or $Mg^{2+}/Al^{3+}/Mn^{2+}$ at 6:2:1. For preparing Mn–Al, in 100 ml of aqueous solution containing $Mn(NO_3)_2 \cdot 4H_2O$ (0.3 mol), $Al(NO_3)_3 \cdot 9H_2O$ (0.1 mol) were mixed. The 0.2 mole NaOH and 0.05 mole Na_2CO_3 were dissolved in 150 mL deionized water until pH value reached 10, and then added slowly into Mn/Al solution. The resulting precipitated LDH powder was separated by filtration and drying at 110°C.

The Mg–Al–Mn powder was also prepared similarly by co-precipitating of 0.3 mole Mg^{2+} and 0.1 mole Al^{3+} and 0.05 mole Mn^{2+} in deionized H_2O treated with 0.2 mole NaOH and 0.05 mole Na_2CO_3.

The activated carbon (AC) powder and granule used in this study were from Merck (Germany). The Mn–Al/AC was also prepared by co-precipitation method. First, 10 g of raw AC powder or granule and 150 mL deionized water with 0.3 mole of Mn^{2+} and 0.1 mole of Al^{3+} were mixed and stirred for 1h at 70°C. Then, alkaline solution mentioned above was added slowly. After stirring, the same method was used to obtain Mg–Al–Mn/AC.

2.2. Experimental apparatus

The test apparatus was installed specifically for measuring mercury removal in the fix-bed reactor scale. A mixture of 1 g of mercury sorbents sample and 3 g of quartz sand was packed into an inch-diameter quartz tube with a 200 mesh aperture carrier disc. A 1kW reactor was provided in a temperature-controlled heat with a furnace chamber surrounding the reactor. For delivering precise concentrations, a mercury permeation tube (Valco, USA) was housed in a temperature controlled oven and constantly emitted mercury with total gas flow rate of 1 $L{\cdot}min^{-1}$ during the experiments. The Hg^0 concentration at the outlet of the reacting bed was continuously monitored with an on-line mercury analyzer (EMP-2, NIC, Japan). The test apparatus was illustrated in figure 1.

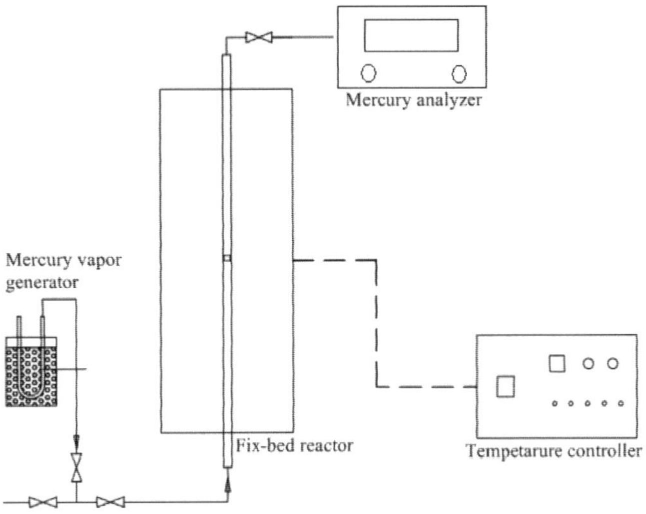

Figure 1. Schematic of the test apparatus.

2.3. Testing procedure

The study carried out experiments to measure the adsorption breakthrough time and the equilibrium adsorption capacity of the mercury sorbents. In the tests, about 1g different mercury sorbents were added into quartz reactor tube. Mercury flow was maintained through all testing procedure at a gas hourly space velocity (GHSV) of 19736 h^{-1}. Sorbents were taken at ambient to 400°C temperature conditions and measured the pre-breakthrough absorption capacity. The breakthrough is defined at 90% mercury removal and the screening tests stop until the breakthrough.

3. Results and discussion

3.1. Surface characterisation of sorbents

JEOL S4800 (Japan) scanning electron microscope (SEM) with a finely focused electron beam was used to obtain high-resolution microstructure of the sorbents. And the energy dispersive X-ray spectrometry (EDS) was used to examine the distribution of the elements on the surface. Figure 2 shows the SEM images of Mg–Al–Mn, Mn–Al LDH, virgin AC (V–AC), Mg–Al–Mn/AC and Mn–Al/AC. It is clear that V–AC has developed pore structure in figure 2(a) and Mg–Al–Mn/AC and Mn–Al/AC in figures 2(b) and 2(c) showed that the active component (Mn) illustrating brighter spot can be evenly distributed in the AC surface. The SEM images of Mg–Al–Mn and Mn–Al LDH in figure 2(d) and 2(e) show that the samples have a clear plate-like morphology, which was typical for the LDH lamella morphology. The LDH lamella structures are homogeneously deposited across the entire substrates in Mg–Al–Mn and Mn–Al LDH samples.

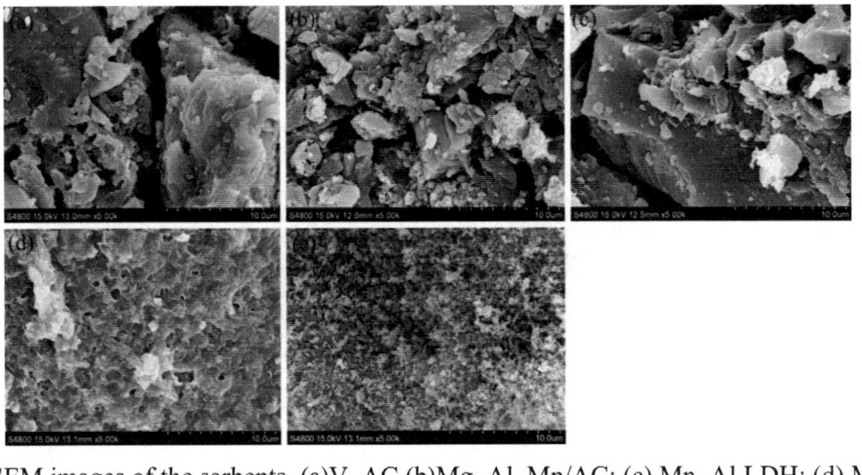

Figure 2. SEM images of the sorbents, (a)V–AC (b)Mg–Al–Mn/AC; (c) Mn–Al LDH; (d) Mg–Al–Mn LDH (e) Mn–Al LDH.

Table 1 presented the BET surface areas, pore structure and pore size results of Mg–Al–Mn and Mn–Al LDH sorbents. When comparing with Mg–Al–Mn and Mn–Al, it demonstrated that surface area, pore volume and pore size almost the same as 11.45 m^2/g, 0.02 mL/g, 1.92 nm to 12.21 m^2/g, 0.04 mL/g and 1.93 nm, respectively. When measured with Mn, it was varied from 1.52 to 15.1 wt.% MnO in Mg–Al–Mn and Mn–Al by EDS analysis.

Table 1. BET surface content and element composition of sorbents

Sorbents	Surface area, m^2/g	Pore volume, ml/g	Pore size, nm	MOx, wt.%
Mg–Al–Mn	11.45	0.023	1.92	MnO = 1.52
Mn–Al	12.21	0.039	1.93	MnO = 15.1

3.2. Mercury removal efficiency of different kinds sorbents

Reaction temperature plays an important role on removal behavior and reaction characteristics of sorbents and mercury. The performances of Mg–Al–Mn and Mn–Al under different adsorption temperatures (25–400°C) were analyzed to figure out the temperature effect. The results show that the removal efficiency has a trend of increasing at first and then dropping with the rise of reaction temperature. Figure 3 shows Hg^0 absorption equivalent data under different absorption temperature of Mg–Al–Mn and Mn–Al. The mercury removal testing under the temperature of 300°C has the highest removal efficiency. Under this temperature, the maximum absorption equivalents of Mg–Al–Mn and Mn–Al are up to 247 and 90.9 μg/g. But when the temperature reaches 400°C, the absorption equivalent of Mg–Al–Mn and Mn–Al decreases to 43.5 and 153.2 μg/g. However, the temperature effect on various physicochemical properties of LDHs, AC, Mn/AC and LDHs/AC need to be ascertained (referenced and/or tested) to propose the various mechanisms involved in these processes.

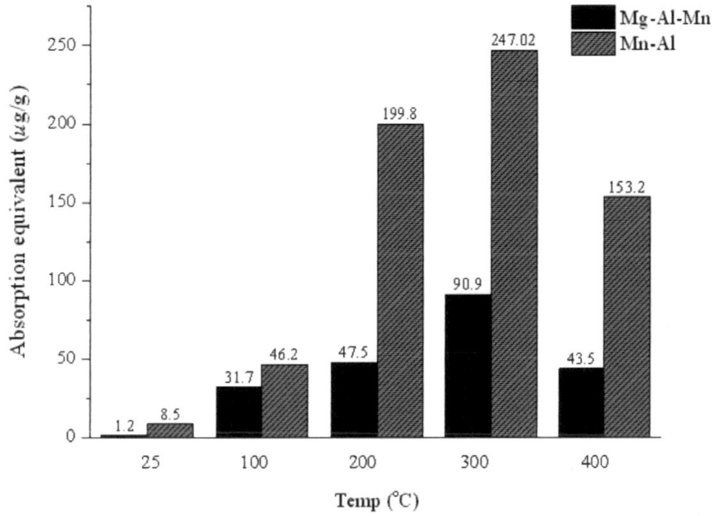

Figure 3. Hg^0 adsorption equivalent under different adsorption temperature of Mg–Al–Mn and Mn–Al.

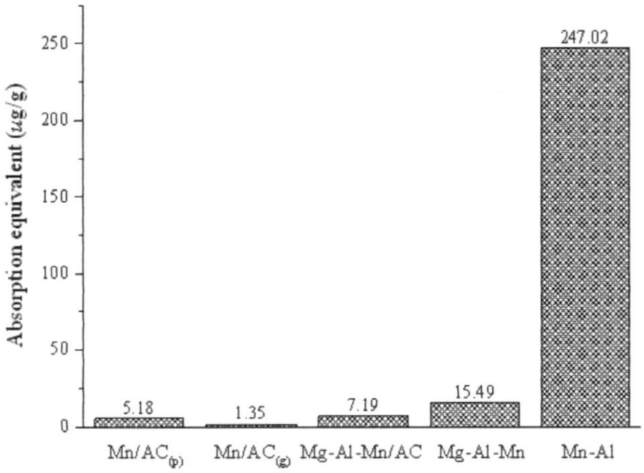

Figure 4. Hg^0 adsorption equivalent under 300°C temperature of different sorbents.

On the other way, Figure 4 shows the mercury removal efficiency on different kinds of sorbents at 300°C. It is showed that the Mn-modified AC performs poor efficiency in the mercury removal process. The worst efficiency is likely due to the less of Mn content formed on the AC. Furthermore, Mn–Al has the best removal efficiency for the removal of element mercury, followed by Mg–Al–Mn.

The mercury removal efficiency decrease in the following sequence: Mn–Al > Mg–Al–Mn > Mg–Al–Mn/AC ≒ Mn/AC$_{(p)}$> Mn/AC$_{(g)}$. Moreover, it is believed that manganese oxide is the most significant component for the removal of mercury. Therefore, the studies presented in the paper were carried out that the mercury removal tests by adding transition metals to sorbents are an important step that greatly increases the mercury removal efficiency of sorbents.

4. Conclusions

This study has found that the elevated temperature removal of mercury via manganese oxide contained sorbents testing in fix-bed reactor. These results demonstrated that the mercury removal efficiency of different kinds of manganese oxide-containing sorbents decreasing in the following sequence: Mn–Al > Mg–Al–Mn > Mg–Al–Mn/AC ≒ Mn/AC$_{(p)}$> Mn/AC$_{(g)}$. Characterization of surface properties of samples was indicated that manganese oxide embedded on the materials which were the most effective component for the removal of mercury at elevated temperature conditions. Mercury removal testing under the temperature of 300 °C has the highest removal efficiency, and the maximum absorption equivalent of Mg–Al–Mn and Mn–Al were 247 and 90.9 μg/g, and then gradually decreasing after 400 °C. These results represent the feasibility of using such mercury sorbents for Hg^0 removal under elevated temperature conditions, and the detailed mechanism is needed to be further studyed.

Acknowledgments

The research was supported by the research program of National Science Council of R.O.C, Taiwan under Contract No. MOST 104-3113-E-008-001.

References

[1] Kilgroe J D, Sedman C B, Srivastava R K, Ryan J V, Lee C W and Thorneloe S A 2002 *Interim Report Including Errata Dated 3-21-02* (EPA-600/R-01-109)
[2] Miller S J, Dunham G E, Olson E S and Brown T D 1998 *The 1998 Conference on Air Quality: Mercury, Trace Elements, and Particulate Matter* McLean VA **19–21** Paper A4-3
[3] Galbreath K C and Zygarlicke C J 2011 *Fuel Process. Technol.* **65-66** 289-310
[4] Vidic R D and Siler D P 2001 *Carbon* **39** 3-14
[5] Granite E J, Freeman M C, Hargis R A, O'Dowd W J and Pennline H W 2007 *J Environ Manage* **84** 628-34
[6] Bailey S E, Olin T J, Bricka R M and Adrian D D 1999 *Water Res.* **33** 2469-79
[7] Zhao Y, Mann M D, Pavlish J H, Mibeck B A F, Dunham G E and Olson E S 2006 *Environ. Sci. Technol.***40** 1603-8
[8] Presto A A and Granite E J 2006 *Environ. Sci. Technol.* **40** 5601-9
[9] Zhuang Y, Thompson J S, Zygarlicke C J and Pavlish J H 2007 *Fuel* **86** 2351-9
[10] Xin L, Zhouyang L, Jinsoo K and Youp L J 2013 *App. Catal. B: Environ* **132-133** 401-7
[11] Zhang L, Zhuo Y, Du W, Tao Y, Chen C and Xu X 2012 *Ind. Eng. Chem. Res.* **51** 5292-8
[12] Khan A and O'Hare D 2001 *J. Mater. Chem* **12** 3191-8

IOP Conf. Series: Materials Science and Engineering **103** (2015) 012043 doi:10.1088/1757-899X/103/1/012043

Using magnetorheological fluids in an innovative hybrid bicycle damper

Y J Shiao[1] and T S Nguyen

Department of Vehicle Engineering, National Taipei University of Technology, Taipei 106, Taiwan

E-mail: yshiao@mail.ntut.edu.tw

Abstract. Magnetorheological fluids are capable of changing their viscosity quickly. This can provide good controllability and fast dynamic response. A conventional passive suspension system with air spring or hydraulic damper has simple design and financial benefit for bicycles, but its operation is uncontrollable and non-adaptive. This paper presented a semi-active hybrid bicycle suspension system which combines conventional air spring and a new magneto-rheological damping brake together to reduce vibration of a bicycle. A multi-layer magneto-rheological brake and linkage mechanism are connected to bike fork to form the adaptive damping part of the innovative hybrid suspension system. The simulation results proved that the semi-active suspension system can reduce bike vibration effectively.

1. Introduction

Vehicular suspension system is used to absorb shock impact from roads, and can be classified into categories of passive, semi-active and active devices according to their performances. This paper presented a hybrid bicycle suspension system which combines conventional air spring and a magneto-rheological damping brake together to reduce vibration of a bicycle.

Conventional front suspension system in a bike uses air spring as the main source of damping force. The damping force depends on moving speed of piston, and the force is not controllable to effectively reduce bike vibration. Thus the control force cannot be adaptive to different road conditions. A suspension system with controllable force and extended force range is required to provide more comfortable riding to bike rider.

Magneto-rheological devices (MR devices) and their applications get attentions these years because of their control flexibility and fast response [1]. MR devices were commonly designed as linear translation dampers to be applied in vehicular suspension system. But such kind of damper needs large space to be installed. Instead of the linear damper, a rotary MR brake is used in this research to provide similar damping force for bike suspension.

2. Hybrid bike suspension system

The proposed hybrid suspension system combines an air spring and a MR brake, as shown in figure 1, to provide damping force together. An MR brake can provide large angular resistance torque with smaller volume compared to linear translation damper. Since MR brake provides resistance force to

[1] Address for correspondence: Y J Shiao, Department of Vehicle Engineering, National Taipei University of Technology, Taipei 106, Taiwan. E-mail: yshiao@mail.ntut.edu.tw.

against angular motion, not translation motion, a slider-crank mechanism is needed to convert the translation motion of bike fork into angular motion of MR brake. This hybrid suspension system has advantages of simple design, acceptable cost, good controllability, fast response and high accuracy.

The in-line slider crank (as shown in figure 2) consists of three rotating joints and one sliding joint. The linear motion of slider drives the rotation of crank. Since the slider travels along a straight line which passes through the base joint of the crank, a symmetric back-and-forward slider motion is created. Thus, the in-line slider crank mechanism has advantages of harmonic motion and easy assembly over other four-bar linkage mechanisms.

Figure 1. Hybrid bike suspension system.

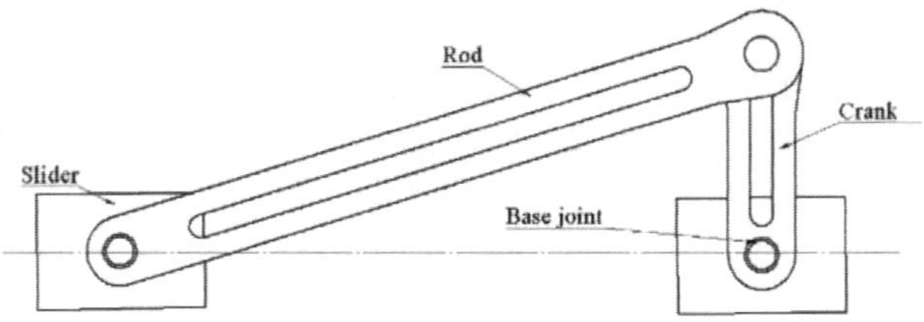

Figure 2. Slider-crank mechanism.

In the designed system in figure 1, the slider is the front fork of a bicycle and the pillar joint of rod connects to the top of front fork. The crank is the rotating part of MR brake which is fixed to the head tube of a bike. Parameters of this slider-crank mechanism are listed in table 1. For the sake of smooth movement of slider and crank, an appropriate proportion of crank length to rod length is from 1:3 to 1:5.

Table 1. Parameter of designed slider-crank.

parameters	crank length (mm)	rod Length (mm)	maximum stroke (mm)
Value	40	130	120

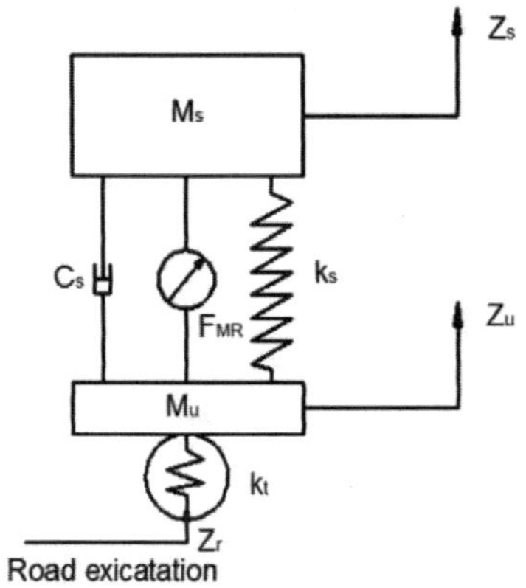

Figure 3. Model of slider-crank mechanism for hybrid suspension system.

To understand the dynamics of this hybrid suspension system, a dynamic model was built as shown in figure 3. The variables z_s, z_u and z_r are the displacements of sprung mass, unsprung mass and road excitation respectively. The sprung mass m_s and unsprung mass m_u correspond to the upper and lower weights of bike fork, while k_s and k_t are the stiffness of air spring and bike tire. Parameter c_s represents the viscous damping of suspension system, and F_{mr} is the controllable force from MR brake.

3. MR brake design

The term "magnetorheological fluid" (MR fluid) was firstly proposed by Rabinow in 1948. MR fluid, which is one of intelligent materials, can rapidly change its state from liquid state into quasi-solid state after applying an external magnetic field. Devices with MR fluid commonly work in three working modes: valve mode, direct-shear mode and squeeze mode (figure 4). For applications of braking, an MR brake works under the direct shear mode for its operation. Because of the characteristics of fast response, good controllability and quiet operation, MR brake is gradually adopted in many fields for braking, deceleration or precise motion control.

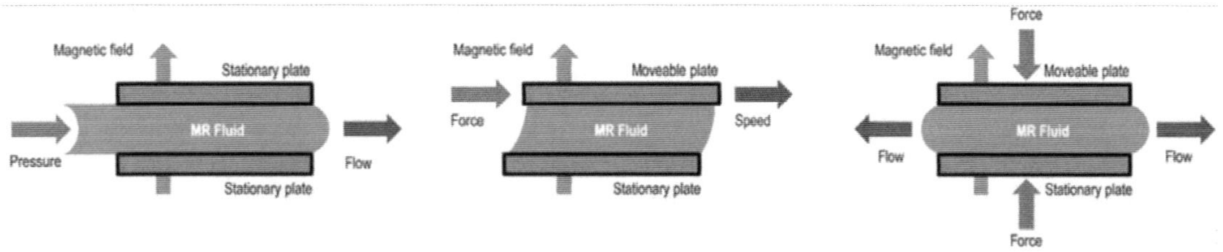

Figure 4. Three operational modes of MR fluid [2].

The magnitude of brake torque of an MR brake is directly related to the effective acting area between inner brake surface and MR fluid. In order to enhance brake force of an MR brake, its acting area of MR fluid needs to be increased. However, increment of the active area usually accompanies increase of unwanted brake size and weight. It is a target of this research to find a way to increase the effective active area of MR fluid and to keep the small size of MR brake at the same time. Therefore, a multi-layer disk structure instead of conventional single-layer disk structure is considered in this research.

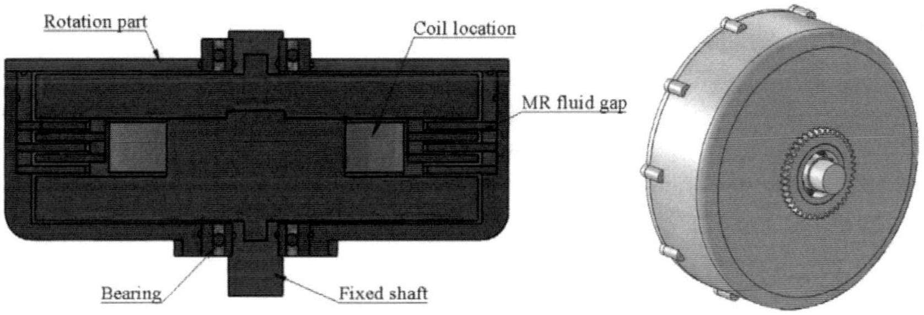

Figure 5. New multi-layer MR brake.

Figure 5 shows the configuration of a multi-layer MR brake [3]. This MR brake has a stationary part which is enclosed by a rotary casing. Coils are wounded around the axle of stationary part to produce magnetic field after applying current to the coils. Two and three discs are infixed into the stationary part and rotary casing respectively, and gaps between the corresponding discs are filled with MR fluid. Then braking force is produced due to the viscous force between MR fluid and brake inner surface. Magnetic simulation of the MR brake, shown in figure 6, shows that strong magnetic flux penetrates the discs and MR fluid in gaps to produce strong viscous force and corresponding brake torque. The magnitude of brake force or brake torque varies as input current changed. Figure 7 shows the simulated relation of brake force and input current. Approximately linear relation of force and current can be observed in the figure.

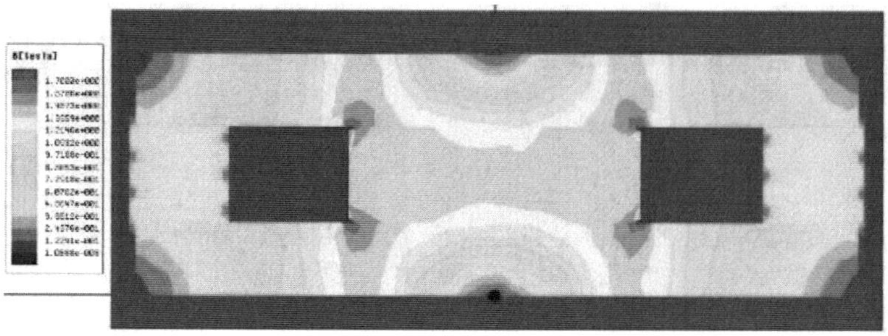

Figure 6. Magnetic simulation of the MR brake.

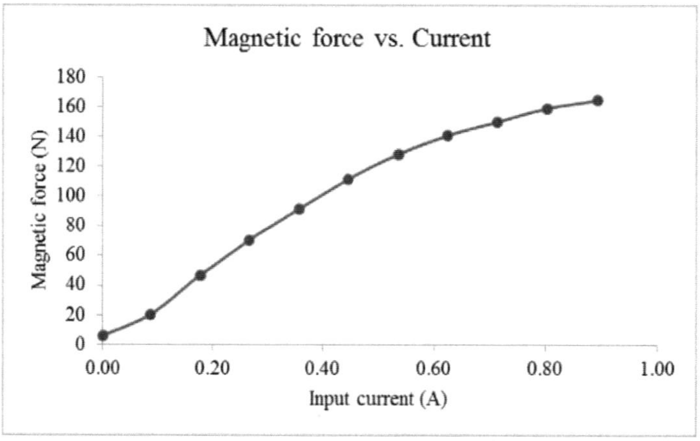

Figure 7. Relation of magnetic force and input current.

4. Vibration control

Because of structure similarity, a model of quarter car's suspension model, shown in figure 8, can be adopted in this research. The sprung mass m_s denotes the total weight applied to the upper head tube, including the weight of MR brake. The unsprung mass m_u is the weight of fork, shock absorber and wheel. Stiffness of spring and tire are k_s and k_t respectively.

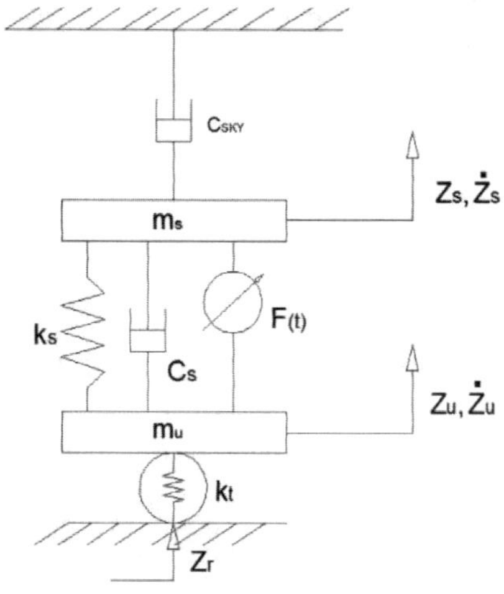

Figure 8. Skyhook control algorithm.

There were various studies proposed to control a semi-active suspension system. In most of studies and reports, skyhook method is highlighted as an optimal control for semi-active suspension system [4]. Figure 8 illustrates model of sky hook control for this two-order suspension system. The sprung mass is linked to the sky to reduce the vertical oscillation of the chassis.

Equations for the Skyhook model are presented as follows:

$$m_u \ddot{z}_u(t)+c_s[\dot{z}_u(t)-\dot{z}_s(t)]+k_s[z_u(t)-z_s(t)]+k_t[z_u(t)-z_r(t)]=-F(t) \quad (1)$$

$$m_s \ddot{z}_s(t)+c_s[\dot{z}_s(t)-\dot{z}_u(t)]+k_s[z_s(t)-z_u(t)]=F(t) \quad (2)$$

where $\dot{z}_s$ and $\dot{z}_u$ are velocities; $\ddot{z}_s$ and $\ddot{z}_u$ are the accelerations of the sprung and unsprung masses respectively. *F(t)* is the controllable force from MR brake.

Among all controllers, PID controller is the most common form of feedback control. It was an essential element of early governors, and then became the standard tool when process control emerged in 1940s. With advantages of easy implementation, high accuracy and fast response, PID is used widely in process and dynamic control today. The control law of a PID controller is shown in equation (3):

$$u(t)=K(e(t)+\frac{1}{Ti}\int_0^t e(t)dt+T_d\frac{de(t)}{dt} \quad (3)$$

Block diagram of the closed-loop control utilizing PID controller is illustrated as shown in figure 9. The system error e, which is the difference of reference signal r and output signal y, is used as input of PID controller. Then the control signal u will be computed and input to the plant. The steady state error always possesses in proportional control. Increasing of gain P will result in decreasing of system steady state error and increasing of oscillation in output. Adopting an integral gain I can effectively reduce steady state error. The increase of derivative gain D will enhance the response speed and reduce output overshoot.

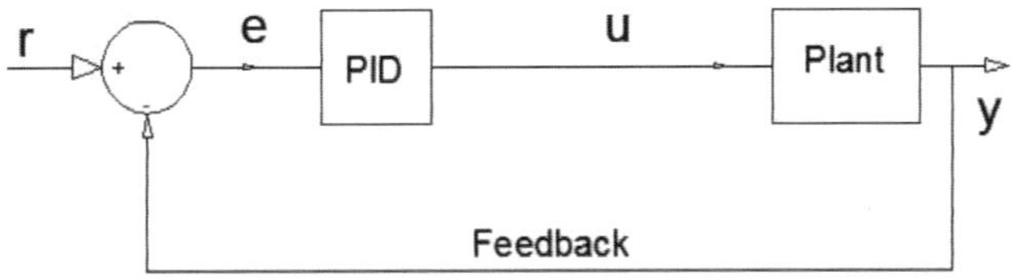

Figure 9. Closed-loop with PID controller.

To simplify the dynamic model, some assumptions are set as following: The tire is as similar as a linear spring with the spring rate k_t, and it always contacts with road surface; Behaviors of spring and damper are linear; The friction in the joints are neglected. As the results, the model of this hybrid suspension is nearly a two-order suspension system. Considering the weight distribution of a bicycle, there is about 30% to 35% weight applied to the front wheel and 65% to 70 % to rear wheel. Consequently, if the total weight of bike and cyclist is 100 kg, then the weight applied to the front wheel is about 30 kg to 35 kg. The weight, which mentioned above, is the sprung weight of this model. Table 2 lists the parameters for simulation.

Table 2. Simulation parameters [5].

Parameters	Value	Unit
Sprung mass (m_s)	32.5c	Kg
Unsprung mass (m_u)	5.5	Kg
Spring rate (k_s)	2950	N/m
Tire stiffness (k_t)	35055	N/m
Damping coefficient (c_s)	105	Ns/m

Damping performance of the suspension system needs to be controlled according to the motion of suspension travel. Intuitively, the MR brake's force F_{brake} increases while it has the same direction with the relative velocity between sprung and unsprung masses. Otherwise, the force F_{brake} is set to zero. This semi-active condition [4] is expressed in equation (4):

$$F_{brake} = \begin{cases} F_{brake}, F_{brake}(\dot{z}_s - \dot{z}_u) > 0 \\ 0, F_{brake}(\dot{z}_s - \dot{z}_u) \leq 0 \end{cases} \quad (4)$$

Vibration control performances of bike fork with hybrid MR suspension system are evaluated under bumping road condition. The road condition setting in this simulation is shown in figure 10 and equation (5) [6]:

$$z_r = a\frac{1-cos(8\pi t)}{2} \quad (5)$$

where $a = 6$ cm for $0.5 \leq t \leq 0.75$, $a = 3$ cm for $3.0 \leq t \leq 3.25$ and 0 otherwise.

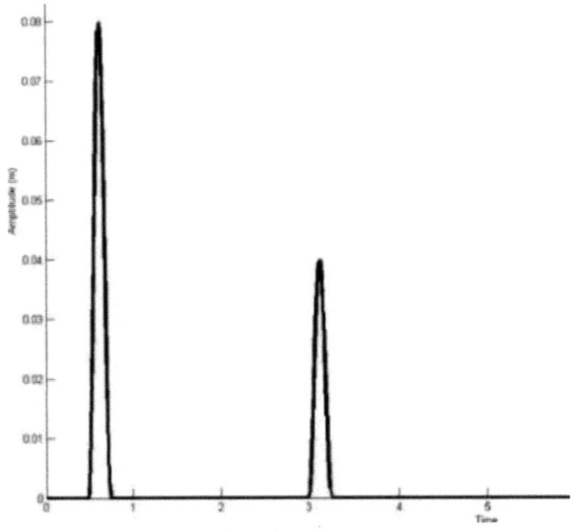

Figure 10. Road's profile for simulation.

For designing the PID controller, Ziegler-Nichols tuning method was employed to find the gain parameters K_P, K_I, and K_D with value of 25.5, 11.7, and 0.15, respectively.

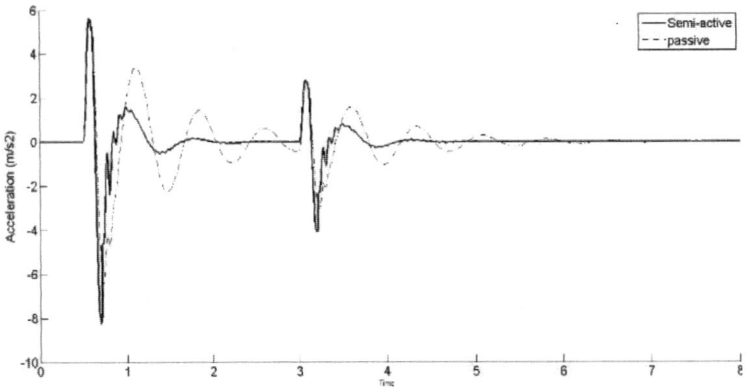

Figure 11. Accelerations responses of hybrid suspension system.

5. Results and discussion

The comparisons of semi-active and passive suspension behavior are depicted as shown in figures 11 to 13. The amplitudes and vibration frequency of the displacement, acceleration are both smaller significantly when we adopted MR brake as a variable actuator. The combination of skyhook model and PID controller is not only an effective solution for the control of semi-active suspension system in vehicles, but also in bicycles.

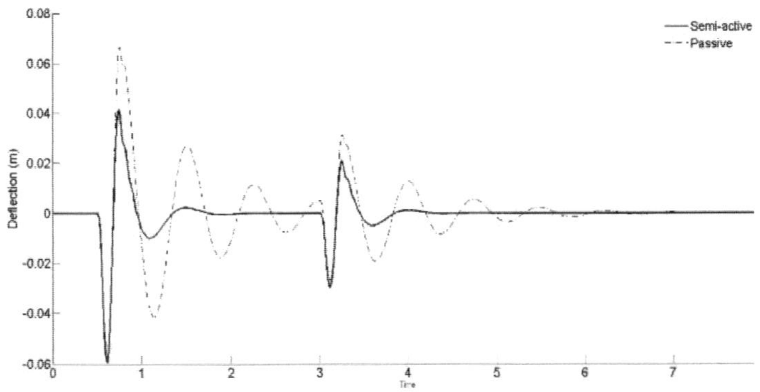

Figure 12. Displacements of hybrid suspension system.

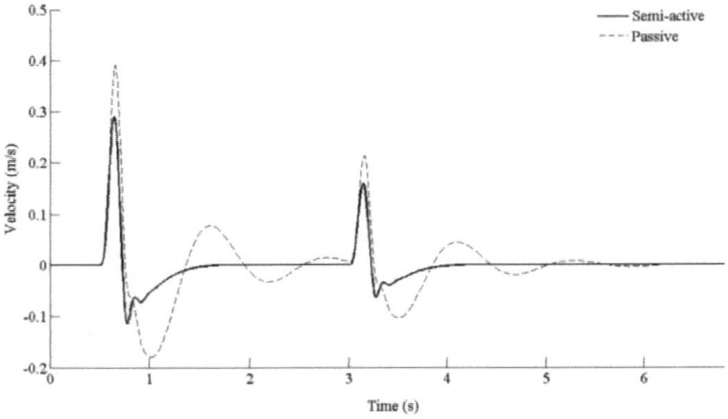

Figure 13. Velocities of hybrid suspension system.

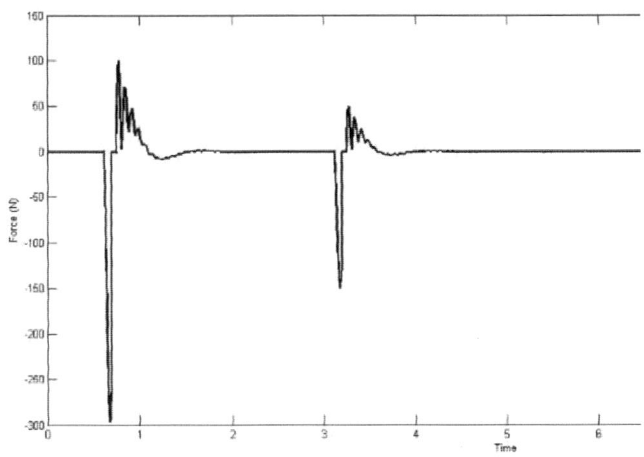

Figure 14. Controllable force.

The results shown in figure 14 indicate that the control force had to vary in a wide range in a short time. According to the magnetic force in figure 7, this MR brake can provide a rapid variable force that is bigger than the required force. It means that MR brake is obviously adaptable for application on front bike suspension.

6. Conclusions

In this paper, we introduced a new hybrid suspension system which consists of a multi-layer MRB with compact size and large brake torque. The simulation results proved that the semi-active suspension system can reduce bike vibration effectively. With these advantages revealed in this paper, the new suspension has the potentials to replace the passive bike's suspensions in the market and provides cyclists with comfortable riding.

References

[1] Shiao Y and Nguyen Q A 2014 Structural analysis and validation of the multi-pole magnetorheological brake for motorcycles. *Procedia Eng.76* pp 24-34 http://dx.doi:10.1016/j.proeng.2013.10.002.

[2] Vicente J D, Klingenberg D J and Hidalgo-Alvarez R 2011 Magnetorheological fluids: A review *Soft Matter.* **7(8)** 3701

[3] Shiao Y, Lai C H and Lin T Y 2015 Development of light weight magneto-rheological brake for adaptive knee orthosis *Proc. Int. Conf. of Engineering Technology Innovation (Kaohsiung, Taiwan)* pp 128-134

[4] Seong M S, Choi S B and Sung K G 2006 *Control Strategies for Vehicle Suspension System Featuring Magnetorheological (MR) Damper.* (Vib Anal Control - New Trends Dev) pp 97-114. http://dx.doi:17687.

[5] Gauderman Ph D and Morrison J 2007 *Owner's Manual* (Giant Bicylce, Inc.)

[6] Conde E C, Carbajal F B, Papaloapan U and Bonita C L 2011 Generalized PI control of active vehicle suspension systems with MATLAB *Applications of MATLAB in Science and Engineering.* ed T Michalowski (Rijeka, Croatia: InTech) http://www.intechopen.com/books/applications-of-matlab-in-science-and-engineering/generalized-pi-control-of-active-vehicle-suspension-systems-with-matlab.

IOP Conf. Series: Materials Science and Engineering **103** (2015) 012044 doi:10.1088/1757-899X/103/1/012044

Electrochemical way of molybdenum extraction from the Bimetallic systems of Mo-W

L K Kudreeva, M K Nauryzbaev, A P Kurbatov, D H Kamysbaev, A O Adilbekova[1] and Z S Mukataeva

Department of Chemistry and Chemical Technology, Al-Farabi Kazakh National University, Almaty 050040, Kazakhstan

E-mail: Akbota.Adilbekova@kaznu.kz

Abstract. Electrochemical dissolution of molybdenum and tungsten was investigated in water-dimethylsulfoxide (DMSO) media at different concentrations of lithium chloride and magnesium perchlorate. The terms of efficient extraction of molybdenum from bimetallic systems of Mo-W have been determined. The polarization curves of the electrooxidation of molybdenum in the solution of 0.25 M LiCl in the DMSO at the different rates of rotations and the scan rate equal to 50 mV/s were obtained. In the presence of the addition of water at the potential of 0.1-0.75 V the small area of polarizability occurs, then with increasing potentials above 1.5 V there is a sharp increase of the oxidation current. Comparison of the current values of anodic dissolution of molybdenum and tungsten showed that the rate of anodic dissolution of molybdenum significantly exceeds the rate of anodic dissolution of tungsten. In the case of molybdenum, the dissolution process is limited by diffusion, in the case of tungsten - by the passive film formation on the electrode surface.

1. Introduction

The chemical etching process is widely used in contemporary lamp production for molybdenum extraction from bodies of incandescent, wherein the dissolution of molybdenum from the Mo-W bimetallic systems takes place [1]. At the etching process the highest rate of dissolution of molybdenum can be achieved at the lowest weight loss of spirals, and the lowest consumption of pickling liquids. The solution consisting of a mixture of the nitric acid (56%) and the sulfuric acid (78%) in the volume ratio 1: 3 was used for the etching process [2].

The nitric acid well oxidizes the surface of the metal molybdenum till MoO_3. But the molybdenum oxide obstructs the access of acid to the metal surface and thereby it slows the dissolution of the metal. The sulfuric acid dissolving MoO_3 and accelerates the dissolution of metallic molybdenum. It is important to keep the ratio of acids because the decrease of the amount of nitric acid leads in loss weight of tungsten and the reduction of sulfuric acid slows the rate of etching. The end of the evolution of nitrogen oxide brown fumes accompanying the process of etching indicates the completion of the reaction. After washing with running water, the spirals were immersed in the boiling solution of 5% sodium hydroxide to neutralize the acids and for purification from oxides formed on the spiral surface by acids. Then the spirals was washed in hot running water and dipped 2-3 times into a bath with acetone or alcohol and dried in the oven at a temperature not higher than $100^{o}C$.

[1] Address for correspondence:A O Adilbekova, Department of Chemistry and Chemical Technology, Al-Farabi Kazakh National University, Almaty 050040, Kazakhstan. E-mail: Akbota.Adilbekova@kaznu.kz.

When etching molybdenum core, the tungsten is covered with an oxide film which well soluble in alkali. The loss in weight is determined by comparing the mass of spirals etched and mechanically removing the core. The weight loss of spirals should not exceed 3-5%.

The spirals dried after etching must have a metallic gray surface without traces of oxides and stains. The dark colour of the spirals indicates the incomplete removing of the molybdenum.

The process of chemical dissolution is environmentally dangerous, as it leads to the formation of the strongly acidic molybdenum containing solutions of the approximate composition (wt. %): H_2SO_4 - 18, HNO_3 - 8, Mo (VI) - 1,6-3, and also these solutions can contain SO_2 and nitrogen oxides [1]. The concentration of Mo (VI) in the washing water reaches 8 mg/l. The maximum allowable concentration of Mo(VI) is equaled to 0.0012 mg/l. It is impossible to achieve the maximum allowable concentration of [Mo(VI)] at the waste water treatment [1].

Thus, the disadvantages of this method of etching molybdenum cores are:

- formation and evolution of nitrogen oxides,
- the high aggressiveness of etching solutions,
- the etching process is virtually ungovernable,
- the low efficiency of molybdenum capture from spent solution (50%). In this connection, the problems of environmental, technological and resource-saving character take place.

Currently, there are a significant number of ways to remove the core samples from the tungsten [3-14]. However, they have some disadvantages: they are expensive, difficult to operate (number of them requires adjustment of the technological process of spiral production), decrease the rate of dissolution of the molybdenum.

The purpose of the work is to study the possibility of the selective dissolution of molybdenum from the bimetallic system of Mo-W in the water - dimethylsulfoxide (DMSO) solution of lithium chloride and magnesium perchlorate. In this work the anodic dissolution of molybdenum and tungsten was carried out in the solution of 0.5 M $Mg(ClO_4)_2$ and 0.25 M LiCl in the solutions of water - DMSO. The electrolyte is devoid of the most of the drawbacks pointed above at the process of chemical etching [15].

2. Experimental

The study of the electrooxidation of molybdenum and tungsten was carried out by means of the potentiostat-galvanostat *AUTOLAB PG101/101 M* (Switzerland) in the thermostatic three-electrode cell.

The rotating and stationary molybdenum and tungsten rods pressed in Teflon were used as a working electrode (anode). Before each experiment, the electrodes were polished with help of the micron sandpaper to a mirror finish and washed with working solution.

When using a stationary electrode, the measurements were performed in another three-electrode glass cell. The cell before the experiments was washed with hot soda solution, the large amount of water, and then with bi-distilled water. Then the cell was steamed and dried under vacuum. Immediately before the test the cell was rinsed with the working solution. The platinum plate was used as an auxiliary electrode and the silver chloride electrode was used as a reference electrode placed in a capillary with the saturated electrolyte solution of LiCl in the DMSO.

Before each experiment the setup was prepared by accordance with the requirements. The polarization of the electrode was carried out in the range of potentials from 0.00 till +2.0 V in depending on the working electrode and the conditions of the electrochemical process (all potentials were given with respect to the silver chloride electrode).

3. Results and discussion

Figure 1 demonstrates the polarization curves of the electrooxidation of molybdenum in 0.25 M LiCl in DMSO solution at the different speeds of rotations and the scan rate of 50 mV/s. In the presence of the water addition at the potential of 0.1 - 0.75 V the small area of polarizability occurs. Then the increasing of the potentials above 1.5 V results in the extreme increase of the oxidation current because the lowest oxides of molybdenum (δ and γ-phase) are formed [15] which have a high electronic conductivity. With increasing of scanning speed, the speed of the electrochemical dissolution of Mo increases.

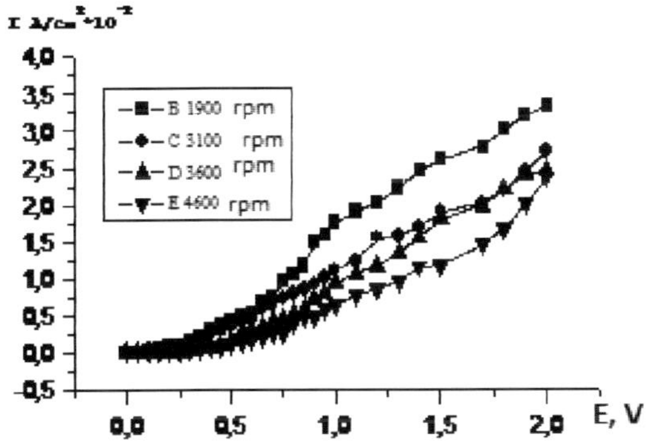

Figure 1. The polarization curves of the anodic dissolution of molybdenum in 0.25 M LiCl in DMSO.

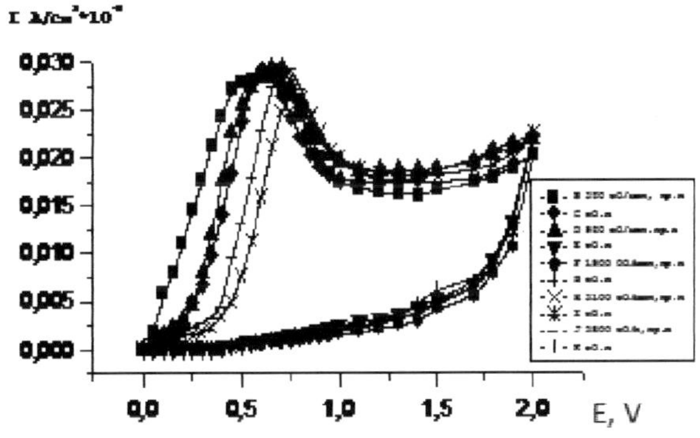

Figure 2. The polarization curves of the anodic dissolution of tungsten in 0.25 M LiCl in DMSO.

Figure 2 shows the polarization curves of tungsten electrooxidation in 0.25M LiCl at the different rates of rotation. It is shown from the figure 2 that the growth of potential leads to the sharp increase of the current which reaches a maximum at 0.5 - 0.7 V. Then in a relatively small potential range from 1.1 V to 1.5 V the rate of oxidation of the metal reduces. The oxidation current of tungsten decreases and then from 1.45 V to 2 V, the oxidation rate slightly increases. The area of the passive state potential is limited by super passivation accompanied by the formation of oxygen compounds of WO_3 (α-phase). Hence the maximum at potential is approximately equaled to + 0.7 V and the subsequent decrease of the speed of the anodic dissolution of tungsten indicates the full coating of the surface by oxides of poorly conducting film from the α-phase [12].

Comparison of the current values of anodic dissolution of molybdenum and tungsten showed that the rate of anodic dissolution of molybdenum significantly exceeds the rate of anodic dissolution of tungsten. In the case of molybdenum, the dissolution process is limited by diffusion, in the case of tungsten - by the passive film formation on the electrode surface [11].

Using data obtained the values of the coefficient of efficiency of anodic dissolution of molybdenum in the presence of tungsten at the different potentials and different concentrations of LiCl were calculated according to the formula:

$$K_{эфф} = i_{Mo}/i_{W},$$

where i_{Mo} и i_W - the density of current of the anodic dissolution of Mo and W, respectively, at the same values of potentials.

Figures 3 and 4 show that the increase of the potential the concentration of LiCl differently influents on the process (Figure 3), but in comparison with the perchlorate solutions, the coefficient of effectiveness of the anodic dissolution of molybdenum is higher in a solution of LiCl. The influence of water additions in [LiCl] solutions (Figure 4) was investigated. As can be seen from the Figure 4 the increase of water concentration causes the growth of the effectiveness coefficient of the anodic dissolution of Mo at the both concentrations of LiCl. The most efficient extraction of molybdenum from the bimetallic systems of Mo-W proceeds at 0.5 M LiCl in 5.2 M DMSO solution, the concentration of water 32.2 mol/l in the potential range of 1 - 2.2 V.

The most effective electrolyte composition is equaled to 0.25 M or 0.5 M LiCl in 5.2 M DMSO and 32.2 M H_2O. Therefore, we conducted the electrolysis in the selected area of concentrations of the components. The electrolysis was carried out in the glass electrochemical cell with separated spaces.

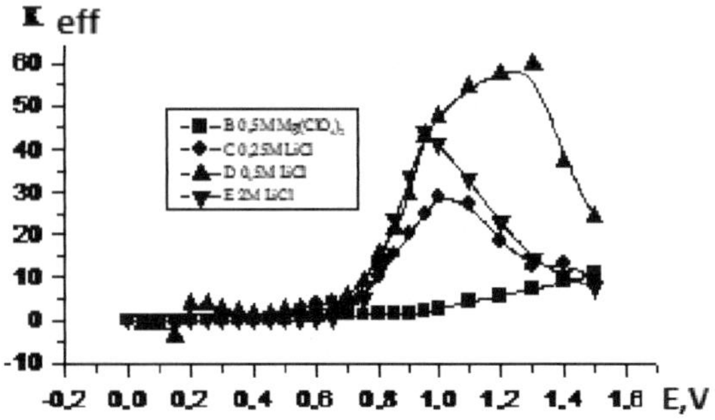

Figure 3. The dependence of the effectiveness coefficient of anodic dissolution of molybdenum on potential at the different concentrations of LiCl and 0.5 M $Mg(ClO_4)_2$.

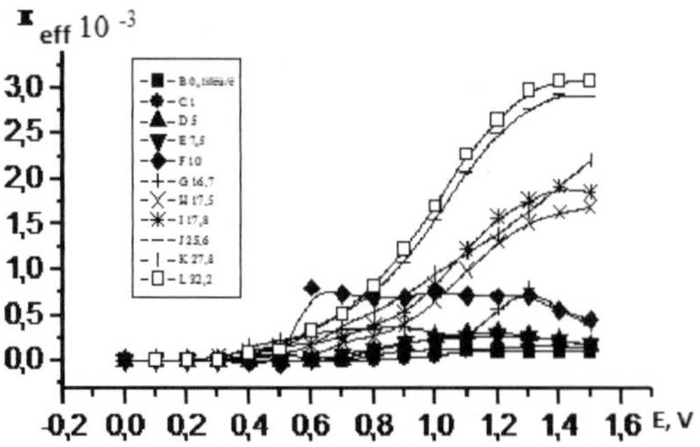

Figure 4. The dependence of the effectiveness coefficient of anodic dissolution of the molybdenum on potential of 2M LiCl at the different concentrations of water.

The indifferent metal (Pt) was cathode, and the tungsten - molybdenum spiral wound around the molybdenum rod was used as a anode. The process was carried out at the room temperature and a current density of 220 mA / cm^2. The amount of metal has passed into solution was adjusted by the difference between the initial and final masses of the anode.

Also for the determination of element contents after the anodic dissolution of the bimetallic system the solution was analyzed by atomic absorption method and mass spectrometric methods.

Agilent 7500 (Agilent Technologies) is the mass spectrometer with inductively coupled plasma. This apparatus is characterized by the ability to detect trace amounts of micro impurities (up to the level of ng/g) and by rapid investigation of the quantitative content of elements (over than 70) for information on the composition of the samples studied.

The Chemstation Software of Agilent 7500 automates the analyses and promotes the accurate decoding of the data. The atomic absorption analysis of Mo and W on the AAS1N device in nitrous oxide-acetylene flame was carried out at the wavelength of 285.2 nm. The content of metals was identified by means of the calibration curve. The calibration was obtained using the accurate weight of molybdenum with metal purity of 99.99%.

The table 1 shows the results for various electrolyte solutions.

Table 1. The results of analysis of the electrolyte solution after the electrolysis.

Electrolyte composition	Concentration in solution, mg/l	
	Mo	W
0.25 MLiCl, 32.2 M H_2O, 5.2 M DMSO	24,400	0,0214
0.5MLiCl, 32.2 M H_2O, 5.2M DMSO	98,247	0,0218

Based on the data presented in the table 1, it can be concluded that in the selected electrolyte is a significant dissolution compared with molybdenum and tungsten concentration of this region is the most suitable for the selective dissolution. Thus, in an electrolyte containing 0.25 M LiCl the concentration of molybdenum passed into the solution on ~ 4000 times higher than the concentration of tungsten, and in 0.5 M LiCl solution is higher by 1000 times.

4. Conclusions

The most effective extraction of molybdenum from the Mo-W bimetallic systems was observed at the next electrolyte composition: 0.25 M or 0.5 M LiCl; 5.2 M DMSO; 32.2 M H_2O in the range of potentials from 1 to 2.2 V.

The dependence of the coefficient of efficiency of the anodic dissolution of molybdenum on the potential value was determined with the presence of the different electrolytes. The ratio of rates of the anodic dissolution of Mo and W is 4000: 1 (in the terms f the experiment). The results obtained give the possibility to use the proposed solutions as an electrolyte mixture for the selective dissolution of molybdenum in the presence of tungsten.

The proposed method is different from the known ones and it significantly increases the selectivity of dissolution and simplifies the process.

References

[1] Yapryntseva O A, Kolosnitsyn V S, Yatsyk N A and Krasnogorskaya N N 2002 Anodic dissolution of molybdenum and tungsten in solutions of sodium sulfate *Russian Journal of Applied Chemistry* **75(4)** 678-9

[2] Ulmishek L G 1958 *The Production of Electrical Incandescent Lamp* (Leningrad)

[3] Binder L, Rathwallner G, Mollay B and Nauer G 2004 The electrochemical dissolution of molybdenum *Proc. 55th of the International Society of Electrochemistry Meeting* (Greece: Thessaloniki) S7FP3

[4] Binder L 1988 The electrochemical dissolution of molybdenum *Proceedings 39th of the International Society of Electrochemistry Meeting* (UK: Glasgow) 547

[5] Armstrong R, Bell M and Metcalfe A 1977 The anodic dissolution of molybdenum in alkaline solutions electrochemical measurements *Journal of Electroanal. Chem.* **84** 61-72

[6] Abbas Q and Binder L 2011 The electrochemical dissolution of molybdenum in non-aqueous media*Int. Journal of Refractory Metals and Hard Materials* **29** 542-6

[7] Hull M 1972 On the anodic dissolution of molybdenum in acidic and alkaline electrolytes *Journal of Electroanal. Chem.* **38** 143-57

[8] Anikeev K M, Ginzburg A A and GandelsmanI L 1966 *A Method for Removing of Molybdenum from Tungsten Spiral of Electron Tubes* (USSR: AC 185064, MPK C 22B 34/30, Bulletin 16)

[9] Jurkiewicz J N and Kolenkov K Y 1969 *The Electrolytic Separation of Tungsten and Molybdenum in the Processing of the Tungsten-molybdenum Alloys* (USSR: AC 223347, MPK C 22B 34/00, Bulletin 10)

[10] Gurevich E A, Polyakov B I, Tsyganov G A, Zakarchevny DI and Kalkov A A 1971 *The Electrolytic Separation of Molybdenum with Refractory Metals* (USSR: AC 290948, MPK C 22B 49/00, Bulletin 3)

[11] Muta Myusudzaki and Fidzikay Kutato Intedzetto 1989 *Method of Selective Dissolution of Molybdenum* (Japan: AC 145734, MPK C 22B 34/36)

[12] Zworykin A Y, Perelman F M and Rode E YA 1950 *Method for Recovering of Tungsten and Molybdenum from Ores and Concentrates* (USSR: AC 116335 MPK C 22B 34/00)

[13] Alkatsev M I and Guriev V R 2002 *Method of Recycling Carbide* (USSR: AC 2189402 MPK B 25B 1/00)

[14] Kozlov N B, Noskov A V, Lilin S A and Zuev A V 1999 About the selectivity of electrochemical dissolution of molybdenum and tungsten *Russian Journal of Applied Chemistry* **72(1)** 101-4

[15] Kudreeva L K 2006 Electrooxidation of molybdenum and tungsten in water - dimethyl sulfoxide solution of electrolytes (Kazakhstan:Al-Farabi Kazakh National University, Ph.D thesis)

IOP Conf. Series: Materials Science and Engineering **103** (2015) 012045 doi:10.1088/1757-899X/103/1/012045

Theoretical and experimental investigations of thickness-stretch modes in 1-3 piezoelectric composites

Z T Yang[1], D P Zeng, M He and H Wang

College of Biomedical Engineering, State Key Laboratory of Ultrasound Engineering in Medicine Co-founded by Chongqing and MOST, Chongqing Municipal Key Laboratory of Ultrasound Engineering in Medicine, Chongqing Medical University, Chongqing, 400016, China

E-mail: yangzengtao@163.com

Abstract. Bulk piezoelectric ceramics operating in thickness-stretch (TSt) modes have been widely used in acoustic-related devices. However, the fundamental TSt waves are always coupled with other modes, and the occurrence of these spurious modes in bulk piezoelectric ceramics affects its performance. To suppress the spurious modes, 1-3 piezoelectric composites are promising candidates. However, theoretical modeling of multiphase ceramic composite objects is very complex. In this study, a 1-3 piezoelectric composite sample and a bulk piezoelectric sample are fabricated. The electrical impedance of these two samples are compared. A simple analytical TSt vibration mode from the three dimensional equations of linear piezoelectricity is used to model the performance of 1-3 piezoelectric composites. The theoretical results agree well with the experimental results.

1. Introduction

Bulk piezoelectric ceramics operating in thickness-stretch (TSt) modes have been widely used in ultrasonic transducers, resonators, filters, vibration sensors and other acoustic wave devices. Strictly speaking, pure TSt motions for bulk piezoelectric plates are only possible in unbounded plates. However, piezoelectric devices have finite dimensions of thickness and radius or length in practical applications. Besides, due to in-plane field variation, Poisson's effect and edge effects, the fundamental TSt waves in piezoelectric plates are always coupled to the in-plane extensional wave and the symmetric thickness-shear (TSh) wave [1, 2]. The occurrence of these spurious modes in bulk piezoelectric ceramics plates is undesired, as it may interfere with the operation of the transducers.

To reduce the coupling effect and to suppress the spurious modes, many novel structured materials have been invented. 1-3 piezoelectric composites consisting of active piezoelectric rods embedded in a polymer phase, which combines the advantages of both ceramic and polymer phases, have been widely used in transducer applications [3, 4]. Compared with bulk piezoelectric ceramics, 1-3 piezoelectric composites show many advantages: lower acoustic impedance (Z), a higher coupling coefficient (k_t), a lower mechanical quality factor (Q_m) and dielectric loss ($\tan\delta$) and better design

[1] Address for correspondence: Z T Yang, College of Biomedical Engineering, State Key Laboratory of Ultrasound Engineering in Medicine Co-founded by Chongqing and MOST, Chongqing Municipal Key Laboratory of Ultrasound Engineering in Medicine, Chongqing Medical University, Chongqing, 400016, China. E-mail: yangzengtao@163.com.

flexibility [5, 6]. One example is that Or and Chan [7] have used piezoelectric composite rings to suppress the undesirable radial and wall thickness modes in transducer construction. Nowadays, 1-3 piezoelectric composites can replace bulk piezoelectric materials in bio-medical transducers, underwater applications, and micropositioning systems [8].

Modeling the dynamic behavior of transducers made of 1-3 piezoelectric composites is important in real applications. It is also helpful for the technical improvement of ultrasound-related fields. However, theoretical modeling for multiphase ceramic composite objects is very complex. In the past, many models have been introduced for investigating the physical and electrical properties of 1-3 composites [5, 9, 10]. In [5, 9], 1-3 composites can be treated as an effective homogeneous medium. However, details of the thickness vibration expression and electric independence were not derived in the paper [10]. In addition, the mechanical losses of 1-3 composite piezoelectrics were neglected, and the type of thickness mode was not discerned in the analysis [9]. The finite element method (FEM), which had received considerable attention in recent years, was used to analyze and optimize the 1-3 piezoelectric composites, providing reliable descriptions of the behavior of real devices [4, 11]. However, FEM was always computationally expensive and time consuming. Thus, an analytical solution from which the behaviors of the system can be seen explicitly and directly is still needed.

In this paper, both 1-3 piezoelectric composite samples and bulk piezoelectric samples are fabricated. The electrical impedances between these 2 samples are compared. Then, the paper presents a simple analytical TSt vibration mode from the three dimensional equations of linear piezoelectricity to model the performance of 1-3 piezoelectric composite plates. The theoretical results agree well with the experimental results.

2. Impedance comparison between a 1-3 piezoelectric composite disc and a bulk piezoelectric disc

To illustrate the superior performance of 1-3 piezoelectric composites, two kinds of samples are fabricated and compared. As shown in figure 1, one disc is made of the 1-3 piezoelectric composite. The other is made of bulk piezoelectric ceramics. In this work, both samples are operated at a fundamental frequency of 0.8 MHz. The bulk piezoelectric disc is fabricated using lead zirconate titanate (PZT-5A, [12]). The 1-3 piezoelectric composite disc is fabricated using PZT-5A and epoxy. The PZT-5A unit rods used as the active component are embedded in a passive epoxy. The aspect ratio (ratio between the thickness and the lateral dimension of the PZT unit cell) of the sample is about 3, larger than 2 mentioned in [13], in order to model the thickness mode sufficiently.

Note that the fundamental TSt mode is most useful in applications [14]. The electrical impedances near the fundamental frequency of the 1-3 piezoelectric composite disc and the bulk PZT disc are compared in figure 2. The electrical impedance of the two samples is measured by an impedance analyzer (Agilent 4294A). As shown in figure 2, the impedance of the 1-3 piezoelectric composite disc

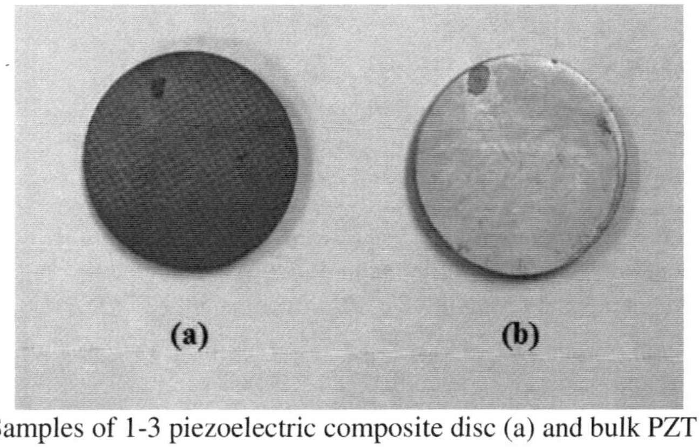

Figure 1. Samples of 1-3 piezoelectric composite disc (a) and bulk PZT disc (b).

at its anti-resonance frequency is much higher than that of the bulk PZT disc. In addition, many spurious modes occur at the low-frequency region of the bulk PZT disc, which would interfere with the performance. Moreover, figure 2 shows that the coupling coefficient (k_t) of the 1-3 piezoelectric composite disc is much higher than that of the bulk PZT disc.

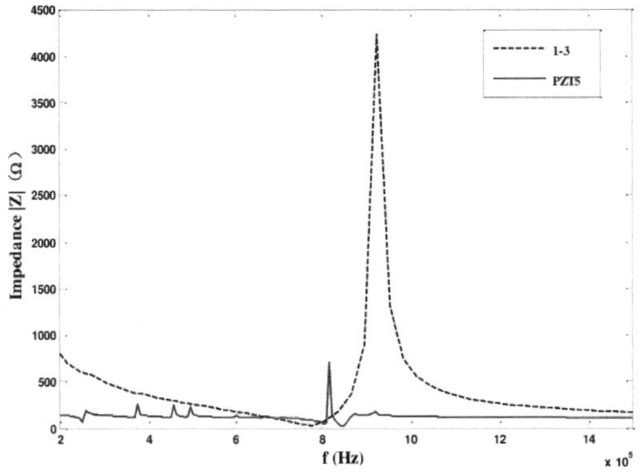

Figure 2. Electrical impedance comparison between sample (a) and (b).

3. Modeling of 1-3 piezoelectric composites operating in TSt modes

1-3 piezoelectric composites, as shown in figures 1(a) and 3, are formed by embedding piezoelectric ceramic rods in a passive polymer medium. The poling direction of the PZT rods is along the longitudinal direction. As mentioned in [5, 9], 1-3 composites can be treated as an effective homogeneous medium with a set of effective material properties. Thus, an analytical model in which the 1-3 composites are replaced by a bulk ceramic plate is proposed. Consider the bulk ceramic plate with a thickness of 2 h poled along the x_3 axis (see figure 3). The plate is bounded by two planes at $x_3 = \pm h$ that are traction-free and electroded. A time-harmonic voltage is applied across the plate thickness.

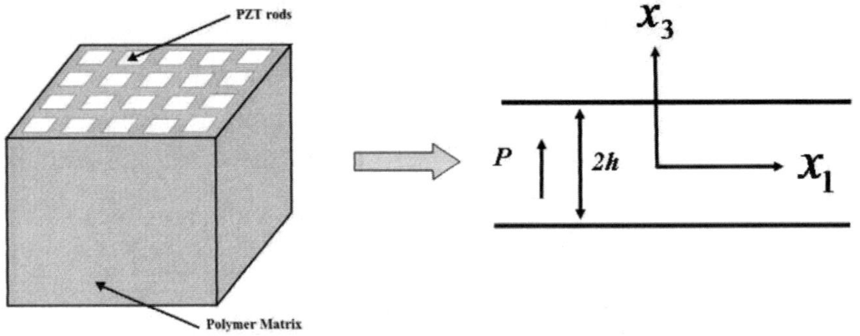

Figure 3. Schematic representation of 1-3 piezoelectric composite plate and its theoretical model.

The governing equations of three-dimensional linear piezoelectricity [15] and the boundary conditions for the 1-3 piezoelectric composite plate shown in figure 3 can be written as [16]:

$$T_{ji,j} = \rho \ddot{u}_i, \quad D_{i,i} = 0,$$

$$T_{ji,j} = c_{ijkl} S_{kl} - e_{kij} E_k, \quad D_i = e_{ikl} S_{kl} + \varepsilon_{ik} E_k,$$

$$S_{ij} = (u_{i,j} + u_{j,i})/2, \quad E_i = -\phi_{,i},$$
$$T_{3j} = 0, \quad x_3 = \pm h, \tag{1}$$
$$\phi(x_3 = h) - \phi(x_3 = -h) = Ve^{i\omega t}.$$

where u_i is the mechanical displacement vector, T_{ij} is the stress tensor, S_{ij} is the strain tensor, E_i is the electric field vector, D_i is the electric displacement vector, and ϕ is the electric potential. c_{ijkl}, e_{kij} and ε_{ij} are the effective elastic, piezoelectric and dielectric constants of the 1-3 piezoelectric composite plate, respectively. We denote the effective mass density by ρ. We use the Cartesian tensor notation, the summation convention for repeated tensor indices and the convention that a comma followed by an index denotes partial differentiation with respect to the coordinate associated with the index. A superimposed dot represents the time derivative.

For 1-3 piezoelectric composites poled in the x_3 direction, consider the so-called thickness-stretch (TSt) vibration of the plate only depending on u_3. We assume the solutions are in the following forms:

$$u_3 = u(x_3)e^{i\omega t}, \; u_1 = u_2 = 0, \; \phi = \phi(x_3)e^{i\omega t}, \tag{2}$$

Modes described by Equation (2) are allowed by the three-dimensional equations of linear piezoelectricity.

4. Theoretical and experimental results and discussion

To describe the TSt vibrations of the disc, the effective constitutive relationship for this homogeneous medium was introduced in [5].

$$c_{33} = v[c_{33}^E - \frac{2\bar{v}(c_{13}^E - c_{12}^P)^2}{v(c_{11}^P + c_{12}^P) + \bar{v}(c_{11}^E + c_{12}^E)}] + \bar{v}c_{11}^P,$$
$$e_{33} = v[e_{33}^E - \frac{2\bar{v}e_{31}^E(c_{13}^E - c_{12}^P)}{v(c_{11}^P + c_{12}^P) + \bar{v}(c_{11}^E + c_{12}^E)}], \tag{3}$$
$$\varepsilon_{33} = v[\varepsilon_{33}^E + \frac{2\bar{v}(e_{31}^E)^2}{v(c_{11}^P + c_{12}^P) + \bar{v}(c_{11}^E + c_{12}^E)}] + \bar{v}\varepsilon_{11}^P,$$
$$\rho = v\rho^E + \bar{v}\rho^P.$$

where v is the volume fraction of the piezoceramics in the composite, and $\bar{v} = 1 - v$ is the volume fraction of the epoxy. The material parameters with a superimposed E and P inside are for the ceramic phase and polymer phase, respectively. The corresponding material parameters for both PZT-5A and epoxy are given in table 1.

Table 1. Material parameters for PZT-5A and epoxy.

Piezoceremics	PZT-5A [12]	Epoxy	Araldite D
c_{11}^E (10^{10} N/m^2)	12.1	c_{11}^P (10^{10} N/m^2)	0.8
c_{12}^E (10^{10} N/m^2)	7.59	c_{12}^P (10^{10} N/m^2)	0.44
c_{13}^E (10^{10} N/m^2)	7.54	$\varepsilon_{11}^P / \varepsilon_0$	4
c_{33}^E (10^{10} N/m^2)	11.1	ρ^P (Kg/m^3)	1150

e_{31}^{E} (C/m^2)	-5.4	$\varepsilon_0(10^{-12}$F/m)	8.85
e_{33}^{E} (C/m^2)	15.8		
ε_{31}^{E} (10^{-8}C/Vm)	0.811		
ε_{33}^{E} (10^{-8}C/Vm)	0.735		
ρ^{E} (Kg/m^3)	7750		

The calculated electrical impedance of the disc is compared with the one obtained experimentally by the impedance analyzer (Agilent 4294A). The comparison between the experimental and theoretical electrical impedances is shown in figure 4. There is excellent agreement between the experimental electrical impedance and that obtained by the theoretical TSt vibration we proposed. This means that the 1-3 piezoelectric composite can operate in nearly pure TSt vibration. Other modes can be suppressed near the fundamental TSt resonance. The model we present could be used as a guideline for transducer design.

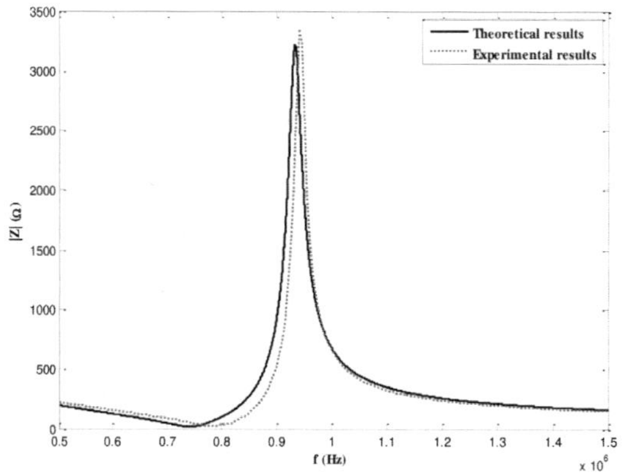

Figure 4. The comparison of electrical impedances between the experimental and theoretical results.

5. Conclusions

This paper presents the theoretical and experimental studies of the TSt modes of the 1-3 piezoelectric composite. An analytical solution of TSt vibration from the three dimensional equations of linear piezoelectricity to analyze the performance of 1-3 piezoelectric composites is performed. The electrical impedance obtained by the theoretical TSt vibration agrees well with experimental results. This means that the 1-3 piezoelectric composite can operate in nearly pure TSt vibration. This simple model of TSt vibration is sufficient to analyze 1-3 piezoelectric composites with larger aspect ratios quickly and efficiently. The results could be used as a guideline for transducer design.

Acknowledgment

This work is supported by the National Natural Science Foundation of China (Grant number 81201102, 81127901, 81401424) and Natural Science Foundation Project of CQ CSTC (Grant number cstc2012jjA10075). Additionally, this work is also supported by the Plan of Top Young Innovative Talents of Chongqing City.

References

[1] Huang R, Lee P C Y, Lin W S and Yu J D 2002 Extensional, thickness-stretch and symmetric thickness-shear vibrations of piezoceramic disks *IEEE Trans. Ultrason, Ferroelectric.* **49** 1507-15

[2] Lin S Y 1998 Coupled vibration analysis of piezoelectric ceramic disk resonators *Journal of Sound and Vibration* **218** 205-17

[3] Newnham R E, Skinner D P and Cross L E 1978 Connectivity and piezoelectric-pyroelectric composites *Materials Research Bulletin* **13** 525-36

[4] Hossack J A and Hayward G 1991 Finite-element analysis of 1-3 composite transducers *IEEE Trans. Ultrason. Ferroelectr.* **38** 618-29

[5] Smith W A and Auld B A 1991 Modeling 1-3 composite piezoelectrics: Thickness-mode oscillations *IEEE Trans. Ultrason. Ferroelectr.* **38** 40-7

[6] Janas V F and Safari A 1995 Overview of fine-scale piezoelectric ceramic/polymer composite processing *J. Am. Ceram.* **78** 2945-55

[7] Or S W and Chan H L W 2001 Mode coupling in lead zirconate titanate/epoxy 1-3 piezocomposite rings *Journal of Applied Physics* **90** 4122-9

[8] Topolov V Y and Bowen C R 2009 *Electromechanical properties in composites based on ferroelectrics* (London: Springer)

[9] Chan H L W and Unsworth J 1989 Simple model for piezoelectric ceramic/polymer 1-3 composites used in ultrasonic transducer applications *IEEE Trans. Ultrason. Ferroelectr.* **36** 434-41

[10] Gururaja T R, Schulze W A and Cross L E 1985 Piezoelectric composite materials for ultrasonic transducer applications, part I: resonant modes of vibration of PZT rod-polymer composites *IEEE Transactions on Sonics and Ultrasonics.* **32** 481-98

[11] Ballandras S, Wilm M, Edoa P F, et al. 2003 *Jounal of Applied Physics* **93** 702

[12] Berlincourt D A, Curran D R and Jaffe H 1964 Piezoelectric and piezomagnetic materials and their function in transducers *Physical Acoustics* vol 1-A ed Mason W P (New York: Academic Press) p 247

[13] Hayward G and Bennett J 1996 Assessing the influence of pillar aspect ratio on the behaviour of 1-3 connectivity composite transducers *IEEE Trans. Ultrason. Ferroelectr.* **43** 98-108

[14] Huang D J and Yang J S 2014 On the propagation of long thickness-stretch waves in piezoelectric plates *Ultrasonics* **54** 1277-80

[15] Tiersten H F 1969 *Linear Piezoelectric Plate Vibrations* (Boston: Springer)

[16] Yang J S 2006 *Analysis of piezoelectric devices* (Singapore: World Scientific)

IOP Conf. Series: Materials Science and Engineering **103** (2015) 012046 doi:10.1088/1757-899X/103/1/012046

A study of factors affecting properties of AM/AMPS/NVP terpolymeric microspheres prepared by inverse suspension polymerization

J F Jiang, Q Zhao, M Q Lin[1], Y F Wang, S M Dang, F F Sun

Research Institute of Enhanced Oil Recovery, China University of Petroleum, Beijing 102200, China

E-mail: linmq@cup.edu.cn

Abstract. Terpolymeric microspheres were synthesized by the inverse suspension polymerization of functional monomers including AMPS, NVP, and AM. The morphology and size of the obtained microspheres were measured by scanning electron microscopy (SEM) and optical microscopy. Furthermore, the swelling performances of the obtained microspheres were measured with alaser particle analyzer (LPA), and the thermal stability of the microspheres obtained was measured by differential thermal analysis (DSC-TG) and high temperature experiments involving microsphere/water dispersion. The results revealed that the extreme value of the microsphere size distribution decreased from 280 μm to 20 μm as the stirring rate increased from 175 rpm to 500 rpm. At temperatures below 25°C, the maximum achieved swelling ratio of the microspheres was 21, and the thermal stability of the terpolymer microspheres was significantly higher than that of the dipolymer microspheres. The terpolymer/water dispersions were kept at 120°C for 19d before any damage was observed.

1. Introduction

In deep profile control of oil fields, polymer microspheres are able to travel through cracks and channels in high-pressure are a sand get to deeper parts, thus achieving control along the profile. This capability can be attributed to ideal particle size, good swelling performance after absorbing water, deformation performance, thermal resistance, and shear resistance [1-3]. As profile control is such an important aspect of oil mining, studies on factors affecting the properties of polymer microspheres are of great significance [4-8].

Various lab experiments and practical applications performed in oil fields have indicated that polymer microspheres have a significant enhancement effect on oil recovery factors [9-10]. Deep profile control achieved by polymer microspheres can only be successful in cases where the particle size matches well with the size of the reservoir fractures and channels; polymer microspheres of various sizes can be synthesized using different polymerization methods [11-13]. Previous studies on the swelling performance, heat resistance, salt resistance, and strength of acrylamide-based polymer microspheres revealed that these microspheres showed reasonable swelling performance and stability in water at 90°C for more than 10 d [14-17]. Moreover, the particle size decreased as salt

[1] Address for correspondence: M Q Lin, Research Institute of Enhanced Oil Recovery, China University of Petroleum, Beijing 102200, China. E-mail: linmq@cup.edu.cn.

concentration and shear rate increase [18]. Although many studies have been carried out involving polymer microspheres, few reports on their thermal resistance have been published. In this article, copolymer microspheres were prepared by the inverse suspension polymerization of AM and two monomers (AMPS and NVP) with good thermal resistance [19]. Additionally, the effects of the AMPS:NVP ratio, stirring speed, concentrations of cross-linking agent and emulsifier on particle size, swelling performance, and thermal resistance were investigated.

2. Experimental

2.1. Materials

The materials used for the experiments of this research included the following items, which are listed with their grade and company of origin: AM (Acrylamide), analytically pure, Beijing Modern East Fine Chemical Co. Ltd.; AMPS (2-acrylamido-2-methyl propanesulfonic acid), industrial grade, Shandong Quanxin Chemical Industry Co. Ltd.; NVP (N-vinyl pyrrolidone), analytically pure, TCI-Tokyo Chemical Industry Co. Ltd.; NMBA (N, N-methylenebis acrylamide), chemically pure, Tianjin Guangfu Fine Chemical Industry Institute; KPS (potassium peroxodisulfate), chemically pure, Tianjin Guangfu Fine Chemical Industry Institute; Cyclohexane, analytically pure, Beijing Yili Fine Chemical Co. Ltd.; Span 80 (dehydrated sorbitan fatty acid ester), analytically pure, Beijing Yili Fine Chemical Co. Ltd.; absolute ethanol, analytically pure, Beijing Yili Fine Chemical Co. Ltd.; Methylene Blue, analytically pure, Beijing Yili Fine Chemical Co. Ltd.; D.I. water, made in lab.

2.2. Methods

2.2.1. Preparation of terpolymeric microspheres. A certain amount of span 80 was mixed with cyclohexane and stirred until clear and a transparent oil phase was observed; monomers, NMBA, and KPS were dissolved in water to obtain a water phase. Then, the oil phase was poured into a three-neck flask in a thermostatic waterbath. The stirring speed and waterbath temperature were adjusted accordingly and the water phase was added dropwise into the flask under a constant pressure. After several hours, the mixture was filtrated and the obtained products were washed with absolute ethanol and dried at 50°C for 24 h. Finally, pale yellow copolymer microspheres were obtained.

2.2.2. Measurement of microsphere size. The morphologies of the microsphere powder and microspheres after swelling were checked by microscope (Olympus BX-41). The size and surface morphology were checked by scanning electron microscopy (SEM, SIRION 200).

2.2.3. Measurement of swelling performance. The particle sizes before and after water absorption were measured by laser particle analyzer (MACROTRAC S3500).

2.2.4. Measurement of thermal resistance. The thermal resistances of the microspheres were measured by simultaneousthermalanalyzer (SDT Q600) with a heating rate of 10°C/min (up to 600°C). The microspheres were dispersed in water and placed in a high-temperature oven. The samples were checked by microscope and SEM at 24 h intervals.

3. Experimental results

3.1. Effects of monomer ratio on the properties of microspheres

Seven kinds of microspheres were obtained by adjusting the monomer ratio (AM: AMPS: NVP), and the properties of these microspheres were evaluated.

3.1.1. Swelling behaviour. The initial average diameters, the average diameters after swelling in de-ionized water for 8 h, 24 h, and 72 h, and the swelling ratios of the seven kinds of spheres were measured, and are summarized in table 1. The microscopic images of the terpolymer microspheres (monomer ratio AM: AMPS: NVP=7:1.5:1.5) before and after 24-hour swelling are shown in figure 1.

Table 1. Size distribution of microspheres with different ratios of monomers.

Monomer ratio AM: AMPS: NVP	Initial size (d) /μm	Size after 8h swelling/μm	Size after 24 h swelling /μm	Size after 72 h swelling/μm	Volumetric swelling (D^3/d^3)
7:0:3	76.46	99.61	124.26	126.17	4.29
7:0.5:2.5	67.04	97.15	120.67	119.35	5.83
7:1:2	60.45	91.52	116.75	115.96	7.21
7:1.5:1.5	55.69	100.24	139.23	139.85	15.63
7:2:1	51.83	103.74	139.49	141.14	19.49
7:2.5:0.5	49.28	132.58	206.98	206.26	74.09
7:3:0	46.62	165.46	209.78	210.52	91.11

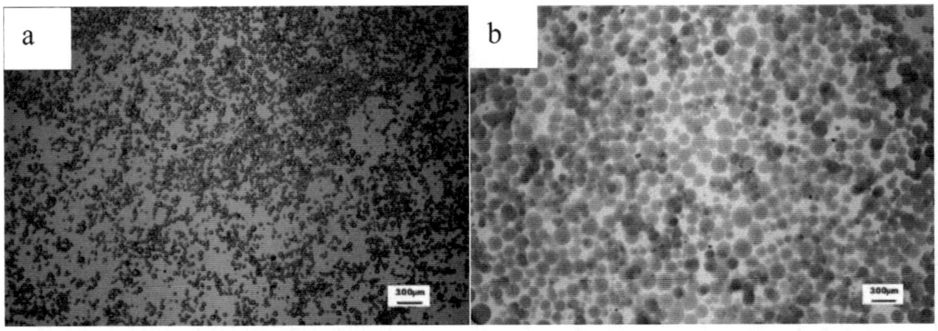

Figure 1. SEM images of terpolymer microspheres (a) before swelling; and (b) after swelling. The swelling temperature was 30°C and the microsphere concentration was 20 mg/L.

Figure 1(a) shows a SEM image of the dry-powdered terpolymer microspheres. As can be seen, the initial particle diameters were less than 90μm. Figure 1(b) shows a SEM image of the microspheres after 24 h swelling and stained with methylene blue. As can be seen, the diameters of the spheres at this stage were in the range of 100-200 μm. After swelling, the microspheres maintained a spherical morphology and the diameters increased significantly.

As shown in table 1, as the percentage of AMPS increased, the diameters of the microspheres decreased, while the swelling capability increased. After 8 h swelling, the diameters of the microspheres increased drastically. After 24 h swelling, the diameters of the microspheres increased further. However, after 72 h swelling, the diameter of the microspheres was similar to that of the microspheres after 24 h swelling, demonstrating that the spheres reached their swelling limit after 24hand the diameter of the microspheres did not increase further.

3.1.2. Thermal resistant properties in aqueous solution. Thermal stability durations of the microspheres in aqueous solution at 120°Cwere measured, and are shown in table 2. As can be seen, as the proportion of AMPS increased and the proportion of NVP decreased, the thermal stability durations of the microspheres in the aqueous solution first increased and then decreased. The thermal resistant duration was optimized (19 d) at AMPS: NVP = 1:1. The thermal resistances of the AM/NVP and AM/AMPS copolymer microspheres were relatively poor, with a thermal resistant duration of 8 d and 7 d, respectively. Therefore, it was indicated that optimum thermal resistance could only achieved with the AM/AMPS/NVP terpolymer microspheres (with a proper monomer ratio).

IOP Conf. Series: Materials Science and Engineering **103** (2015) 012046 doi:10.1088/1757-899X/103/1/012046

Table 2. Thermal stability of microspheres with different ratios of monomers.

Monomer ratio (AM:AMPS:NVP)	7:0:3	7:0.5:2.5	7:1:2	7:1.5:1.5	7:2:1	7:2.5:0.5	7:3:0
Duration /d	8	11	16	19	15	10	7

3.1.3. Analysis of TG-DSC curves. The TG-DSC curves of the dry-powdered microspheres are shown in figure 2 and table 3. As shown in figure 2(a), the weight loss curves of the microspheres with monomer ratios (AM: AMPS: NVP) of 7:0:3, 7:0.5:2.5, and 7:3:0 were significantly different from the other curves, especially in the range of 200-400°C. At the same temperature, weight loss percentages of the microspheres with monomer ratios of 7:0:3 and 7:0.5:2.5 were lower than that of the other microspheres, while microspheres with a monomer ratio of 7:3:0 exhibited the lowest weight loss percentages. Figure 2(b) shows the DSC curves of different microspheres. As can be seen, the intensities and positions of all endothermic peaks of the microspheres with monomer ratios of 7:1:2 and 7:1.5:1.5 were similar to each other. The DSC curves of the microspheres with monomer ratios of 7:2:1 and 7:2.5:0.5 were similar at temperatures lower than 360°C. In the range of 150-230°C, all curves had similar endothermic peaks at the same temperature, indicating that all the microspheres decomposed in a similar way in this range.

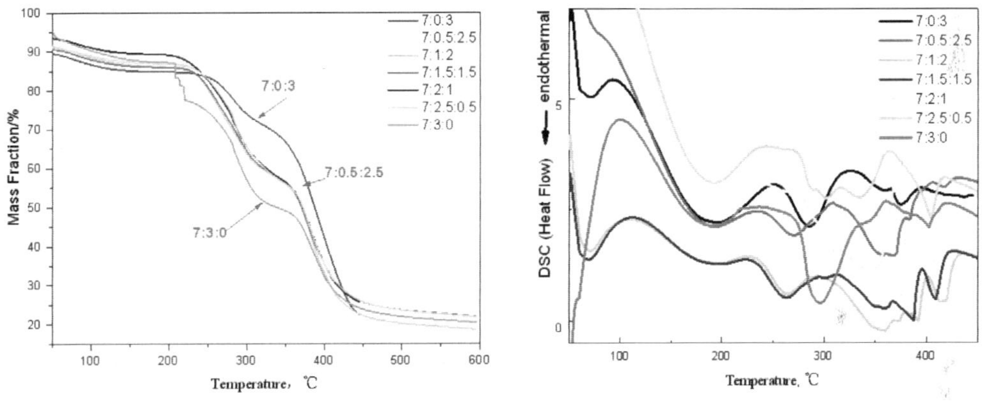

Figure 2. TG and DSC curves of microspheres with different ratios of monomers.

Table 3. Various temperatures of microspheres with different ratios of monomers.

Monomer ratio	GTT /°C	IDD /°C	Degradation temperatures corresponding to different weight loss					
			10	20	30	40	5060	
7:0:3	139	197.5	295.26	356.76	379.04	395.31	412.21	433.88
7:0.5:2.5	/	193.5	270.05	310.89	368.63	384.57	401.22	431.60
7:1:2	147.8	197.5	262.34	292.49	348.34	376.80	393.07	420.42
7:1.5:1.5	148.4	199.6	265.37	295.24	352.66	377.65	398.46	446.39
7:2:1	/	193.9	262.45	289.57	329.79	373.62	391.46	419.43
7:2.5:0.5	130	192.6	272.46	296.08	341.34	377.78	396.03	438.13
7:3:0	140	191.5	226.25	278.46	298.97	363.44	388.03	419.14

[a] GTT: Glass transition temperature.
[b] IDT: Initial degradation temperature.

As shown in table 3, the initial glass transition temperatures of the microspheres with monomer ratios (AM: AMPS: NVP) of 7:1:2 and 7:1.5:1.5 were significantly higher than that of the other microspheres, indicating these microspheres showed better thermal stability. The temperature corresponding to a weight loss of 60% for the microspheres with a monomer ratio of 7:1.5:1.5 was the highest (446.39°C), while that for the microspheres with a monomer ratio of 7:3:0 was the lowest

(419.14°C). Therefore, it was concluded that the microspheres with a monomer ratio of 7:1.5:1.5 demonstrated the optimized thermal resistance, while the microspheres with a monomer ratio of 7:0:3 had the worst thermal resistance in the practical temperature range of oil fields.

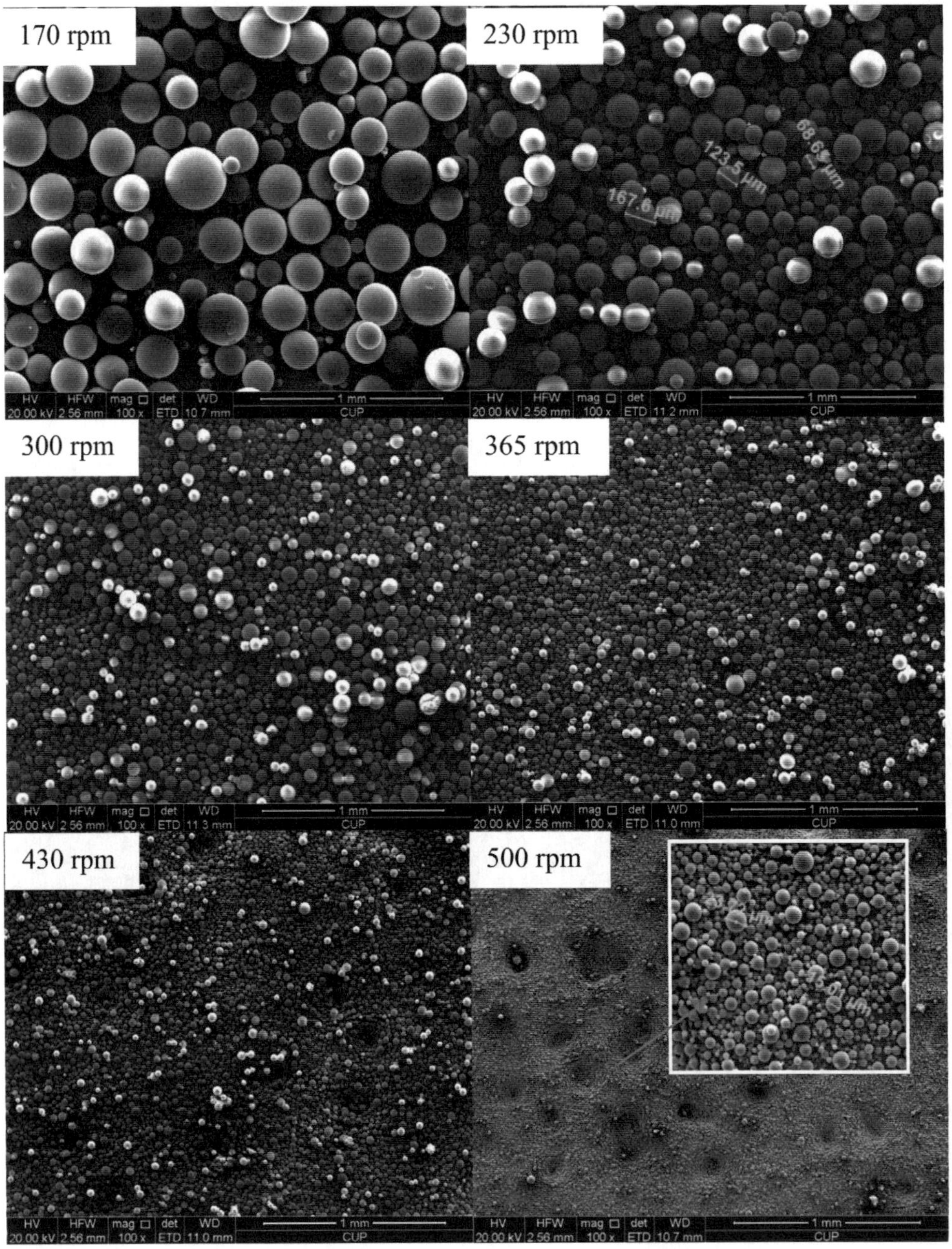

Figure 3. SEM images of microspheres prepared by different stirring rates.

3.2. Effects of stirring speed on the properties of microspheres

3.2.1. Initial diameter. With the other experimental conditions kept constant, six kinds of microspheres with different diameters were obtained by adjusting the stirring speed (170-500 r/min), and SEM images of these samples are shown in figure 3. As can be seen, the diameter of the microspheres decreased as the mechanical stirring speed increased. As shown in table 4, the average initial diameter of the microspheres stirred at 175 rpm was 230.90 μm, while that of the microspheres stirred at 500 rpm was 11.14 μm, indicating that the stirring speed had a significant effect on the diameter of the microspheres. A possible reason is that in the inverse suspension system, external forces and the proper amount of emulsifier are needed to form stable water-in-oil droplets, and the stirring speed directly affects the size of the droplets containing monomers, initiators, and crosslinking agents, thus affecting the diameter of the polymer microspheres.

Table 4. Initial average size of microspheres with different stirring rates.

Stirring rate /rpm	175	230	300	365	430	500
Initial size /μm	230.90	136.00	69.81	41.71	30.68	11.14

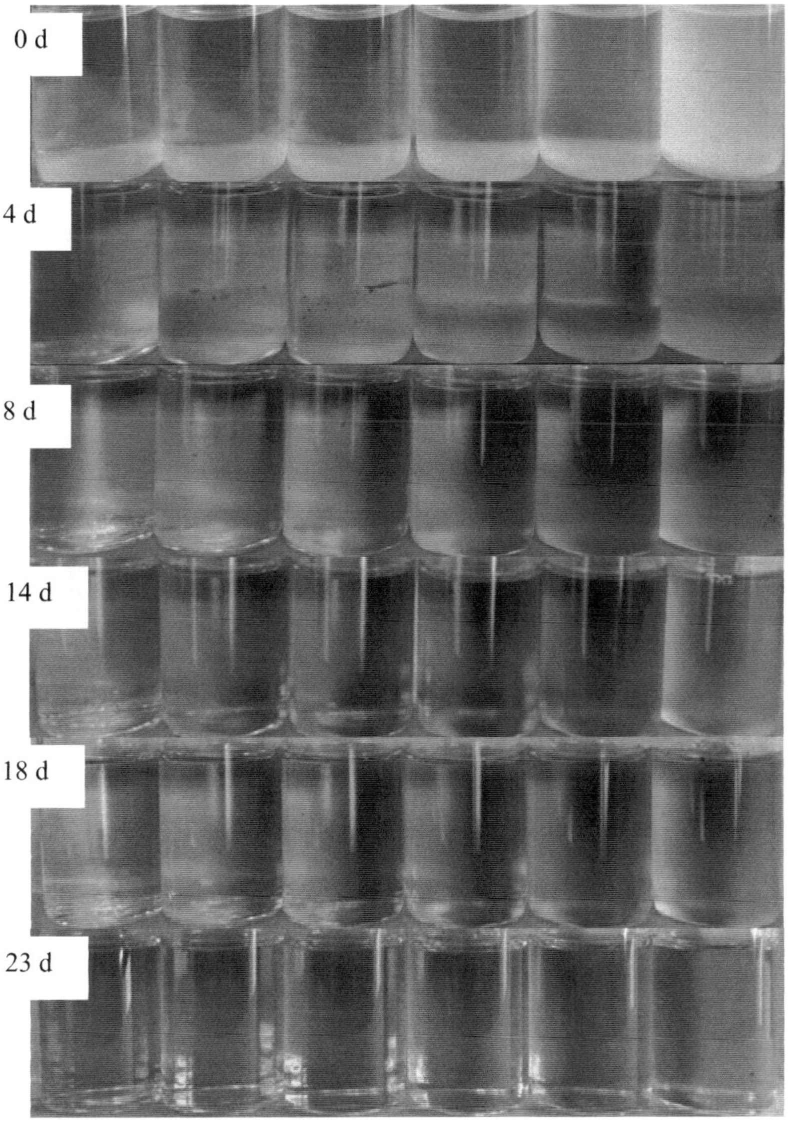

Figure 4. Thermal stability (in water solution) of microspheres prepared by different.

3.2.2. Thermal stability in aqueous solution. Microspheres obtained at different stirring speeds were transferred into 120°C water for different durations and monitored. As shown in figure 4 (stirring speed increasing from left to right), the layered interfaces between the microspheres and water were obvious in the first four bottles, with the microspheres precipitated at the bottom. For the samples obtained at a stirring speed of 430 rpm, some suspended microspheres were observed, while most of the microspheres precipitated at the bottom. In the samples obtained at a stirring speed of 500 rpm, a turbid liquid different from the others was observed due to the small diameter of the microspheres. As the duration increased, microsphere swelling was observed due to water absorption. After 8 d, the microspheres obtained at 175 rpm occupied nearly all of the solution. In the next 48 h, the microspheres swelled further and filled the solution from left to right. This status was maintained for 5 d. After 14 d, decomposition of the microspheres obtained at 175 rpm was observed and the diameter of the microspheres started to decrease. Then, the decomposition continued from left to right. In the last 23 d of the experiment, the microspheres decomposed completely and the solutions were transparent.

The specific thermal resistant durations of the different samples are summarized in table 5. The number of days decreased as the microsphere diameters increased. This could be attributed to the increase in the amount of AMPS contained in a single microsphere and the amount of negative groups produced, which enhanced the interaction forces between polymer segments. Additionally, segments containing fewer negative groups were more fragile at elevated temperatures, resulting in accelerated decomposition of the microspheres.

Table 5. Thermal stability (in duration) of microspheres prepared by different stirring rates (water temperature = 120°C and microsphere concentration = 1 g/L).

Stirring rate /rpm	175	235	300	365	430	500
Stability duration/d	18	18	19	20	22	23

3.3. Effects of crosslinking agent amounts on the properties of microspheres

3.3.1. Initial diameter and swelling performance. A series of microspheres were prepared by adjusting the amount of crosslinking agents used (0.05%-2%).The average diameters of all the microspheres obtained before and after swelling are summarized in table 6. As the amount of crosslinking agent increased from 0.1% to 2%, the average diameter of the microspheres decreased from 63.63 μm to 47.95 μm. With the amount of crosslinking agents at 0.05%, the average diameter of the microspheres was 47.91 μm, indicating a negative correlation between the amount of crosslinking agent and the diameter of the microspheres. However, no obvious trends were observed for the average diameter of microspheres after swelling. With the amount of crosslinking agent in the range of 0.33%-1.5%, the swelling performance of the microspheres degraded.

Table 6. Initial size and swelling performance of microspheres with different amounts of cross-linker.

NMBA content/%	0.05	0.1	0.33	1	1.5	2
Initial size/μm	47.91	63.63	60.06	56.3	54.26	47.95
Size after swelling/μm	150.19	174.66	186.49	156.85	122.8	156.32
Swelling ratio	30.75	20.61	29.91	21.64	11.59	34.63

3.3.2. Thermal resistance

- TG-DSC curves

The TG curves of the microspheres obtained with different amounts of crosslinking agents are shown in figure 5. As can be seen, the different curves are highly similar to each other, and the decomposition and weight loss percentages at certain temperatures are almost identical, indicating that the amount of crosslinking agent had a negligible effect on the thermal resistance of the dry-powdered microspheres.

- Thermal resistance performance in high temperature aqueous solution

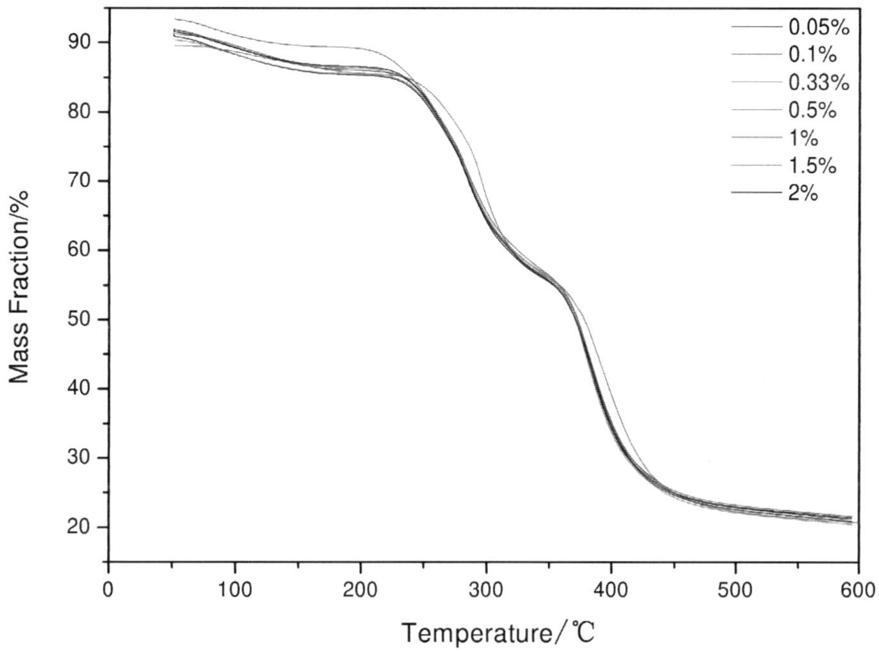

Figure 5. TG curves of microspheres prepared by different amounts of cross-linker.

Microspheres obtained with different crosslinking agent amounts (0.05%-2%) were dispersed in aqueous solution, and the thermal stability durations of these disperse systems were measured. As shown in table 7, the thermal stability durations of the microspheres dispersed in aqueous solution stayed constant at 18-19 das the amount of crosslinking agent was increased. This result was consistent with TG curves, thus verifying the conclusion that the effect of crosslinking agent amount on the thermal resistance of the microspheres was negligible.

Table 7. Thermal stability of microspheres prepared by different amounts of LX cross-linker.

NMBA content/%	0.05	0.1	0.33	0.5	1	1.5	2
Stability duration /d	18	19	18	19	19	18	18

3.4. Effects of emulsifier amount on the properties of microspheres

With the other experimental conditions kept constant, seven kinds of microspheres were obtained by adjusting the amount of emulsifier (1.6%-20%). The size distributions and thermal resistance durations in aqueous solution of all microspheres were obtained and are shown in table 8. It was observed that the average diameter of the dry-powdered microspheres decreased from 82.25 μm to 64.74 μm as the emulsifier amount increased from 1.6% to 20%; the emulsifier amount had a negligible effect on the thermal resistance duration. The results demonstrated that the diameter of the microspheres decreased when the emulsifier amount increased, while the thermal resistances of these microspheres were not affected.

Table 8. Initial size and thermal stability (in duration) of microspheres prepared by different amounts of emulsifier.

Emulsifier content /%	1.6	3.2	6.4	10.0	13.2	16.4	20
Initial size /μm	82.25	72.01	72.46	70.19	69.25	66.69	64.74
Stability duration /d	19	17	18	19	19	17	19

4. Conclusion

After performing the experiments described in this paper, the following conclusions could be drawn:

- The monomer ratio (AM: AMPS: MVP) had a significant effect on the properties of the microspheres obtained by inverse suspension polymerization. With the AM content fixed, the initial diameter of the microspheres decreased and the swelling performance was enhanced by increasing the amount of AMPS and decreasing the amount of NVP. After 24 h swelling, saturation was observed and the diameter of the microspheres remained constant thereafter. As the amount of AMPS increased and the amount of NVP decreased, the thermal resistance durations of the microspheres obtained in aqueous solution first increased and then decreased. With a monomer ratio of 7:1.5:1.5, the microspheres exhibited optimized thermal resistance (19 d). The TG curves suggested that the microspheres with a monomer ratio of 7:1.5:1.5 had an optimized thermal resistance, while those of the microspheres with a monomer ratio of 7:0:3 showed the lowest thermal resistance.
- The stirring speed significantly affected the initial diameter and the thermal resistance of the microspheres prepared by inverse suspension polymerization, as well as the thermal resistance of microspheres. With the formula fixed and stirring speed increased from 170 to 500 rpm, the initial average diameter of the microspheres decreased from 230.90 μm to 11.14 μm, while the thermal resistance durations increased from 18 d to 23 d.
- The effects of the crosslinking agent amount on the initial diameter, swelling performance, and thermal resistance of the microspheres were negligible. As the amount of crosslinking agents increased from 0.05%to 0.33%, the volumetric expansion ratio of the microspheres first decreased and then increased. As the amount of crosslinking agents increased from 0.33% to 1.5%, the swelling performance of the microspheres degraded initially and then improved. As the amount reached 2%, the swelling of the microspheres obviously increased until the maximum amount of crosslinking agents was added.
- The amount of emulsifier had a notable effect on the initial diameter of the microspheres, but not on the thermal resistance. As the amount of emulsifier increased, the initial diameter of the microspheres decreased from 82.25 μm to 64.74 μm, although no obvious trends were observed.

Acknowledgment

This research is supported by National Natural Science Foundation of China No. 51274211 and Major National Science and Technology Projects (2011ZX05009).

References

[1] Frampton H, Morgan J C, Cheung S K. Development of a Novel Waterflood Conformance Control System, *Spe/doe Symposium on Improved Oil Recovery*, 2004, Tulsa. *SPE* 89391:233-235.
[2] James P, Frampton H, Brinkmam J. Field Application of a New in-Depth Waterflood Conformance Improvement Tod[R]. SPE 84897, 2003: 27-29.
[3] Choi S K, Ermel Y M, Bryant S L. Transport of a pH sensitive Polymer in Porous Me- dia for Novel Mobility-control Application[J]. SPE 99656, 2006: 324-326.
[4] Sohn O, Sim S J, Lee D H. Preparation and Swelling Properties of Poly(potassiumacryl-ate-co-acrylamide).*Superabsorbent Particles*. Polymer-Korea, 2004, (1):18-23.
[5] Mudiyanselage T K and Neckers D C 2008 Highly absorbing superabsorbent polymer *J PolymSci Part A: PolymChem* **46** 1357-64

[6] Smith J E 1986 Quantitative evaluation of polyacrylamide crosslinked gels for use in enhanced oil recovery *International ACS Symposium* (Anaheim, California) pp 9-12
[7] Smith J E, Mack J C. Gels Correct In-Depth Reservoir Permeability Variation. *Oil and Gas Journal*, 1997, 94(34): 33-39.
[8] Smith J E. Performance of 18 Polymers in Aluminium Citriate Colloidal Dispersion Gels. *SPE*28989, 1995: 461-470.
[9] Fernández I J, PDVSA-Intevep. Evaluation of Cationic Water-Soluable Polymers with Improved Thermal Stability. *SPE International Symposium on Oilfield Chemistry,* 2005, Texas, SPE93003: 1-13.
[10] Mack J C, Smith J E. In-Depth Colloidal Dispersion Gels Improve Oil Recovery Efficiency. *SPE* 27780, 1994: 527-539.
[11] Zhao Q, Lin M Q, Dang S M and Hua Z 2014 Acrylamide/2-Acrylamido-methyl-propanesusonic-acid/1-Vinyl-2-pyrrolidinone terpolymeric microspheres by orthogonal experiments *Journal of Asia Chemistry* **26** 5615-8
[12] Fielding Jr R C, Gibbons D H, Legrand F P. In-Depth Drive Fluid Diversion Using an Evolution of Colloidal Dispersion Gels and New Bulk Gels-An Operational Case History of North Rainbow Ranch Unit. *SPE* 27773, 1994: 433-444.
[13] Dong Z X and Li M Y 2003 Deformation of linked polymer coils *J Chem.* **11** 686-90
[14] Candau F and Leong Y S 1982 Inverse microemulsion polmerization *Phys. Chem.* **86** 193-5
[15] Jaroslav B, Sakurako K and Keiji F 2000 Preparation of partly hydrophobized, crosslinked polyacrymide particles by perpolymerization of pcrylamide/N,N-methylenebisacrylamide/ styrene in inverse microemulsion *Polym Int.* **49** 358-66
[16] Puig L, Sanchez D J C and Villacampa M 2001 Microstructured polyacrylamide pydrogels prepared via pnverse picroemulsion polymerization *Colloid Interface Sci.* **235** 278-82
[17] Chauveteau G, Tabary R, Renard M. Disproportionate Permeability Reduction by Soft Preformed Microgels. *SPE* 89390, 2004: 1-8.
[18] Frampton H, Morgan J C, Cheung S K. Development of A Novel Waterflood Conformance Control System. *SPE* 89391, 2004: 1-9.
[19] Lin M Q, Zhao Q, Dang S M and Wang Y F 2015 Preparation and properties ofterpolymericmicrospheres for deep profile control in oilfields *Materials Reseach Innovations* **19** S2-1

Estimating surface hardening profile of blank for obtaining high drawing ratio in deep drawing process using FE analysis

C. J. Tan[1,2,4], A. Aslian[1,2], B. Honarvar[3], J. Puborlaksono[1,2], Y. H. Yau[1] and W. T. Chong[1]

[1]Department of Mechanical Engineering, Faculty of Engineering, University of Malaya, 50603 Lembah Pantai, Kuala Lumpur, Malaysia
[2]Centre of Advanced Manufacturing and Material Processing (AMMP Centre), University of Malaya, 50603 Lembah Pantai, Kuala Lumpur, Malaysia
[3]Department of Electrical Engineering, Faculty of Engineering, University of Malaya, 50603 Lembah Pantai, Kuala Lumpur, Malaysia

E-mail: tancj1@yahoo.com/ tancj@um.edu.my

Abstract. We constructed an FE axisymmetric model to simulate the effect of partially hardened blanks on increasing the limiting drawing ratio (LDR) of cylindrical cups. We partitioned an arc-shaped hard layer into the cross section of a DP590 blank. We assumed the mechanical property of the layer is equivalent to either DP980 or DP780. We verified the accuracy of the model by comparing the calculated LDR for DP590 with the one reported in the literature. The LDR for the partially hardened blank increased from 2.11 to 2.50 with a 1 mm depth of DP980 ring-shaped hard layer on the top surface of the blank. The position of the layer changed with drawing ratios. We proposed equations for estimating the inner and outer diameters of the layer, and tested its accuracy in the simulation. Although the outer diameters fitted in well with the estimated line, the inner diameters are slightly less than the estimated ones.

1. Introduction

Industries have been forming high strength steel and aluminium alloy sheets into lightweight auto parts. However, the formability of the sheets is low at room temperature. Vollertsen and Lange developed the Tailored Heat Treated Blanks technology to improve the formability of the sheets.[1] They applied laser-induced heating to locally soften the blank holding portion of the high strength aluminium alloy blank. They then drew the partially softened blank into a cylindrical cup, and the LDR of the cup increased from 2.0 to 2.6. The soft portion of the cup aged back to its original hardness after seven days. However, this method is only applicable for precipitation hardenable alloys, and the laser heat treated area is large, resulting in low processing speed. Laser transformation hardening has improved wear characteristics, hardness, life performance of tools and dies in the industries with minimum part distortion.[2] It locally heats the surface of a thick specimen, transforming the surface layer to an austenite structure, leaving bulk material unchanged. Due to self-quenching, rapid cooling of the austenite forms the hard and brittle martensite. Kovalenko and Golovko obtained uniform hardened depth by high power diode laser due to its top-hat beam profile.[3] Since heat dissipation from bulk materials hardens the specimen surface, surface hardening of thin specimens is difficult. So and Ki reported that they couldn't obtain surface hardening for AISI 1020 sheets having 1.3 mm thickness and less.[4] However, Ki et al. developed a heat-sink assisted laser transformation hardening method.[5] They increased the hardened width and depth of a 2 mm thick DP590 dual-phase steel sheet to 5 mm and 1.6 mm, respectively. The hardness of the target section increased from about 190 HV to 350 HV. For large

[4] Address for correspondence: C. J. Tan, Department of Mechanical Engineering, Faculty of Engineering, University of Malaya, 50603 Lembah Pantai, Kuala Lumpur, Malaysia. E-mail: tancj1@yahoo.com/ tancj@um.edu.my

stamped parts, the surface hardening speed can be increased by increasing the track width of the laser beam. Seifert et al. reported a track width of more than 50 mm at once is possible with the multi-kilowatt range high power diode laser.[6] Since hardened depths decrease with high laser travel speed and low laser power, it is important to study the minimum dimensions of the hard layer at different hardness levels for high production rate and low equipment cost.[7] Nguyen and Merklein evaluated various heat-treatment layouts in deep drawing process of fast-hardenable aluminium alloy sheets into cross-shape drawn cups using finite element analysis.[8] However, they didn't report the detailed dimensions of the hardened layer and its effects on increasing the LDR.

In this study, we calculated the deep drawing process of cylindrical cups from partially hardened blanks using ABAQUS Standard Ver. 6.10. In the axisymmetric model, we partitioned an arc-shaped hard section into a flat blank following the shape reported by Ki et al.[5] We obtained the values for the inner diameter, outer diameter and the minumum depth of the hard layer at 2 different hardness levels for successful drawings of cups at different drawing ratios. When the drawing ratios exceeded the limit, we observed localised necking around the bottom corner of the cup in the simulation. We determined the LDR of the cup based on the successful drawing right before necking.

2. Increase in LDR with partially hardened blank

Figure 1 shows the comparison between the cylindrical cup formed from the partially hardened blank and the one formed from the normal blank for the same drawing ratio. We observed high stress concentrations in the hard layer with increasing punch stroke. Due to the high tensile strength and the uniform distribution of stresses in the hard layer, we observed no necking in the cup. However, necking occured around the bottom corner of the cup formed from the normal blank for the same drawing ratio due to stress concentrations.

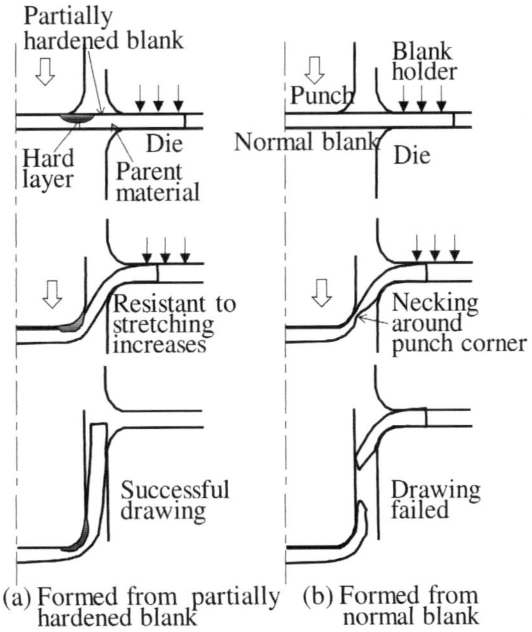

Figure 1. Comparison between cylindrical cup formed from partially hardened blank and one formed from normal blank for same drawing ratio.

3. Conditions for FE simulation

Figure 2 shows the simulation conditions and the FE model for the deep drawing process of a cylindrical cup. We modeled the blank as solid deformable and all the tools as analytical rigid. We treated the model as axisymmetry in the simulation. We fixed the initial blank thickness and the diameter of the blank at 2 mm and 80 mm, respectively. R_p, R_d and C remained constant. We determined the drawing ratio (DR) from the ratio of the diameter of the blank to the diameter of the punch. We reduced the diameters of the punch and the die by 1 mm for each increment in DR. The clearance between the punch and die remained constant at 3 mm in each increment. We started the test at DR=2.11 using D_p=38 mm as the LDR for the DP590 blank is equal to 2.11. We obtained successfully drawn cups for each increment in DR using different hard layers and hardened depths until we observed localized necking. We applied a constant blank holding force of 11 kN for all DRs. The cross section of the hard layer consists of a horizontal line at the top and an arc at the bottom. We constructed the arc from 3 points i.e. D_1, D_2 and t_h. The position of t_h is in between D_1 and D_2.

Target inner diameter of hard layer, $D_1 = D_p - 2(R_p) - t_0$

Target outer diameter of hard layer, $D_2 = D_p + C + \frac{t_0}{2}$

Width of hard layer, $w_h(\text{mm}) = \frac{D_2 - D_1}{2}$

Thickness of hard layer, $t_h(\%)$ = percentage of t_0

Where,

t_0= initial blank thickness (mm)

D_p = diameter of punch (mm)

D_d = diameter of die (mm)

R_p = punch corner radius (mm)

R_d = die corner radius (mm)

C = Clearance between punch and die

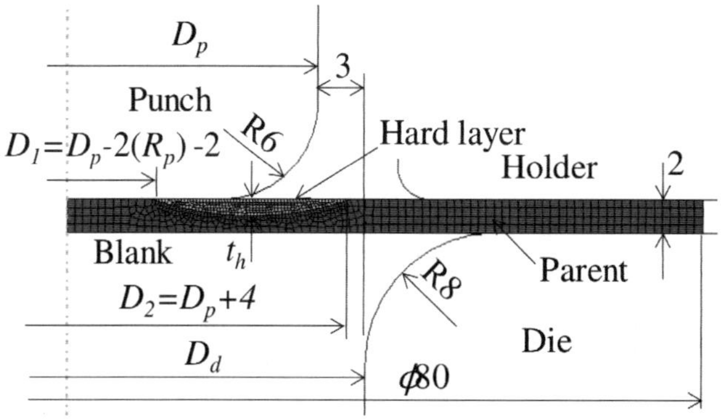

Figure 2. Simulation conditions and FE model for deep drawing process of cylindrical cup.

We modeled all contact surfaces following Coulomb's friction law, and assumed a penalty value of 0.1 for the coefficient of friction. Penalty friction limits motion between the surfaces to an elastic slip. We divided the blank into 4-node quadrilateral elements with reduced integration and hourly glass control. We defined an average global seed size of 0.5 mm for the parent material and 0.2 mm for the hard layer. We assumed the blank including the hard layer to be homogeneous and isotropic in the model. During the simulation, we first applied the blank holding force at the flange portion. A punch moving downward then pushed the blank completely into the die cavity.

We assumed the parent material as DP590 and the hard layer as DP780 or DP980 in the simulation. We adopted the flow stress curves reported by Sung et al. to define the plastic deformation behaviors for DP590, DP780 and DP980.[9] We estimated the hardness for these materials from the Ultimate Tensile Strength (UTS) using the following correlation reported by Martin and Gejza.[10]

$$\text{UTS} = 2.77 \times \text{Hardness (HV)} + 92.754 \quad (1)$$

The estimated Vickers hardness for DP590, DP780 and DP980 are equal to 186 HV, 255 HV and 328 HV, respectively. The hardness values set for the hard layers i.e. DP780 and DP980 are less than the maximum hardness obtained by Ki et al. i.e. 350 HV in the laser transformation hardening process of DP590 sheets having the same blank thickness.[5)] Hence, hardness values set for the hard layers in this study are within the process limit, and it is feasible in reality.

4. Results obtained from simulation

4.1. Relationships between drawing ratios and dimensions of hard layers

We obtained the dimensions for the two different types of hard layers from successfully drawn cups using different DRs. Figure 3 shows the relationship between the thickness of the hard layer and the drawing ratio. The LDRs for DP980 and DP780 are equal to 2.50 and 2.42, respectively by taking into consideration of the process limit as reported by Ki et al. i.e. the maximum hardened depth is equal to 1.6 mm or 80% of the initial blank thickness.[5] t_h increased with DR for both types of hard layers to enhance resistance to failure around the punch corner. The amount of increase in hardened depth for DP780 is more than the one for DP980 due to its lower tensile strength.

Figure 4 shows the comparison between the calculated and the target inner and outer diameters of the hard layer for successful drawing at each DR. D_2 for both layers fitted well with the upper target line. However, D_1 for both layers are less than the lower target line when the drawing ratio is greater than 2.3. The stress concentration moved from the punch corner toward the punch bottom for high DR. Reduction in D_1 increased the resistance to failure of the blank.

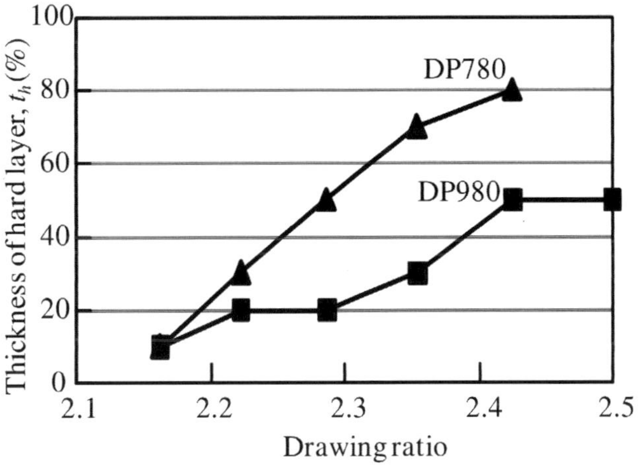

Figure 3. Minimum thicknesses of hard layer for successful drawing at each drawing ratio.

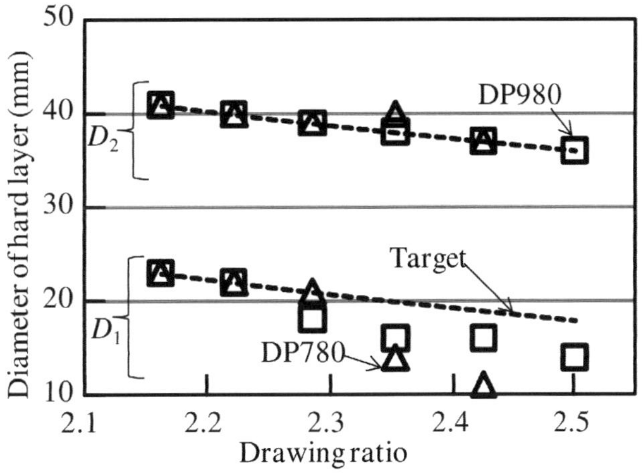

Figure 4. Comparison between calculated and target inner and outer diameters of hard layer for successful drawing at each drawing ratio.

4.2. Increase in LDR using hard layer equivalent to DP980 or DP780

Figure 5 shows the successfully and failed drawn cups obtained from the simulation for different partially hardened blanks indicating the increase in LDR. Without the hard layer, the LDR for DP590 blank is 2.11, in agreement with the one reported by Wang et al. i.e. 2.19 for the same steel grade.[11] We observed a maximum stress value of 623 MPa around the upper portion of the cup due to the circumferential compression in the flange portion during the drawing process. The LDR for DP590 blank having a hard layer of DP980 is 2.50. We observed a maximum stress value of 849 MPa in the hard layer for increase in DR due to stretching caused by the punch during the drawing process. Since the highest stress value is lower than the ultimate tensile strength of the hard layer, we prevented necking in the hard layer, and successfully obtained a fully drawn cup. Further increase in hardened depth for DR=2.58 shifted the stress concentration point from the punch corner to the shoulder of the cup leading to necking in this zone. The LDR for DP590 blank having a hard layer DP780 is equal to 2.42. We observed a maximum stress value of 816 MPa in the hard layer. The stresses for most of the elements in the hard layer are less than the ultimate tensile strength. We prevented necking owing to the high resistance to failure of the hard layer. We set the limit for the hardened depth at 80%, taking into the consideration of the process limit. Further increase in DR shifted the stress concentration point toward the upper punch corner. Due to the arc-shape of the hard layer, the hardened depth in this point is insufficient, resulting in localized necking.

Figure 6 shows the relationship between the average change in wall thickness around the cup corner and the percentage of hardened depth for different drawing ratios. For successful drawing, the amount of increase in hardened depth for DP780 is larger than the one for DP980 for increasing DR. The resistance to the failure of the cup increases with the tensile strength of the hard layer resulting in high LDR. We measured the average change in wall thickness from 5 points around the corner of the cups. The amount of thinning around the corner for DP780 is higher than the ones for DP980 except for drawing ratio 2.35 due to the 2.5 times increase in hardened depth. The partially hardened blank has potential application in multi-stage stamping process of wheel disks as the inner corner of the wheel disk tends to fail in the fatigue tests. [12-13]

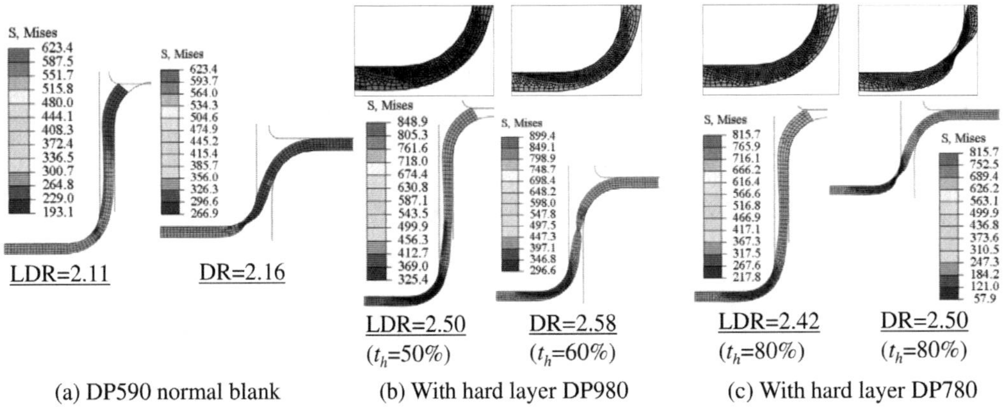

Figure 5. Successfully and failed drawn cups obtained from simulation for different partially hardened blanks indicating increase in LDR.

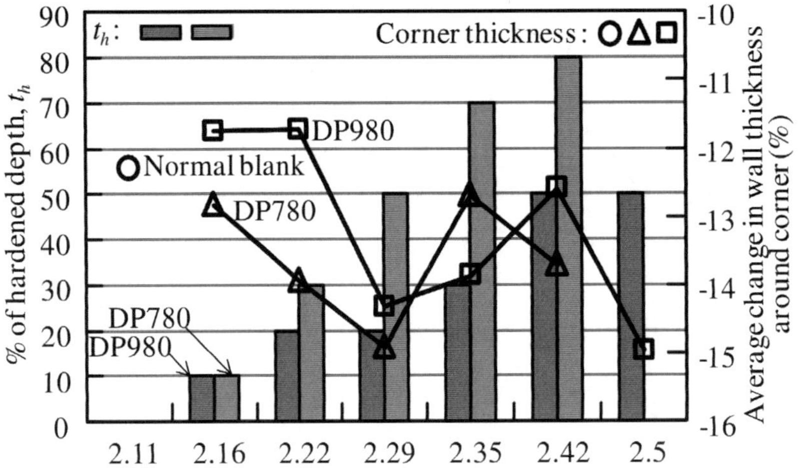

Figure 6. Relationship between average change in wall thickness around cup corner and % of hardened depth for different drawing ratios.

5. Conclusions

We estimated the hardening profiles of the partially hardened blank for increasing LDR of cylindrical cups using FE analysis. In the simulation, we successfully increased the LDR for DP590 blank from 2.11 to 2.50 by increasing the target surface hardness from 186 HV to 328 HV using w_h= 11 mm and t_h= 50 %. However, the LDR reduced to 2.42 when the increase in hardness becomes small i.e. to 255 HV using w_h= 13 mm and

t_h= 80 %. When the hardness level of the target surface decreased, we increased the hardened width, w_h and the hardened depth, t_h to obtain similar LDR. The proposed equations for estimating the inner and outer diameters of the hardening width for successful drawing fitted well with the simulation results although the inner diameter requires additional width at high DRs.

Acknowledgement
The authors would like to thank University of Malaya for providing the research grant support under the project no. RP021A-13AET.

References

[1] F. Vollertsen and K. Lange: 'Enhancement of drawability by local heat treatment', CIRP Ann. Manuf. Techn., 1998, 47, 181–184.
[2] V.S. Kovalenko, A.D. Verhoturov, L.F. Golovko and I.A. Podehernyaeva: 'Laser surface hardening and electric spark surface hardening of materials', JSLR, 1988, 263-274.
[3] V.S. Kovalenko, L.F. Golovko: 'The role of dimensional factors and absorption efficiency in laser surface hardening', Opt. Laser. Eng., 1990, 12-1, 55-65.
[4] S. So and H. Ki: 'Effect of specimen thickness on heat treatability in laser transformation hardening', Int. J. Heat. Mass. Transfer., 2013, 61, 266–276.
[5] H. Ki, S. So, S. Kim: 'Laser transformation hardening of carbon steel sheets using a heat sink', J. Mater. Process. Technol., 2014, 214, 2693–2705.
[6] M. Seifert, S. Bonß, B. Brenner, E. Beyer: 'High power diode laser beam scanning in multi-kilowatt range', Proc. 23rd Int. Congr. on 'Applications of lasers & electro-optics', San Francisco, California, USA, October 4 - 7, 2004.
[7] D. B. Purushothaman, B. Gengusamynaidu, R. B. Karupuudaiyar: 'Experimental studies on the microstructure and hardness of laser transformation hardening of low alloy steel', T. Can. Soc. Mech. Eng., 2012, 36-3, 241-258.
[8] H. Nguyen and M. Merklein: 'Improved formability of aluminum alloys using laser induced hardening of tailored heat treated blanks', Phys. Procedia, 2012, 39, 318 – 326.
[9] J. H. Sung, J. H. Kim, R.H. Wagoner: 'A plastic constitutive equation incorporating strain, strain-rate, and temperature', Int. J. Plast., 2010, 26, 1746–1771.
[10] G. Martin and R. Gejza: 'Correlation between hardness and tensile properties in ultra-high strength dual phase steels – short communication', Mater. Eng. - Mater. Inz., 2011, 18, 155-159.
[11] W. Wang, C. He, Z. Zhao, X. Wei: 'The limit drawing ratio and formability prediction of advanced high strength dual-phase steels', Mater. Des., 2011, 32, 3320–3327.
[12] C. J. Tan, K. Mori, Y. Abe: 'Forming of tailor blanks having local thickening for control of wall thickness of stamped products', J. Mater. Process. Technol., 2008, 202-1-3, 443-449.
[13] C. J. Tan, Y. Abe, K. Mori, T. Nonaka, O. Ebihara, Increase of wall thickness around corner of multi-stage drawn cup with flange using conical punches, Key Eng. Mater., 2007, 340-341, 761-766.

Experimental Study and Dynamic Modeling of Metal Rubber Isolating Bearing

Ke Zhang[1], Yanguo Zhou[2,4] and Jian Jiang[3]

[1]National Key Laboratory on Ship Vibration & Noise, China Ship Development and Design Center, Wuhan, China
[2]Department of Engineering Mechanics, School of Civil Engineering, Wuhan University, Wuhan, China
[3]Institute of Seismology, CEA, Wuhan, China

E-mail: zhouyanguowhu@126.com/ asdff3456@163.com

Abstract. In this paper, dynamic shear mechanical properties of a new metal rubber isolating bearing is tested and studied. The mixed damping model is provided for theoretical modeling of MR isolating bearing, the shear stiffness and damping characteristics of the MR bearing can be analyzed separately and easily discussed, and the mixed damping model is proved to be an rather effective approach. The test results indicate that loading frequency bears little impact over shear property of metal rubber isolating bearing, the total energy consumption of metal rubber isolating bearing increases with the increase in loading amplitude. With the increase in loading amplitude, the stiffness of the isolating bearing will reduce showing its "soft property"; and the type of damping force gradually changes to be close to dry friction. The features of "soft property" and dry friction energy consumption of metal rubber isolating bearing are very useful in practical engineering application.

1. Introduction

Metal rubber material is a kind of new functional structural material featuring net structure by staggered connection of metal wires at its inside. In macroscopic aspect, it has rubber-like macromolecular structure and elasticity; in microcosmic aspect, it is characteristic of mutual dry friction effect of its internal metal wires. With the excellent mechanical properties of vibration damping, the metal-wire spring rolls inside the metal rubber material and structure will rub, slip, extrude and deform under the action of external load to consume much energy, which leads to playing the role of damping. As a full-metal product, the metal rubber features strong adaptability to the environment, resistance to high or low temperature, resistance to corrosion; thus it is the optimal substitute of traditional rubber material. Due to the excellent vibration damping and isolation of metal rubber material under special conditions such as high or low temperature and corrosive environment, vibration isolators made of metal rubber have extensive application value in industries of machinery, military, aerospace, ocean ships[1-8].

The isolating bearing made of laminated rubber is generally used for current building vibration isolation and requires some improvement due to the endurance issues such as ageing, corrosion, fatigue. However, the metal rubber vibration isolator is applying metal rubber material to replace the traditional rubber, which is a kind of new material isolating bearing. In addition, as a nonlinear hysteresis characteristic material with variable stiffness and damping, the dynamic model of metal rubber material is an important research issue. In this paper, a new vibration isolator is developed and manufactured with stainless steel metal rubber, then its dynamic shear performance is tested and studied, followed by dynamic modeling and theoretical analysis, which can be referenced by the application of metal rubber material in practical engineering design[9-12].

[4] Address for correspondence: Yanguo Zhou, Department of Engineering Mechanics, School of Civil Engineering, Wuhan University, Wuhan, China. E-mail: zhouyanguowhu@126.com/ asdff3456@163.com.

2. Experimental details and results

In this paper, 1Cr18Ni9Ti austenitic stainless steel wire with the wire diameter of 0.2mm is employed to make a spiral roll for workblank preparation, and then a cylindrical metal rubber isolating bearing is made after the process of cold press moulding within a round model, as shown in Fig. 1. The test specimen features external diameter of 30mm and internal diameter of 20mm with moulding density of 0.23.

Figure 1. MR isolating bearing.

Figure 2. Test specimen and machine.

In the test, the shear performance of the manufactured isolating bearing is tested on Hongda HT9711 electronic dynamic material tester, with the loading form of displacement control, as shown in Fig.2. The hysteresis curves of the metal rubber isolating bearing with different loading frequencies and loading amplitudes are tested, and to make sure that the deformation of the test specimen is within the restorable deformation scope, the loading strain amplitudes are 5%, 10%, 15% and 20% respectively, and the loading frequencies are 0.1Hz, 0.5Hz, 1.0Hz and 3.0 Hz respectively.

The shear stress-strain hysteresis curves of the test specimen with the strain amplitude of 10% and 15% at loading frequencies of 0.1Hz, 0.5Hz, 1.0Hz and 3.0Hz are presented in Fig.3 and Fig.4, from which it can be seen that the testing hysteresis curves with the same strain amplitude almost coincide despite of the different loading frequencies, which shows that the shear property of metal rubber isolating bearing vary little with variation of loading frequency. It means that the MR isolating bearing can exhibit some good performances when used in the engineering design within a wide range of loading frequency or seismic loading frequency. Therefore, as a typical test, one of the loading frequency (i.e. 1.0Hz) is selected for the following loading tests.

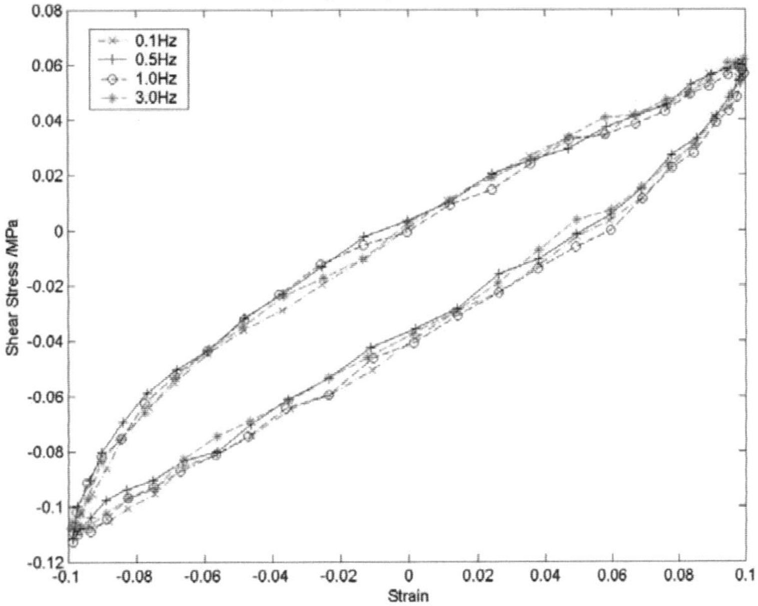

Figure 3. Hysteresis curves of test specimen with strain amplitude of 10%.

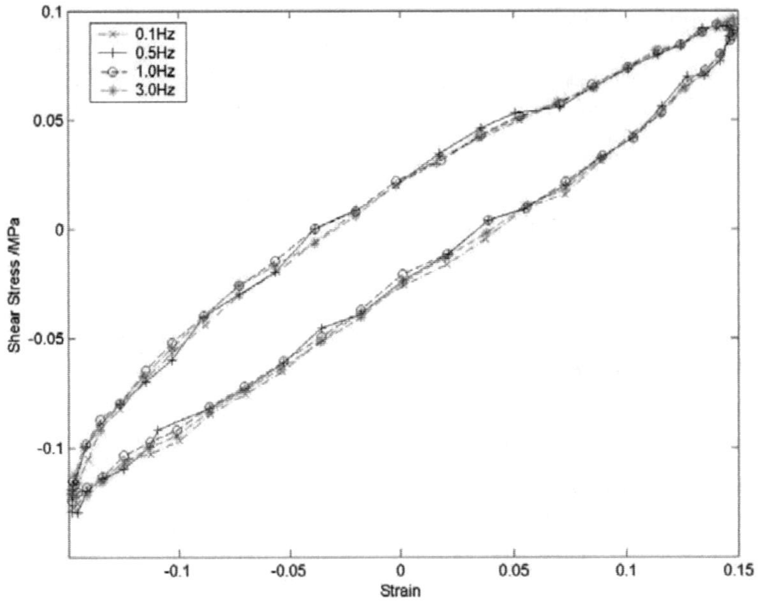

Figure 4. Hysteresis curves of test specimen with strain amplitude of 15%.

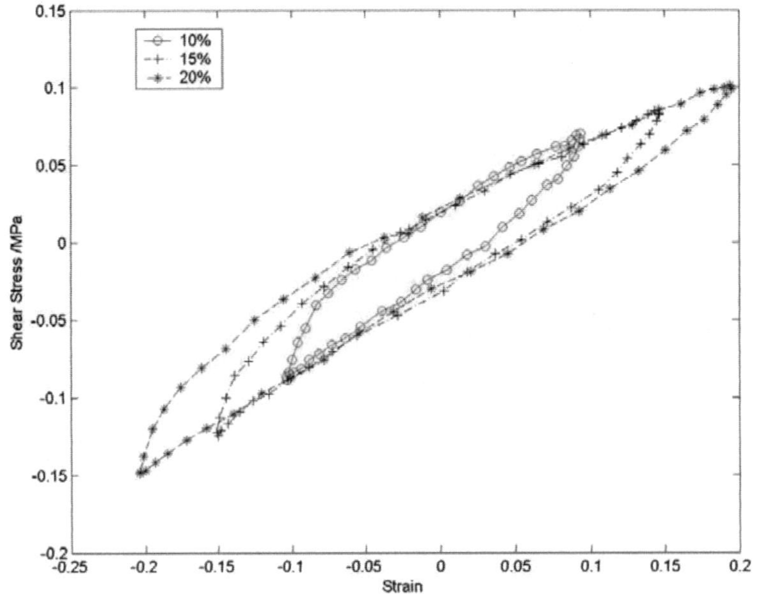

Figure 5. Hysteresis curves of test specimen with strain amplitudes of 10%, 15% and 20%.

For the influence of the strain amplitude on the hysteresis behavior of MR bearing, the shear stress-strain hysteresis curves of the test specimen with loading strain amplitudes of 10%, 15% and 20% respectively are illustrated in Fig.5, from which it can be seen that with the increase of the loading amplitude, the shear stiffness seems decreasing, and the area enveloped by the curve seems growing, which means the total energy consumption of metal rubber isolating bearing increases accordingly.

The internal mechanism is that the relative slippage between metal wires is very small when the amplitude is small, and the mutual friction is not yet fully produced; however, with the increase of the loading amplitude, the obvious slippage appears between metal wires, and the total energy consumption increases accordingly. These properties will be very useful for the isolating bearing in practical engineering.

By the curve in Fig.5, it can be basically understood the general characteristics of MR bearings and its mechanism, however, the above characteristics is not clearly shown from the experimental curve with just basic data processing.

For a better understanding and comparison of the behavior of different MR bearing, the following theoretical modeling approach will be applied, and the overall resilience force will be decomposed into two parts, then the shear stiffness and damping characteristics of the MR bearing were discussed and analyzed seperately.

3. Theoretical modeling and analysis

The metal rubber material is featuring nonlinear hysteresis property of variable stiffness and variable damping, and it is very complicated in the constitutive relation, it is hardly described by any available mathematic model. Fortunately, some advisable theoretical modeling approaches are provided in reference [13,14]. In this study, the mixed damping model will be adopted for its theoretical modeling. Dynamic modeling and theoretical analysis for shear property of the isolating bearing are presented according to the experimental results, and the damping mechanism of the isolating bearing is discussed after the reconstruction of the displacement resilience force curve.

The following mathematic expression of resilience will be used and for identification of specific model parameters:

$$F_f = F_k + F_c = k_1 x + k_3 x^3 + k_5 x^5 \pm c\sqrt{(A^2 - x^2)^\alpha} \tag{1}$$

Where, F_f is the resilience force, F_k is the nonlinear spring force, F_c is the nonlinear damping force. A is the amplitude of vibration, x is the vibration displacement, k_1, k_3, k_5 is the coefficient of spring force, c is called the damping coefficient, α is called the damping composition coefficient, which shows the sensitivity of damping force with the velocity.

The above model has clear physical significance, where the first three items are spring forces and the last item is damping force. There are five undetermined parameters to be identified with the following parameter separation and identification algorithm. The identification process features distinct physical significance and is not difficult to identify all the parameters with high precision, which are detailed as the following four steps: 1) Data Processing: classify the measured experimental data to obtain displacement-resilience hysteresis curve; 2) Decomposition: break the resilience force into two parts of spring force and damping force for separate parameter identification; 3) Curve fitting: apply the least square method to nonlinear spring force for parameter identification; 4) Curve fitting: apply the least square method to nonlinear damping force for parameter identification.

To test the effectiveness of the above mentioned identification algorithm, take the hysteresis curves in Fig.3 as an example. Parameters of the hysteresis loop with frequency of 1.0Hz and strain amplitude of 15% (shown in Fig. 3) are identified and the hysteresis curves are reconstructed, which are shown in Fig.6 and Fig.7. From Fig.6 it can be seen that the decomposed spring force and the decomposed damping force is clearly shown, and from Fig.7 it can be seen that the theoretical curves coincides with experimental curves satisfiedly, indicating the adopted method here has good accuracy and clear physical significance.

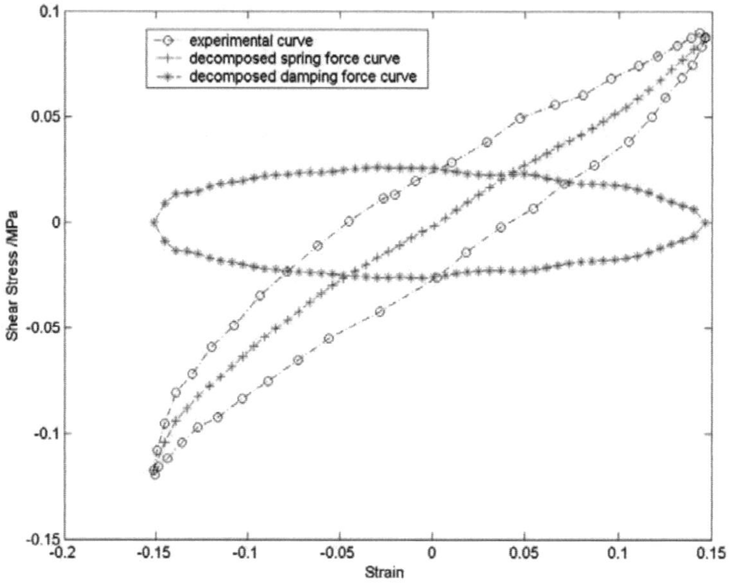

Figure 6. Decomposition of hysteresis resilience forece into spring force and damping force.

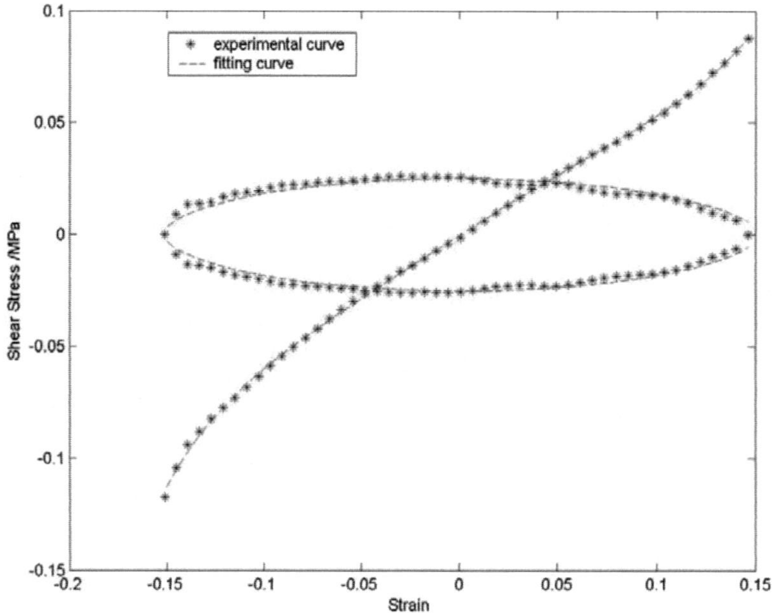

Figure 7. Curve fitting of the spring force and damping force.

According to the above theoretical modeling methods, the parameter identification and the reconstruction of hysteresis loop curves with the frequency of 1.0Hz and the strain amplitude of 10%, 15% and 20% respectively are presented in Fig.8 and Fig.9. It shows that the theoretical results can better simulate the experimental results, indicating that this approach is rather effective in all cases. Then the behavior of MR bearing can be understood and comparatively analyzed clearly.

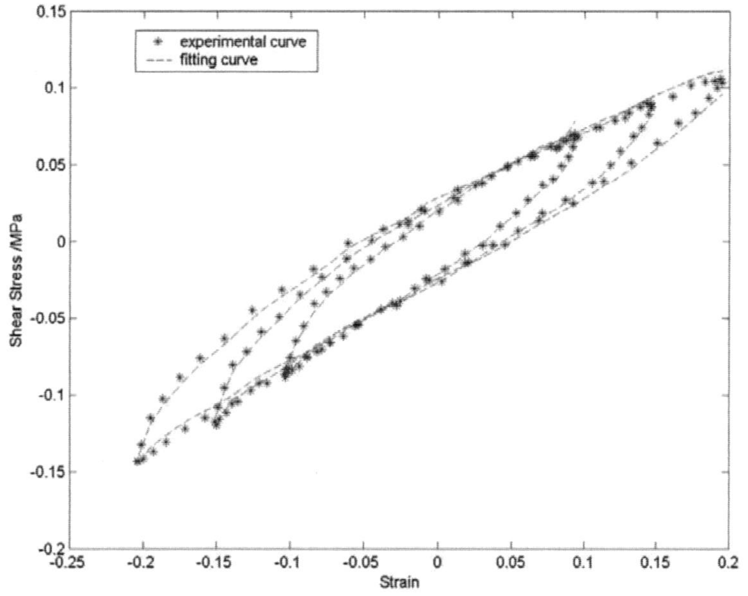

Figure 8. Reconstruction of hysteresis curve with frequency of 1.0hz and strain amplitude of 10%, 15% and 20% respectively.

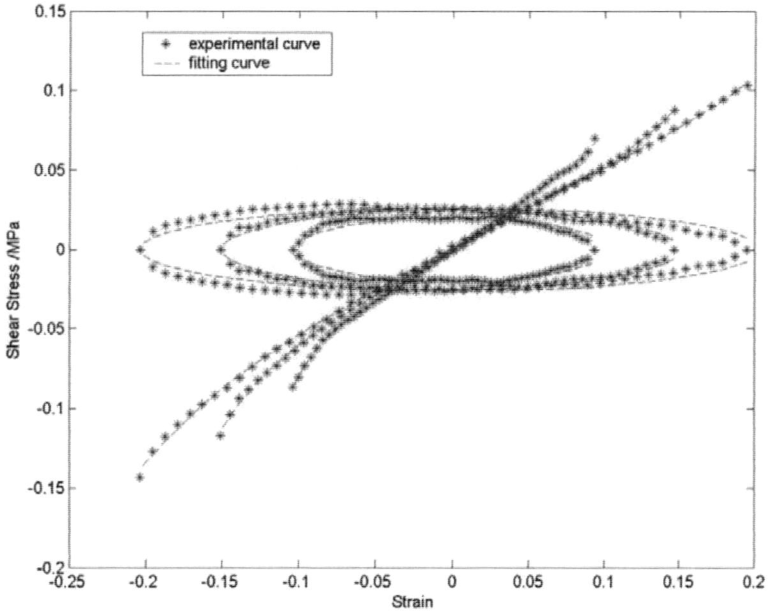

Figure 9. Reconstruction of spring force and damping force curves with frequency of 1.0Hz and strain amplitude of 10%, 15% and 20% respectively.

From the results of parameter identification in Fig.9, it can be easily concluded that the identification method features distinct physical significance, the nonlinear spring force parameters vary regularly, for a specific loading frequency, the shear stiffness decreases with the increase in amplitude, which indicates the shear stiffness of MR isolating bearing reduces with the increasing loading amplitude. This property can be expressed by so-called "soft property".

From the results of identification of nonlinear damping force, it also can be seen easily that the process of damping force identification features distinct physical significance, the enveloped damping area increases much with the increase in loading amplitude, indicating that the damping energy consumption of the system increases greatly with the increase in loading amplitude.

In addition, as can be seen from the decomposition of the hysteresis curve, with the increase in loading amplitude, the damping force curve changes gradually from an ellipse to a rectangle close to that of dry friction damping. The reason can also be explained that the relative slippage and the mutual friction between metal wires will be more obvious with the increase of the loading amplitude, and the type of damping mechanism will be transformed into dry friction damping.

The features of "soft property" and dry friction energy consumption of metal rubber isolating bearing with increase in loading amplitude are very useful in practical engineering application.

4. Conclusions

A new kind of isolating bearing is developed and manufactured with stainless steel metal rubber, of which shear mechanical properties are tested and studied. The test result indicates that loading frequency bears little impact over shear property of metal rubber isolating bearing, i.e. its shear property hardly vary with variation of loading frequency.

Dynamic modeling and theoretical analysis of isolating bearing shear property are presented based on the test results, which indicates that the mixed damping model can satisfiedly describe the shear hysteresis dynamic property of metal rubber isolating bearing, with good accuracy and clear physical significance.

The results of modeling and analysis show that the shear stiffness of MR isolating bearing reduces with the increasing loading amplitude, expressed as "soft property"; with the increase in loading amplitude, the type of damping force changes gradually close to that of dry friction, which is very useful in practical structural engineering designing.

Acknowledgment

This work was financially supported by the National Science Foundation of China (Grant No. 51108357) and the Fundamental Research Funds for the Central Universities (Grant No. 210-275818).

References

[1] Tinker M L, Cuthins M A. Damping phenomenon in a wire rope vibration isolator. *Journal of Sound and Vibration*, 1992, 157(1): 7-18.

[2] Li Yuming, Zheng Jian, Bai Hongbai. Dynamic mechanical model of metal rubber material. *Journal of Materials Research*, 2003, 17(5):499-504.

[3] Kou Hongning, Liu Guoquan,Yang Jianchun. Microstructure and mechanical properties of stainless steel wires used for metal rubber. *Journal of Aeronautical materials*, 2006, 26(4): pp. 32-33.

[4] Jiang Hongyuan, Ao Hongrui, Xia Yuhong. Research on manufacturing technology and application of metal rubber. *Mechanical Technologist*, 2001, 35(2): 35-36.

[5] Ertas B H, Luo H G. Nonlinear dynamic characterization of oil-free wire mesh dampers. *Journal of Engineering for Gas Turbines and Power*, 2008, 130(3): 032503-1-8.

[6] Zarzour M, Vance J. Experimental evaluation of a metal mesh bearing damper. *Journal of engineering for gas turbines and power*, 2000, 122(2): 326-329.

[7] Zhang D Y, Scarpa F, Ma Y H. Compression mechanics of nickel-based superalloy metal rubber. *Materials Science and Engineering*: A, 2013, 580: 305-312.

[8] Lee Y B, Kim T Y, Kim C H. Effects of mesh density on static load performance of metal mesh gas foil bearings. *Journal of Engineering for Gas Turbines and Power*, 2012, 134(1): 012502-1-7.

[9] San A L, Chirathadam T A, Kim T H. Measurement of structural stiffness and damping coefficients in a metal mesh foil bearing. *Journal of Engineering for Gas Turbines and Power*, 2010, 132(3):032503.

[10] Zhao Yagebai, Cong Jianxin. Shear mechanical properties of stainless steel metal rubber under actions of static and dynamic loads. *Journal of Northeast Forestry University*, 2011, 29(8);96-100.

[11] Zhou Yanguo, Zhang Jing, He Yong. Design and Performance Studying of Vibration Isolator Using Metal Rubber, *China Civil Engineering Journal*, 2010,(1): 352-355.

[12] MAO Chenxi, ZHAO Yagebai, LI Suchao, LI Hui, Mechanical behavior of stainless steel metallic pseudo-rubber for three-directional isolation. *World Earthquake Engineering*,2010,26(4): 66-72

[13] Zhou Yanguo, Qu Wenzhong. Several practical approaches for nonlinear dynamic modeling of metal rubber. *Noise and Vibration Control*, 2013, 33(1):31-36.

[14] Li Dongwei, Bai Hongbai, Yang Jianchun. Modeling of nonlinear system with hysteresis characteristics. *Chinese's Journal of Mechanical Engineering*, 2005, 41(10): 205-209.

Tribological behavior of HM1 steel fabricated by precision spray forming under high temperature

Y Q Cheng [1], P Zhang, M D Zhu and Y S Sun

School of Materials and Energy, Guangdong University of Technology, Guangzhou, 510006, China

E-mail: Y.Q.Cheng@163.com

Abstract. In this study, we investigated the tribological behavior of HM1 steel fabricated by precision spay forming (PSF). WE used block ring friction test for our investigation, at various temperature, which was compared with that of the as-cast specimen. The results indicate that the wear rate and the friction coefficient of the PSFed specimen are reduced compared to that of the as-cast specimen. Attribution to these results is the fine grain, the eliminated segregation of elements, and the uniformly distributed matrix material elements for the PSFed specimen. SEM morphology of wear scar shows that the mainly wear mechanism of the as-cast specimen is adhesive wear, while the wear mechanism of the PSFed specimen is mainly abrasive wear.

1. Introduction

The procedure of hot work dies created different types of damage such as wear, plastic deformation, thermal and mechanical cracking, gross cracking to occur during the process. More than 70% of the failure for the hot work die is caused by wear [1]. Therefore, it is important to reduce the wear and decrease the friction in order to retard the wear failure. It is usually better to have the wear resistance of hot work die fabricated by the material with more uniform microstructure and fine grain size [2, 3].

The materials processed by spray forming have many advantages. This includes, uniform microstructure and fine grain size and non-segregation, which has a better wear resistance than that of the same materials processed by other methods [4~8]. Xu et al [4] studied the microstructures and wear resistance of the Cr12MoV steel and the results indicated that the spray formed steel exhibited more homogenous distribution of alloy element and finer grains than that of the conventional cast steel. The wear resistance of the spray formed steel improved near 20% compared to that of the conventional specimen. Rodenburg et al [5] found that the wear rate of spray formed M3:2 high-speed steel was lower than the specimens produced by powder metallurgy, which potentially is attributed to the carbide size distribution.

Within the last few decades, PSF was developed as an effective way to process steel into near-net-shape mould & die and tooling [9~11], which could reduce the cost and time compared with the conventional machining. The service life of a PSFed forging die was improved over 25% than the conventional die [10, 11]. In addition, the dimension tolerances are usually within ±0.05mm in the length of 200 mm with a satisfactory surface roughness, which is used directly without machining [10, 11].Therefore, it is important to study the tribological behavior of PSFed steel in order to promote the

[1] Address for correspondence: CHENG Yong Qi, School of Materials and Energy, Guangdong University of Technology, Guangzhou, 510006, China. E-mail: Y.Q.Cheng@163.com.

application of the hot work dies fabricated by PSF. The influence of PSF on the tribological behavior of HM1 steel is investigated under high temperature using dry sliding in order to explore the method to improve the wear resistance of HM1 steel.

2. Experimental procedure

2.1. Materials

The materials used in this study were fabricated using the self-developed PSF device with the type of PSF-20KG. The scrapped HM1 steel forging die was used as the raw materials, which was induction melted in the medium frequency furnace under a nitrogen atmosphere with slag refining after eliminating the oil, oxide, etc. The chemical composition of the raw materials is listed on Table 1.

Table 1. The chemical composition of HM1 steel (wt. %)

Composite	C	Cr	Mo	W	V	Si	Mn
	0.41	3.05	2.8	1.28	1.02	0.72	0.28

During the procedure of PSF, the superheated temperature of the metallic solution is about 100°C. A ring-seam constraint straight spray nozzle was used with the atomization pressure of 0.6MPa under the industrial nitrogen. The spray deposition height was 140mm, and the ceramic plate was used as the receiver that was preheated to the temperature of about 600°C. The reference specimen was produced by the ceramic mold casting under the same melting conditions. For the PSF specimen, the average density was up to 99.5%, which we used the method of drainage to measure [10].

2.2. Wear and wear scar presentation and measurement

The circular specimens with a dimension of 60mm(Diameter)×10mm(Height) are electro-discharged machined from the precision spray deposited ingot and the casting ingot. Prior to testing, all specimens were polished using diamond pastes, alcohol cleaning, drying, and the ultrasonic cleaning was used for 15 min. The rotating dry sliding wear test in the nitrogen atmosphere was carried out on HT-1000 type friction, while the wear testing carried out at a linear speed of 0.4m/s for 40 minutes under the load of 200N in a nitrogen atmosphere at the high temperature of 420 °C, 520 °C,620 °C, and 720°C, respectively. In addition, a C-Si ceramic ball with a diameter of 4mm was used as the counterpart. The wear rate was measured by multi-functional materials surface topography tester. All tests were repeated at least three times to calculate the average values. Worn surface morphology and energy spectrum analysis were carried on Hitachi S-3400II type scanning electron microscope.

3. Results and discussion

3.1. Friction behavior

Fig.1 shows a group of typical relationship between friction coefficient and the times of both types of specimens. It is observable that the friction behavior is similar with each other under different temperatures for all of the specimens. After the stabilizing of the wear test around 15min into the friction and wear procedure, the friction coefficient of both types of specimens has an obvious fluctuation.

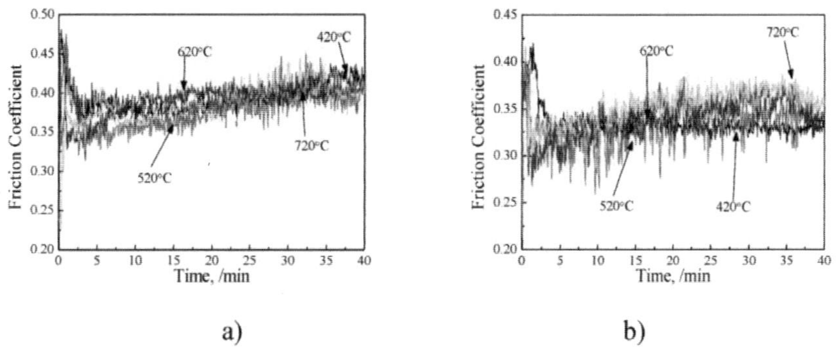

Figure 1. The friction coefficient curves of a) As-cast; b) PSFed HM1 steel under different temperature.

3.2. *Wear behavior*

Fig.2 illustrates the wear rate and the friction coefficient of HM1 steel for the as-cast specimen and PSFed specimen. It is observable that the wear rate of the PSFed specimens are lower than that of the as-cast specimens under various temperatures, which indicates that the wear resistance of HM1 steel could be enhanced by PSF. The friction coefficient of the PSFed specimens is also lower than the as-cast specimens. The results show that the wear of materials failure could be reduced by PSF. In addition, the wear rate for both types of specimen increases slightly when the temperature is lower than 620°C, whereas it rises at the temperature of 720°C. This implies that the wear mechanism was potentially changed at 620 °C. These also indicate that the working temperature for HM1 steel should be not above of 620°C in order to decrease the wear loss.

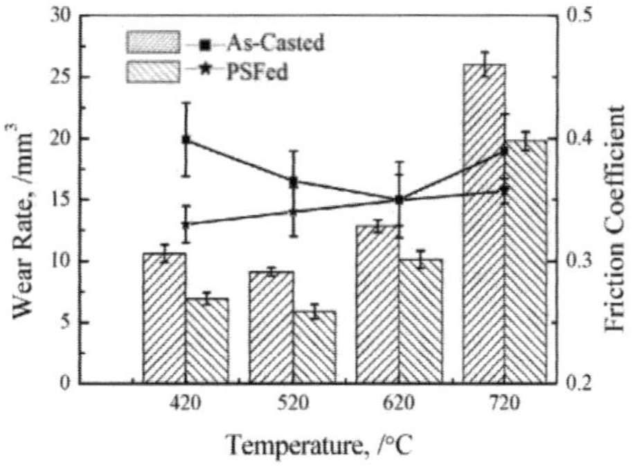

Figure 2. The wear rate and the friction coefficient of HM1 steel under different temperatures.

For the PSFed specimens, the friction coefficient changes slightly, which indicates that the temperature has little effect on the friction coefficient for both types of specimens under the experimental conditions.

3.3. *Wear mechanisms*

Fig.3 shows the typical SEM morphology of the wear scar of both types of specimens at the temperature of 620°C. In addition, the wear scar of the as-cast specimen exhibits a large number of adhesive material morphology (Fig.3 a), which indicates that the adhesion wear occurs under the experimental condition. While for the PSFed specimen, the wear scar surface morphology is observably different from the as-cast specimen and the amount of white hard particles presents on the wear scar surface. This illustrated that the abrasive wear occurred.

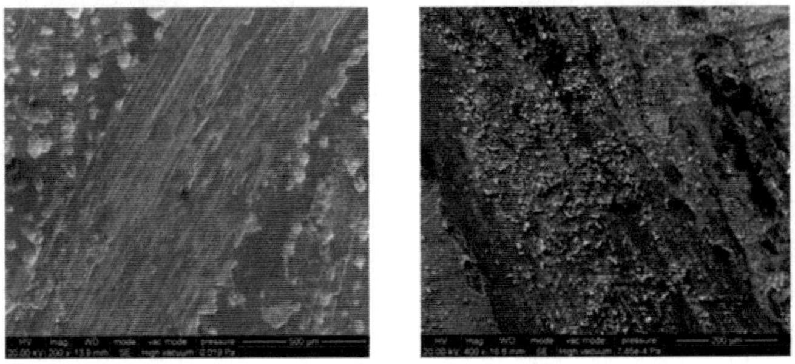

a) b)

Figure 3. SEM morphology of the wear scar of both types of specimens under the temperature of 620°C. a) As-cast; b) PSFed

Fig.4 shows the EDS linear scan spectrums of the worn scar at the testing temperature of 620°C. It is observable that the Fe and O contents for both types of specimens on the wear surface fluctuate greatly and the content of O element increases at the adhesion complexes. While the content of O and Fe elements tends to stabilize in the original region. All of these observations indicate that the oxidation wear occurred during the friction procedure. Furthermore, a lot of heat will generate under the testing condition with a sliding dry friction, especially at an elevated temperature. Therefore, the surface oxidation phenomenon of the specimens occurs although the testing was carried out in a nitrogen atmosphere. In addition, the content of O element on the worn surface for the as-cast specimen is higher than the PSFed specimen. According to the results of EDS quantitative analysis, it indicates that the oxidation wear could be reduced for the specimen produced by PSF.

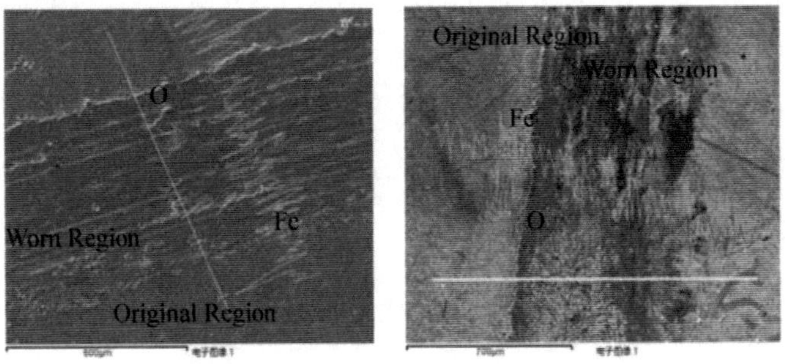

a) b)

Figure 4. EDS linear image of the worn surface for both types of specimens at 620°C. a) As-cast; b) PSFed

3.4. Discussion

We concluded from the experimental results that the wear resistance of PSFed specimens can be reduced compared with that of the as-cast specimens. Fig.5 shows the microstructure of as-cast HM1 steel and PSFed specimen. In Fig.5 a, the grain of as-cast specimen is coarse with the average grain size of approximately 100μm. In addition, the coarse intergranular connected net carbide is observable. The microstructure of PSFed specimen is more uniform with fine grains and fine carbide dispersed. The average grain size was approximately 15μm. . Therefore, compared with the as-cast specimen, the improved wear resistance of the PSFed specimen can be attributed to the uniform microstructure and fine grains.

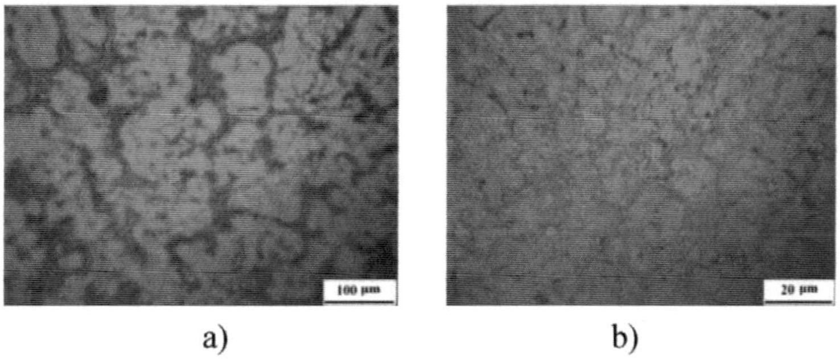

Figure 5. Microstructure of HM1 steel a) As-cast; b) PSFed.

Fig.6 shows the X-ray diffraction patterns of both the as-cast HM1 steel and the PSFed. As observed from the diffraction pattern, the martensite and Cr_7C_3 type carbide and residual austenite are found in both types of specimens. The relative content of the austenite for the as-cast samples and the PSFed samples are solved by X-Ray diffraction quantitative analysis method. The arithmetic mean of the relative volume fraction for the austenite is 14% for the as-cast samples and 8% for the PSFed samples, which solves for six sets of data. It might be attributed to the relatively high amounts of Cr, Mo, W, V and other alloy elements for the PSFed specimen are solid solution during rapid solidification, which should improve the wear resistance of the PSFed specimen.

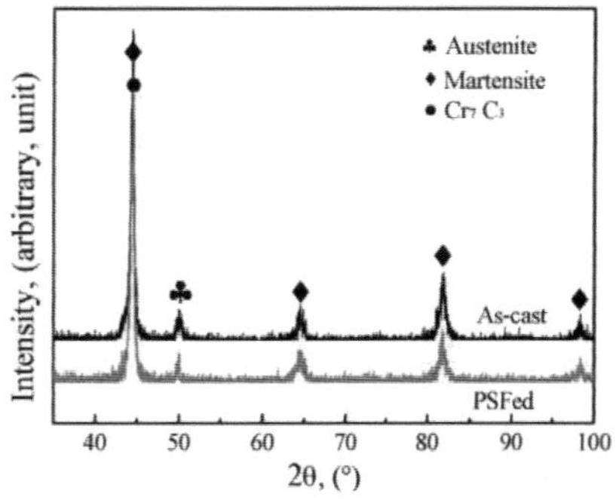

Figure 6. X-ray diffraction pattern of as-cast and PSFed HM1 steel.

Therefore, in order to improve the wear resistance of HM1 steel, the material preparation process should focus on the content of MC type high hardness carbide with the fine grain and non-segregation. This was optimized around the PSF and the subsequent heat treatment process of steel [12].

4. Conclusions

The wear resistance and wear mechanism of HM1 steel fabricated by casting and PSF at high temperature are investigated in order to explore the influential factors to improve their working performance. The following conclusions can be drawn from this study:

- The wear rate of PSFed HM1 steel is lower than the as-cast specimen under the experimental conditions with different high temperatures, which indicates that the wear resistance can be enhanced by PSF.
- Observably, the wear rate of as-cast HM1 steel and PSFed specimen are increased at the temperature of 620°C, which implies that the wear mechanism may change at 620 °C.
- At the temperature of 620°C, the adhesion wear for the as-cast specimen occurs, and this is

also where the abrasive wear for the PSFed specimen is exhibited. The attributions to the changes are to the fine grain, eliminated segregation of elements, and uniformly distributed matrix material elements for the PSFed specimens.

Acknowledgments
The authors of this study acknowledge the financial support from the National Natural Science Foundation of China (51205065), the Public Research and Capacity Building Project of Guangdong Province (2014A010105043), and the Natural Science Foundation of Guangdong Province (S2012010009730).

References

[1] C. Choi, A. Groseclose, T. Altan: 'Estimation of plastic deformation and abrasive wear in warm forging dies', J Mater Process Technol, 2012, 212, 1742-1752.
[2] S.Q. Wang, T. Zhu, Y.S. Mao: 'Wear resistance and wear mechanism of hot-forging die steels', Mater Sci & Technol,2012, 20, 140-144.
[3] K. Kadirgama , K. Abou-El-Hossein, S.M. Noor: 'Tool life and wear mechanism when machining Hastelloy C-22HS', Wear, 2011, 270, 258-268.
[4] H.B. Xu, X.G. Jian, J.S. Wu: 'Microstructures and Wear Resistance of the Spray Formed and Rolled Cr12MoV Steel', Acta Metal Sinica, 2001, 37, 889-892.
[5] C. Rodenburg, W.M. Rainforth: 'A quantitative analysis of the influence of carbides size distributions on wear behaviour of high-speed steel in dry rolling/sliding contact', Acta Mater, 2007, 55,2443-2454.
[6] Q.L. Liu, Y.Z. Liu, W.H. Xiao: 'Friction and Wear Properties of High Alloy Tool Steel HGSF01 Produced by Spray Forming', Mater Mech Eng, 2010, 6, 67-70.
[7] T.T. Matsuo, C.S. Kiminami, W.J. Botta: 'Sliding wear of spray-formed high-chromium white cast iron alloys', Wear, 2005, 259, 445-452.
[8] N. Raghukiran, R. Kumar: 'Processing and dry sliding wear performance of spray deposited hyper-eutectic aluminum–silicon alloys', J Mater Process Technol, 2013, 213, 401-410.
[9] Y.F. Yang, S.P. Hannula: 'Soundness of spray formed disc shape tools of hot-work steels', Mater Sci Eng A, 2004, 383, 39-44.
[10] Y.F. Yang, S.P. Hannula: 'Development of precision spray forming for rapid tooling', Mater Sci Eng A, 2008, 477, 63-68.
[11] Y.Q. Cheng, W. Yu, Y.S. Sun, P. Zhang: 'Microstructure and Properties of HM1 Hot Forging Die Processed by Precision Spray Forming', Special Cast Nonfer Alloy, 2012, 32, 127-130.
[12] P. Sarojrani, D. Karunakar, P.K. Jha: 'Developments in investment casting process—A review', J Mater Process Technol, 2012, 212, 2332-2348.

IOP Conf. Series: Materials Science and Engineering **103** (2015) 012050 doi:10.1088/1757-899X/103/1/012050

Shear strength and structural behaviours of HPRWO with web openings with circular steel tubes

S J Lee[1,3] and J M Park[2]

[1]Department of Architecture, Chosun College of Science & Technology, Gwangju 501-744, South Korea
[2]Department of Architecture Interior, Kyungbuk College, Yeongju-si 750-712, South Korea

E-mail: lsj3948@cst.ac.kr

Abstract. The purpose of this study was to evaluate the shear strength and structural behavior of reinforced-concrete beams with web openings (hereinafter, "HPRWO"), where the web openings are reinforced with circular steel tubes. The experiments were conducted under the monotonic loading condition. Based on the structural experiments involving HPRWO, ultimate load ratio (d_0,d_0/h, etc.), ductility, load-deflection curve, and failure mode comparisons were made for evaluation purposes. This study utilized the design formula for predicting the shear strength proposed by the previous studies and formulas to determine the appropriate shear strength for HPRWO. The results of the experiments confirmed that the rigidity, ductility, and other properties of the HPRWO specimens reinforced with circular steel tubes, fiber, and admixtures were superior to those of the unreinforced HPRWOs. With Mansur's formula, a noticeable tendency for the increase in d_0 and the sectional area of the web openings to lead to the overestimation of $V_u/V_{u,cal}$ was found. The $V_u/V_{u,cal}$ value was found to be more in line with the experiment results based on the AIJ formula compared with the results obtained using other formulas.

1. Introduction

In the field of architectural and civil engineering, there has been a tendency of late for buildings to be built in such a way as to avoid the conventional standardized forms, thus incorporating non-standardized forms, various new styles, and state-of-the-art equipment and systems. The end result is ultra-high-rise buildings. That said, the taller the buildings are, the greater the dead space on each of the floors. Such increase in the dead space is playing a major role in increasing the total construction costs. The possible solutions to the said problems include the use of HPRWO, where perforation is introduced to the beam members along their vertical axes, wherein various facility-ducting services (e.g., HVAC systems, electrical equipment, and hygienic systems) and piping are installed in the depths of the beams [1]. At present, the HPRWO system is roughly divided into (a) HPRWO with steel frames and (b) HPRWO with steel-concrete reinforcement. The strength formula from the previous decades frequently used the conditions subject to the application of both bending based on balance and the shearing force. Since 1990, two suggestions have been made: (a) the design formula for nominal strength, which calculates the maximum shearing force in the unreinforced rectangular openings while plastic collapse is achieved due to the Vierendeel phenomenon [2] in the openings and the maximum

[3] Address for correspondence: S. J. Lee, Department of Architecture, Chosun College of Science & Technology, Gwangju 501-744, South Korea. E-mail: lsj3948@cst.ac.kr.

moment from bending; and (b) the three-dimensional correlative equation for the maximum strength while both shearing force and moment are being applied [3]. In South Korea, the related research on the dynamic behavior of H-beams with web openings was started in 1978. Since then until 2003, a number of papers have been published, though lacking specified codes, relying too much on the structural designer's experience while designing or constructing reinforced stiffeners at the sites, and showing insufficient research on the structures subject to repeated loading [4-5]. As regards the HPRWO with steel-concrete reinforcement, American researchers have published an extensive body of research outcomes, starting with the research in HPRWO focusing on RC bending, shearing, and torsion. In 1984, the study on rectangular HPRWOs subject to a combination of bending and torsion gave impetus to the research in this area, followed by the publication of many studies and proposed formulas until the early 2000s [6-8]. In South Korea, the outcomes of the researches on reinforced HPRWOs started to be published in 1985. Since then, the experiments on and analysis results of the structural performance of web openings have been unsatisfactory [9].

Garnet is simply used for physical filtering to remove minute turbid and floating matters. It has been reported that experiments using garnet contributed to improving the strength and brittleness of polymer mortar [10]. Based on the outcomes of previous studies [11], this study aimed to examine topics such as analysis of crack development in specimens and investigation of failure mechanisms using parameters affecting the structural performance of HPRWOs, including web opening area ratio, width-thickness ratio of steel tubes, and mixing proportions of fiber and admixtures. In addition, this study examined the web openings' load-deflection curve, ultimate load ratio $(V_u / V_{u,hprc})$, and ductility, and evaluated the results of the previous studies and compared them with the results of the current studies in terms of the applicability of the formula for calculating the shear strength proposed by previous researchers, and their design formulas regarding the shear strength of HPRWO bending members.

2. Test sample overview of high-performance RC beams with web openings

Table 1 and Figure 1 show the experimental parameters of the eight specimens that were used for the bending tests under monotonic loading. A total of eight specimens were produced (two HPRWOs without reinforcing circular steel tubes and six HPRWOs with circular steel tubes), along with detailed drawings of such specimens. The principal parameters that were used in the tests were fiber, admixtures, web opening area ratio, web opening configuration, and width-thickness ratio. Testing was conducted to identify the structural performance of the new reinforcement method with circular steel tubing used to reinforce the openings, instead of the conventional reinforcement method using rebar. The proportioning strength of the concrete that was used for the openings was 44.7 MPa for high-strength concrete. The test set-up and LVDT location are listed in Figure 1.

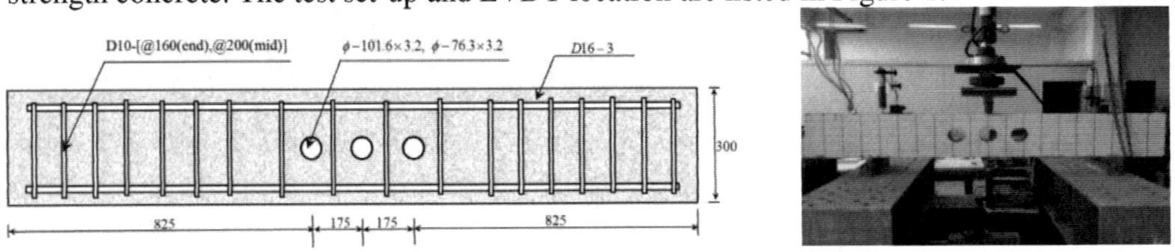

Figure 1. Test specimen and loading system.

Table 1. Details of HPR specimens.

Specimens	b	h	Main -bar	Additive	Fiber	Circle Steel Tube	Web Opening			
							Area	Number	Location	D/t
HPRC3	150 mm	300 mm	D16	None	None	None	0.333	3	M	None
HPRGNC3				Garnet	Ny	76.3×3.2				21.8
HPRGPC3					PP					
HPRGSFC3					SF					
HPRC4				None	None	None	0.444			None

HPRFNC4				Fly-ash	Ny	101.6× 3.2				29.8
HPRFPC4					PP					
HPRFSFC4					SF					

Note: HPRGNC:HPR(High Performance RC Beams with Web Opening), G(Garnet), F(Fly-ash), N{Fiber; N(Nylon), P(Polypropylene), SF(Steel Fiber)}, C(Circle Steel Tube), 3(0.333), 4(0.444)

3. HPRWO test results and crack development

Figure 2 confirms that in general, the specimens whose HRPWO sectional area was 0.333 showed a more extensive crack development than the specimens whose HPRWO sectional area was 0.444. Specimens HPRGNC3, HPRGPC3, and HPRGSFC3, with different fibers incorporated, exhibited the initial crack development pattern (i.e., bending cracks) that intensively progressed from the lower edges of the HPRWO in the center to the areas surrounding the openings as the load increased. Furthermore, since the occurrence of crack development on the loading points due to the concrete crush, the cracks had gradually expanded. With the HPRGSFC3 specimen including steel fiber, shear cracks were observed to be developing at a location 32 cm away from the opening, which was by far the closest; the range of failure was found to be the smallest. Before and after the yield, shear cracks comprised the majority of the cracks observed; towards the end of the testing, however, a combination of various cracking patterns was visible, as confirmed. Specimen HPRGSFC3 exhibited vertical fractures in the lower edges of the HPRWO in the center, but it showed no buckling or rupture of the steel tubes. Specimens HPRGPC3 and HPRGNC3 showed a composite mode around the openings, which was the dominant pattern, while specimen HPRGSFC3 showed a mixture of diagonal and tangential modes. As for the specimens whose opening size was larger than the depth of the openings (i.e., HPRFNC4, HPRFGC4, and HPRFSFC4), the initial cracks were found to be similar to those of the previous specimens; their shear cracks were found within the 30, 20, and 25 cm range on each side of the openings, respectively. Little cracking or failure was observed in the $\ell/4$ area (edge) of the total span, although extensive bending cracks were found around the openings in the center. Additionally, fracture was observed on the surface of the concrete at the loading locations, due to crush. The range of such fracture tended to be focused on the distributions smaller than the steel tube's $D/t = 21.8$. Specimen HPRFNC4 showed shear cracks before and after the yield, as well as the composite mode failure around the openings. Specimens HPRFPC4 and HPRFSFC4 exhibited crack and fracture patterns that include a combination of diagonal and tangential modes.

Based on the above findings, the circular steel tubing provided as a reinforcement on the web openings showed crack control effects for the specimens, and offered expected merits as an alternative material for the conventional reinforcement (i.e., steel reinforcement and steel bars) used for the openings.

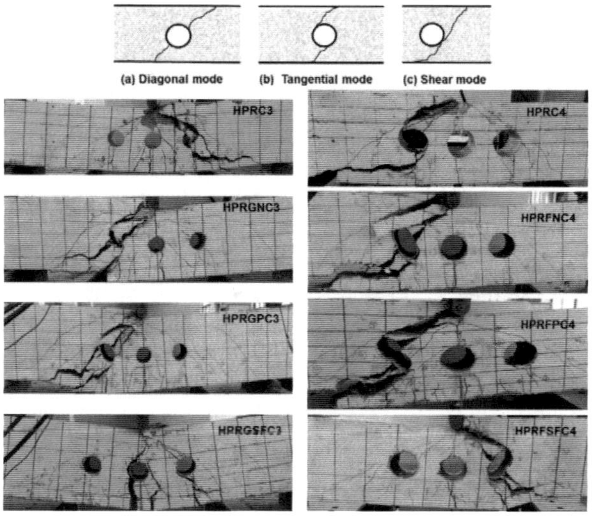

Figure 2. Failure mode types and failure modes of specimens.

4. Load-deflection curve

Table 2 and Figure 3(a) and (b) list the maximum loads for the specimens for testing the web openings, and the load-deflection curve measured from the specimens' center LVDTs. The overall behavior exhibited by the specimens was the typical bilinear one, and the figures and table confirmed that the C4 series showed somewhat lower values than the C3 series in terms of the rigidity of the members. Regarding their behavior after the achievement of the maximum strength, the unreinforced openings exhibited a rapid drop in strength after the achievement of the maximum strength, whereas the reinforced openings had a somewhat gradual decrease in strength. No significant difference was observed from the center part's deflection to the yield load at each stage of loading, but from the yield load to the maximum strength, the deflection in the reinforced openings, compared with that in the unreinforced openings, showed a relatively large increase. For the C3 series, except for HPRGPC3, the increase was between 11.72 and 55.21%; the C4 series showed increases of 13.27 to 25.44%. In terms of the maximum strength, specimen HPRGSFC3, compared with the unreinforced specimens, exhibited a 4.57% increase in strength, but the rest of the C3 series specimens showed somewhat small strength values. The C4 series, compared with specimen HPRC4, had 14.59-23.48% increases in strength (Figure 3(b)).

5. Ultimate load ratio ($V_u/V_{u,hprc}$) and the relationship between d_0 and d_0/h

Table 3 and Figure 4 (a), (b) show the relationship between the web openings' ultimate load (V_u) and ultimate load ratio ($V_u/V_{u,hprc}$). Overall, as the depth of the openings (d_0) and the d_0/h value increased, V_u decreased while $V_u/V_{u,hprc}$ showed an increase. In particular, specimen HPRFSFC4 showed a significant increase. Compared with the unreinforced openings, specimens HPRFSFC3 and HPRFSFC4 showed 4.8 and 30.6% increases in the $V_u/V_{u,hprc}$ value, respectively. The presumed causes of such increase include the following: (a) contrary to the expectation that an increase in the area ratio of the steel tubes used as reinforcing materials would naturally decrease the ultimate load value, the yield (V_y) and ultimate load (V_u) values, after the initial loading, were found to be far greater; and (b) compared with the ultimate load value of the specimens produced with nylon and polypropylene, the specimens produced with steel fiber supposedly underwent a split in strength among their components (i.e., reinforcing steel tubes, steel fiber, and stirrups) from the point where the yield load was achieved until the point where the ultimate load was reached.

6. Ductility

The specimens' ductility performances can be expressed as:

$$\Delta = \Delta_u / \Delta_y \tag{1}$$

In the above formula, Δ_y refers to the yield displacement, and Δ_u to the displacement in the limit strength. Their values were determined when the member strength was approximately 80% less than the maximum strength. The criteria reflect the member's non-elastic deformation yield strength. Figure 4 (c) shows the comparison of the circular steel tubes' width-thickness ratios (D/t) and ductility rates. With the unreinforced specimen HPRC3, the ductility was found to be 6.349, while the specimens with reinforced openings ($D/t = 21.8$) exhibited ductility values of between 4.683 and 11.579. It was found that the ductility of the beams that had web openings did not show a significant difference from the other specimens' if reinforced with steel tubes and fiber. Additionally, the results of this testing were found to be similar to those from previous studies; it was understood that greater effects could be achieved by providing web openings in the moment area rather than in the composite or shear mode area.

Compared with the unreinforced specimens with web openings, specimens HPRGNC3 and HPRGPC3 showed a tendency for their ductility to be lower (by 26.3 and 10.1%, respectively), although most specimens were found to excel in ductility compared with the unreinforced beams with web openings.

Table 2. Test results of HPR.

Specimen	V_i		V_y		V_u		V_u/ $V_{u,HPRC}$	Ductility
	Load (kN)	Deflection (mm)	Load (kN)	Deflection (mm)	Load (kN)	Deflection (mm)		
HPRC3	27.58	1	142.34	6.3	167.25	21.1	100*	6.349
HPRGNC3	24.02	0.7	143.23	6.4	161.91	23.9	96.81	4.683
HPRGPC3	27.58	0.9	141.45	7.0	162.79	19.7	97.3	5.714
HPRGSFC3	1.78	0.2	145.00	5.7	175.25	47.1	104.8	11.579
HPRC4	7.12	0.2	122.7	6.0	130.7	8.5	100**	1.667
HPRFNC4	12.45	0.6	141.0	6.3	153.01	9.8	117.0	2.698
HPRFPC4	4.45	0.4	139.6	5.9	156.57	10.3	119.7	3.729
HPRFSFC4	26.68	0.7	153.0	5.7	170.80	11.4	130.6	3.947

Note: *-HPRC3, **-HPRC4, V_i-Initial load), V_y-Yield load, V_u-Maximum load

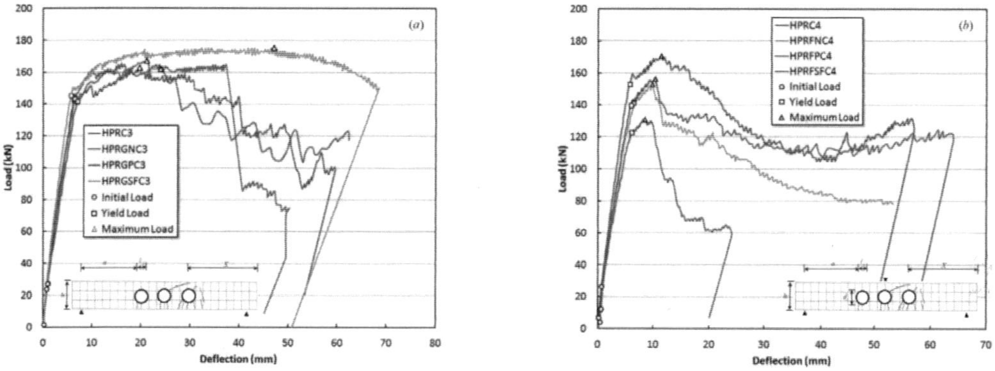

Figure 3. (a), (b) Experimental curves of load versus deflection.

Table 3. Comparisons between tests and shear strength.

Specimen	V_u (kN)	$V_{u,cal}$ (kN)			$V_u/V_{u,cal}$		
		①	②	③	V_u/①	V_u/②	V_u/③
RCB	221.5	152.4/15%	183.4/12%	152.4/15%	1.453	1.207	1.453
HPRC3	167.3	167.9/9.9%	198.4/8.4%	101.7/16%	0.996	0.843	1.645
HPRGNC3	161.9	167.9/9.6%	198.4/8.2%	101.7/16%	0.964	0.816	1.592
HPRGPC3	162.8	167.9/9.6%	198.4/8.2%	101.7/16%	0.969	0.821	1.601
HPRGSFC3	175.3	167.9/10%	198.4/8.8%	101.7/17%	1.044	0.884	1.724
HPRC4	130.7	167.9/7.8%	188.2/6.9%	86.3/15%	0.778	0.694	1.514
HPRFNC4	153.0	167.9/9.1%	188.2/8.1%	86.3/18%	0.911	0.813	1.773
HPRFPC4	156.6	167.9/9.3%	188.2/8.3%	86.3/18%	0.933	0.832	1.815
HPRFSFC4	170.8	167.9/10%	188.2/9.1%	86.3/20%	1.017	0.908	1.979

① ACI; ② AIJ; ③ Mansur

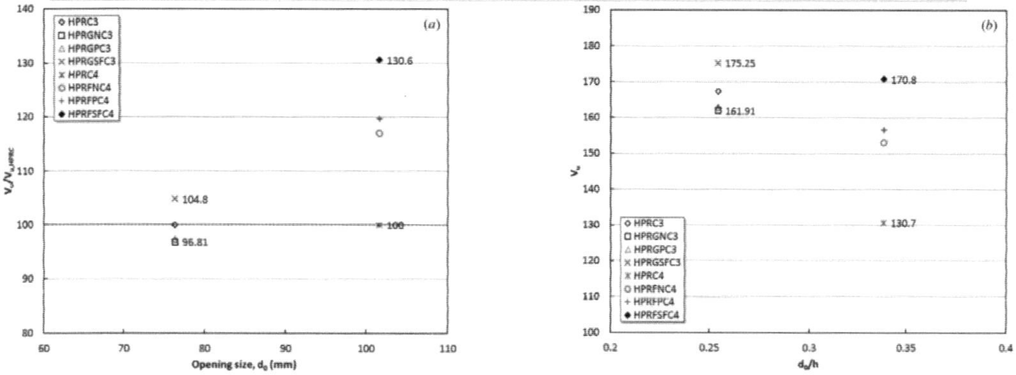

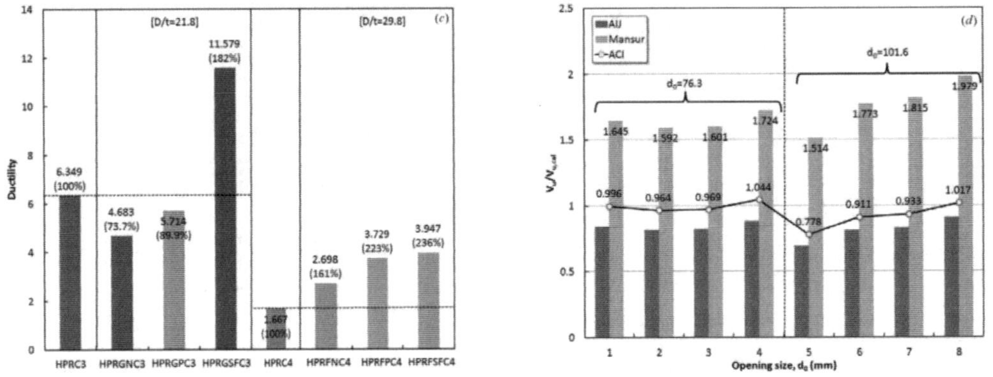

Figure 4. Test results of HPRWOE: (a) relation between d_0 and $V_u/V_{u,hprc}$; (b) relation between d_0/h and V_u; (c) displacement ductility; (d) relation between d_0 and $V_u/V_{u,cal}$.

In particular, the steel-fiber-infused HPRGSFC3 and HPRFSFC4 exhibited a considerably superior ductility of 82 and 136%, respectively, compared with the unreinforced web openings, with particularly superior ductility found in the C4 series, where the sectional area of openings and D/t had increased. Overall, an average of 25.7% increase in ductility was found in the reinforced openings compared with the unreinforced openings. The related observations include the following: (a) the fiber (i.e., nylon, polypropylene, steel fiber) and admixtures (i.e., garnet, fly ash) incorporated into the production of the specimens affected the ductility performance of the specimens; and (b) compared with the previous researchers' use of steel reinforcement or rebar as reinforcing materials to control the crack and increase the rigidity and ductility around the web openings, the use of steel tubing as reinforcing material for the openings was equally effective.

7. Design items and analysis of HPRWO

Regarding the design practices for the typical steel-framed beams with web openings, the formula proposed by Darwin [3], as described in Chapter 1, a three-dimensional formula for the maximum strength under conditions subject to both shear and moment, has been frequently used. That said, the most commonly and frequently used design practice for HPRWOs at present is the additional strength theorem. This method is based on the conventional design method ($V_c + V_s$) of the ACI [12] shear design. The formula of AIJ (Architectural Institute of Japan) [13] is also an additional strength theorem addressing the stress withstood by concrete and steel, respectively. In addition to these theorems, the formula proposed by Mansur [14], which applies various parameters for web openings, was used to compare, analyze, and examine the shear strength compared with the test values.

7.1. ACI 318

Two categories were employed: (a) the concrete-distributed load (V_c) for the concrete resisting against shearing prior to crack development; and (b) the steel-distributed load as the web reinforcement resists against shearing (V_s). The nominal shear strength [12] is provided in formula (2).

$$V_{u,cal} = V_c + V_s, \tag{2}$$

where $V_c = \left(\dfrac{\sqrt{f_{ck}}}{6}\right) b_w d,\ V_s = \left(\dfrac{A_v f_{yt} d}{S}\right) \le 0.67\sqrt{f_{ck}}\, b_w d$.

7.2. AIJ

The next formula is the one that is specified in AIJ's Standard for Structural Calculation of Reinforced Concrete Structures (1988) [13], which was proposed and standardized by Hirosawa. This formula specifies the evaluation of the shear strength ($V_{u,cal}$) of beams, including the fine perforation, and is similar to the conventional method mentioned earlier. With this formula, as shown below, the total shear resistance is achieved by adding the values for the steel reinforcement horizontally crossing the

45° angle failure section that passes the concrete and the center line of the web openings. K_u is the value falling between 0.72 and 1, which addresses the sizing effects within the shearing. It is the function of the effective depth d.

$$V_{u,cal} = \left\{ \frac{0.092K_p(f_{ck}+17.7)}{\frac{M}{Vd}+0.12} \times \left(1-\frac{1.61d_0}{h}\right) + 0.846\sqrt{\overline{\rho_w} f_{yv}} \right\} bd_v \tag{3}$$

where K_p is $K_p = 0.82\left(\frac{100A_s}{bd}\right)^{0.23}$ and d_0 is the diameter of the circular web openings or of the circular circumference. With rectangular web openings, it is either $h/3$ or a value smaller than $h/3$. In addition, h is the total depth of the beams and is expressed as $M/(Vd) \le 3$. In formula (3), $\overline{\rho_w}$ refers to the ratio of the web reinforcement placed within the vertical distance $d_v/2$ from the center of the web openings. $\overline{\rho_w}$ is defined as follows:

$$\overline{\rho_w} = \frac{A_v(\sin\alpha + \cos\alpha)}{bd_v} \tag{4}$$

Here, d_v refers to the distance between the upper and lower reinforcements; A_v to the sectional area of the web reinforcement (vertical stirrups and diagonal reinforcement); α to the angle of inclination for the web reinforcement; and f_{yv} to the yield strength of the web reinforcement. The first part of formula (3) refers to the shear resistance of concrete; it is assumed to decrease in proportion to the depth of the web openings. The second part of the formula refers to the shear resistance by web reinforcement.

7.3. Mansur

The formula proposed by Mansur [14] was based on ACI's shear design method, as follows:

$$V_{u,cal} = V_c + V_s \tag{5}$$

$$V_c = \left(\frac{\sqrt{f_{ck}}}{6}\right) b_w(d-d_0) \tag{6}$$

$$V_s = V_{sv} + V_{sd} = \left(\frac{A_v f_{yv}}{s}\right)(d-d_0) + A_d f_{yd}\sin\alpha \tag{7}$$

Formula (6) refers to the shear resistance in the concrete while V_{sv} in formula (7) refers to the case where shear reinforcement was provided with vertical stirrups, and V_{sd} to the shear reinforcement using diagonal reinforcement. In addition, A_d refers to the total sectional area of the diagonal reinforcement through the fracture section; α to the angle of inclination of the diagonal reinforcement; and f_{yd} to the yield strength of the diagonal reinforcement.

8. Influence of the web opening depth (d_0) vs $V_u/V_{u,cal}$

Overall, as the d_0 value increased, the $V_u/V_{u,cal}$ value tended to increase as well in the design formulas, except for the ACI formula. The $V_u/V_{u,cal}$ value, in particular, in the Mansur-proposed formula, increased to the maximum value (23.5%) in specimen HPRFSFC4 compared with the unreinforced specimens (Figure 4 (d)). There was a noticeable tendency of overestimation with Mansur's formula, whereas the AIJ and ACI specified formulas tended to underestimate or approximate the value of 1. The comparison of the test results for the unreinforced specimens and the values $d_0 = 76.3$ and $d_0 = 101.6$ confirmed that the design strength values increased to 0.37 and 18.41%, respectively. With the ACI formula, as shown in Table 3, $V_u/V_{u,cal}$ showed a lower value

(0.973) on average. This result was based on the fact that in the design formula for concrete (V_c) and steel reinforcement (V_s), the effects on d_0 were not incorporated. Also, as shown in the calculated strength ($V_{u,cal}$), there was little change in the strength value, which could indicate that the multi-parameter conditions of the specimens that were used in this study had not been reflected properly. Thus, similar results were found in the strength comparison data. The AIJ formula exhibited a tendency in which an increase in d_0 led to a moderate increase in $V_u/V_{u,cal}$, as shown in the analysis, with no significant change. This is thought to be attributable to the fact that the effects of the shear resistance design area of the HPRWOs $\{1-(1.61d_0/h)\}$ were properly incorporated into the concrete. As for $d_0 = 101.6$ (i.e., a 24.9% increase from the d_0 value), the $V_u/V_{u,cal}$ value was found to have increased by approximately 1.3%.

9. Conclusions

- The results of the comprehensive examination of the data from the tests conducted on the HPRWOs showed that when providing facility systems in a specimen that combines a steel tubing width-thickness ratio (*D/t*) of 21.8 and a mixture of steel fiber and garnet, benefits, including increased strength and crack-controlling effects, can be expected.
- Compared with the unreinforced specimens, the steel-tube-reinforced HPRWPOs did not underperform in terms of ductility; rather, all the specimens in the category of the C4 series were found to excel in performance. It was confirmed that the smaller the d_0 value was, and with garnet and steel fiber incorporated into the make-up of the specimens, the better the performance became.
- The ACI formula is unable to reflect the parameters of the tests conducted for this study, thus making it difficult to achieve accurate comparative results. In the case of the shear strength evaluation based on Mansur's formula, the increases in the d_0 value led to noticeable decreases in shear strength. As for the AIJ standardized formula, the results were by far the most satisfactory, with an insufficient incorporation of the effects resulting from the increased $\mathrm{D/t}$ of the steel tubing and those resulting from the reinforcement of the web openings with steel tubing. Such insufficiency makes it difficult to apply the AIJ formula to the specimens for this study as it is.

Acknowledgement
This research was supported by the Basic Science Research Program of the National Research Foundation of Korea (NRF-2014R1A1A2059498) and funded by the Ministry of Education, Science and Technology.

References

[1] H. R. Daniel and A. E. Mcmullellen, "Torsion in concrete beams containing and opening", *Journal of the Structural Division*, ASCE, 141 (1962).
[2] K. F. Chung, T. C. Liu, and A. C. H. Ko, "Investigation on vierendeel mechanical in steel beams with circular web openings", *Journal of Constructional Steel Research*, No. 57, pp. 467-490 (2001).
[3] D. Darwin, "Steel and composite beams with web openings", *Steel Design Guide Series 2*, AISC Inc, pp. 1-63 (1990).
[4] J. S. Kim, E. S. Kim, Y. S. Yang and J. S. Park, "An experimental study on the ultimate strength of H-shape steel beams with multiple web openings", *Journal of Korean Society of Steel Construction*, No.5 (3), pp. 147-159 (1993).
[5] E. T. Lee, Y. J. Lee, H. J. Shim and K. H. Chang, "Local buckling behavior of steel members with web opening under cyclic loading", *Journal of Korea Architecture*, No.19 (11), pp. 15-22 (2003).
[6] R. G. Redwood and U. Minoru, "Critical loads for webs with holes", *Journal of the Structural Division*, 105, pp. 2053-2076 (1979).
[7] M. A. Mansur and P. Paramasivam, "Reinforced concrete beams with small opening in bending

and torsion", *ACI Journal*, pp. 180-185 (1984).
[8] S. Kakusho and S. Hayashi, "Experimental and analytical study on reinforcing method of reinforced concrete beams with web opening", *Proceeding of the Eight World Conference on Earthquake Engineering*, No.6, pp. 21-28 (1984).
[9] K. H. Yang, H. C. Eun and H. S. Chung, "The prediction on the shear strength of reinforced concrete deep beams with rectangular web openings", *Journal of Korea Architecture*, No.19 (12), pp. 49-56 (2003).
[10] B. Shirai, "Properties of strength of the polymer cement mortar using garnet sand", *Journal of Structural and Construction Engineering*, pp. 865-866 (1996).
[11] S. J. Lee, "A study on the shear strength and structural behaviors of RC beams with multiple web openings reinforced with square steel tube", *Journal of Korea Architecture*, No.28 (2), pp. 29-36 (2012).
[12] ACI Committee 318-05, "Building code requirements for reinforced concrete and commentary (ACT 318-05/ACI 318 R-05)", American Concrete Institute (2005).
[13] Architectural Institute of Japan, "AIJ standard for structural calculation of reinforced concrete structures", (1988).
[14] M. A. Mansur, "Effect of openings on the behavior and strength of R/C beams in shear", *Cement and Concrete Composites*, No.20, pp. 477-486 (1998).

Effects of specimen width and rolling direction on the mechanical properties of beryllium copper alloy C17200

W C Lee[1] and Z R Liu

Department of Mechanical Engineering, National Taiwan University of Science and Technology, Taipei, 10607, Taiwan, Republic of China

Email: wclee@mail.ntust.edu.tw

Abstract. The objective of the research was to study the effects of specimen width and rolling direction of beryllium copper alloy, C17200, on the mechanical properties of yield strength and Young's modulus. The experimental results showed that the reduction of the specimen width from 12.5 mm to 2.5 mm did not affect the yield strength but reduced the Young's modulus by 4%. Also, the change of rolling direction affected both the yield strength and the Young's modulus. When the tension direction is parallel to the rolling direction, the maximum yield strength was obtained. The results can help predict the behavior of small-scale beryllium copper products more accurately.

1. Introduction

The current trend for electronic products is towards a small scale; therefore, the sizes of the components in these products need to be reduced as well. When the sizes of the metal parts are reduced, the mechanical properties may change. Research has shown that specimen size does affect the mechanical properties of some metals. Kals *et al.* [1] found that for CuNi18Zn20 material, the yield strength decreased with reductions in specimen thickness and specimen width. For example, when the thickness of a specimen was reduced from 1.0 mm to 0.1 mm and the width reduced from 20 mm to 2 mm, the yield strength decreased about 15.8%. The study by Reaulea *et al.* [2] on pure aluminum reported a similar phenomenon. Under the condition that the grain size was less than 126.5 μm, the tensile strength and yield strength were reduced with the reduction of the thickness of the specimen. Dubos *et al.* [3] found that when the ratio of the specimen's thickness to the grain size was less than a critical value, highly pure copper became softer in terms of its mechanical properties. Yang and Lu [4] reported similar results, in that the critical value was not a constant but dependent on the grain size. Michael *et al.* [5] used brass in their study and found a different phenomenon. When the grain size of the specimen was 40 μm and the thickness was 0.1 mm, the tensile tests for five different widths showed that the yield strength decreased with the decrease of the width, but the amount of reduction was small. Therefore, they concluded that the change of width did not affect the flow stress.

Why does specimen size matter in some metals? Miyazaki *et al.* [6] thought that the surface constraint force was less than the interior constraint force when metals were subjected to loads. The interior of the metal takes more of the load, so when the thickness of the specimen was reduced, the constraint force, as well as the flow stress, was reduced. Engel *et al.* [7] explained that the hardness of

[1] Address for correspondence: W C Lee, Department of Mechanical Engineering, National Taiwan University of Science and Technology, Taipei, 10607, Taiwan, Republic of China. E-mail: wclee@mail.ntust.edu.tw.

the grain on the free surface was less than that inside the metal. When the grain sizes were the same, the smaller the diameter of the specimen, the more the surface grains. Because the hardness of the surface grain was less than that of the inside grain, the flow stress decreased.

The rolling direction of the metal is also an important factor when designing electrical spring contacts made of coiled sheet metal. To make the subsequent assembly process easier, the longitudinal direction of the spring contacts is usually perpendicular to the rolling direction of the material. Research has shown that there is no fixed relationship between the mechanical properties of the materials and the angle between the rolling direction and the longitudinal direction in tensile tests. For simplicity, in the following we have used the rolling direction (RD) to represent the angle between the rolling direction and the longitudinal direction in the tensile tests. To illustrate this phenomenon, Leong *et al.* [8] used annealed α–brass in the tensile tests to obtain the yield strength at 0° and 90°. The average yield strength at 90° was greater than that at 0° by 5.3%. Stachowicz *et al.* [9] tested 85–15 brass for the yield strength at 0°, 45°, and 90°. The brass was categorized as M85–1, M85–2, M85–3 and M85–4 based on the annealing time. For M85–1 and M85–2, the yield strength was highest at 0°. However, for M85–3 and M85–4, the yield strength was highest at 90°. Therefore, when the annealing conditions were different, the relationship between the rolling direction and yield strength changed as well.

In the connector industry, beryllium copper is a commonly used material for electrical contacts because of its good electrical and mechanical properties. As the size of beryllium copper contacts continues to be reduced, we need to understand if there are any effects on the mechanical properties so that we can predict the behavior of the contacts better. Therefore, one of the objectives of this research was to investigate the change of the yield strength and Young's modulus of the beryllium copper specimens when the width was reduced; another objective was to investigate those of the beryllium copper specimens at rolling directions of 0°, 45°, and 90°.

2. Experimental methods

The beryllium copper used in this study was C17200–25–1/2 H (TD02). It went through solution heat treatment, then a rolling process and, finally, the annealing condition. The material contained 1.8–2% beryllium. The grain sizes of all the specimens were close and ranged from 6.2 to 9.5 μm; therefore, we did not consider the effect of grain size.

Regarding the design of the experiment, the experimental factors that we selected were width, thickness, and rolling direction. We changed one factor at a time to observe the changes in the yield strength and the Young's modulus. Because materials with different thicknesses may be from different batches, we did not study further the effect of thickness. When conducting the tensile tests, we used four samples for each condition. The statistical method, analysis of variance (ANOVA), was employed to study the effects due to various widths and rolling directions.

The universal testing machine used was INSTRON 3365. Because the specimens were made of thin metal sheets, we followed ASTM E 8M-01 [10] to design the specimen. The gauge length was 50 mm, and the width was 12.5 mm. Then we modified the design by reducing its width to 7.5 mm and 2.5 mm, respectively. There were four different material thicknesses, 0.05 mm, 0.10 mm, 0.15 mm and 0.20 mm, used in the experiments and rolling directions of 0°, 45°, and 90°. The specimens were prepared by using wire electrical discharge machining. The tests were performed by following ASTM E 345-93. [11] The strain rate was set at 0.01 mm/mm/min to obtain the specimen's yield strength.

3. Results and discussion

3.1. *Effect of specimen width and rolling direction on the yield strength*

It was found that the yield strength changed little when the width was reduced from 12.5 mm to 2.5 mm. By using the specimen with a thickness of 0.05 mm and a rolling direction of 0°, we plotted the experimental results as shown in Figure 1 We then used ANOVA to analyse the data further. The ANOVA result is listed in Table 1. The confidence level adopted for the analysis was 95%. From the

IOP Conf. Series: Materials Science and Engineering **103** (2015) 012051 doi:10.1088/1757-899X/103/1/012051

F-ratios and the p-values in Table 1, we found that the p-value associated with the width factor was 0.133, which was greater than the α risk of 0.05; therefore, the width was not a significant factor for yield strength. Figure 2 shows the main effect plots associated with the effects of width, thickness and rolling direction on the yield strength. It can be seen that the yield strength remained about the same for the different widths, ranging between 580 MPa and 582 MPa. This phenomenon was similar to the conclusion made by Michel *et al.* [5] concerning brass.

As shown in Table 1, the p-value associated with the rolling direction factor was minimal, and less than the α risk at 0.05. Therefore, the rolling direction was a significant factor for yield strength. It can be seen from Figure 2 that the average yield strength was 611 MPa when the rolling direction was 0°, which was about 7% more than the average yield strength for rolling directions of 45°or 90°. This result was different from the results for copper alloy and aluminum alloy obtained in previous research. The reason may be related to the composition of the beryllium copper and the annealing conditions.

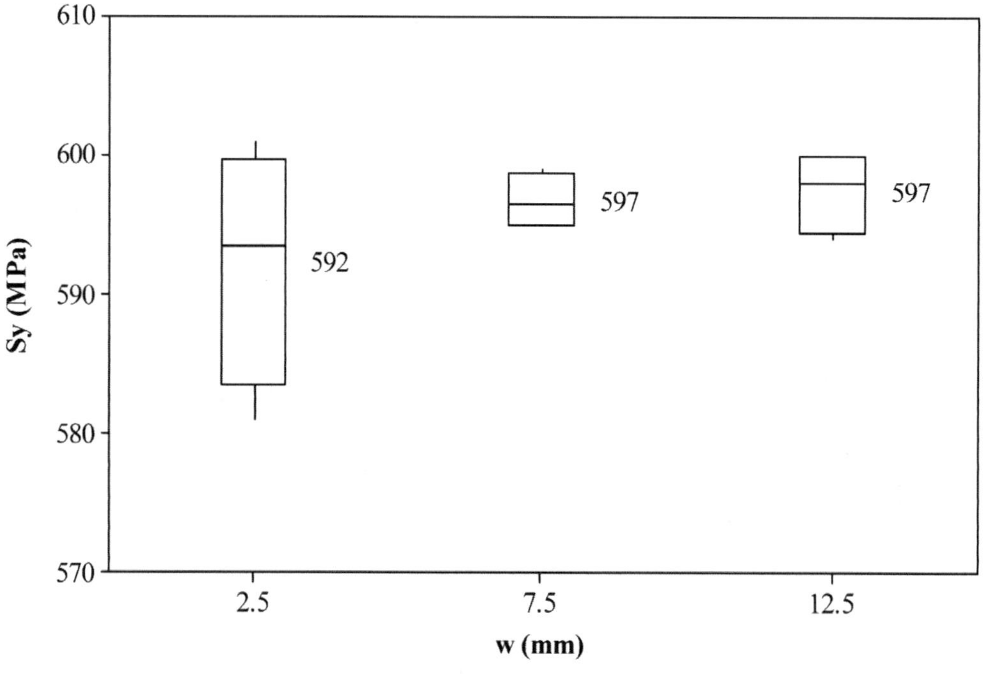

Figure 1. Yield strength associated with various widths for the thickness of 0.05 mm and the rolling direction of 0° (RD: rolling direction, t: thickness, Sy: yield strength, w: width).

Also, we observed that materials with different thicknesses can have obviously different average yield strengths, which were as high as 602 MPa and as low as 566 MPa. This variation was consistent when the results were compared with the inspection report provided by the vendor of the material. So, the variations in the yield strength of materials with different thicknesses could be due to their different metal compositions and different manufacturing conditions, with the variation not only a result of the thickness.

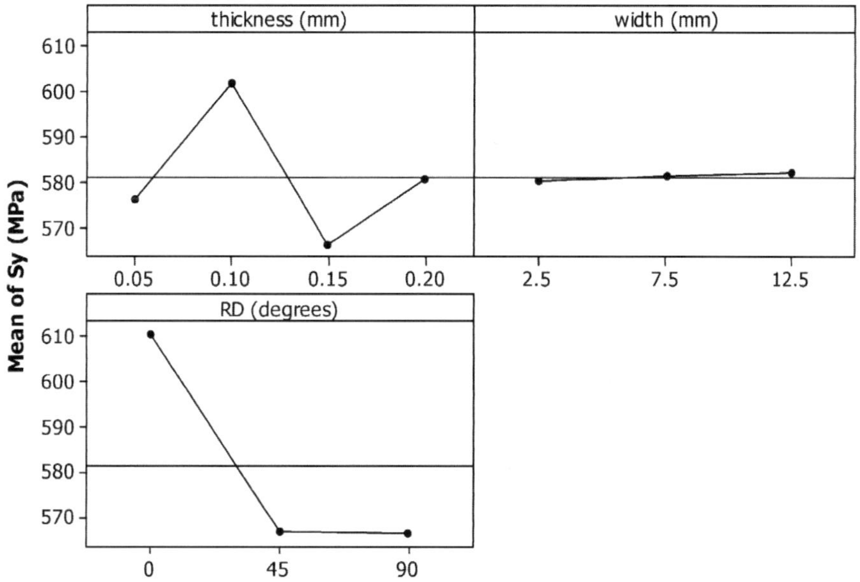

Figure 2. Main effect plots of thickness, width and rolling direction regarding the yield strength.

Table 1. ANOVA results regarding the yield strength.

Source	Sums of Squares	Degrees of Freedom	Mean Squares	F-ratio	p-value
w	76	2	38	2.06	0.13
RD	61457	2	30729	1668.32	0.00
t	24617	3	8206	445.49	0.00
w×RD	77	4	19	1.05	0.39
w×t	387	6	64	3.50	0.00
RD×t	7753	6	1292	70.15	0.00
w×RD×t	182	12	15	0.83	0.63
Error	1989	108	18		

3.2. Effect of specimen width and rolling direction on the Young's modulus

Based on the experimental results, we found that both the width and the thickness affected the Young's modulus. For example, Figure 3 shows the Young's modulus at different widths, when the thickness was 0.05 mm and the rolling direction was 0°. As the width decreased, the Young's modulus also decreased. As the ANOVA results listed in Table 2 show, the p-values of the width and the rolling direction were minimal, which meant that the width and the rolling direction did affect the Young's modulus. In addition, the F-ratio of the thickness was 3.87 and the p-value of the thickness was 0.01, which meant that the thickness also affected the Young's modulus; however, the effect was not as obvious as for the width or rolling direction.

The effects of thickness, width, and rolling direction on the Young's modulus were plotted in Figure 4. It can be seen that the average Young's modulus was 125 GPa when the width was 12.5 mm. Compared to the average Young's modulus of 120 GPa when the width was 2.5 mm, the Young's modulus showed a decrease of about 4.4%.

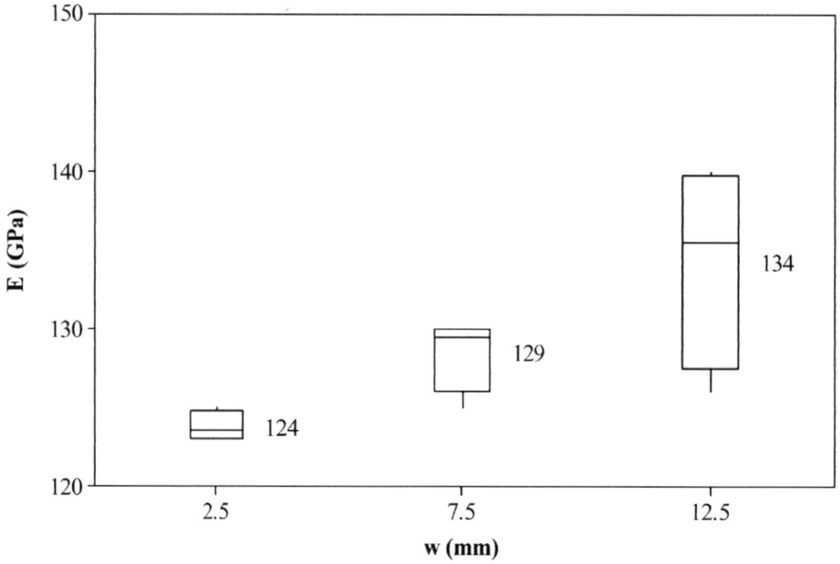

Figure 3. Young's modulus associated with various widths for the thickness of 0.05 mm and the rolling direction of 0°.

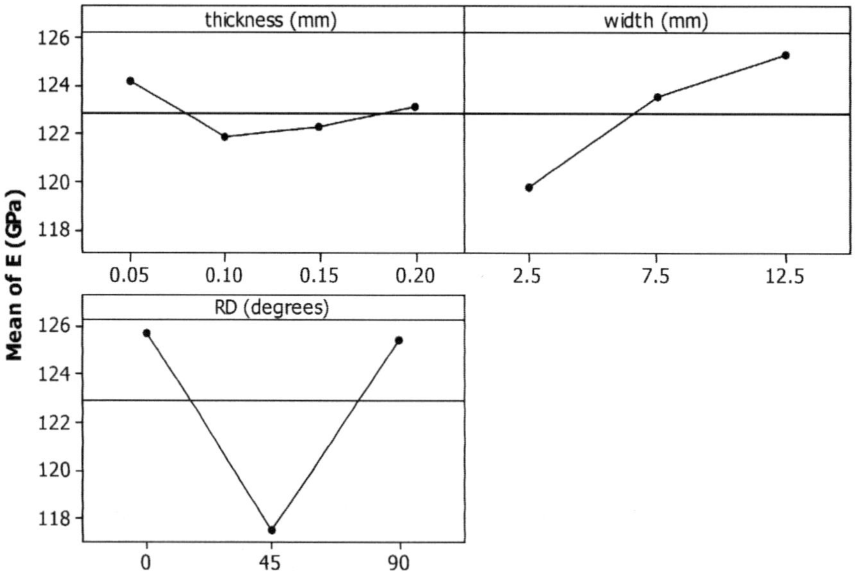

Figure 4. Main effect plots of thickness, width and rolling direction regarding the Young's modulus.

Regarding the rolling direction, the average Young's modulus was 126 GPa and 125 GPa for a rolling direction of 0° and 90°, respectively. The difference here was negligible. However, the average Young's modulus was 118 GPa when the rolling direction was 45°, which was about 6.5% less than that for a rolling direction of 0°. It can also be seen in Figure 4 that there was no obvious difference in the Young's modulus when the thickness varied. The range was from 122 GPa to 124 GPa.

Table 2. ANOVA results regarding the Young's modulus.

Source	Sums of Squares	Degrees of Freedom	Mean Squares	F-ratio	p-value
W	773.17	2	386.58	41.29	0.00
RD	2066.38	2	1033.19	110.34	0.00
T	108.74	3	36.25	3.87	0.01
w×RD	27.96	4	6.99	0.75	0.56
w ×t	106.44	6	17.74	1.90	0.09
RD×t	249.57	6	41.60	4.44	0.00
w×RD×t	171.93	12	14.33	1.53	0.12
Error	1011.25	108	9.36		

4. Conclusions

In our study, for the material C17200–25–1/2H (TD02), when the widths ranged from 2.5 mm to 12.5 mm, the thicknesses ranged from 0.05 mm to 0.20 mm and the rolling directions ranged from 0° to 90°; the change of the width of the specimen did not affect the yield strength, but did affect the Young's modulus. When the width was reduced from 12.5 mm to 2.5 mm, the Young's modulus was reduced by about 4.4%.

On the other hand, under the same condition, the change of the rolling direction of the specimen did affect both the yield strength and the Young's modulus. When the rolling direction was 0°, the yield strength and the Young's modulus were at their maximum. When the rolling direction was 45°, the yield strength and the Young's modulus were at their minimum. The difference for the yield strength was about 7%, and the difference for the Young's modulus was about 6.5%. When the rolling direction was 90°, the yield strength was close to that at 45° and the Young's modulus was close to that at 0°.

Based on the above results, when a finite element software tool is used to analyse a tiny component made of beryllium copper, if the longitudinal direction of the component is perpendicular to the rolling direction, it is necessary to consider lowering the allowable stress. If the width of the component is as little as 2.5 mm, we need to consider reducing its Young's modulus. By so doing, the accuracy of the analysis results should be improved.

References

[1] Kals T A and Eckstein R 2000 Miniaturization in sheet metal working. *J. Mater. Process. Tech.* 103, 95-101.
[2] Raulea L V, Goijaerts A M, Govaert L E, and Baaijens F P T 2001 Size effects in the processing of thin metal sheets', *J. Mater. Process. Tech.* 115, 44-48.
[3] Dubos P A, Hug E, Thibault S, Gueydan A, and Keller C 2013 Strain path influence on size effects during thin sheet copper microforming. *Int. J. Mater. Prod. Tec.* 47, 3-11.
[4] Yang L and Lu L 2013 The influence of sample thickness on the tensile properties of pure Cu with different grain sizes *Scripta Mater.* 69, 242-245.

[5] Michel J F and Picart P 2003 Size effects on the constitutive behaviour for brass in sheet metal forming *J. Mater. Process. Tech.*141, 439-446.
[6] Miyazaki S, Shibata K, and Fujita H 1979 Effect of specimen thickness on mechanical properties of polycrystalline aggregates with various grain sizes. *Acta Metall.* 27, 855-862.
[7] Engel U and Eckstein R 2002 Microforming - From basic research to its realization *J. Mater. Process. Tech.* 125-126, 35-44.
[8] Leong W and Nourbakhsh S 1989 Tensile behavior of cold rolled α-brass *Scripta Metall.* 23, 1011-1014.
[9] Stachowicz F 1989 Effects of microstructure on the mechanical properties and limit strains in uniaxial and biaxial stretching. *J. Mech. Work. Technol.* 19, 305-317.
[10] ASTM, E8M-01 2003 Standard Test Methods for Tension Testing of Metallic Materials. [Metric]
[11] ASTM 2003 E345-93, Standard Test Methods of Tension Testing of Metallic Foil.

IOP Conf. Series: Materials Science and Engineering **103** (2015) 012052 doi:10.1088/1757-899X/103/1/012052

High-temperature tribological properties of NiCoCrAlY-WSe_2-BaF_2·CaF_2 solid lubricant coatings prepared by plasma spraying

X H Chen, X J Yuan[1], J Xia and Z H Yu

Xi'an Research Institute of Hi-Tech, Shaanxi province, China 710025

E-mail: yxj2003@263.net

Abstract. In this paper, NiCoCrAlY-WSe_2-BaF_2·CaF_2 solid lubricant coatings were produced on a substrate by plasma spray and investigated at the high temperature, such as 500℃ and 800℃. The structure of the coatings was characterized using XRD pattern and scanning electron microscopy. The TC1 (83wt% NiCoCrAlY) coating has a low friction coefficient at 500℃, where the WSe_2 is a good solid lubricant. The TC2 (65wt% NiCoCrAlY) coating has the low friction coefficient (0.279) at 800℃, due to the formation of BaCrO4 on the surfaces. As a result, the TC2 coating has the optimal tribological property in the wide temperature.

1. Introduction

High-temperature self-lubricant coating which has well mechanical and tribological properties at 25~1000℃ was a hotspot in the industry [1-3]. Recently, several lubricant coatings were researched. Where, in the solid lubricant coatings, the silver and barium-calcium fluoride eutectic was added to the binder and act as high-temperature lubricants. The composite coatings of PM212/PS212, PS300, PS400 were composed of a good matrix (NiCr-Cr_3C_2, or NiCr-Cr_2O_3) and Ag with BaF_2·CaF_2 eutectic solid lubricants [4-7]. Research of the coatings indicates that the lubricious glaze formed in the surface of coatings to achieve low friction and wear rate of a high wide temperature range. However, at room temperature, the wear rate of the coatings is even high that the lubricious glaze was destroyed. Above 800℃, the oxidation and the decline of the mechanical property degraded its wear resistance [8-11].

So, in order to obtain optimal friction property of lubricant coatings at wide temperature range, the NiCoCrAlY-WSe_2-BaF_2·CaF_2 coatings were fabricated on the substrates of ASTM4145 alloy steel. The tribological properties of the composite coatings were tested at temperature 500℃ and 800℃.

2. Experimental procedure

NiCoCrAlY-WSe_2-BaF_2·CaF_2 composite powders with a different content was listed in Tab.1, which involve NiCoCrAlY, BaF_2, CaF_2, WSe_2, Y, and BN with a mean grain size of 5-80μm. The solid lubricant coatings were deposited by HT-80 atmospheric plasma spray system (HT-80 Plasma Spray System, Beijing Aerospace Zhenbang Pml Precision Mechanism Ltd.). The substrate is ASTM5145 alloy steel.

The tribological property was tested on a CSM high-temperature tribometer (CSM Co. Rue de la Gare4 Galileo Center Ch-2034 Peseux, Switzerland). For pre-testing, the sample disk (Φ30mm×8mm) surfaces were cleaned with acetone. A commercial Al_2O_3 ceramic ball, which was 6mm in diameter, was also used. The test was performed with a normal load of 10 N and a sliding velocity of 0.24 m/s for 60 min. The test temperature was set as 500°C and 800°C, respectively.

[1] Address for correspondence: X J Yuan, Xi'an Research Institute of Hi-Tech, Shaanxi province, China 710025. E-mail: yxj2003@263.net.

The microstructure of worn surface at different temperature for NiCoCrAlY-WSe_2-$BaF_2 \cdot CaF_2$ composite coatings were characterised using scanning electron microscopy (SEM; JSM-5600LV; JEOL, Tokyo, Japan), energy dispersal spectroscopy (EDS; Kevex Corp, Foster City, CA), and XRD spectrum (XRD; Philips X' Pert-MRD X-ray diffract-meter, 40 kV, 30 mA) with Cu Ka radiation over an angular range of $10° \leq 2\theta \leq 90°$.

Table 1. Compositions of the spray NiCoCrAlY-WSe_2-$BaF_2 \cdot CaF_2$ composite particle.

No.	NiCoCrAlY	BaF_2	CaF_2	WSe_2	Y	BN
TC1	83%	4.67%	2.33%	7%	3%	-
TC2	65%	12.3%	9.8%	5.9%	3%	4%

3. Results and discussions

3.1. High-temperature tribological property of the NiCoCrAlY-WSe_2-$BaF_2 \cdot CaF_2$ composite coatings

Friction coefficients of TC1 and TC2 coatings at 500℃ and 800℃ were listed in Tab.2. From the Tab.2, it can be seen that the average friction coefficient of TC1 coatings are about 0.440 at 500℃ and 0.461 at 800℃. The friction coefficient of the TC2 coatings was 0.748 and 0.279 at 500℃ and 800℃, respectively. Where, the TC2 coating at 500℃ has the higher friction coefficient because $BaF_2 \cdot CaF_2$ cannot possess tribology property for the brittle structure. However, the experiment result shows that WSe_2 is an optimal solid lubricant phase at 500℃. At 800℃, the friction coefficient was decreased to 0.279, And, the wear rate of TC1 solid lubricant coatings are 22.5×10^{-5}(mm^3/N·m) and 5.47×10^{-5}(mm^3/N·m) at the different temperature, that of TC2 solidi lubricant coatings are 11.6×10^{-5}(mm^3/N·m) and 2.54×10^{-5}(mm^3/N·m), respectively. Especially, the low tribological property of TC2 coatings was obtained.

Table 2. Friction coefficient of TC1 and TC2 coatings at different temperature.

Sample No.	Temperature (℃)	Friction coefficient	Wear Rate ($\times10^{-5}$(mm^3/N·m))
TC1	500	0.440	22.5
TC1	800	0.461	5.47
TC2	500	0.748	11.6
TC2	800	0.279	2.54

At 500℃, The XRD results show that there was not oxidized in the NiCoCrAlY-WSe_2-$BaF_2 \cdot CaF_2$ lubricant coatings; most of phase in the coatings keep the original statement. Comparably, at 800℃, the BaF_2 and CaF_2 peaks does not exist in the XRD pattern instead of the $BaCrO_4$. All of them were proved as the good solid lubricants. Especially, the $BaCrO_4$ with the baritone structure has a good lubricating property at high temperature.

3.2. Worn surface of the NiCoCrAlY-WSe_2-$BaF_2 \cdot CaF_2$ composite coatings at high temperature

Fig.2 shows the worn surfaces of the TC1 coatings. From Fig.2(a, b), at 500°C, the surfaces are rough, some dispersed particles and grooves existed in the wear region, which indicates that wear could depend on the delamination. From Fig. 2(c, d) (at 800°C), a relatively smooth surface can be taken, where some oxidation leads to the formation of the glaze layer.

The worn surfaces of TC2 coatings were shown in the Fig.3. It was noted that the detachment in the rich $BaF_2 \cdot CaF_2$ areas indicates that the wear mechanism is micro-fracture and delamination at 500℃ (shown in Fig.3 (a b)). At 800℃, relatively smooth surface form, many fine grooves, and deformed surfaces are present along slide direction, which takes on the micro-plough with deformation and delamination (shown in Fig.3(c, d)).

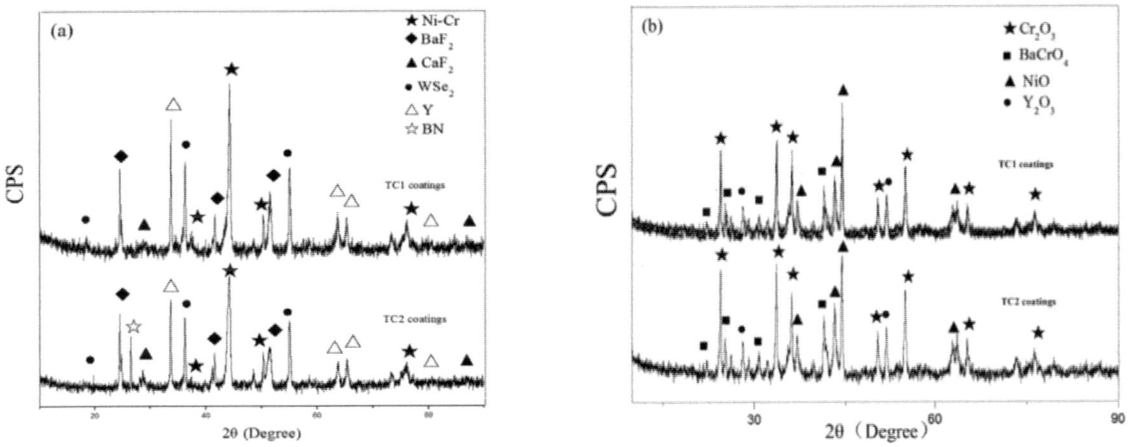

Figure 1. XRD patterns of the NiCoCrAlY-WSe_2-$BaF_2{\cdot}CaF_2$ composite coatings (a) 500℃; (b) 800℃.

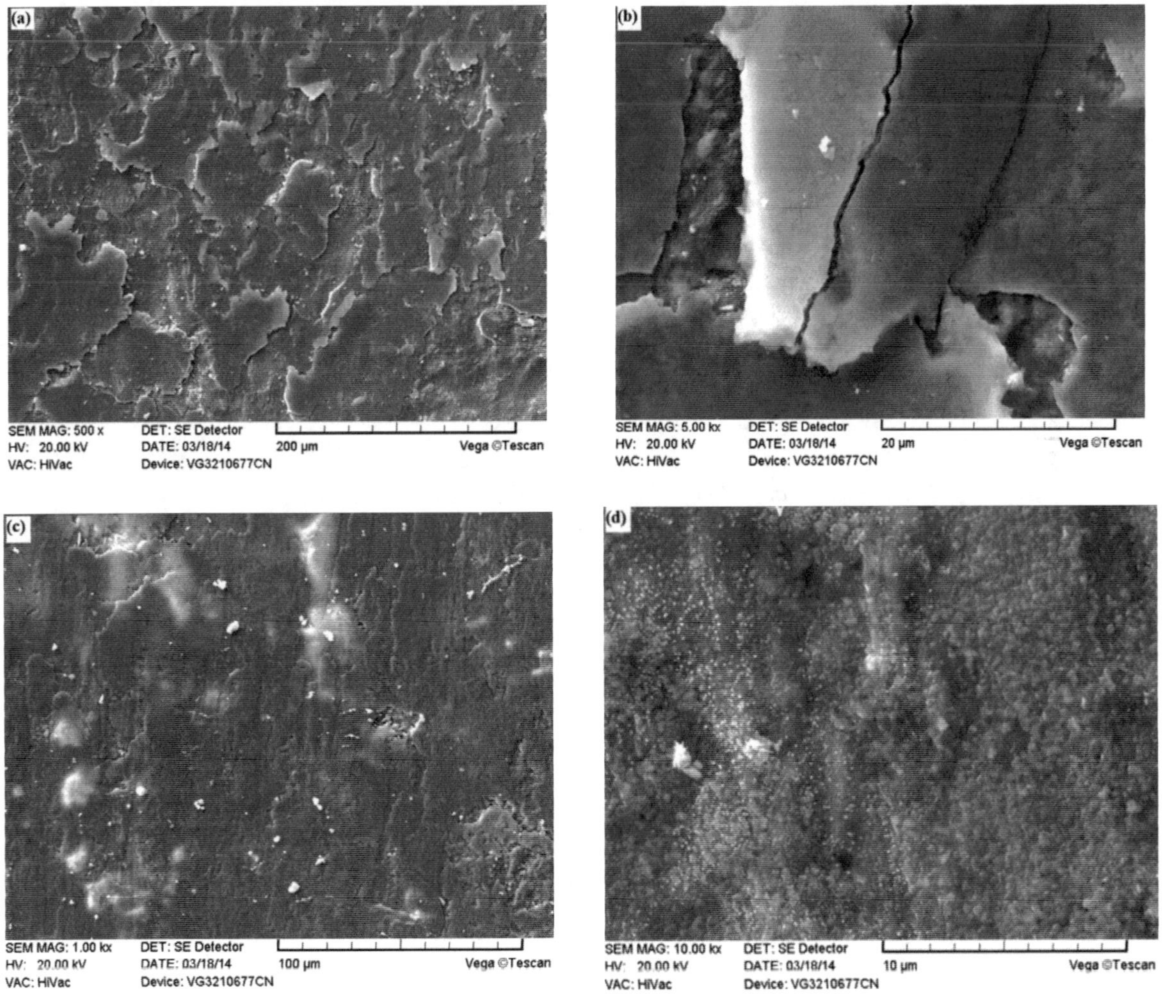

Figure 2. Worn surface of TC1 coatings at high temperature: (a, b) 500℃; (c, d) 800℃.

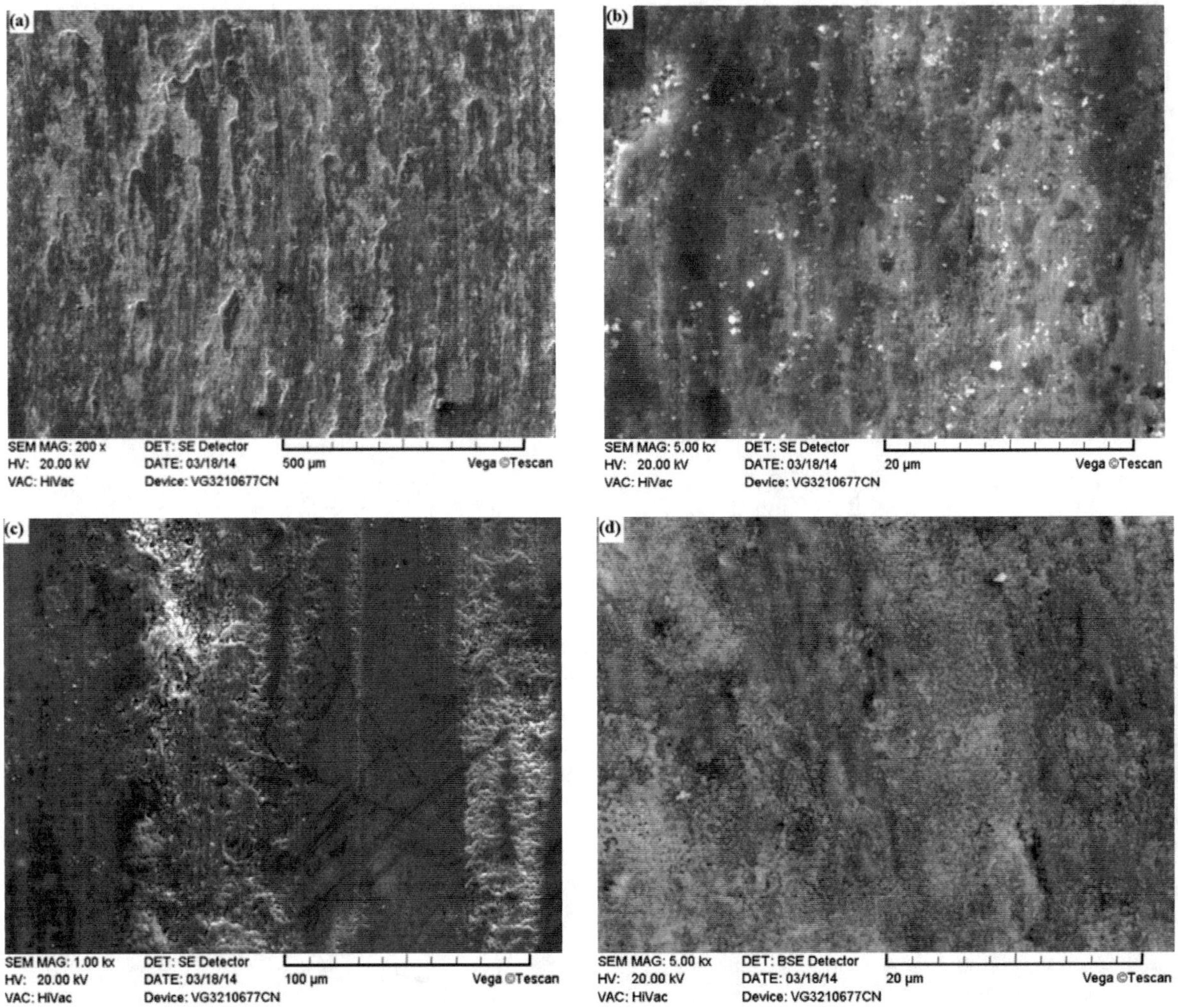

Figure 3. Worn surface of TC2 coatings at high temperature: (a, b) 500℃; (c, d) 800℃.

3.3. High-temperature wear mechanisms of the NiCoCrAlY-WSe_2-BaF_2·CaF_2 composite coatings

The results demonstrate that the TC1 coatings have better self-lubricating properties at 500℃, and the TC2 coatings have the better performance at friction and wear-resistant properties at 800℃. When the temperature is low (such as 500℃), the WSe_2 acts as the lubricant phase in the coatings, TC1 coating has more, so it has a low average friction coefficient at 500℃. When the temperature comes to 800℃, the WSe_2 is disabled; the BaF_2 and CaF_2 play the role instead. The phase changes on the surfaces of TC1 coatings at 800℃, they are the oxidant, and the film is on the surface of coatings so as to wear easily. The debris on the sliding surfaces was effected on the friction pair, and the structure characteristic of the coatings.

The self-lubricant property of the NiCoCrAlY-WSe_2-BaF_2·CaF_2 composite coatings in a wide temperature was consisted of BaF_2·CaF_2 and WSe_2. At the lower temperature, the WSe_2 act as solid lubricants. At high temperature, the TC2 coating has the highest friction coefficient at 500℃. At 800℃, the TC2 coating has the lowest friction coefficient. Where, the new production, $BaCrO_4$, provides an excellent tribological property at high temperature [12]. Moreover, the compactness of the coatings at high temperature (at 800℃) is bad contrast to them at 500℃for the decomposition of WSe_2. So, there are some pores in the coatings.

4. Conclusions

- The high-temperature tribological properties of NiCoCrAlY-WSe_2-BaF_2·CaF_2 composite coatings prepared by Plasma Spraying were researched. The TC2 (65wt% NiCoCrAlY) coating shows a better friction property at high temperature.

- At 800℃, the production, $BaCrO_4$, provides an excellent lubricating property due to the tribo-chemical reaction at high temperature.
- The low friction coefficient of the TC2 coatings at 800℃ is due to the balance between the mechanical and lubricity of the worn surface.

Acknowledgments
The authors acknowledge the financial support of the National Natural Science Foundation of China (NSFC) under Grant No. 51405497.

References
[1] Lu, J.J., Xue, Q.J., Wang, J.B., Ouyang, J.L.: The effect of CeF_3 on the mechanical and tribological properties of Ni-based alloy. *Tribol. Int.*, 1997(9): 659–662.
[2] Shengyu Zhu, Qinling Bi, Jun Yang, Weimin Liu, Effect of Fluoride Content on Friction and Wear Performance of Ni_3Al Matrix High-Temperature Self-Lubricating Composites, *Tribol. Lett.*, 2011, 43(3): 341-349.
[3] Er-yong Liu. Tribological Properties of Adaptive Ni- Based Composites with Addition of Lubricious Ag_2MoO_4 at Elevated Temperatures, *Tribol. Lett.*, 2012, 47(1): 21-30.
[4] Sliney H.E., Dellacorte C., Lukaszewicz V.: The tribology of PS212 coatings and PM212 composites for the lubrication of titanium 6A1-4V components of a Stirling engine space power system. *Tribol. Trans.* 1995(38): 497–506.
[5] DellaCorte C., Laskowski J.: Tribological evaluation of PS300: a new chrome oxide-based solid lubricant coating sliding against Al_2O_3 from 25 to 650 ℃. *Tribol. Trans.* 1997(40): 163–167.
[6] Dellacorte, C.: The effects of substrate material and thermal processing atmosphere on the strength of PS304: a high temperature solid lubricant coating. *Tribol. Trans.* 2003(46): 361–368.
[7] DellaCorte, C., Fellenstein, J. A.: The effect of compositional tailoring on the thermal expansion and tribological properties of PS300: a solid lubricant composite coating. *Tribol. Trans.* 1997(40) 639–642.
[8] Mumm, D., Evans, A., Spitsberg, I.: Characterization of cyclic displacement instability for a thermally grown oxide in a thermal barrier system. *Acta Mater.*, 2001(49): 2329–2340.
[9] Bi Qinling, Shengyu Zhu, and Weimin Liu. High Temperature Self-Lubricating Materials, Tribol. in Eng., 2013. http://dx.doi.org/10.5772/55645.
[10] Erdemir, A.: A crystal-chemical approach to lubrication by solid oxides. *Tribol. Lett.* 2000 (8): 97–102
[11] Zhu, S..Effect of particle size on tribological behavior of Ni_3Al matrix high temperature self-lubricating composites, *Tribol. Int.*, 2011,44:1800-1809.
[12] Ouyang, J.H.. Spark-plasma-sintered $ZrO_2(Y_2O_3)$-$BaCrO_4$ self-lubricating composites for high temperature tribological applications, *Ceram. Int.*, 2005, 31(4):543-553.

Institute of Physics
Dirac House, Temple Back
Bristol BS1 6BE UK

ISSN: 1757-8981
ISBN 978-1-5108-1812-5

9 781510 818125

CURRAN ASSOCIATES INC.
proceedings
.com